Lightweight and Sustainable Materials for Automotive Applications

Lightweight and Sustainable Materials for Automotive Applications

Edited by
Omar Faruk
Jimi Tjong
Mohini Sain

CRC Press
Taylor & Francis Group
Boca Raton London New York

CRC Press is an imprint of the
Taylor & Francis Group, an **informa** business

CRC Press
Taylor & Francis Group
6000 Broken Sound Parkway NW, Suite 300
Boca Raton, FL 33487-2742

First issued in paperback 2019

ISBN-13: 978-1-4987-5687-7 (hbk)
ISBN-13: 978-0-367-87665-4 (pbk)

Visit the Taylor & Francis Web site at
http://www.taylorandfrancis.com

and the CRC Press Web site at
http://www.crcpress.com

To My Beloved Wife
Shaila Shumi
and
My Beloved Daughter
Ornela Suhiya

Omar Faruk

To My Beloved Wife
Jasmin Reyes Tjong
and
My Beloved Daughter
Dr. Vehniah Kristin Tjong

Jimi Tjong

Contents

Preface

World automotive manufacturers are required to decrease CO_2 emissions and increase fuel economy while assuring driver comfort and safety. In recent years, there has been rapid development in lightweight and sustainable materials application in the automotive industry to meet the above-mentioned criteria. Researchers are seeking to develop vehicle lightweighting strategies that will allow them to cost-effectively meet fuel economy targets, and increasingly shifting their focus to incorporating mixed-material solutions at mass produced scales. The global automotive lightweight materials market is expected to grow at a CAGR of 8.1% during 2014–2019. The major drivers of the global automotive lightweight materials market are government regulations on fuel economy and emission controls, increasingly stringent safety regulations, high vehicle production, increasing use of lightweight materials, higher gasoline prices and fuel economy, and replacement of traditional materials. Aluminum, magnesium, and carbon fiber are emerging to be the frontrunners in achieving lightweight results.

Besides these materials, composite materials, thermoplastic, thermoset polymer, rubber material, nanocomposite materials, and foaming materials are also significantly employed as lightweight materials in automotive applications. The appeal of bio-based parts for use in the next generation of vehicles is increasing, and there are compelling factors such as stringent fuel economy, emissions regulations, and sustainability at work. There is rising demand for lighter-weight materials, which can translate into huge energy and cost savings for manufacturers. Bio-fiber reinforced composite materials are not only lightweight, they also reduce dependence on nonrenewable resources such as conventional petroleum-based polymeric plastics, which are fossil-fuel materials. Biocomposite materials are made from renewable resources and the latest generation can achieve price-performance competitiveness, potentially lowering costs for both manufacturers and consumers. These factors have prompted the Ford Motor Company and other automakers to invest in bio-based material research. However, there are challenges to overcome before these materials can be more widely adopted.

This book provides critical reviews and the latest research results and applications of various lightweight and sustainable materials in automotive applications. It also provides current applications and future trends of lightweight materials in the automotive area. In recent years, enormous research and commercialization have been performed with lightweight and sustainable materials. There are few books published on lightweight materials in automotive applications mainly focusing on metallic lightweight materials. To date, no book has been published that focuses on lightweight materials including metal, plastic, composites, bio-fiber, bio-polymer, carbon fiber, glass fiber, nano materials, rubber materials, and foaming materials. Therefore, this book will be a significant guide to academia, researchers, and industries that are involved in this current in-demand research and commercialization area.

Omar Faruk
Jimi Tjong
Mohini Sain

Editors

Dr. Omar Faruk works at the Powertrain Engineering Research & Development Centre of Ford Motor Company, Canada. He is also an adjunct professor at the Centre for Biocomposites and Biomaterials Processing, University of Toronto, Canada. He earned his PhD in mechanical engineering from the University of Kassel, Germany, and was previously a Visiting Research Associate at Michigan State University. He has more than 75 publications to his credit including 12 book chapters which have been published in different international journals and conferences. He also edited two books, titled *Biofiber Reinforcement in Composite Materials* and *Lignin in Polymer Composites*, published by Woodhead Publishing Ltd and Elsevier Ltd, respectively. In addition, he is an invited reviewer for 69 international reputed journals, government research proposals, and book proposals.

Dr. Jimi Tjong is the Technical Leader and Manager of the Powertrain Engineering, Research & Development Centre of Ford Motor Company, Canada. He earned his PhD in mechanical engineering from the University of Windsor, Canada, and he has worked for more than 30 years at Ford Motor Company. His principal field of research and development encompasses optimizing automotive test systems for cost, performance, and full compatibility. It includes the development of test methodology and cognitive systems, calibration for internal combustion engines, alternative fuels, bio fuels, lubricants and exhaust fluids, lightweight materials with the focus on aluminum, magnesium, bio materials, batteries, electric motors, super capacitors, stop/start systems, HEV, PHEV, BEV systems, nano sensors and actuators, high performance and racing engines, nondestructive monitoring of manufacturing and assembly processes, advanced gasoline and diesel engines focusing on fuel economy, and performance and cost opportunities. He has published and presented numerous technical papers in the above fields internationally. Dr. Tjong is also an adjunct professor at the University of Windsor, McMaster University and the University of Toronto in Canada. He mentors graduate students in completing course requirements as well as career development coaching.

Prof. Mohini Sain specializes in advanced nano-cellulose technology, biocomposites, and bio-nanocomposites at the University of Toronto, Faculty of Forestry. He is cross-appointed to the Department of Chemical Engineering and Applied Chemistry. He is a fellow of the Royal Society of Chemistry, UK, and a Fellow of the Canadian Engineering Society. He is also an adjunct professor at the University of Guelph and the University of Lulea, Sweden; an honorary professor at the Slovak Technical University and the Institute of Environmental Science at the University of Toronto; and collaborates with American and European research institutes and universities. Prof. Sain is the recipient of several awards, most recently the Plastic Innovation Award and KALEV PUGI Award for his innovation and contribution to the industry. He is the author of more than 400 papers and is designated as a "high-cited" researcher by Reuter Thompson. Prof. Sain hugely contributed to the society at large by translating research to commercialization. He has tens of patents, 30 technology transfers to industry, and created new companies for making products ranging from packaging to automotive to building construction to packaging materials. He is also the co-author of the world's first book on cellulose nanocomposites, cellulose for electronic devices, and has co-edited a number of books on renewable advanced materials. Prof. Sain's role as a pioneer in creating nonprofit organizations is highly meaningful for society at large.

Contributors

Masoud Akhshik
Centre for Biocomposites and
  Biomaterials Processing
University of Toronto
Toronto, Ontario, Canada

Abdullah Al Mamun
Adler Pelzer Group
HP Pelzer Holding GmbH
Witten, Germany

Ahmet T. Alpas
Engineering Materials Program
Department of Mechanical, Automotive
  and Materials Engineering
University of Windsor
Windsor, Ontario, Canada

M. Montserrat Alvarez Grima
ARLANXEO Performance Elastomers
Geleen, the Netherlands

Sandeep Bhattacharya
Engineering Materials Program
Department of Mechanical, Automotive
  and Materials Engineering
University of Windsor
Windsor, Ontario, Canada

Nathaniel Brown
Department of Automotive Engineering
Clemson University
Greenville, South Carolina

Katherine Dean
Industrial Interfaces Team,
  Commonwealth Scientific and
  Industrial Research Organisation
  (CSIRO), Manufacturing Flagship
Melbourne, Australia

Hans-Josef Endres
Institute for Bioplastics and
  Biocomposites
University of Applied Sciences and Arts
Hanover, Germany

Omar Faruk
Centre for Biocomposites and
  Biomaterials Processing
University of Toronto
Toronto, Ontario, Canada

Maik Feldmann
Institute for Materials Engineering,
  Polymer Engineering
University of Kassel
Kassel, Germany

Wojciech (Voytek) S. Gutowski
Industrial Interfaces Team,
  Commonwealth Scientific and
  Industrial Research Organisation
  (CSIRO), Manufacturing Flagship
Melbourne, Australia

and

Professor, Chinese Academy of
  Sciences
Institute of Applied Chemistry
Changchun, China

Hans-Peter Heim
Institute for Materials Engineering,
  Polymer Engineering
University of Kassel
Kassel, Germany

Frank Henning
Fraunhofer Institute for Chemical
   Technology (ICT)
Pfinztal, Germany

and

Karlsruhe Institute of Technology (KIT)
Karlsruhe, Germany

Philip Hough
ARLANXEO Performance Elastomers
Geleen, the Netherlands

Birat KC
Centre for Biocomposites and
   Biomaterials Processing
University of Toronto
Toronto, Ontario, Canada

Joyce Kersjes
ARLANXEO Performance Elastomers
Geleen, the Netherlands

Mark T. Kortschot
Department of Chemical Engineering
   and Applied Chemistry, Advanced
   Materials Group
University of Toronto
Toronto, Ontario, Canada

D. Sameer Kumar
Department of Mechanical Engineering
R.V.R. & J.C. (Rayapati Venkata
   Rangarao & Jagarlamudi
   Chandramouli) College of Engineering
Guntur, India

Sheng Li
Industrial Interfaces Team,
   Commonwealth Scientific and
   Industrial Research Organisation
   (CSIRO), Manufacturing Flagship
Melbourne, Australia

Mohammad Ali Nikousaleh
Institute for Materials Engineering,
   Polymer Engineering
University of Kassel
Kassel, Germany

Numaira Obaid
Department of Chemical Engineering
   and Applied Chemistry, Advanced
   Materials Group
University of Toronto
Toronto, Ontario, Canada

Srikanth Pilla
Department of Automotive Engineering
   and Department of Materials Science
   and Engineering
Clemson University
Greenville, South Carolina

Sai Aditya Pradeep
Department of Automotive Engineering
   and Department of Materials Science
   and Engineering
Clemson University
Greenville, South Carolina

Ashok Rajpurohit
Fraunhofer Institute for Chemical
    Technology (ICT)
Pfinztal, Germany

and

Karlsruhe Institute of Technology (KIT)
Karlsruhe, Germany

Mohini Sain
Faculty of Forestry
Centre for Biocomposites and
    Biomaterial Processing
University of Toronto
Toronto, Ontario, Canada

and

Adjunct Professor
King Abdulaziz University
Abdulla Sulayman
Jeddah, Saudi Arabia

C. Tara Sasanka
Department of Mechanical Engineering
R.V.R. & J.C. (Rayapati Venkata
    Rangarao & Jagarlamudi
    Chandramouli) College of Engineering
Guntur, India

T. Palanisamy Sathishkumar
Faculty of Mechanical Engineering
School of Building and Mechanical
    Sciences
Kongu Engineering College
Erode, Tamilnadu, India

Moyeenuddin Ahmad Sawpan
Composites Materials Research
Pullton Composites Ltd
Gisborne, New Zealand

Arne Schirp
Fraunhofer Institute for Wood Research
Wilhelm-Klauditz-Institut
Braunschweig, Germany

Srishti Shukla
Department of Automotive Engineering
Clemson University
Greenville, South Carolina

and

Department of Mechanical Engineering,
    Indian Institute of Technology
    (Banaras Hindu University)
Varanasi, India

Ahmed Sobh
Centre for Biocomposites and
    Biomaterials Processing
University of Toronto
Toronto, Ontario, Canada

Jimi Tjong
Centre for Biocomposites and
    Biomaterials Processing
University of Toronto
Toronto, Ontario, Canada

Niels van der Aar
ARLANXEO Performance Elastomers
Geleen, the Netherlands

Martin van Duin
ARLANXEO Performance Elastomers
Geleen, the Netherlands

Marjan van Urk
ARLANXEO Performance Elastomers
Cologne, Germany

Weidong Yang
Industrial Interfaces Team,
    Commonwealth Scientific and
    Industrial Research Organisation
    (CSIRO), Manufacturing Flagship
Melbourne, Australia

Xiaoqing Zhang
Industrial Interfaces Team,
    Commonwealth Scientific and
    Industrial Research Organisation
    (CSIRO), Manufacturing Flagship
Melbourne, Australia

1 Natural Fiber Reinforced Thermoplastic Composites

*Omar Faruk, Birat KC, Ahmed Sobh,
Jimi Tjong, and Mohini Sain*

CONTENTS

1.1 INTRODUCTION

The automotive industry is currently experiencing environmental, legislative, and consumer pressure to improve the environmental sustainability of passenger vehicles. Just one of the approaches being taken to address this issue is the reconsideration of materials used in the automotive application.

The incorporation of natural fibers as a reinforcing agent in thermoplastic polymer composites has gained increasing applications both in many areas of engineering and technology. Several natural fiber–based thermoplastic composite materials have been developed using modified synthetic strategies to extend its application in the automotive area. The advantage of using natural fibers is more economical and ecologically favorable. By substituting 50% of synthetic fibers with natural fiber composites in automotive applications, 3.07 million tons of CO_2 emissions and 1.19 million m^3 of crude oil could be reduced in North America alone [1].

In Europe, the second largest application sector for biocomposites is the automotive interior sector. The production and use of 90,000 tons of natural fiber composites in the automotive sector in 2012 could expand to over 350,000 tons of natural fiber composites in 2020 in the automotive sector [2]. Among this 90,000 tons of natural fiber composites, thermoplastic matrices were used for approximately 55% and the rest was thermoset matrices. There is a trend in growth of injection molded natural fiber composites, which is mainly with a thermoplastic matrix. Nowadays, the global

1

players are offering their new developed materials indicating sustainable lightweight thermoplastic solutions for the automotive sector.

The composites' shape, surface appearance, environmental tolerance, and overall durability are dominated by the matrix while the fibrous reinforcement carries most of the structural loads, thus providing macroscopic stiffness and strength. The polymer market is dominated by commodity plastics with 80% consuming materials based on nonrenewable petroleum resources. The effects of the incorporation of natural fibers in petrochemical-based thermoplastic matrixes were extensively studied. Polypropylene (PP), polyethylene (PE), polystyrene (PS), polyvinyl chloride (PVC), thermoplastic polyurethane (TPU), polyamide (PA), and acrylonitrile butadiene styrene (ABS) were used for thermoplastic matrices.

The increasing number of publications during recent years including reviews and books reflect the growing importance of these new biocomposites. Pereira et al. [3] represented a statistical analysis of scientific publications in the area of natural fiber composites with thermoplastic and thermoset matrices (Figure 1.1). It is seen that the number of publications about natural fiber composites with thermoplastics increased significantly from 2000 to 2014 compared to natural fiber composites with thermoset matrices.

This chapter presents a literature review examining the effect of natural fiber reinforcement on the mechanical, physical, and rheological properties, photodegradation, weathering, flame retardancy, processability, and morphological properties of natural fiber reinforced thermoplastic composites for a better understanding to increasing the implementation of these materials in automotive applications day by day.

1.1.1 POLYPROPYLENE COMPOSITES WITH NATURAL FIBERS

Hao et al. [4] investigated the effects of processing conditions (processing time, processing temperature, and processing pressure) on the mechanical properties of kenaf/polypropylene nonwoven composites and kenaf fiber content was 50% by weight (Figure 1.2).

It was found that temperature and time are the most significant processing factors. The performance of sound absorption and sound insulation was also investigated and it is mentioned that an adhesive-free sandwich structure provided excellent sound absorption and insulation performance.

The flexural modulus of the kenaf and polypropylene composite was influenced by a number of kenaf layers, heating time, and kenaf weight fraction prepared by pressing [5]. It is seen that the maximum flexural modulus of the composite is optimized by kenaf weight fraction, which increased with increasing bulk density.

The influence of acetylation on the structure and properties of flax fibers was investigated. Modified flax fiber reinforced PP composites were also prepared [6]. The effect of acetylation on the degree of polymerization and crystallinity of flax fiber is illustrated and it was observed that the degree of polymerization slowly decreased while the degree of acetylation increased to 18%. Once reaching an acetyl content of 18%, the degree of polymerization decreased rapidly due to the vigorous degradation of cellulose. It was also notable that the degree of crystallinity increased a little bit regarding the degree of acetylation because of the removal of lignin and

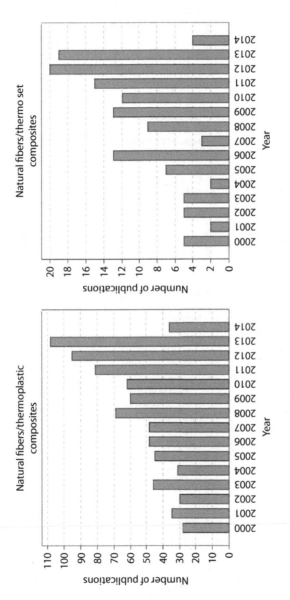

FIGURE 1.1 Number of publications about natural fiber composites with thermoplastic and thermoset matrices in the last years from Web of Science database.

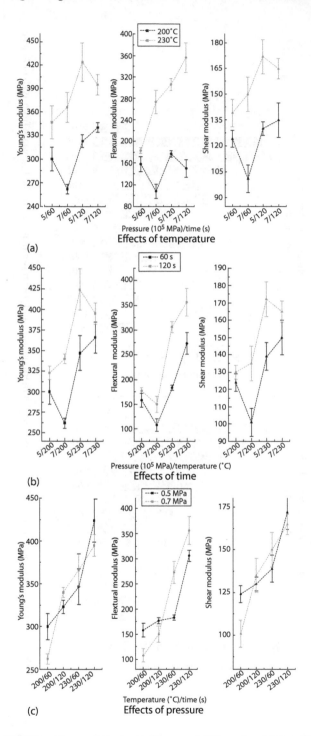

FIGURE 1.2　Influence of manufacture conditions on (a) Young's modulus, (b) flexural modulus, and (c) in-plane shear modulus of kenaf fiber reinforced PP composites.

extractible. After that, the degree of crystallinity of the cellulose decreased with respect to the degree of acetylation owing to the decomposition of acetylated amorphous components on the cellulose surface. The influences of the degree of acetylation of flax fibers on the moisture absorption properties are illustrated. At 95% RH, the 3.6%, 12%, 18%, and 34% degree of acetylated flax fibers showed about 14%, 18%, 27%, and 42% lower moisture absorption properties than untreated flax fiber, respectively. The moisture absorption properties decreased proportionally with increasing acetyl content of fibers due to the reduction of hydrophilicity of the fibers.

Bledzki et al. [7] have investigated the reinforcement of four types of natural fibers that are most commonly used in the plastic industry. Polypropylene biocomposites were reinforced with softwood, abaca, jute, and kenaf fibers having the same matrix-to-fiber content (60/40 wt%). Figure 1.3 shows that the heat deflection temperature was increased by a factor of two for softwood and abaca and for jute and kenaf was further improved approximately 20°C. The increase in HDT of the composites is mainly influenced by fiber loading and fiber-matrix interfacial strength and interfacial strength is also strongly related to the geometry of the fibers and particularly their aspect ratio.

Bledzki et al. [8,9] examined the mechanical properties of abaca fiber reinforced PP composites regarding different fiber lengths (5, 25, and 40 mm) and different compounding processes (mixer-injection molding, mixer-compression molding, and direct compression molding process). It was observed that with increasing fiber length (5–40 mm), the tensile and flexural properties showed an increasing tendency although not a significant one. Among the three different compounding processes compared, the mixer-injection molding process displayed a better mechanical performance (tensile strength is around 90% higher) than the other processes.

When abaca fiber PP composites were compared with jute and flax fiber PP composites, abaca fiber composites had the best notched Charpy and falling weight

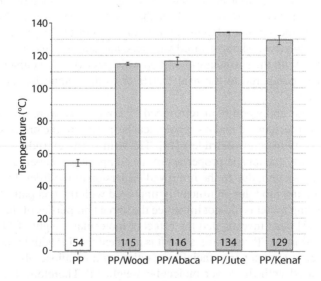

FIGURE 1.3 Heat deflection temperature of PP biocomposite.

impact properties. Abaca fiber composites also showed higher odor concentration compared to jute and flax fiber composites.

Bledzki et al. [10] investigated PP composites with enzyme-treated abaca fibers. The surface morphologies of enzyme-treated and untreated abaca fibers demonstrated that the untreated fiber surface is rough, containing waxy and protruding parts. The surface morphology of treated fibers is observed that if the waxy material and cuticle in the treated surface are removed, the surface becomes smoother. Fibrillation is also known to occur when the binding materials are removed from the surface of the treated fibers. Fiber surface damage was also observed for naturally digested fibers that occur in natural digestion systems (NDSs). The effects of enzyme treatment of abaca fiber on the tensile and flexural strength are shown. The tensile strength of enzyme-treated abaca composites was found to have increased 5%–45% due to modification. NDS modified abaca fiber composites showed little improvement in comparison to unmodified abaca fiber composites. The tensile strength of fungamix modified composites increased by 45% as compared to unmodified fiber composites. A conventional coupling agent (MA/PP) has a positive effect on the tensile strength. It was found to improve by 40%.

The tensile and bending properties of jute/PP unidirectional composites with 20% jute fiber (Vf) content show remarkable improvement when compared to virgin PP [11]. The improvements in the mechanical properties are broadly related to the wettability of resin melts into fiber bundles, interfacial adhesion, orientation, and uniform distribution of matrix fibers and the lack of fiber attrition and attenuation during the tubular braiding process (Figure 1.4).

Espert et al. [12] have investigated the water absorption properties of coir, sisal fibers reinforced PP composites in water at three different temperatures: 23, 50, and 70°C. The tensile properties of the composites were decreased, showing a great loss in mechanical properties of the water-saturated samples compared to the dry samples.

Ljungberg et al. [13] reported the nanocomposite films of isotactic PP reinforced with cellulose whiskers were prepared for the first time, which is highly dispersed with surfactant and compared with either bare or grafted aggregated whiskers (aggregated cellulose whiskers, grafted whiskers, and surface coated whiskers). The nanocomposites with the surfactant-modified whiskers exhibited enhanced ultimate properties when compared to the neat matrix or to the composites containing the other filler types.

Rice husk flour and wood flour reinforced PP composites were investigated with the focus on the mechanical and morphological properties of the processing systems (single-screw or twin-screw extruder) [14]. The twin-screw extruding system influenced significantly the tensile properties of the composites compared to the single-screw extruding system, due to the improved dispersion of the filler. In addition, the tensile strength and modulus significantly improved with the compatibilizing agent. But the processing systems did not influence the impact properties of the composites.

Qiu et al. [15] have investigated the effect of molecular weight of PP on the cellulose fiber reinforced PP composites and it is reported that PP with higher molecular weight revealed a stronger interfacial interaction with cellulose fibers in the composites, compared with the lower molecular weight PP. Therefore, the composites

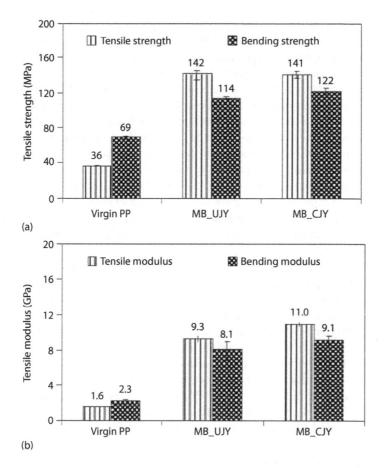

FIGURE 1.4 Mechanical properties of jute/PP unidirectional composites using microbraid jute yarns (a) tensile strength and (b) tensile modulus (CJY: coated jute yarns; MB: microbraid; UJY: uncoated jute yarns).

derived from the higher molecular weight of PP exhibited stronger tensile strength at the same cellulose content.

Another research has been performed on the optimization of fiber treatments for PP/cellulosic fiber composites [16]. The grafting method of a PP chain by an ester bond capable of co-crystallizing or entangling with the fibers was used. While the modification of cellulose fiber was done by isocyanates for improvement in the rupture, initiation strength was observed by PP modification of fibers resulting in the crack propagation strength.

In the study by Van de Valde and Kiekens [17], unidirectional (UD) and multi-directional (MD) composites of flax and PP were analyzed at various processing conditions. Superior mechanical properties were found in UD composites of boiled flax combined with MAA/PP. In an unexpected result, they found the treatment of flax did not provide any additional property enhancement for MD composites.

Cantero et al. [18] investigated the effects of chemical treatments on fiber wettability and mechanical behavior of flax/PP composites. Two types of flax fibers (natural flax and flax pulp) were compounded with PP to produce the composite material. The chemical treatments used were maleic anhydride (MA), maleic anhydride–polypropylene copolymer (MAPP), and vinyl trimethoxy silane (VTMO). Effects of chemical treatments on fiber surface morphology were characterized using infrared spectroscopy. Surface energy values were determined using two techniques for the different types of flax fibers used (Figure 1.5). For the long fiber, a dynamic contact angle method was used, whereas for the irregular pulps a capillary rise method was implemented. Small discrepancies in the wettability values were attributed to the different methods of measurement. Nevertheless, all treatments resulted in a reduced polar component of fiber surface energy. In consideration of mechanical behavior, MAPP-treated fiber composites exhibited the highest mechanical properties, and the other chemical treatments resulted in virtually no improvement as from the untreated case.

Keener et al. [19] also reported on the use of maleated coupling agents to strengthen and improve composite performance. Maleated polyolefins are widely used for two main reasons: (1) economic feasibility and (2) effectiveness in improving interfacial adhesion. Optimization of composite performance was achieved in this study through careful selection of a maleated coupling agent with the appropriate balance between its molecular weight and maleic anhydride content.

Arbelaiz et al. [20] studied fiber treatments and their effects on fiber thermal stability and crystallization of flax/PP composites. The treatments included the use of maleic anhydride, maleic anhydride PP copolymer, vinyltrimethoxy silane, and alkalization. Thermal stability of flax fibers was assessed using thermogravimetry (TG) analysis, and kinetic parameters were determined with DSC experiments and

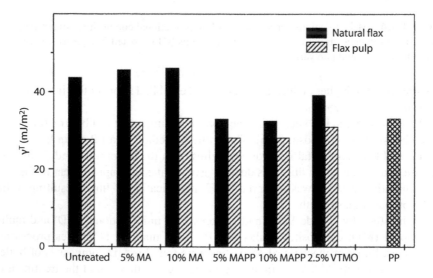

FIGURE 1.5 Effect of fiber treatments on total surface energy of natural flax fibers and pulps.

the Kissinger method. In terms of thermal stability results, all chemical treatments enhanced the flax fiber degradation temperature limit. With respect to crystallization, the study indicated an increased crystallization rate of PP with the addition of flax fiber.

Bos et al. [21] focused on improving the mechanical properties of flax/PP composites. They concluded that compatibilization of the fiber/matrix interphase is essential in improving composite tensile strength and stiffness. However, once the interfacial properties are optimized, the internal fiber structure becomes the weakest point.

In the efforts by Angelov et al. [22], pultrusion techniques were used to manufacture flax/PP composites of rectangular cross-sectional profiles. The study focused on the optimization of several processing parameters including preheating temperature, die temperature, and pulling speed. These parameters were of particular interest due to their relevance in future scale-up to industrial type production. Quality and characterization of the manufactured profiles were assessed to establish the effects, if any, of pultrusion processing parameters. Mechanical performance was characterized by Charpy impact and three-point bending tests.

In the report by Madsen and Lilholt [23], influences of porosity on physical and mechanical properties of UD composites were discussed. A winding machine was first used to align the flax yarns onto a square metal frame. Then the PP matrix foils were introduced by film stacking and the process was then completed by vacuum heating, and pressure consolidation procedures. The results indicated that the porosity fractions increased with fiber weight fractions. In relation to mechanical performance, the axial stiffness ranged from 27 to 29 GPa and axial strength was from 251 to 321 MPa. An improved model was developed to predict composite tensile properties using the "rule-of-mixtures," including parameters of composite porosity and anisotropy of fiber properties.

Mtuje et al. [24] attempted to exploit and improve the tensile properties of cannabis in order to achieve comparable performance to glass fibers. MAPP was used as the compatibility agent which decreased the hydrophobic nature of PP and improved the dispersion and interfacial qualities of both the fiber and matrix. They were able to produce composites with tensile strength as high as 80% of glass fiber/PP composites.

Pracella et al. [25] prepared composites of hemp and PP through batch mixing. Addition of compatibilizers, fiber surface functionalization, and modification to the polymer matrix were all explored for improvement in fiber/matrix interactions. Hemp fibers were functionalized by melt grafting with glycidyl methacrylate (GMA). As expected, when compared to plain PP, all composites achieved higher tensile modulus and lower elongation at break (Table 1.1). Additionally, the authors found that compatibilization with PP-g-GMA (10 phr) effectively increased composite stiffness and attributed to an improved fiber–matrix interfacial adhesion.

Wambua et al. [26] conducted yet another study that focused on comparing the performance of natural fibers to glass fibers. In this work, they found kenaf, hemp, and sisal composites to have comparable tensile strength and modulus properties, but hemp outperformed kenaf with respect to impact strength (Figure 1.6). Coir fiber composites were reported to have achieved the worst mechanical performance with the exception of impact strength which was greater than that of jute and kenaf. The

TABLE 1.1

Tensile Properties of PP/Hemp and PP/Hemp/PP-g-GMA Composites

Sample (wt.%/wt.%/phr)	Tensile Modulus (GPa)	Stress at Max (MPa)	Stress at Break (MPa)	Elongation at Break (%)
PP	1.8 ± 0.2	31.8	19.8	18.0
PP/Hemp 90/10	2.9 ± 0.3	27.0	25.4	4.0
PP/Hemp 80/20	2.8 ± 0.2	25.7	24.8	2.8
PP/Hemp/PP-g-GMA 90/10/5	1.9 ± 0.2	27.9	27.2	3.5
PP/Hemp/PP-g-GMA 90/10/10	3.1 ± 0.3	27.8	26.9	2.6
PP/Hemp/PP-g-GMA 80/20/5	2.5 ± 0.5	23.1	22.2	2.2
PP/Hemp/PP-g-GMA 80/20/10	3.0 ± 0.5	25.7	25.3	2.3

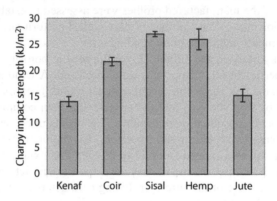

FIGURE 1.6 Charpy impact strengths of fiber reinforced polypropylene composites.

general conclusion made was that in most cases, the specific properties of natural fiber composites were comparable to those of glass.

Pickering et al. [27] aimed to optimize the growth time and digestion parameters for production of New Zealand hemp fiber for use in PP composites. A growing period of 114 days produced fibers with an optimal tensile strength of 857 MPa and a Young's modulus of 58 GPa. The strongest composite, with tensile strength and Young's modulus of 47.2 MPa and 4.88 GPa, respectively, consisted of PP with 40 wt% fiber and 3 wt% MAPP.

Although most studies focused on fiber surface modifications, Doan et al. [28] chose to investigate the effect of MAPP coupling agents on composite properties. They were able to improve the adhesion between PP and jute fibers significantly by the addition of 2 wt% MAPP to PP matrices, which resulted in improved composite mechanical performance.

Thermal degradation of natural fibers during processing is critical to avoid. Fung et al. [29] aimed to report on the processing of sisal fiber reinforced PP composites by injection molding. Tensile properties for high temperature (210°C) and low temperature (180°C) PP specimens were compared. The specimens that were injection molded at 210°C showed serious darkening as well as degradation. It was also observed that the emission of empyreuma odor occurred and the odor stayed for a few months without disappearing. In the case of low-temperature specimens that were injection molded at 180°C, they were light in color and without odor (Figures 1.7 and 1.8). Their results suggested that with 10 wt% of sisal fibers, Young's modulus increased by approximately 150% and tensile strength by about 10%. More importantly, they found that reinforcement efficiency for specimens produced at low temperature was slightly higher than high temperature samples.

Joseph et al. [30,31] attempted to expand the knowledge base on the use of sisal fibers in PP composites. In their study, they manufactured sisal/PP specimens by melt-mixing and solution-mixing methods. Effects of fiber content and chemical treatment on composite thermal properties were discussed. Composites with treated fibers showed superior properties to the untreated system. From the DSC measurements, crystallization temperature and crystallinity of the PP matrix increased with fiber content. These results were explained by the nucleating effect on the fiber surface which formed regions of trans-crystallinity.

Interested in the effects of compatibilizing agents, Han-Seung Yang et al. [32] chose rice-husk flour as a particle-reinforcement for PP. Mechanical and morphological properties were studied as a function of compatibilizing agent content. To prepare the samples, four loading levels of filler (10, 20, 30, and 40 wt%) and three levels of compatibilizer (1, 3, and 5 wt%) were used. Results indicated a decrease in tensile strength with increased filler loading, but tensile properties improved significantly with compatibilizing agent.

Sain et al. [33] reported on the flammability of PP, rice-husk filled PP, and the flame retarding effect of magnesium hydroxide (Figure 1.9). They found that magnesium hydroxide was effective in reducing the flammability of the composites by 50%, but with adverse effects on mechanical properties. Even with the help of coupling agents, the addition of flame-retardant negatively impacted tensile and flexural properties resulting in a 17% and 15% reduction, respectively, as compared to the composite without flame retardant.

Suarez et al. [34] performed experiments to understand the fracture mechanisms found in sawdust/PP composites. The discussion mainly focused on the morphological analysis of the fracture, tensile properties of the composites, and the effect of MAPP content on the mechanical properties. Fixed processing conditions were used to prepare composites of PP or PP plus MAPP and saw dust coated with 22.4 wt%

FIGURE 1.7 Tensile bars injection molded using the high temperature (HT-210°C) and low temperature (LT-180°C) settings.

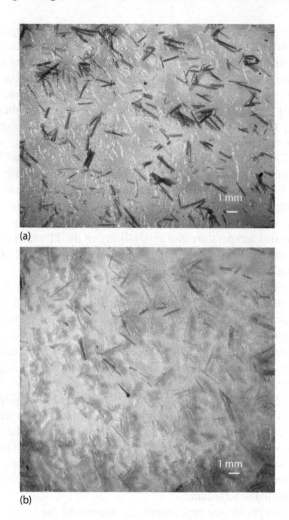

FIGURE 1.8 Photographs showing the degradation (and the related darkening) of the sisal fibers in the tensile bars injection molded with (a) high temperature (HT-210°C); and (b) low temperature (LT-180°C) settings.

MAPP. Scanning electron microscope (SEM) images showed the improved adhesion of sawdust to the PP matrix with increasing MAPP content. However, increasing the MAPP content of composites with up to 10% MAPP did not show any additional enhancement in tensile properties.

In an effort to improve the interfacial interaction between lufa fiber (LF) and PP matrix, Demir et al. [35] investigated the use of three different coupling agents: (1) (3-aminopropyl)-triethoxysilane (AS), (2) 3-(trimethoxysilyl)-1-propanethiol (MS), and (3) MAPP. They discussed the effects of the coupling agents on the mechanical properties in addition to the water sorption characteristics and morphology. They concluded that tensile strength and Young's modulus increased with coupling agents. Also, water absorption decreased with treatment which was attributed

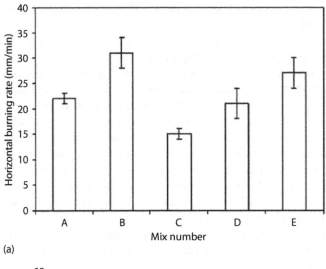

(a)

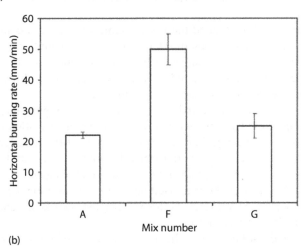

(b)

Formulation of the mixes

Ingredients	Mix number						
	A	B	C	D	E	F	G
PP	100	47.5	47.5	47.5	47.5	47.5	47.5
Sawdust	–	50	25	25	25	–	–
Rice husk	–	–	–	–	–	50	25
E-43 (coupling agent)	–	2.5	2.5	2.5	2.5	2.5	2.5
Magnesium hydroxide	–	–	25	20	20	–	25
Boric acid	–	–	–	5	–	–	–
Zinc borate	–	–	–	–	5	–	–

(c)

FIGURE 1.9 Horizontal burning rate of mixes A–G (a) and (b). Formulation of series (c).

to a better adhesion between fiber and matrix. Optimal mechanical properties were obtained with MS treatment.

Wheat straw fibers have great potential for use in injection molded applications with PP. Panthapulakka et al. [36] prepared fibers by mechanical and chemical processes and completed a full characterization with respect to chemical composition, morphology, and physical, mechanical, and thermal properties. The researchers also prepared composites of PP with 30% wheat straw fibers and their mechanical properties were evaluated. They suggested that chemically prepared fibers exhibited better mechanical, physical, and thermal properties. Composite properties were significantly enhanced compared to virgin PP. However, composite strength properties were less for the specimens with chemically prepared fiber as compared to mechanically prepared fibers. This was attributed to poor dispersion under the processing conditions used. They concluded that wheat straw fiber has potential as a reinforcing material for thermoplastic composites.

Jang and Lee [37], also interested in flame retardants for natural fiber composites, investigated the use of various retardants such as Saytex8010, TPP, Mg(OH)$_2$, and antimony trioxide.

Rozman et al. [38] investigated the use of lignin powder as a compatibilizer in coconut fiber/PP composites. Composites with lignin compatibilizers were found to have higher flexural properties relative to the control composites. However, increasing the lignin content resulted in reduced tensile properties. Through SEM, the improvement of interfacial characteristics was made evident.

Shibata et al. [39] conducted a study on key factors affecting the flexural modulus of kenaf/PP composites. The key factors involved in the investigation were (1) number of kenaf layers, (2) heating time, and (3) kenaf weight fraction. Increasing both the number of kenaf layers and heating time, the composite flexural modulus was improved. These results were explained by an improved homogeneous PP dispersion in the composite board from the increased number of layers allowing better contact between kenaf and PP. Moreover, a better wetting between kenaf and matrix was achieved with the longer heating time.

Mwaikambo et al. [40] illustrated the use of kapok/cotton fabric as reinforcement for PP and MAPP resins. Chemical treatment with acetic anhydride and sodium hydroxide proved to significantly enhance the thermal properties of the plant fibers, whereas poor results were observed when testing composite mechanical properties. Mercerization and weathering improved the toughness of the materials compared to the acetylated fiber composites. This was documented as an indication of the increased plasticization and ductility of the fibers and composites. Acetylated fiber composites showed less rigidity compared to other treatments. Tensile modulus was improved with the modified PP, but toughness was compromised due to the brittle nature of MAPP.

Tensile and flexural properties of pineapple leaf fiber/PP composites were reported by the work of Arib et al. [41]. The primary goal of the work was to evaluate the mechanical behavior as a function of fiber volume fraction. Results from the experiments indicated that the tensile strength and modulus of composites increased with fiber content as predicted with the rule of mixtures.

Czigany [42] conducted an evaluation on the mechanical properties of hybrid composites. The materials used were basalt, hemp, glass, and carbon fibers as reinforcement with a PP matrix. A carding, needle-punch, and press process was used to manufacture the hybrid composites. To enhance interfacial adhesion, the fibers were treated with maleic acid anhydride and sunflower oil. The mechanical and microscopic testing and analysis performed in the study proved that the surface-treated fibers enhanced the hybrid composite performance.

Using the injection molding process, Ruksakulpiwat et al. [43] prepared vetiver/PP composites with various fiber weight fraction and length. Comparing composite mechanical properties to the virgin matrix, tensile strength and Young's modulus were higher in the case of the composite, but elongation at break and impact strength were lower. With increasing fiber content, composite viscosity, heat distortion temperature, crystallization temperature, and Young's modulus all increased. Contrarily, the decomposition temperature, tensile strength, elongation at break, and impact strength decreased. Improved mechanical properties were observed with the treated vetiver grass.

Compatibilizing agents, maleic anhydride (MA)-grafted polypropylene (MAPP) and MA-grafted polyethylene (MAPE) were tested for their effects on thermal properties of bio-flour-filled/PP, and low density polyethylene (LDPE) composites. Kim et al. [44] found that thermal stability, storage modulus, glass transition temperature, and loss modulus peak temperature increased with increasing MAPP and MAPE content. The composite melting temperature was not significantly affected, but the crystallinity increased with MAPP and MAPE content. Enhanced thermal properties were attributed to the improved interfacial characteristics between flour and matrix.

The behavior of bamboo/PP and bamboo-glass/PP hybrid composites under hygrothermal aging and mechanical fatigue conditions was reported by Liao et al. [45]. Implementing the MAPP coupling agent, the moisture absorption and tensile strength degradation was suppressed in both composite systems.

Abu-Sharkh and Hamid have investigated the date palm leaves reinforcement with PP and UV stabilizers to form composite materials [46]. It is found that PP/date palm leaves fiber composites are found to be much more stable than PP under the severe natural weathering conditions of Saudi Arabia and in accelerated weathering conditions. It is mentioned that compatibilizing has a deteriorating effect on the composites due to the lower stability of the maleated polypropylene.

Hemp fiber reinforced PP composites exhibited interesting recyclability [47]. The obtained results prove that the mechanical properties of hemp fiber/PP composites remain well preserved, despite the number of reprocessing cycles. The Newtonian viscosity decreases with cycles, indicating a decrease in molecular weight and chain scissions induced by reprocessing. The decrease of fiber length with reprocessing could be another reason for the decrease of viscosity.

The mechanical properties of corona-treated hemp fiber/PP composites were characterized by tensile and compressive stress–strain measurements [48]. The treatment of compounds (fibers or matrix) leads to a significant increase in tensile strength. The modification of cellulosic reinforcements rather than PP matrix allows greater improvement of the composites properties with an enhancement of 30% of Young modulus. The observation of the fracture surfaces enables discussion about adhesion improvement.

Mohanty et al. [49] used MAPP as a coupling agent for the surface modification of jute fibers. It was found that a fiber loading of 30% with a MAPP concentration of 0.5% in toluene and 5 min of impregnation time with 6 mm average fiber lengths resulted in the best possible outcomes. An increase in flexural strength of 72.3% was observed for the treated composites.

The influence of MAH/PP on the fiber–matrix adhesion in jute fiber-reinforced PP composites and on the material behavior under fatigue and impact loadings was investigated [50]. The fiber–matrix adhesion was improved by treating the fibers' surface with the coupling agent MAH/PP. It was shown that a strong interface is connected with a higher dynamic modulus and reduction in stiffness degradation with increasing load cycles and applied maximum stresses. The specific damping capacity resulted in higher values for the composites with poorly bonded fibers. Furthermore, the stronger fiber–matrix adhesion reduced the loss-energy for non-penetration impact tested composites about roughly 30%. Tests, which were performed at different temperatures, showed higher loss energies for cold and warm test conditions compared to those carried out at room temperature. The post-impact dynamic modulus was roughly 40% after 5 impact events and was 30% lower for composites with poor and good fiber–matrix adhesion, respectively.

Thermoforming has proven to enable the successful fabrication of kenaf fiber reinforced PP sheets into sheet form [51]. The optimal fabrication method found for these materials was at the compression molding process, which utilizes a layered sifting of a micro-fine PP powder and chopped kenaf fibers. The fiber content (30 and 40 wt%) provided adequate reinforcement to increase the strength of the PP matrix. The kenaf/PP composites compression molded in this study proved to have superior tensile and flexural strength when compared to other compression molded natural fiber composites such as other kenaf, sisal, and coir reinforced thermoplastics. With the aid of the elastic modulus data, it was also possible to compare the economic benefits of using kenaf composites instead of other natural fibers and E-glass. The manufactured kenaf maleated PP composites have a higher modulus/cost and a higher specific modulus than sisal, coir, and even E-glass.

Thus, they provide an option for replacing existing materials with a higher strength, lower cost alternative that is environmentally friendly. Hybrid composites of wood flour/kenaf fiber and PP were prepared to investigate the hybrid effect on the composite properties [52]. The results indicated that while nonhybrid composites of kenaf fiber and wood flour exhibited the highest and lowest modulus values, respectively, the moduli of hybrid composites were closely related to the fiber to particle ratio of the reinforcements. With the help of the hybrid mixtures equation, it was possible to predict the elastic modulus of the composites better than when using the Halpin–Tsai equation.

Magnesium hydroxide and zinc borate were incorporated into sisal/PP composites as flame retardants [53]. Adding flame retardants into sisal/PP composites reduced the burning rate and increased the thermal stability of the composites. No synergistic effect was observed when both magnesium hydroxide and zinc borate were incorporated into the sisal/PP composites. In addition, the sisal/PP composites exhibited insignificant differences in shear viscosity at high shear rates indicating that the types of flame retardants used in this study had no impact on the processability of the composites. The sisal/PP composites which flame retardants were added to exhibited tensile and flexural properties comparable to those of the sisal/PP composites, which flame retardants had not been added to.

Ramie fiber reinforced PP composites were fabricated using a hybrid method of melt-blending and injection molding processes [54]. Different ramie fiber/PP composites were fabricated by varying the fiber length, fiber content, and method of fiber pretreatment. The results exhibited increases in fiber length and fiber content and also show increased tensile strength, flexural strength, and compression strength noticeably in turn. Yet, they also result in negative influences on the impact strength and elongation behavior of the composites.

The development of composites for ecological purposes (eco-composites) using bamboo fibers and their basic mechanical properties was evaluated [55]. The steam explosion technique was applied to extract bamboo fibers from raw bamboo trees. The experimental results showed that the bamboo fibers (bundles) had a sufficient specific strength, equivalent to that of conventional glass fibers. The tensile strength and modulus of PP-based composites increased about 15% and 30% when using steam-exploded fibers. This increase was due to good impregnation and a reduction of the number of voids, in comparison to composites using fibers that were mechanically extracted.

The effects of MAH/PP coupling agents on rice-husk flour reinforced PP composites were evaluated [56]. The tensile strengths of the composites decreased as the filler loading increased, but the tensile properties were significantly improved with the addition of the coupling agent. Both the notched and unnotched Izod impact strengths remained almost the same with the addition of coupling agents. A morphological study revealed the positive effect of a compatibilizing agent on interfacial bonding.

The compression and injection molding processes were performed in order to evaluate which is the better mixing method for fibers (sugarcane bagasse, bagasse cellulose, and benzylated bagasse) and PP matrixes [57]. The injection molding process performed under vacuum proved to work best. Composites were obtained with a homogeneous distribution of fibers and without blisters. However, the composites did not have good adhesion between the fiber and the matrix according to their mechanical properties.

KC et al. [58] illustrated an application of Taguchi method to optimize injection molding process parameters of sisal–glass fiber hybrid PP biocomposite. Six parameters that influence the flow and cross-flow shrinkage such as injection pressure, melt temperature, mold temperature, holding pressure, cooling time and holding time were selected as variables and two hybrid biocomposites were used with different content of sisal (SF) and glass fiber (GF); SF20GF10 and SF10GF20. For the experimental design, L18 orthogonal array with a mixed-level design and signal-to-noise (S/N) of smaller-the-better was used. Figure 1.10 illustrates the S/N ratios for each factor and their levels. Optimal factor levels for flow shrinkage of both hybrid biocomposites were injection pressure 90 bars, melt temperature 210°C, mold temperature 40°C, cooling time 40 s, and hold time 6 s. However, optimal holding pressure was 70 bar and 50 bars for SF20GF10 and SF10GF20, respectively. In comparison with the flow shrinkage, the response of each factor on x-flow shrinkage

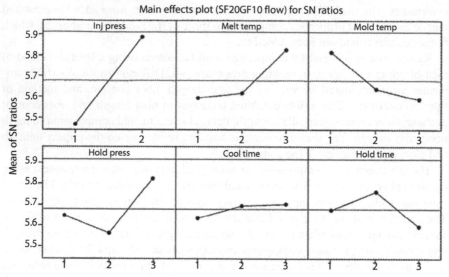

Signal-to-noise: smaller is better
(a)

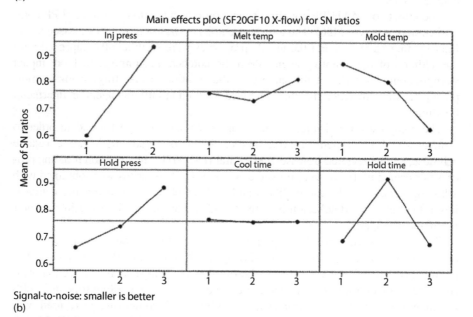

Signal-to-noise: smaller is better
(b)

FIGURE 1.10   Main effects plots of S/N ratios for flow (a and c) and x-flow shrinkage (b and d) of SF20GF10 and SF10GF20.   (*Continued*)

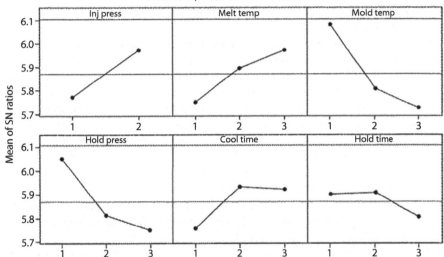

Signal-to-noise: smaller is better

(c)

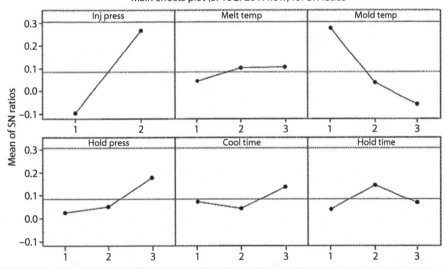

Signal-to-noise: smaller is better

(d)

FIGURE 1.10 (CONTINUED) Main effects plots of S/N ratios for flow (a and c) and x-flow shrinkage (b and d) of SF20GF10 and SF10GF20.

was similar. Moreover, S/N ratio for x-flow shrinkage was smaller than the ratio for flow shrinkage possibly due to the lower influence of injection molding parameter on x-flow shrinkage.

Various investigations of flax fiber/polypropylene composites have been completed. These studies focus on many different variables, including: comparison between natural fiber thermoplastic mat (NMT) and glass fiber thermoplastic mat (GMT) [59], the influence of fiber/matrix modification and glass fiber hybridization [60], the influence of surface treatment on interface by glycerol triacetate, thermoplastic starch, -methacryl oxypropyl trimethoxy-silane and boiled flax yarn [61], comparison of matrices (PP and PLA) on the composite properties [62], and the influence of processing methods [63]. Buttler [64] presented the feasibility of using flax fiber composites in the coachwork and bus industry.

Jute fiber reinforced PP composites were evaluated regarding the effect of matrix modification [65], the influence of gamma radiation [66], the effect of interfacial adhesion on creep and dynamic mechanical behavior [67], the influence of silane coupling agent [68,69], and the effect of natural rubber [70].

The influence of treated coir fiber on the physicomechanical properties of coir/PP composites [71,72], and the effect of fiber physical, chemical, and surface properties on the thermal and mechanical properties of coir/PP composites were investigated and evaluated [73].

The using of rice husk as filler for rice husk/PP composites [74], the thermogravimetric analysis of rice-husk filled HDPE and PP composites [75], the effect of different concentrations and sizes of particles of rice-husk ash in the mechanical properties of rice husk/PP composites [76], the effect of different coupling agent on rice husk/copolymer PP composites [77], the effects of oil extraction, compounding techniques and fiber loading [78], and the effect of matrix modification [79] on the mechanical properties of oil palm empty fruit bunch filled PP composites were examined. In addition, a comparative study of oil palm empty fruit bunch fiber/PP composites and oil palm derived cellulose/PP composites [80] were studied.

1.1.2 POLYETHYLENE COMPOSITES WITH NATURAL FIBERS

Flax fiber was reinforced with recycled high density polyethylene manufactured by a hand lay-up and compression molding technique [81]. Figure 1.11 illustrates a significant increase in toughness, the impact energy of the composite by some 50 kJ/m^2 over that of recycled HDPE. It is indicated that the flax fiber plays an important role in the toughening processes. It is also mentioned that such large toughness enhancement has come from a number of deformation and failure mechanisms acting in the notch tip process zone (type 1 toughening mechanisms) and in the crack wake zone (type 2 toughening mechanisms).

Thermal diffusivity, thermal conductivity, and specific heat of flax fiber/HDPE biocomposites were evaluated [82]. The thermal conductivity, thermal diffusivity, and specific heat of the flax/HDPE composites decreased with increasing fiber content, but the thermal conductivity and thermal diffusivity did not change significantly at temperatures in the range (170–200°C) studied (Table 1.2). The specific heat of the biocomposites increased gradually with temperature.

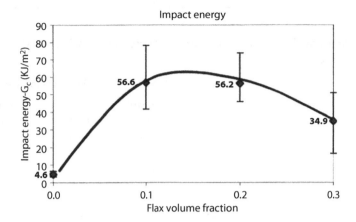

FIGURE 1.11 Toughness of flax/HDPE composites showing a peak at about Vf = 0:15.

Traditionally the mechanism of moisture absorption is defined by diffusion theory, but the relationship between the microscopic structure-infinite 3D-network and the moisture absorption could not be explained. Wang et al. [83] introduced the percolation theory and a percolation model was developed to estimate the critical accessible fiber ratio—the moisture absorption and electrical conduction behavior of composites. At high fiber loading when fibers are highly connected, the diffusion process is the dominant mechanism; while at low fiber loading close to and below the percolation threshold, the formation of a continuous network is key. Hence, percolation is the dominant mechanism. The model can be used to estimate the threshold value, which can, in turn, be used to explain moisture absorption and electrical conduction behavior. The composite started showing conductivity after it absorbed approximately 50% of maximum moisture. After this, conductivity increased quickly with further moisture absorption. The pattern of the increment of electrical conductivity suggests a diffusion process of moisture absorption.

The hygrothermal weathering properties of rice hull reinforced HDPE composites were studied by Wang et al. [84]. The samples (50% rice hull and 50% HDPE) absorbed 4.5% moisture after 2000 h of exposure to a relative humidity (RH) of 93% at 40°C. The walls of the samples swelled significantly (7.1%) in thickness and ultimately developed about 5 mm of longitudinal bowing based on 61 cm long boards. Both expansion and bowing were partially recovered after another 2000 h exposure to 20% RH at 40°C. Deformation results in increased moisture contents. Temperature also played a significant role in causing direct thermal expansion or contraction and by affecting the rate and the amount of moisture adsorption.

Sisal fiber reinforced PE [85] and HDPE [86] composites were examined regarding their interfacial properties, isothermal crystallization behavior, and mechanical properties. Sisal fiber treatment with stearic acid increased the interfacial shear strength by 23% compared to untreated fibers in sisal/PE composites (Figure 1.12). Permanganate ($KMnO_4$) and dicumyl peroxide (DCP) roughen the fiber surface and introduced mechanical interlocking with the HDPE. Sisal fiber reinforced HDPE shows a stable debonding process with Permanganate and DCP treatment.

TABLE 1.2

Specific Heat and Thermal Diffusivity of HDPE and Flax Fiber–HDPE Biocomposites

Materials	Temperature (°C)	Specific Heat (kJ/kg°C)		Thermal Diffusivity (m²/s)	
		Mean	Standard Deviation	Mean	Standard Deviation[b]
10% fiber biocomposite[a]	170	2.44	0.01	1.68×10^{-7}	7.45×10^{-9}
	180	2.47	0.02	1.66×10^{-7}	7.49×10^{-9}
	190	2.50	0.04	1.64×10^{-7}	7.62×10^{-9}
	200	2.52	0.04	1.63×10^{-7}	7.69×10^{-9}
20% fiber biocomposite[a]	170	2.28	0.02	1.59×10^{-7}	1.51×10^{-8}
	180	2.29	0.01	1.58×10^{-7}	1.50×10^{-8}
	190	2.32	0.01	1.56×10^{-7}	1.48×10^{-8}
	200	2.34	0.01	1.55×10^{-7}	1.47×10^{-8}
30% fiber biocomposite[a]	170	2.25	0.02	1.40×10^{-7}	1.34×10^{-8}
	180	2.27	0.02	1.39×10^{-7}	1.32×10^{-8}
	190	2.29	0.03	1.38×10^{-7}	1.32×10^{-8}
	200	2.30	0.03	1.37×10^{-7}	1.31×10^{-8}
	170	2.48	0.04	1.82×10^{-7}	4.40×10^{-9}
	180	2.50	0.04	1.81×10^{-7}	4.30×10^{-9}
	190	2.52	0.04	1.79×10^{-7}	4.30×10^{-9}
HDPE	200	2.54	0.06	1.78×10^{-7}	5.00×10^{-9}

[a] Flax fiber–HDPE biocomposite.

[b] Standard deviation of thermal diffusivity (SDα) was calculated according to the equation

$$SD_\alpha = \sqrt{\left(\frac{\partial \alpha}{\partial k}SD_k\right)^2 + \left(\frac{\partial \alpha}{\partial \rho}SD_\rho\right)^2 + \left(\frac{\partial \alpha}{\partial C_p}SD_{C_p}\right)^2}$$

where SD_k = standard deviation of thermal conductivity, SD_ρ = standard deviation of density, and SD_{C_p} = standard deviation of specific heat.

Choudhury [87] showed that silane treated sisal fiber reinforced HDPE exhibits an unstable debonding process. A high crystallization rate and short crystallization half-time were observed for sisal/HDPE composites compared to neat HDPE (Figure 1.13). This demonstrates the strong nucleating abilities of sisal fibers. With increasing fiber content, the crystallization activation energy of the composites decreased.

The effects of extrusion processing temperature on the various properties of kenaf fiber/high-density polyethylene (HDPE) composites were investigated. It is concluded with its best performance of thermo-mechanical and tensile properties for low (LPT) and (HPT) high processing temperatures [88]. It is illustrated that at HPT, pure HDPE and kenaf fiber/HDPE composites show a higher tensile modulus compared to those processed at LPT and the tensile modulus values increase with increasing fiber loading when processed at HPT (Figure 1.14).

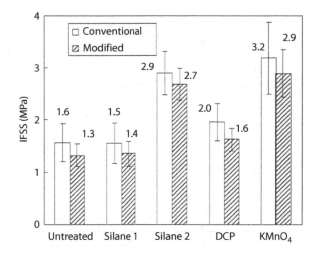

FIGURE 1.12   Interfacial shear strength of sisal fiber reinforced HDPE composites.

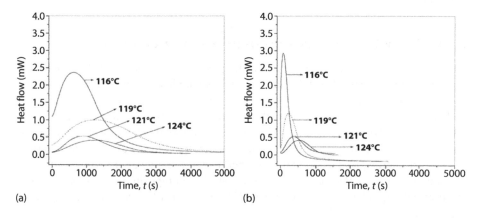

FIGURE 1.13   Heat flow curves of (a) neat HDPE and (b) 80:20 HDPE/sisal composite during isothermal crystallization.

Abdelmouleh et al. [89] illustrated the mechanical properties of the cellulose reinforced polyethylene composites and it is found that mechanical properties increased with increasing the average fiber length. It is also mentioned that cellulose fibers treated with g-methacryloxypropyltrimethoxy silane (MPS) and g-mercaptopropyl-trimethoxy silanes (MRPS) displayed good mechanical performances and with hexadecyltrimethoxy-silanes (HDS) bearing merely aliphatic chain only a modest enhancement of composite properties.

Georgopoulos et al. [90] have used residues from eucalyptus wood, ground corn-cob, and brewery's spent grain as reinforcing fillers with LDPE matrix. It is found that all natural fillers lead to a decrease in tensile strength of the LDPE but Young's modulus increased due to the higher stiffness of the fibers.

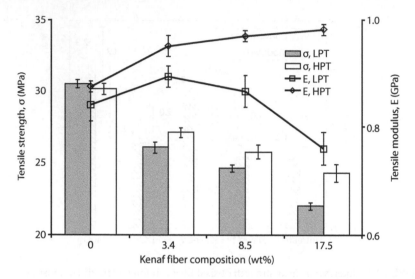

FIGURE 1.14 The tensile modulus and tensile strength of pure HDPE and kenaf fiber/ HDPE composites.

1.1.3 Other Thermoplastic Composites with Natural Fibers

Only limited studies were reported regarding the usage of PS and PVC as matrixes for natural fibers. The thermal and dynamic mechanical behavior [91], as well as the rheological [92] and mechanical properties [93] of PS composites reinforced with sisal fibers, were studied. The PS composites reinforced with agave fiber were also studied regarding their fiber surface modification [94]. The impact properties of PVC composites reinforced with bamboo fibers [95], the influence of fiber type (i.e., bagasse, rice straw, rice husk, and pine fiber) and loading level of styrene-ethylene-butylene-styrene (SEBS) block copolymer on PVC composite properties [96], and the thermal degradation of abaca fiber [97] reinforced PVC composites were also evaluated.

Zheng et al. [98] evaluated the influence of surface treatments of bagasse fiber (BF) with benzoic acid and the mechanical properties of BF-polyvinyl chloride (PVC) composite. It is represented that the ratio of PVC/BF, the content of benzoic acid, and processing temperature had a significant effect on the mechanical properties of the composite, which was examined by the orthogonal optimal method (Figure 1.15). It is seen that the interaction graphs of tensile strength against the three variables, which illustrates the changes in tensile strength with different factors, and the tensile strength clearly increased as the content of BF and the content of the modifier increased.

The influence of fiber content on mechanical properties of kenaf fibers reinforced poly (vinyl chloride)/thermoplastic polyurethane poly-blend (PVC/TPU/KF) composites were investigated [99]. It was found that the tensile strength, impact strength, and tensile strain decrease while tensile modulus increases with an upsurge in fiber content. In addition, high impact strength was observed with 40% fiber content. It is

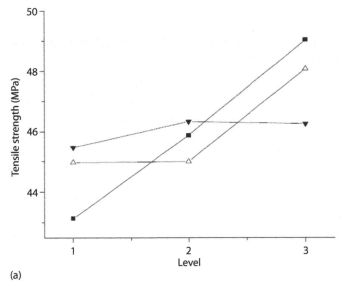

(a)

Factors and levels of the orthogonal optimal method

Levels	Factors		
	BF content (%) (factor A)	Benzoic acid content (%) (factor B)	Fiber-treatment temperature (°C) (factor C)
1	15	3	140
2	25	5	150
3	35	10	160

(b)

FIGURE 1.15 (a) Interaction graphs of tensile strength against three variables. (■) BF content, (△) benzoic acid content, (▼) treatment temperature. (b) Factors and levels of the orthogonal optimal method.

also mentioned that thermal degradation took place in three steps. Composites, as well as the matrix, had a similar stability in the first step and at the second step, the matrix showed a slightly better stability than the composites. Finally, at the last step, the composites showed better stability than the matrix (Table 1.3).

Sapun et al. [100] investigated the mechanical properties of soil buried kenaf fiber reinforced thermoplastic polyurethane (TPU) composites. It was found that tensile strength of kenaf fiber reinforced TPU composite dropped to 16.14 MPa from 28.68 MPa (before soil burial test) after 80 days of soil burial test (Figure 1.16). It was also observed that there was no significant change in flexural properties of soil buried kenaf fiber reinforced TPU composite specimens.

Curaua fibers were reinforced with PA-6 by using a co-rotating twin-screw extruder [101]. The fiber contents of 20 wt%, 30 wt%, or 40 wt% and fiber lengths of 0.1 or 10 mm were studied. Fibers were treated with N_2 plasma or washed with NaOH solution to improve their adhesion to PA-6. The tensile and flexural properties

TABLE 1.3

Mass Loss Percentage of Pure PVC/TPU, PVC/TPU/KF 20%, PVC/TPU/KF 30%, PVC/TPU/KF 40%, and Kenaf Fiber

	Weight Loss (%)						
Sample	$T_{100.83}$ °C	$T_{252.5}$ °C	$T_{299.16}$ °C	$T_{351.66}$ °C	$T_{398.33}$ °C	$T_{450.83}$ °C	$T_{503.33}$ °C
Pure PVC/TPU	0.58	5.26	51.10	65.17	74.69	84.38	90.89
PVC/TPU/KF 20%	0.14	4.52	47.77	62.10	69.48	78.39	84.78
PVC/TPU/KF 30%	0.81	5.84	49.58	61.78	68.25	75.79	81.61
PVC/TPU/KF 40%	0.89	6.66	51.75	63.19	69.09	76.46	82.13
Kenaf fiber	8.54	10.98	20.96	50.62	76.50	79.98	82.86

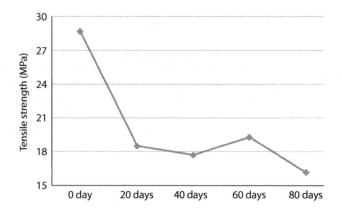

FIGURE 1.16 Tensile strength of soil buried kenaf reinforced TPU composites.

of this composite are better than unfilled but are still lower than those of glass fiber reinforced polyamide-6. Table 1.4 shows that the lower fiber content and the shorter fiber length improve the tensile strength of the composite and the higher fiber content and the shorter fiber length improves the tensile modulus of the composite.

A natural fiber blend from a mixture of kenaf, flax, and hemp fibers was added to nylon 6 using melt mixing to produce compounded pellets [102]. Injection molded natural fibers/nylon 6 composites were prepared with varying concentrations of natural fibers (from 5 to 20 wt%). Due to the addition of natural fiber blend, the tensile and flexural properties of the nylon 6 composites were significantly increased. The natural fiber blend weight fraction of 20% showed the maximum strength and modulus of elasticity of the nylon 6 composites. The incorporation of natural fibers lowered the Izod impact strength of composites without any surface treatments and a coupling agent. The melt flow index of the composites was decreased with increasing natural fiber blend concentration. It is found that the tensile and flexural modulus of elasticity is in accordance with the rheological data from the MFI measurements. It is mentioned that the higher mechanical results with a lower density of a natural

TABLE 1.4

Tensile Strength (σ_{max}), Elastic Modulus (E), and Strain at Break (ε_b) of the PA 6-Curaua Fiber Composites (Short Fiber: 0.1 mm, Long Fiber: 10 mm)

	Tensile Tests		
Sample	σ_{max} (MPa)	E (GPa)	ε_b (%)
PA-6	44 (±3)	0.9 (±0.3)	35 (±13)
PA-6 + 20 wt% short fiber	34 (±10)	1.9 (±0.1)	3 (±2)
PA-6 + 30 wt% short fiber	34 (±6)	2.1 (±0.1)	3 (±1)
PA-6 + 40 wt% short fiber	30 (±3)	2.4 (±0.2)	1.7 (±0.2)
PA-6 + 20 wt% long fiber	30 (±3)	1.8 (±0.1)	3 (±1)
PA-6 + 30 wt% long fiber	32 (±8)	2.1 (±0.2)	2 (±1)
PA-6 + 40 wt% long fiber	26 (±3)	2.8 (±0.1)	1.4 (±0.2)

fiber blend can be used as a sufficient reinforcing material for low-cost, eco-friendly composites in the automotive industry.

High performance thermoplastic matrices such as polyamides (PA 6 and PA6,6) instead of the commonly used polyolefins were used to develop natural fiber composites for substituting glass fibers without renouncing their mechanical properties [103]. Flax, jute, pure cellulose, and wood pulps have been melt compounded with different polyamides to analyze the effect of fiber content on mechanical properties. Thermogravimetrical analysis was performed to determine the thermal behavior of the different fibers to know the boundary for processing at high temperatures since the melting points of the polyamides are much higher than those of polyolefins and this could lead to a higher degradation of the natural fibers. Flexural and tensile properties indicated that natural fibers are effective in reinforcing with polyamide 6 and polyamide 6,6 because an increase in both strength and modulus has been observed for all the fibers with respect to the unreinforced matrices. It is also observed that longer mixing times seemed to improve fiber dispersion since mechanical properties slightly increased and both torque and viscosity of composites were reduced by increasing the mixing time.

Banana fiber reinforced composites based on high density polyethylene/nylon-6 blends were prepared via a two-step extrusion method [104]. The coupling agents maleic anhydride grafted styrene/ethylene–butylene/styrene triblock polymer (SEBS-g-MA) and maleic anhydride grafted polyethylene (PE-g-MA) were used to enhance impact performance and interfacial bonding between banana fibers and the resins. It is observed that the presence of SEBS-g-MA enhanced the strengths and moduli of HDPE/Nylon-6 based composites compared with corresponding HDPE based composites.

The flexural strength and modulus have continuously improved up to 48.2% banana fiber content at a fixed weight ratio of PEg-MA to banana fiber, while impact toughness was lowered gradually (Table 1.5). At 29.3 wt.% banana fiber content, the experimental tensile modulus and the predicted tensile modulus by the Hones–Paul model for 3D random fiber orientation agreed well. It is also mentioned that the

TABLE 1.5

Effect of Banana Fiber Loading on Mechanical Properties of Banana Fiber Filled Composites Based on HDPE/Nylon-6/SEBS-g-MA (80/20/6, w/w) Blend

BaF (wt.%)	Tensile Strength (MPa)	Tensile Modulus (GPa)	Izod Notched Impact Strength (kJ/m²)	Flexural Strength (MPa)	Flexural Modulus (GPa)
29.3	25.5 ± 0.6	1.15 ± 0.05	8.66 ± 0.43	31.7 ± 1.0	1.29 ± 0.05
38.8	34.5 ± 0.5	2.13 ± 0.06	7.02 ± 0.18	47.1 ± 1.0	2.31 ± 0.03
48.2	33.8 ± 0.4	2.63 ± 0.11	6.45 ± 0.34	52.2 ± 2.4	2.55 ± 0.13

Note: The weight ratio of PE-g-MA to BaF was fixed at 7.5 wt.%.

randomly oriented fiber models underestimated experimental data at higher fiber levels. The presence of SEBS-g-MA had a positive influence on reinforcing effect of the Nylon-6 component in the composites, which was observed. The fractionated crystallization of the nylon-6 component in the composites was induced by the addition of both SEBS-g-MA and PE-g-MA.

Thitithanasarn et al. [105] tried to improve the thermal stability of jute fibers with the use of flexible epoxy surface coating that could facilitate processing with engineering thermoplastics. Compression molded jute fabric and Polyamide 6 (PA6) composites were manufactured, and the thermal decomposition characteristics of the jute fabric were evaluated by using thermogravimetric analysis. It is seen that thermal degradation resistance of jute fabric was improved by coating with flexible epoxy resin and the flexural modulus was improved with increasing curative concentration.

Neher et al. [106,107] investigated the palm fiber (PF) reinforcement in acrylonitrile butadiene styrene (ABS) matrix, which was prepared by employing an injection molding machine (IMM). Three sets of samples were prepared by injection molding process for three different wt% (5%, 10%, and 20%) of fiber contents. The mechanical (tensile strength, flexural stress, micro hardness, and Leeb's rebound hardness), physical (bulk density and water absorption) properties, surface morphology, microstructure (if it is crystalline or noncrystalline), and new bond formation after preparation of the composites were measured. The observed result reveals that the tensile strength (TS) and flexural stress (FS) were decreased with increasing fiber contents in the PF-ABS composites except 10% fiber content and it is clear that the palm fiber, ABS, and PF-ABS composites are amorphous in nature. It is also mentioned that there is no new bond formed after the addition of palm fiber in ABS polymeric matrix to create PF-ABS composites measured by FTIR. Figure 1.17 represents that micro hardness of palm fiber-ABS composites shows more or less the same hardness for different loads [106].

Saba et al. [108] reviewed the dynamic mechanical properties of natural fiber reinforced polymer composites. Tables 1.6 and 1.7 illustrate the reported work on the dynamic mechanical analysis of natural fibers reinforced thermoplastic polymer composites and natural fibers reinforced hybrid thermoplastic polymer composites.

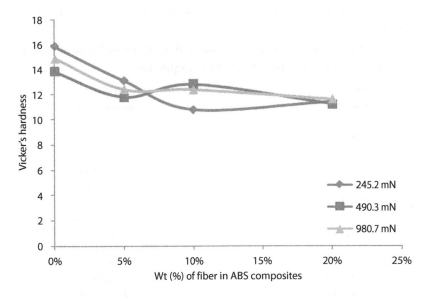

FIGURE 1.17   Vicker's hardness versus wt (%) of fiber in ABS composites for different loads.

1.1.4 NATURAL FIBER–THERMOPLASTIC COMPOSITES IN AUTOMOTIVE

Nowadays, especially in Europe, many automotive components (interior and exterior) are made from bio-fiber reinforced composite materials which are mainly based on polypropylene with reinforcing jute, flax, hemp, and kenaf fibers.

Ford Motor Company in Germany is using kenaf fibers imported from Bangladesh in their model Ford Mondeo and the door panels of the Ford Mondeo are manufactured by kenaf reinforced PP composites. Ford North America is also using kenaf fiber reinforced PP composites for the bolsters inside the doors of its new Escape SUV model. In both parts, kenaf fiber content is 50 wt% with PP. Ford Motor has mentioned that using 50 wt% kenaf fiber in one part reduced 25% of weight of that part compared to the current one and improves fuel economy. Ford is also using wheat straw derived PP composites for the interior storage bins in the Ford Flex model and coconut fiber reinforced PP composites for the load floor in the Focus BEV model. Hemp fiber reinforced PP composites were implemented in a Ford Montagetraeger.

General Motors has introduced flax fiber reinforced PP composites in their trim and shelving components in the Chevrolet Impala model. The Lexus CT200h has implemented bamboo fibers with PE in its luggage compartment, speakers, and floor mats. A number of Mercedes models have flax fiber reinforced PE composites in the engine and transmission cover. Underbody panels made by compression molded flax fiber reinforced PP composites for the A Class Daimler Chrysler cars. Open Corsa introduced the inner door panels manufactured by flax fiber reinforced PP composites. Flax fiber reinforced PP composites were also used for Freightliner century COE C-2 heavy trucks as well as rear shelf trim panels for model 2000 Chevrolet Impala, implemented by Cambridge Industry in Michigan.

TABLE 1.6

Reported Work on Dynamic Mechanical Analysis of Natural Fibers Reinforced Thermoplastic Polymer Composites

Reinforcement	Matrix
Short coir fiber	Natural rubber
Kenaf fiber	HDPE
Short hemp fiber	Polypropylene
Short sisal fiber	Polystyrene
Wood flour	Polypropylene
Pineapple leaf fiber	Polypropylene
Short hemp fiber	Polypropylene
Hemp fiber	Polypropylene
Jute fiber	Polypropylene
Sisal fiber	Rubber seed oil polyurethane
Short jute fiber	Polypropylene
Oil palm microfibril	Acrylonitrile butadiene rubber
Oil palm microfibril	Natural rubber
(MAPE) modified jute fiber	HDPE
Doum palm fiber	Polypropylene
Chicken feathers	Poly(methyl methacrylate)
Alfa fiber	Polyvinylchloride
Short henequen fiber	Polyethylene
Unidirectional and twill 2/2 flax fiber	Polypropylene
Modified jute fiber	Polypropylene
Oil palm fiber	LLDPE
Treated argan nut shell particles	HDPE
Pineapple fiber	Polyethylene

TABLE 1.7

Reported Work on Dynamic Mechanical Analysis of Natural Fibers Reinforced Hybrid Thermoplastic Polymer Composites

Reinforcement	Thermoplastic Matrix
Pine/agave fiber	High density polyethylene
Sisal/oil palm fiber	Natural rubber
Kenaf fiber/wood flour/rice hulls/newsprint fiber	Polypropylene
Short hemp fiber/glass fiber	Polypropylene
Kenaf, hemp, flax/glass fiber	Polypropylene
Short carbon fiber/kenaf fiber	Natural rubber
Flax/hemp fiber	Polypropylenes
Short bamboo/glass fiber	Polypropylene

1.2 CONCLUSIONS

The petroleum-derived thermoplastics PP and PE are the two most commonly employed thermoplastics in natural fiber reinforced composites. Many automotive components (interior and exterior) are now made from bio-fiber reinforced composite materials which are mainly based on polypropylene with reinforcing bio fibers such as jute, flax, hemp, kenaf, and wood. Day by day, there is great interest in developing natural fiber composites with PA to replace the glass fiber reinforced PA composites parts in the automotive area. It is also mentioned that natural fiber reinforced thermoplastic composites research and their applications significantly increase day by day rather than thermoset matrix, mainly due to their recyclability. Also, the choice of a thermoplastic matrix fits well within the eco-theme of biocomposites, but there are some important limitations on the recyclability and mechanical performance of thermoplastics. Owing to significant weight and cost savings, natural fiber reinforced thermoplastics composites are becoming attractive alternatives to glass fiber reinforced thermoplastic and thermoset composites. However, further research is still required to overcome obstacles such as moisture absorption and increased long-term stability for use as exterior automotive components. Advances in the construction of very large panels, structural design, and cost-effective manufacturing processes are still required.

REFERENCES

1. Pervaiz, M. and Sain, M.M. Carbon storage potential in natural fiber composites. *Resour Conserv Recycl*, 2003, 39, 325–340.
2. Carus, M. and Partanen, A. Biocomposites in the automotive industry. *Bioplastics Magazine*, 2016, 11, 16–17.
3. Pereira, P.H.F., Rosa, M.F., Cioffi, M.O.H., Benini, K.C.C., Milanese, A.C., Voorwald, H.J.C., and Mulinari, D.R. Vegetable fibers in polymeric composites: A review. *Polimeros*, 2015, 25(1), 9–22.
4. Hao, A., Zhao, H., and Chen, J.Y. Kenaf/polypropylene nonwoven composites: The influence of manufacturing conditions on mechanical, thermal, and acoustical performance. *Compos Part B Eng*, 2013, 54, 44–51.
5. Shibata, S., Cao, Y., and Fukumoto, I. Lightweight laminate composites made from kenaf and polypropylene fibres. *Polym Test*, 2006, 25, 142–148.
6. Bledzki, A.K., Mamun, A.A., Lucka-Gabor, M., and Gutowski, V.S. The effects of acetylation on properties of flax fibre and its polypropylene composites. *Express Polymer Letters*, 2008, 413–422.
7. Bledzki, A.K., Franciszczaka, P., Osmanb, Z., and Elbadawi, M. Polypropylene biocomposites reinforced with softwood, abaca, jute, and kenaf fibers. *Industrial Crops and Products*, 2015, 70, 91–99.
8. Bledzki, A.K., Faruk, O., and Mamun, A.A. Influence of compounding processes and fibre length on the mechanical properties of abaca fibre–polypropylene composites. *Polimery*, 2008, 53, 35–40.
9. Bledzki, A.K., Mamun, A.A., and Faruk, O. Abaca fibre reinforced PP composites and comparison with jute and flax fibre PP composites. *Express Polymer Letters*, 2007, 1, 755–762.
10. Bledzki, A.K., Mamun, A.A., Jaszkiewicz, A., and Erdmann, K. Polypropylene composites with enzyme modified abaca fibre. *Composites Science and Technology*, 2010, 70, 854–860.

11. Khondker, O.A., Ishiaku, U.S., Nakai, A., and Hamada, H. A novel processing technique for thermoplastic manufacturing of unidirectional composites reinforced with jute yarns. *Composites: Part A*, 2006, 37(12), 2274–2284.
12. Espert, A., Vilaplana, F., and Karlsson, S. Comparison of water absorption in natural cellulosic fibres from wood and one-year crops in polypropylene composites and its influence on their mechanical properties. *Composites: Part A*, 2004, 35, 1267–1276.
13. Ljungberg, N., Cavaille, J.Y., and Heux, L. Nanocomposites of isotactic polypropylene reinforced with rod-like cellulose whiskers. *Polymer*, 2006, 47(18), 6285–6292.
14. Yanga, H.-S., Wolcotta, M.P., Kimb, H.-S., Kimb, S., and Kim, H.-J. Properties of lignocellulosic material filled polypropylene biocomposites made with different manufacturing processes. *Polymer Testing*, 2006, 25, 668–676.
15. Qiu, W., Endo, T., and Hirotsu, T. Structure and properties of composites of highly crystalline cellulose with polypropylene: Effects of polypropylene molecular weight. *European Polymer Journal*, 2006, 42, 1059–1068.
16. Joly, C., Gauthier, R., and Chabert, B. Physical chemistry of the interface in polypropylene/cellulosic-fibre composites. *Composites Science and Technology*, 1996, 56, 761–765.
17. Van de Velde, K. and Kiekens, P. Effect of material and process parameters on the mechanical properties of unidirectional and multidirectional flax/polypropylene composites. *Composite Structures*, 2003, 62, 443–448.
18. Cantero, G., Arbelaiz, A., Llano-Ponte, R., and Mondragon, I. Effects of fibre treatment on wettability and mechanical behaviour of flax/polypropylene composites. *Composites Science and Technology*, 2003, 63, 1247–1254.
19. Keener, T.J., Stuart, R.K., and Brown, T.K. Maleated coupling agents for natural fibre composites. *Composites: Part A*, 2004, 35, 357–362.
20. Arbelaiz, A., Fernandez, B., Ramos, J.A., and Mondragon, I. Thermal and crystallization studies of short flax fibre reinforced polypropylene matrix composites: Effect of treatments. *Thermochimica Acta*, 2006, 440, 111–121.
21. Bos, H.L., Mussig, J., and van den Oever, M.J.A. Mechanical properties of short-flax-fibre reinforced compounds. *Composites: Part A*, 2006, 37, 1591–1604.
22. Angelov, I., Wiedmer, S., Evstatiev, M., Friedrich, K., and Mennig, G. Pultrusion of a flax/polypropylene yarn. *Composites: Part A*, 2007, 38(5), 1431–1438.
23. Madsen, B. and Lilholt, H. Physical and mechanical properties of unidirectional plant fibre composites—An evaluation of the influence of porosity, *Composites Science and Technology*, 2003, 63, 1265–1272.
24. Mutje, P., Lopez, A., Vallejos, M.E., Lopez, J.P., and Vilaseca, F. Full exploitation of cannabis sativa as reinforcement/filler of thermoplastic composite materials. *Composites: Part A*, 2007, 38, 369–377.
25. Pracella, M., Chionna, D., Anguillesi, I., Kulinski, Z., and Piorkowska, E. Functionalization, compatibilization and properties of polypropylene composites with hemp fibres. *Composites Science and Technology*, 2006, 66, 2218–2230.
26. Wambua, P., Ivens, J., and Verpoest, I. Natural fibres: Can they replace glass in fibre reinforced plastics? *Composites Science and Technology*, 2003, 63, 1259–1264.
27. Pickering, K.L., Beckermann, G.W., Alam, S.N., and Foreman, N.J. Optimising industrial hemp fibre for composites. *Composites: Part A*, 2007, 38, 461–468.
28. Doan, T.-T.-L., Gao, S.-L., and Mader, E. Jute/polypropylene composites, I. Effect of matrix modification. *Composites Science and Technology*, 2006, 66, 952–963.
29. Fung, K.L., Xing, X.S., Li, R.K.Y., Tjonga, S.C., and Mai, Y.-W. An investigation on the processing of sisal fibre reinforced polypropylene composites. *Composites Science and Technology*, 2003, 63, 1255–1258.

30. Joseph, P.V., Kuruvilla, J., and Sabu, T. Effect of processing variables on the mechanical properties of sisal fiber-reinforced polypropylene composites. *Composites Science and Technology*, 1999, 59, 1625–1640.
31. Joseph, P.V., Joseph, K., Thomas, S., Pillai, C.K.S., Prasad, V.S., Groeninckx, G., and Sarkissova, M. The thermal and crystallisation studies of short sisal fibre reinforced polypropylene composites. *Composites: Part A*, 2003, 34, 253–266.
32. Yang, H.-S., Kim, H.-J., Park, H.-J., Lee, B.-J., and Hwang, T.-S. Effect of compatibilizing agents on rice-husk flour reinforced polypropylene composites. *Composite Structures*, 2007, 77, 45–55.
33. Sain, M., Park, S.H., Suhara, F., and Law, S. Flame retardant and mechanical properties of natural fibre–PP composites containing magnesium hydroxide. *Polymer Degradation and Stability*, 2004, 83, 363–367.
34. Suarez, J.C.M., Coutinho, F.M.B., and Sydenstricker, T.H. SEM studies of tensile fracture surfaces of polypropylene sawdust composites. *Polymer Testing*, 2003, 22, 819–824.
35. Demir, H., Atikler, U., Balkose, D., and Thmnoglu, F. The effect of fiber surface treatments on the tensile and water sorption properties of polypropylene-luffa fiber composites. *Composites: Part A*, 2006, 37, 447–456.
36. Panthapulakkal, S., Zereshkian, A., and Sain, M. Preparation and characterization of wheat straw fibres for reinforcing application in injection molded thermoplastic composites. *Bioresource Technology*, 2006, 97, 265–272.
37. Jang, J. and Lee, E. Improvement of the flame retardancy of paper sludge/polypropylene composite. *Polymer Testing*, 2001, 20, 7–13.
38. Rozman, H.D., Tan, K.W., Kumar, R.N., Abubakar, A., Ishak, Z.A. Mohd., and Ismail, H. The effect of lignin as a compatibilizer on the physical properties of coconut fiber polypropylene composites. *European Polymer Journal*, 2000, 36, 1483–1494.
39. Shibata, S., Cao, Y., and Fukumoto, I. Lightweight laminate composites made from kenaf and polypropylene fibres. *Polymer Testing*, 2006, 25, 142–148.
40. Mwaikambo, L.Y., Martuscelli, E., and Avella, M. Kapok/cotton fabric-polypropylene composites. *Polymer Testing*, 2000, 19, 905–918.
41. Arib, R.M.N., Sapuan, S.M., Ahmad, M.M.H.M., Paridahb, M.T., and Khairul Zaman, H.M.D. Mechanical properties of pineapple leaf fibre reinforced polypropylene composites. *Materials and Design*, 2006, 27, 391–396.
42. Czigany, T. Special manufacturing and characteristics of basalt fiber reinforced hybrid polypropylene composites: Mechanical properties and acoustic emission study. *Composites Science and Technology*, 2006, 66, 3210–3220.
43. Ruksakulpiwat, Y., Suppakarn, N., Sutapun, W., and Thomthong, W. Vetiver–polypropylene composites: Physical and mechanical properties. *Composites: Part A*, 2007, 38, 590–601.
44. Kim, H.-S., Kim, S., Kim, H.-J., and Yang, H.S. Thermal properties of bio-flour filled polyolefin composites with different compatibilizing agent type and content. *Thermochimica Acta*, 2006, 451, 181–188.
45. Thwe, M.M. and Liao, K. Durability of bamboo-glass fiber reinforced polymer matrix hybrid composites. *Composites Science and Technology*, 2003, 63, 375–387.
46. Abu-Sharkh, B.F. and Hamid, H. Degradation study of date palm fibre/polypropylene composites in natural and artificial weathering: Mechanical and thermal analysis. *Polymer Degradation and Stability*, 2004, 85, 967–973.
47. Bourmaud, A. and Baley, C. Rigidity analysis of polypropylene/vegetal fibre composites after recycling. *Polymer Degradation and Stability*, 2009, 94, 297–305.
48. Ragoubi, M., Bienaimé, D., Molina, S., George, B., and Merlin, A. Impact of corona treated hemp fibres onto mechanical properties of polypropylene composites made thereof. *Industrial Crops and Products*, 2010, 31, 344–349.

49. Mohanty, S., Nayak, S.K., Verma, S.K., and Tripathy, S.S. Effect of MAPP as a coupling agent on the performance of jute–PP composites. *Journal of Reinforced Plastics and Composites*, 2004, 23, 625–637.

50. Gassan, J. and Bledzki, A.K. Possibilities to improve the properties of natural fiber reinforced plastics by fiber modification–jute polypropylene composites. *Applied Composite Materials*, 2000, 7, 373–385.

51. Du, Y., Zhang, J., Xue, Y., Lacy Jr., T.E., Toghiani, H., Horstemeyer, M.F., and Pittman Jr., C.U. Kenaf bast fiber bundle-reinforced unsaturated polyester composites. III: Statistical strength characteristics and cost–performance analyses. *Forest Products Journal*, 2010, 60, 514–521.

52. Mirbagheri, J., Tajvidi, M., Hermanson, J.C., and Ghasemi, I. Tensile properties of wood flour/kenaf fiber polypropylene hybrid composites. *Journal of Applied Polymer Science*, 2007, 105, 3054–3059.

53. Suppakarn, N. and Jarukumjorn, K. Mechanical properties and flammability of sisal/PP composites: Effect of flame retardant type and content. *Composites Part A: Applied Science and Manufacturing*, 2009, 40, 613–618.

54. Joseph, P.V., Rabello, M.S., Mattoso, L.H.C., Joseph, K., and Thomas, S. Environmental effects on the degradation behaviour of sisal fibre reinforced polypropylene composites. *Composites Science and Technology*, 2002, 62, 1357–1372.

55. Okubo, K., Fujii, T., and Yamamoto, Y. Development of bamboo-based polymer composites and their mechanical properties. *Composites Part A: Applied Science and Manufacturing*, 2004, 35, 377–383.

56. Yang, H.S., Kim, H.J., Park, H.J., Lee, B.J., and Hwang, T.S. Effect of compatibilizing agents on rice-husk flour reinforced polypropylene composites. *Composite Structures*, 2007, 77, 45–55.

57. Luz, S.M., Goncalves, A.R., and Del'Arco Jr., A.P. Mechanical behavior and microstructural analysis of sugarcane bagasse fibers reinforced polypropylene composites. *Composites Part A: Applied Science and Manufacturing*, 2007, 38, 1455–1461.

58. KC, B., Faruk, O., Agnelli, J.A.M., Leao, A.L., Tjong, J., and Sain, M. Sisal-glass fiber hybrid biocomposite: Optimization of injection molding parameters using Taguchi method for reducing shrinkage. *Composites Part A: Applied Science and Manufacturing*, 2016, 83, 152–159.

59. Oksman, K. Mechanical properties of natural fibre mat reinforced thermoplastic. *Applied Composite Materials*, 2000, 7, 403–414.

60. Arbelaiz, A., Fernandez, B., Cantero, G., Llano-Ponte, R., Valea, A., and Mondragon, I. Mechanical properties of flax fibre/polypropylene composites. Influence of fibre/matrix modification and glass fibre hybridization. *Composites Part A: Applied Science and Manufacturing*, 2005, 36, 1637–1644.

61. Duhovic, M., Horbach, S., and Bhattacharyya, D. Improving the interface strength in flax fibre poly(lactic) acid composites. *Journal of Biobased Materials and Bioenergy* 2009, 3, 188–198.

62. Di Landro, L. and Lorenzi, W. Static and dynamic properties of thermoplastic matrix/natural fiber composites—PLA/flax/hemp/kenaf. *Journal of Biobased Materials and Bioenergy*, 2009, 3, 238–244.

63. Harriette, L.B., Mussig, J., and van den Oever, M.J.A. Mechanical properties of short-flax-fibre reinforced compounds. *Composites Part A: Applied Science and Manufacturing*, 2006, 37, 1591–1604.

64. Buttlar, H.B. Natural fibre reinforced construction materials for SMC applications. In: Conference RIKO-2005. 2005. pp. 1–24. http://www.riko.net/download/kwst2005von buttlar.pdf.

65. Doan, T.T.L., Gao, S.L., and Mader, E. Jute/polypropylene composites: I. Effect of matrix modification. *Composites Science and Technology*, 2006, 66, 952–963.
66. Haydaruzzaman, Khan, R.A., Khan, M.A., Khan, A.H., and Hossain, M.A. Effect of gamma radiation on the performance of jute fabrics-reinforced polypropylene composites. *Radiation Physics and Chemistry*, 2009, 78, 986–993.
67. Acha, B.A., Reboredo, M.M., and Marcovich, N.E. Creep and dynamic mechanical behavior of PP–jute composites: Effect of the interfacial adhesion. *Composites Part A: Applied Science and Manufacturing*, 2007, 38, 1507–1516.
68. Wang, X., Cui, Y., Xu, Q., Xie, B., and Li, W. Effects of alkali and silane treatment on the mechanical properties of jute-fiber-reinforced recycled polypropylene composites. *Journal of Vinyl and Additive Technology*, 2010, 16, 183–188.
69. Hong, C.K., Hwang, I., Kim, N., Park, D.H., Hwang, B.S., and Nah, C. Mechanical properties of silanized jute–polypropylene composites. *Journal of Industrial and Engineering Chemistry*, 2008, 14, 71–76.
70. Zaman, H.U., Khan, R.A., Haque, M., Khan, M.A., Khan, A., Huq, T., Noor, N., Rahman, M., Rahman, K.M., Huq, D., and Ahmad, M.A. Preparation and mechanical characterization of jute reinforced polypropylene/natural rubber composite. *Journal of Reinforced Plastics and Composites*, 2010, 29, 3064–65.
71. Haque, M., Hasan, M., Islam, M., and Ali, M. Physico-mechanical properties of chemically treated palm and coir fiber reinforced polypropylene composites. *Bioresource Technology*, 2009, 100, 4903–4906.
72. Islam, M.N., Rahman, M.R., Haque, M.M., and Huque, M.M. Physicomechanical properties of chemically treated coir reinforced polypropylene composites. *Composites Part A: Applied Science and Manufacturing*, 2010, 41, 192–198.
73. Bledzki, A.K., Mamun, A.A., and Volk, J. Barley husk and coconut shell reinforced polypropylene composites: The effect of fibre physical, chemical and surface properties. *Composites Science and Technology*, 2010, 70, 840–846.
74. Kumar, V., Sinha, S., Saini, M.S., Kanungo, B.K., and Biswas, P. Rice husk as reinforcing filler in polypropylene composites. *Reviews in Chemical Engineering*, 2010, 26, 41–53.
75. Kim, H.S., Yang, H.S., Kim, H.J., and Park, H.J. Thermogravimetric analysis of rice husk flour filled thermoplastic polymer composites. *Journal of Thermal Analysis and Calorimetry*, 2004, 76, 395–404.
76. Nascimento, G.C., Cechinel, D.M., Piletti, R., Mendes, E., Paula, M.M.S., Riella, H.G., and Fiori, M.A. Effect of different concentrations and sizes of particles of rice husk ash—RHS in the mechanical properties of polypropylene. *Materials Science Forum*, 2010, 660–661, 23–28.
77. Czel, G. and Kanyok, Z. MAgPP an effective coupling agent in rice husk flour filled polypropylene composites. *Materials Science Forum*, 2007, 537–538, 137–144.
78. Rozman, H.D., Tay, G.S., Kumar, R.N., Abusainah, A., Ismail, H., and Ishak, Z.A.M. Effect of oil extraction of the oil palm empty fruit bunch on the mechanical properties of polypropylene—Oil palm empty fruit bunch-glass fibre hybrid composites. *Polymer—Plastics Technology and Engineering*, 2001, 40, 103–115.
79. Suradi, S.S., Yunus, R.M., Beg, M.D.H., Rivai, M., and Yusof, Z.A.M. Oil palm bio-fiber reinforced thermoplastic composites—Effects of matrix modification on mechanical and thermal properties. *Journal Applied Science*, 2010, 10, 3271–3276.
80. Khalid, M., Ratnam, C.T., Chuah, T.G., Ali, S., and Choong, T.S.Y. Comparative study of polypropylene composites reinforced with oil palm empty fruit bunch fiber and oil palm derived cellulose. *Materials and Design*, 2008, 29, 173–178.
81. Singleton, C.A.N., Baillie, C.A., Beaumont, P.W.R., and Peijs, T. On the mechanical properties, deformation and fracture of a natural fibre/recycled polymer composite. *Composites Part A: Applied Science and Manufacturing*, 2003, 34, 519–526.

82. Li, X., Tabil, L.G., Oguocha I.N., and Panigrahi, S. Thermal diffusivity, thermal conductivity, and specific heat of flax fiber–HDPE biocomposites at processing temperatures. *Composites Science and Technology*, 2008, 68, 1753–1758.
83. Wang, W., Sain, M., and Cooper, P.A. Study of moisture absorption in natural fiber plastic composites. *Composites Science and Technology*, 2006, 66, 379–386.
84. Wang, W., Sain, M., and Cooper, P.A. Hygrothermal weathering of rice hull/HDPE composites under extreme climatic conditions. *Polymer Degradation and Stability*, 2005, 90, 540–545.
85. Torres, F.G. and Cubillas, M.L. Study of the interfacial properties of natural fibre reinforced polyethylene. *Polymer Testing*, 2005, 24, 694–698.
86. Li, Y., Hu, C., and Yu, Y. Interfacial studies of sisal fiber reinforced high density polyethylene (HDPE) composites. *Composites Part A: Applied Science and Manufacturing*, 2008, 39, 570–578.
87. Choudhury, A. Isothermal crystallization and mechanical behavior of ionomer treated sisal/HDPE composites. *Materials Science and Engineering A*, 2008, 491, 492–500.
88. Salleh, F.M., Hassan, A., Yahya, R., and Azzahari, A.D. Effects of extrusion temperature on the rheological, dynamic mechanical and tensile properties of kenaf fiber/HDPE composites. *Compos Part B Eng*, 2014, 58, 259–266.
89. Abdelmouleh, M., Boufi, S., Belgacem, M.N., and Dufresne, A. Short natural-fibre reinforced polyethylene and natural rubber composites: Effect of silane coupling agents and fibres loading. *Composites Science and Technology*, 2007, 67(7–8), 1627–1639.
90. Georgopoulos, S.Th., Tarantili, P.A., Avgerinos, E., Andreopoulos, A.G., and Koukios, E.G. Thermoplastic polymers reinforced with fibrous agricultural residues. *Polymer Degradation and Stability*, 2005, 90, 303–312.
91. Nair, K.C.M., Thomas, S., and Groeninckx, G. Thermal and dynamic mechanical analysis of polystyrene composites reinforced with short sisal fibres. *Composites Science and Technology*, 2001, 61, 2519–2529.
92. Nair, K.C.M., Kumar, R.P., Thomas, S., Schit, S.C., and Ramamurthy, K. Rheological behavior of short sisal fiber-reinforced polystyrene composites. *Composites Part A: Applied Science and Manufacturing*, 2000, 31, 1231–1240.
93. Antich, P., Vazquez, A., Mondragon, I., and Bernal, C. Mechanical behavior of high impact polystyrene reinforced with short sisal fibers. *Composites Part A: Applied Science and Manufacturing*, 2006, 37, 139–150.
94. Singha, A.S., Rana, R.K., and Rana, A. Natural fiber reinforced polystyrene matrix based composites. *Advances in Materials Research*, 2010, 123–125, 1175–1178.
95. Wang, H., Chang, R., Sheng, K.C., Adl, M., and Qain, X.Q. Impact response of bamboo-plastic composites with the properties of bamboo and polyvinylchloride (PVC). *Journal of Bionic Engineering Supplement*, 2008, 5, 28–33.
96. Xu, Y., Wu, Q., Lei, F., Yao, Y., and Zhang, Q. Natural fiber reinforced poly(vinyl chloride) composites: Effect of fiber type and impact modifier. *Journal of Polymers and the Environment*, 2008, 16(4), 250–257.
97. Zainudin, E.S., Sapuan, S.M., Abdan, K., and Mohamad, M.T.M. Thermal degradation of banana pseudo-stem filled unplasticized polyvinyl chloride (UPVC) composites. *Materials and Design*, 2009, 30, 557–562.
98. Zheng, Y.T., Cao, D.R., Wang, D.S., and Chen, J.J. Study on the interface modification of bagasse fibre and the mechanical properties of its composite with PVC. *Composites Part A: Applied Science and Manufacturing*, 2007, 38, 20–25.
99. El-Shekeil, Y.A., Sapuan, S.M., Jawaid, M., and Al-Shuja'a, O.M. Influence of fiber content on mechanical, morphological and thermal properties of kenaf fibers reinforced poly (vinyl chloride)/thermoplastic polyurethane poly-blend composites. *Mater Des*, 2014, 58, 130–135.

100. Sapuan, S.M., Pua, F.L., El-Shekeil, Y.A., and AL-Oqla, F.M. Mechanical properties of soil buried kenaf fibre reinforced thermoplastic polyurethane composites. *Materials and Design*, 2013, 50, 467–470.
101. Santos, P.A., Spinace, M.A.S., Fermoselli, K.K.G., and De Paoli, M.A. Polyamide-6/vegetal fiber composite prepared by extrusion and injection molding. *Composites Part A: Applied Science and Manufacturing*, 2007, 38, 2404–2411.
102. Ozen, E., Kiziltas, A., Kiziltas, E.E., and Gardner, D.J., Natural fiber blend-nylon 6 composites. *Polymer Composites*, 2013, 34, 544–553.
103. Alvarez de Arcaya, P., Retegi, A., Arbelaiz, A., Kenny, J.M., and Mondragon, I. Mechanical properties of natural fibers/polyamides composites. *Polym. Compos.*, 2009, 30, 257–264.
104. Liu, H., Wua, Q., and Zhang, Q. Preparation and properties of banana fiber-reinforced composites based on high density polyethylene (HDPE)/Nylon-6 blends. *Bioresource Technology*, 2009, 100, 6088–6097.
105. Thitithanasarn, S., Yamada, K., Ishiaku, U.S., and Hamada, H. The effect of curative concentration on thermal and mechanical properties of flexible epoxy coated jute fabric reinforced polyamide 6 composites. *Open Journal of Composite Materials*, 2012, 2, 133–138.
106. Neher, B., Bhuiyan, M.R., Kabir, H., Qadir, R., Gafur, A., and Ahmed, F. Study of mechanical and physical properties of palm fiber reinforced acrylonitrile butadiene styrene composite. *Materials Sciences and Applications*, 2014, 5, 39–45.
107. Neher, B., Bhuiyan, M.R., Kabir, H., Qadir, R., Gafur, A., and Ahmed, F. Investigation of the surface morphology and structural characterization of palm fiber reinforced acrylonitrile butadiene styrene (PF-ABS) composites. *Materials Sciences and Applications*, 2014, 5, 378–386.
108. Saba, N., Jawaid, M., Othman, Y., Paridah, M.T. A review on dynamic mechanical properties of natural fibre reinforced polymer composites. *Construction and Building Materials*, 2016, 106, 149–159.

2 Bio-Fiber Thermoset Composites

Ashok Rajpurohit and Frank Henning

CONTENTS

2.1 INTRODUCTION

Bio-fibers, or natural vegetable fibers, have been traditionally used in many applications in textile and allied industries. Recently, these fibers are increasingly determined to have potential to be effectively used in technical applications such as in the form of reinforcements for composite materials. These bio-fibers possess several advantages which further spells out the growing interest of composite manufacturers specifically in automotive and building industries to replace the more commonly used glass fibers. Industry is attempting to decrease the dependence on petroleum-based fuels and products due to the increased environmental consciousness. This is further leading to the need to investigate environmentally friendly, sustainable materials to replace existing ones. In terms of composites, these sustainable materials are in the form of bio-fibers and bio-polymers. In recent years, polymeric biodegradable matrices have appeared in the market as commercial products; but higher costs associated with these matrices is a major drawback thus far. Apart from cost, lower strength and stiffness of these polymer systems make it inappropriate to apply these resins for structural components. Currently the most viable way toward ecofriendly composites is the use of bio-fibers as reinforcement.

Studies have revealed that about 90% of the fibers used as composite reinforcement in Europe were glass fibers and, at most, only 3% were bio-fibers [1]. It has been estimated that substituting synthetic fibers with bio-fibers in automotive composite parts would reduce the material weight by 30% and their cost by 20%. It is reported that the replacement of only 3.5 kg of glass fibers by bio-fibers in each car produced in North America would reduce greenhouse gas emissions by at least 500 million tons per year [1]. This conception of countering greenhouse gas emissions by utilization of bio-fibers has been unequivocally encouraged by several governments worldwide including the European Union and the developing countries in Europe and Asia. The European Union Directive 2000/53/EC in this context requires member countries to reuse and recover at least 95% of the materials used for manufacturing of a vehicle after end-of-life by 2015. Lucintel's report forecasts that the natural fiber composite materials market will grow to $531.3 million in 2016 [2]. Not only in the automotive industry, but the governments from developing countries are also considering diversification of the applications of bio-fibers into various other composite sectors such as civil engineering, and building and construction [3]. Researchers around the world are already working on developing the process and the equipment for bio-fiber composites to replace glass fiber composites along with substantial efforts to develop newer composite products and applications using bio-fibers. Many studies on bio-fiber composites have been reported by various researchers all around the globe [4,5]. Most of them are based on the studies related to the mechanical properties of composites reinforced with short fibers. The components obtained therefore are mostly used to produce nonstructural parts for the automotive industry such as covers, car doors panels, and car roofs [5–7]. Studies dealing with structural composites based on natural reinforcements oriented to housing applications where structural panels and sandwich beams are manufactured out of bio-fibers and used as roofs are also widely reported [7–9]. This chapter aims at highlighting different bio-fibers, their cultivation and manufacturing of reinforcements for composites, and their advantages and drawbacks. Further, it gives insights of thermoset resin systems for different bio-fiber composites manufacturing, along with processes, properties, areas of application, and future scope for these composite systems.

2.2 BIO-FIBER MATERIALS FOR COMPOSITES

2.2.1 INTRODUCTION AND CLASSIFICATION

Fibers can be classified into two main categories: manufactured fibers or synthetic fibers and natural fibers or bio-fibers. These bio-fibers are further categorized into several groups depending on their origin, that is, from plant/vegetable source, from animal source, or from minerals. Some of the other researchers classify bio-fibers into organic and inorganic as the main subcategories and further these are branched into different categories based on their source as mentioned above. One such detailed classification of bio-fibers is diagrammatically shown in Figure 2.1.

When talking about bio-fiber composites, the most widely used category for such applications is from vegetable or plant sources. These include the most prominent one, that is, bast fibers, which include flax, hemp, jute, ramie, and so on. These fibers are deemed to be the most prominently used vegetable fibers due to their wide availability and renewability in a short time with respect to others. The other important sources include seed hair fibers (e.g., cotton), leaf fibers (e.g., abaca and sisal), fruit fibers (e.g., coir), and spear fibers (e.g., bamboo). There are several varieties of fibers that come under the animal fiber and mineral fiber classification, most worthy from this lot are sheep hair, silk, asbestos, and so on. These fibers are not used much in the field of composites and hence are not discussed in this chapter. As it is known that it is the bast or bark that provides the strength and stiffness to the plant, it is logical to try using the fibers coming out of plants' basts for similar composite applications. With respect to bio-fiber thermoset composites, the fibers with a leaf and bast origin proven to be of vast potential are discussed in various review papers. In Table 2.1, an overview of mechanical properties of bio-fibers such as jute, flax, kenaf, hemp, and sisal along with the newer generation of biofibers relevant for thermoset composite application is presented.

2.2.2 STRUCTURE, COMPOSITION, AND EXTRACTION

Bio-fibers or natural fibers primarily consist of cellulose, hemicellulose, and lignin as their main constituents. The mechanical properties of bio-fibers depend on the type of cellulose and the geometry of the elementary cell. A single or elementary plant bio-fiber is a single cell typically of a length from 1 to 50 mm and a diameter of around 10–50 mm [7]. Plant bio-fibers are like microscopic tubes called cell walls surrounding the central lumen. The lumen contributes to the water uptake behavior of the fiber. The fiber consists of several cell walls; hence, structurally natural fibers are considered to be multicellular in nature. These cell walls are formed from oriented reinforcing semi-crystalline cellulose micro-fibrils embedded in a

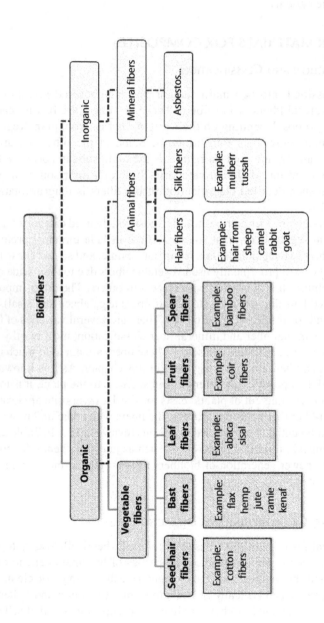

TABLE 2.1

Physical and Mechanical Properties of Relevant Bio-Fibers

Fiber	Density (g cm⁻³)	Diameter (µm)	Tensile Strength (MPa)	Specific Strength (S/ρ)	Tensile Modulus (GPa)	Specific Modulus (E/ρ)	Elongation at Break (%)
Abaca	1.5	–	400	267	12	8	3–10
Bamboo	1.1	240–330	500	454	35.91	32.6	1.4
Banana	1.35	50–250	600	444	17.85	13.2	3.36
Coir	1.2	–	175	146	4–6	3.3–5	30
Flax	1.5	–	800–1500	535–1000	27.5–80	18.4–53	1.2–3.2
Hemp	1.48	–	550–900	372–608	70	47.3	2–4
Jute	1.46	40–350	393–800	269–548	10–30	6.85–20.6	1.5–1.8
Kenaf	1.45	70–250	930	641	53	36.55	1.6
Ramie	1.5	50	220–938	147–625	44–128	19.3–85	2–3.8
Sisal	1.45	50–300	530–640	366–447	9.4–22	6.5–15.2	3–7

Source: John MJ, Thomas S, eds. *Natural Polymers*, Vol. 1, pp. 37–62. Reproduced by permission of The Royal Society of Chemistry. Mussig J, Slootmaker T: In *Industrial Applications of Natural Fibres: Structure, Properties and Technical Applications*, ed. J. Mussig. pp. 44–48. 2010. Copyright Wiley-VCH Verlag GmbH & Co. KGaA. Reproduced with permission. Saheb DN, Jog JP, *Advances in Polymer Technology* 18(4):351–63, 1999.

hemicellulose–lignin matrix of varying composition. Such microfibrils typically have a diameter of about 10–30 nm and are made up of 30–100 cellulose molecules in extended chain conformation and provide mechanical strength to the fiber. The arrangement of fibrils, microfibrils, and cellulose in the cell walls of a plant bio-fiber can be as seen in Figure 2.2.

Among natural fibers the bast fibers, extracted from the stems of plants such as jute, kenaf, flax, ramie, and hemp, are widely accepted as the best candidates for reinforcements of composites due to their good mechanical properties. Hemp and flax [10,11] were shown to have very promising tensile properties for applications where mechanical properties are a requisite, as many authors agree; the two basic parameters that allow characterization of the mechanical behavior of natural fibers are the cellulose content and the spiral angle. In general, the tensile strength of the fibers increases with increasing cellulose content and with decreasing angle of helix axis of the fibers [9,12,13]. The physical and mechanical properties of bio-fiber also depend on the chemical composition of single fibers, the growing conditions, and the extraction methods utilized. Different fibers have different chemical compositions of cellulose, hemicelluloses, lignin, pectin, waxes, water content, and other contents. These can also be seen in Table 2.2. Also, the growing conditions such as soil, water availability, and other climatic conditions along with the harvesting and processing methods affect greatly the outcoming properties of the bio-fibers.

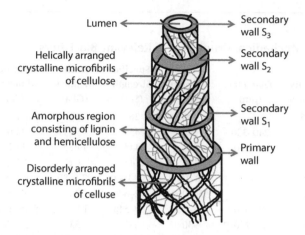

Lumen

Helically arranged
crystalline microfibrils
of cellulose

Amorphous region
consisting of lignin
and hemicellulose

Disorderly arranged
crystalline microfibrils
of celluse

Secondary
wall S$_3$

Secondary
wall S$_2$

Secondary
wall S$_1$

Primary
wall

FIGURE 2.2 Schematic view of structural constitution of natural fiber cell. (Furtado SCR et al. *Natural Fibre Composites: Automotive Applications*. In John MJ, Thomas S, eds., 2012. Reproduced by permission of The Royal Society of Chemistry.)

TABLE 2.2

Mean Chemical Composition of Relevant Bio-Fibers

Fiber	Cellulose	Hemicellulose	Pectin	Lignin	Extractives	Fats and Waxes
Flax	71.2	18.5	2	2.2	4.3	1.6
Hemp	78.3	5.4	2.5	2.9	–	–
Jute	71.5	13.3	0.2	13.1	1.2	0.6
Abaca	70.2	21.7	0.6	5.6	1.6	0.2
Coir	35.6	15.4	5.1	32.7	3	–
Sisal	73.1	13.3	0.9	11	1.3	0.3

Source: Furtado SCR et al. *Natural Fibre Composites: Automotive Applications*. In: John MJ, Thomas S, eds., 2012.

Extraction of bio-fibers from the plant stems can be achieved by various methods. Retting is the starting process for extraction of the bio-fibers from the plant materials; in this process, a controlled degradation of the plant stem is emphasized to allow the fiber to be separated from the woody core and thereby further improving the ease of extraction of the fibers from the plant stems in succeeding steps. Specific amount of time, exposure to moisture, and sometimes mechanical treatment of the stem by a decorticator make the process of retting complete. The retting process can be broadly divided into four different processes, namely, biological, chemical, physical, and mechanical. The biological retting, as the name suggests, separates the fibers by degradation by biological means. These biological means are induced artificially or naturally. Mechanical retting is a simpler and economical process of retting, but it produces lower quality and coarse grades of fiber comparatively. The physical retting process involves the ultrasound retting and the steam explosion retting where the retting process is induced by enhanced enzymatic activities because of steam temperature and pressure. The chemical retting process involves use of water solutions containing sulfuric acid, chlorinated lime, sodium or potassium hydroxide, and soda ash to dissolve the pectin component of the plant material. Chemical retting produces high quality fibers but adds costs to the final product [7,11,14]. Enzymatic retting or retting by microbial means is one of the widely used techniques to extract good quality cellulosic fibers from the plant materials such as jute, hemp, and flax. Researchers [15] also mention the use of a fungal retting process in case of wheat straw before extracting the fibers. The process of fiber separation and extraction has a major impact on fiber yield and final fiber quality. It also influences the structure, chemical composition, and hence the physical and mechanical properties of the fibers.

2.2.3 Pros and Cons of Bio-Fibers

The foremost advantage of bio-fibers is their availability worldwide, depending on the type of fiber and geographical and climatic condition. For example, Europe has hemp and flax fibers in abundance; South America has sisal, abaca, and ramie in abundance; and similarly North America produces large quantities of cotton, hemp, and flax while the Asian subcontinent is home to the largest production areas for several fibers such as bamboo, kenaf, ramie, flax, okra, china reed, and predominantly coir and jute. In addition, bio-fibers have an upper edge considering their superior specific properties over the conventional fibers such as glass for composite applications. These fibers are easy to handle and process as they can be used on the traditional reinforcement making equipment and processes. These fibers are safer as they do not cause any allergies or lung diseases if breathed in or contacted. The composites from bio-fibers also have an advantage as they do not cause much wear to the processing machinery like extruder, pelletizer, or injection molding machines and also are safer during operation in crash performance as they do not show any sharp edge fractures. Bio-fibers provide better thermal and acoustic insulation properties, especially as an automotive interior or construction material part, due to the presence of lumen/void in the fiber. They are lightweight as compared to glass fiber reinforced composites, thereby providing better fuel efficiency or the weight saved, in interior and car trim panels, and can be used in other car components to improve its performance. Furthermore, these fibers provide economic incentive by being a green product; also, they provide a marketing advantage. In the present scenario, the superiority of bio-fibers in the environmental performance, which includes lower energy requirements for production, lower emissions owing to lower weight, and energy and carbon credits from the end of life incineration of bio-fibers, is the main driver for wide interest in bio-fibers in composite applications [7,14]. A schematic representation of properties and specification of different natural fibers and glass fibers is shown in Figure 2.3.

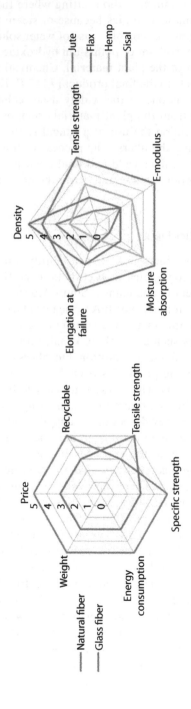

FIGURE 2.3 Strengths of bio-fibers compared to glass fiber composites. (From Lucintel, Natural Fiber Composites Market Trend and Forecast 2011–2016: Trend, Forecast and Opportunity Analysis.)

However, bio-fibers also possess several drawbacks which are considered to be major hindrances for the growth of bio-fibers in technical applications in the field of composites. The foremost of them is the quality consistency of the fibers itself and the reinforcement forms manufactured out of these inconsistent raw materials. Numerous factors such as soil, humidity, sunlight, wind, retting and extraction process, and so on affect the maturity and quality of the bio-fibers produced. These factors (some of them uncontrollable and others even if controlled) produce large variation in fiber quality considering the global parameters [14]. As composite manufacturers expect reproducible mechanical characteristics, the only way to ensure an effective quality level is to take large security coefficients on the fiber properties, which in turn reduces their attractiveness and prevents full exploitation of their advantages. The other challenges that bio-fibers inherently pose are the hydroxyl groups present on the surface of the fiber and the non-crystalline regions of the cell wall which form about 60% of fiber volume fractions of the bio-fibers. The cellulose and the hemicellulose are two out of the three main constituents (the other being lignin) that are hydrophilic and are responsible for moisture retention and further affecting the dimensional stability of the fiber and the composites. The humidity from the air is absorbed by the composite leading to loosening the fiber matrix bond and further weakening the composite structure. Temperature dependencies of bio-fibers and their contribution to mechanical properties are also a noted drawback of the bio-fiber composites with both thermoset and thermoplastic matrices. Studies have reported thermal stability of a typical bio-fiber having a limit as low as 200°C above which the degradation process begins [16,17]. With regard to the above-mentioned drawbacks, many treatments both physical and chemical have been reported to enhance compatibility and mechanical properties of bio-fiber composites; the details can be referred to in Section 2.5 of this chapter.

2.2.4 Reinforcements for Composites

The two basic parameters that provide information on the mechanical behavior of natural fibers are cellulose content and spiral angle. Apart from these, the type and form of reinforcement in which the natural fibers are supposed to be used during composite manufacturing play a very vital role in the final composite properties. Natural fibers are usually in the form of short reinforcements which are used to produce mat fabrics. These discontinuous fibers (chopped) are generally used for a randomly oriented reinforcement (mat) when there is no preferential stress direction. The random orientation of the fibers in the mat leads to a near uniform and considerably low level of stress and strain performance in the composites. The alternative to the use of short fibers is the manufacture of long yarns. Yarns, slivers, and rovings are the long continuous assembly of relatively short interlocked fibers; out of these, rovings and yarns are suitable for use in production of textiles, sewing, crocheting, knitting, weaving, embroidery, and rope making. The fibers twisted at a specific angle to the yarn axis provide the strength to the yarn in its axial direction. Spun yarns are made by twisting or otherwise bonding staple fibers together to make a cohesive thread and may contain a single type of fiber or a blend of various types. Two or more spun yarns, if twisted together, form a thicker twisted yarn. Two or more parallel spun yarns can form a roving. The main advantage of using natural yarns is the possibility to weave them into 2D and 3D fabrics with tailored yarn orientations. The classification of bio-fiber composite reinforcements can be seen in Figure 2.4.

Spun yarns obtained from natural fibers usually produce some short fibers protruding out of the main yarn body. These short fibers are commonly referred to as yarn hairiness. Although not desirable in many cases, the hairiness can lead to better mechanical fiber matrix interface interlocking in composites. Another advantage of natural yarns is the increased surface roughness, which increases interfacial strength due to mechanical interlocking, improving transverse properties. In addition, twisting localizes the micro damages within the yarn leading to higher fracture strength. Besides yarn strength, the amount of twist also affects the inter-yarn impregnation while fabricating reinforced composites. With increased twist level, yarns become more compact making it difficult for the resin to penetrate into the yarn. There is an optimum level of twist, which should be kept as low as possible for the best composite mechanical properties to allow for proper fiber wet-out during composite manufacturing. Figure 2.5 shows the pictorial representation of different types of textile structures made from flax fibers.

Yarns offer a viable and interesting alternative to the use of short fibers as multiple filament yarns can be weaved into 2- or 3-D textiles. Weaving is a textile production method that involves interlacing a set of longer threads, twisted yarn, or rovings (called the warp) with a set of crossing threads (called the weft). This is done on a frame or machine known as a loom, of which there exist a number of types. Some weaving is still done by hand, but the vast majority is mechanized. The main advantage of using weaved fabrics is the possibility to preorient the filaments in designed directions. Natural yarns differ from multifilament or synthetic fibers (i.e., tow) because they are an assembly of short fibers instead of an assembly of aligned continuous fibers. However, the fibers that constitute the yarn have a preferential orientation along a helical trajectory which make use of natural yarns attractive compared to short fibers because in such yarns fibers are mostly along the load direction.

The manner in which the warp and weft threads are interlaced is known as the weave style. Plain weave is the most basic type of textile weaves. The warp and weft are aligned so they form a simple crisscross pattern. The satin weave is characterized by four or more weft yarns floating over a warp yarn, or vice versa, four warp yarns floating over a single weft yarn. Twill is a type of fabric woven with a pattern of diagonal parallel ribs. It is made by passing the weft thread over one or more warp threads and then below two or more warp threads and so on, with a "step" or offset between rows to create the characteristic diagonal pattern. Because of this float structure in twill and satin, these generally drape better than the plain woven fabrics.

Figure 2.6 shows the schematic representation of different types of textile structure weaves.

There are an increasing number of producers of bio-fiber fabrics around the world which are now tailoring their products for composites technology. As mentioned

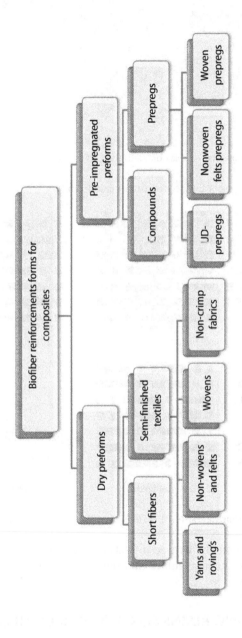

FIGURE 2.4 Reinforcement forms for bio-fiber composites. (From Cristaldi, G., Latteri, A., Recca, G., and Cicala, G. Composites based on natural fiber fabrics. In: Dubrovski, P.D., Ed. *Woven Fabric Engineering*. In-tech Open.)

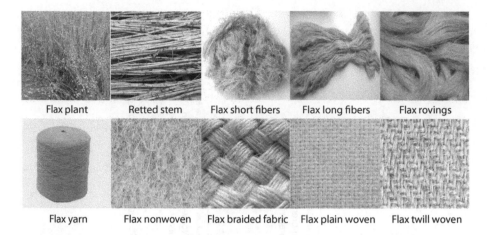

Flax plant   Retted stem   Flax short fibers   Flax long fibers   Flax rovings

Flax yarn   Flax nonwoven   Flax braided fabric   Flax plain woven   Flax twill woven

FIGURE 2.5   Different types of textile structures made from flax fibers.

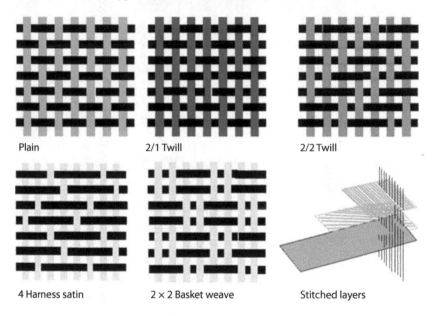

Plain   2/1 Twill   2/2 Twill

4 Harness satin   2 × 2 Basket weave   Stitched layers

FIGURE 2.6   Schematic representation of different types of textile weaves.

before, yarns and rovings can be weaved in 3-D fabrics, even if they are not as widespread as plain ones. To date no commercial example of 3D weaved fabric based on natural yarns is available.

2.3   THERMOSETTING RESINS FOR BIO-FIBER COMPOSITES

Composites, including the ones utilizing bio-fibers, have recently encompassed almost all material domains in day to day use. These include house furnishing, packaging, automotive, transportation, aerospace, sport and leisure, and so on. For these applications until recently synthetic polymers such as thermoplastics, thermoset, and elastomers were widely used as matrix materials. Research efforts in the past couple of decades have been focused on the incorporation of bio-fiber material in synthetic matrices along with process additives and other agents for manufacturing and curing assistance. Thermosetting resins (or thermosets) play an important role in industry due to their high flexibility for tailoring desired properties, leading to their high modulus, strength, durability, and thermal and chemical resistances as provided by high cross-linking density [3–7]. This high cross-linking density in thermosetting resins also makes them poor in their impact resistance and further does not allow them to be reshaped or reformed after curing. To further enhance their performance for industrial applications, thermoset bio-fiber composites are made using a variety of fillers and additives along with the matrix and reinforcement materials.

While bio-fiber thermoplastic composites are currently being researched for a wide range of products and applications, thermoset resin-based bio-fiber composites remain attractive and form a major part of the research and development activities and the output of bio-fiber composite manufacturing industries. Contrary to thermoplastics, which have a high melting point, thermosets are cured at comparatively low temperatures and therefore can be combined with a variety of resin types to design an adequate property profile. These bio-fiber thermoset composites are currently prepared using manual lay-up, pultrusion, and other liquid composite molding techniques. The details of the manufacturing techniques can be found in Section 2.6 in this chapter. By comparison with thermoplastic polymer composites in which fibers are inserted in a polymer melt, followed by forming the composite to the desired shape at an elevated temperature, the use of a thermosetting resin provides better impregnation at lower viscosity and to some extent better reactivity with fillers during the curing processes, leading to finer morphological structures. The most widely used resins for bio-fiber thermoset composites are polyester, epoxies, and vinylesters. These three resins constitute about 90% of the total thermoset resin systems used for bio-fiber composite manufacturing and are discussed in detail in the next section (Table 2.3).

TABLE 2.3
Comparative Analysis of Pros and Cons of Epoxy, Polyester, and Vinyl Ester Resins

Properties	Epoxy Resin	Polyester Resin	Vinylester Resin
Cost	− −	++	+
Ease of usage	−	++	−
Chemical resistance	+	−	++
Environmental resistance	+	−	++
Corrosion resistance	+	−	++
Mechanical properties	++	−	+
Thermal properties	++	−	+
Working time	+	− −	+
Shrinkage	−	++	+

2.3.1 EPOXY RESIN

Epoxy resins are thermosetting materials for which the resin precursors contain at least one epoxy function. These epoxy functions are highly reactive toward diverse functions, such as hydroxyl, leading to extremely versatile materials that can be used for applications ranging from laminated circuit board, wood-based furniture applications to structural carbon fiber composites and adhesives. Nowadays, almost 90% of the world production of epoxy resins is based on the reaction between bisphenol A (2,2-bis(4-hydroxyphenyl)propane and epichlorohydrin, yielding diglycidyl ether of bisphenol A (DGEBA). DGEBA contains one or more bisphenol A moieties depending on the applications. In thermosetting materials, epoxies are currently combined with a large range of reactants, so-called curing agents, such as amines, anhydrides, and amides. Figure 2.7 shows the chemical structure of a typical epoxy resin.

Epoxies are best known for their excellent adhesion, chemical and heat resistance, mechanical properties, and outstanding electrical insulating properties. The chemical resistance of epoxies is excellent against basic solutions. Epoxies are more expensive than polyesters, and their cure times are longer, but their extended range of properties can make them the best choice for critical applications considering better performance/cost ratio. The viscosity of epoxies is a step higher than polyesters or vinyl esters. Attributes of epoxy resins include extremely low shrinkage, good dimensional stability, high temperature resistance, good fatigue, and adherence to reinforcements. Open molding (hand lay-up/spray-up), SMC/BMC, resin transfer molding (RTM), and filament winding are the composite processes widely preferring epoxy resin systems for manufacturing of bio-fiber composites [14,18,19].

2.3.2 POLYESTER RESIN

Polyester resin is a class of unsaturated polyester derived from the polycondensation of a polyol and a polyvalent acid or acid anhydride. Starting from these unsaturated polyester resins, curing reactions are performed through radical or thermal processes in the presence of unsaturated co-monomers such as styrene, leading to polyester resins. Acid anhydrides such as phthalic anhydride or maleic anhydrides are typically used as curing sites all along the polyester backbone, while polyols typically used are di-pentaerythritol, glycerol, ethylene glycol, trimethylolpropane, and neopentylglycol. Unsaturated polyester offers tremendous flexibility and a whole range of polyesters made from different acids, different glycols and monomers with varying properties are available and used in the composite industry. This resin system is the most widely used resin system for bio-fiber thermoset applications. The two principle types of polyester resins that are utilized in composite manufacturing are the orthopthalic polyester resin and the isothaplic polyester resin. The latter is now becoming the preferred choice among the bio-fiber composite manufacturer community because of its superior water resistance and improved heat resistance property. The chemical structure of a typical polyester resin can be seen in Figure 2.8.

Polyesters offer ease of handling, low cost, dimensional stability, as well as good mechanical and chemical resistance and electrical properties. Polyester resins are the least expensive of the resin options, providing the most economical way

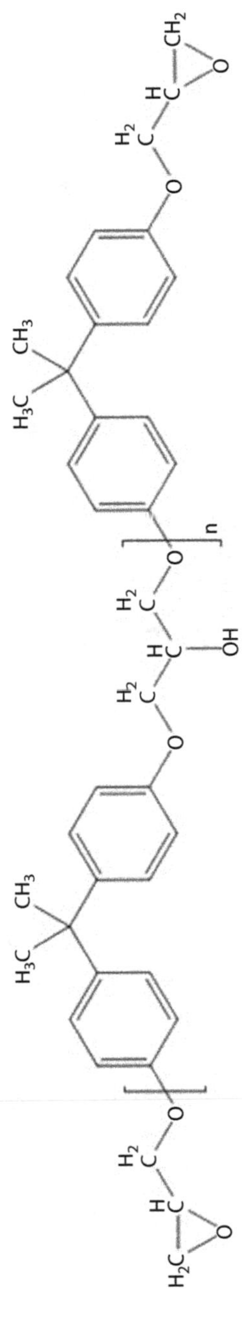

FIGURE 2.7 Chemical structure of a typical epoxy resin.

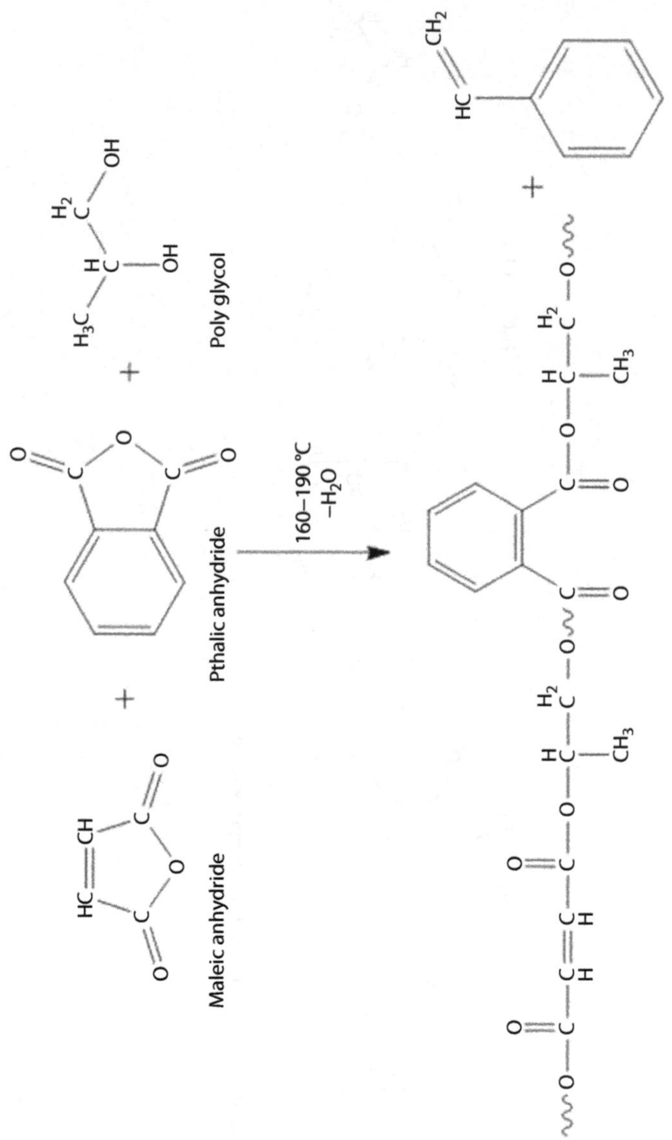

FIGURE 2.8 Chemical reaction for a typical polyester resin.

to incorporate resin, filler, and reinforcement. They are the primary resin matrix used in sheet molding compounds (SMCs) and bulk molding compounds (BMCs) [16,17,19,20]. The low viscosity and raw material cost of polyesters make the additions of filler and reinforcements a matter of practicality. In fact, filler is often called an extender because it extends the value of the resin—reducing the cost of the final composite as much as 50%. Polyester resin (neat) meaning without filler or reinforcements would be a very brittle resin and would never be used without fillers or reinforcements. A wide range of surface qualities are possible to satisfy different application needs, from a cosmetically high surface quality Class A automotive/truck body panel to a rugged underbody utility box.

2.3.3 VINYLESTER RESIN

Vinylester resins are similar in their molecular structure to polyesters, but differ primarily in the location of their reactive sites, these being positioned only at the ends of the molecular chains. As the whole length of the molecular chain is available to absorb shock loadings this makes vinylester resins tougher and more resilient than polyesters. Excellent corrosion-resistant properties have been known to the composite manufacturing industries. Recently, it was established that vinylesters are increasingly specified for aggressive chemical environments and have properties better then polyesters. This is because of the fewer ester groups in the vinylester molecule that are susceptible to water degradation by hydrolysis. The epoxide backbone chemistry, high degree of cross-linking, and higher molecular weight of vinylester resins gives them epoxy-like performance, with the processing ease of polyesters. Bisphenol-A-based epoxy vinyl esters are generally selected for use in environments where composites will be exposed to inorganic acids, alkaline materials, or salt, particularly where greater resin strength is needed to resist vibration loads. Novolac epoxy-based vinylesters contain phenol or cresol and have even higher cross-linking density, providing outstanding thermal stability and mechanical properties for handling a wide range of chemicals, especially organic solvents. The chemical structure of a typical vinylester resin molecule is as shown in Figure 2.9.

The surface quality of the composites manufactured by vinylester resin is not as good as with polyesters. The built-in toughness of vinylesters makes smooth surfaces hard to accomplish. The double bonded nature creates shrinkage, which has a negative impact on surface quality. The double bonded vinyl groups give the entire matrix a toughness that exceeds polyesters by about two times depending on the test used. Next to polyesters, vinylesters are the most economical for traditional fiberglass parts, and can be used with a variety of tooling and processes. Crack resistance

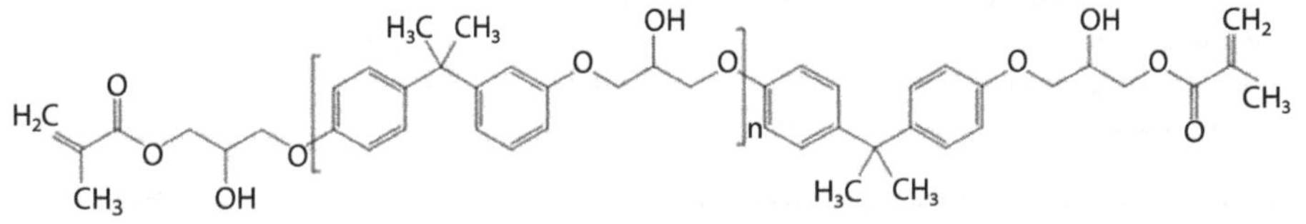

FIGURE 2.9 Chemical structure of a typical epoxy vinyl ester monomer.

of vinylester resins is one of its most important properties that are exploited in fiber reinforced composites [18,21,22].

2.3.4 OTHER THERMOSET RESIN SYSTEMS

Apart from epoxies, polyesters, and vinylesters, a number of other thermoset resin systems are already used for bio-fiber composite applications [18,21]. These include standardized resin systems based on phenolic and polyurethanes. Polyurethanes are one of the most versatile materials, with applications ranging from flexible foam in upholstered furniture to rigid foam as insulation in walls, roofs, and appliances. Polyurethanes represent an important class of thermoplastics and thermosets because their mechanical, thermal, and chemical properties can be tailored by reactions with various polyols and isocyanates, usually catalyzed with tin derivatives. On the other hand, phenolics are preferred for composite applications where stability at higher temperature is required. A comparative analysis of the three most widely used thermoset resins is given in Table 2.4.

The recent significant increase of interest and research in chemicals and polymers from renewable raw materials has led to many new approaches to renewable thermoset technology. Some are based on direct utilization of renewable raw material, such as unsaturated fats and oils, in thermosets; others require fermentation and/or chemical conversion of these raw materials such as sugars to monomers, cross-linking agents and polymers for thermosets. Many researchers and industries are focusing on both of the approaches to develop these so-called bio-thermoset resins [17,23,24]. Some of the examples are linseed, soybean, cottonseed, oilseed radish, and peanut oils. Advantages of these kinds of bio-thermoset resins are

- Improved health and safety in manufacturing
- Better environmental profile of manufacturing processes
- Energy savings in processes replacing aqueous resins with oils

TABLE 2.4
Properties of Epoxy, Polyester and Vinylester Resin Systems

Properties	Unit	Epoxy Resin	Polyester Resin	Vinylester Resin
Density	Mg^3	1.1–1.4	1.2–1.5	1.2–1.4
Young's modulus	GPa	3–6	2–4.5	3.1–3.8
Tensile strength	MPa	35–100	40–90	69–83
Compressive strength	MPa	100–200	90–250	–
Tensile elongation at break	%	1–6	2	4–7
Cure shrinkage	%	1–2	4–8	–
Water absorption at 24 h, 20°C	%	0.1–1.4	0.1–0.3	–
Fracture energy	kPa	–	–	2.5

Source: Ray D, Rout J. *Thermoset Biocomposites*. In: Mohanty AK, Misra M, Drzal LT, eds., 2005.

- Marketing advantage offered by formaldehyde-free resin usage in products
- Wider application of ozone-based technologies in manufacturing
- Sustainable raw material supply and price stability

2.4 LIMITATIONS OF BIO-FIBER THERMOSET COMPOSITES

Currently, more than 95% of the reinforcement materials for fiber reinforced composites are in the form of glass fibers. Bio-fiber composites are limited in applications in several fields including automotive, furnishing, and building and construction [1–3]. This is primarily due to their relatively lower mechanical properties and weak interface characteristics between fiber and matrix, but these properties are being enhanced by new surface treatments, additives, and coatings. Bio-fibers have some advantages over glass and other high performance fibers in aspects such as lower impact on the environment in its production and usage cycle as compared to glass fiber production; fiber content in natural fiber composites is generally higher than that of glass fiber composite for equivalent performance; natural fiber composites are environmentally friendly and reduce more polluting polymer-based contents; light-weight natural fiber composites improve fuel efficiency and reduce emissions when used for automobile components; and end of life incineration of natural fibers results in recovered energy and carbon credits. Barring these advantages, natural fibers are still facing some challenges to improve their properties in moisture absorption, fiber modification, fire resistance, durability, and their quality consistency depends on their location weather conditions [6,7,25,26]. A few important limitations are discussed in the following sections.

2.4.1 INCONSISTENCY

Quality inconsistency of bio-fibers caused by both controllable and uncontrollable factors leads to variation in the mechanical properties of a bio-fiber, further creating complications in the designing and quality controlling aspects of the final bio-fiber composite part. Researchers have acknowledged large variation in the mechanical and other characteristics of bio-fiber composite. As there is a large variation in measured mechanical properties of bio-fibers, they are often used only for low-grade composite applications. Various cross-sectional diameters of fibers may lead to a variation in the mechanical properties of bio-fibers [27,28]. Several factors have been identified to cause variation in the quality and size of natural fiber: geometric location of field, crop variety, harvest seed quality and density, soil quality, fertilizer used, harvesting time, and climate and weather conditions. Apart from these factors, variations are also caused because of fiber production methods, damage due to handling and processing and the differences in drying processes. Considering the above factors, the grade of varieties of bio-fibers is derived for most of the bio-fibers used for technical applications. For example, in India, after several revisions the present jute grading system has eight grades considering six physical parameters like strength, root content, defects, fineness, color, and bulk density. This leads to selection of a very specific grade for a specific application and hence reduces the inconsistency of the fibers and further the end product which in this case is a composite material. Variation in price is also found along with variability in the quality of natural fibers of the plants at the time of their harvest. For consistency in the fibers and hence in the bio-fiber composites, it is necessary to apply quality control checks at each stage of production of bio-fiber reinforcements, and then only utilize the material for composite application which specifically satisfy the quality control features at all the stages.

2.4.2   MOISTURE

It is well known that all bio-fibers are inherently hydrophilic in nature; this nature provides them with a high tendency to absorb water from their surroundings. This hydrophilic property of bio-fibers is strongly dissimilar to other reinforcement such as glass fiber materials for composites which are hydrophobic in nature and hence the bio-fibers have a lower competitive advantage compared to glass and other technical reinforcement materials. When in wet conditions, bio-fiber composites absorb moisture from the atmosphere, resulting in fiber swelling or interface failures which makes the bio-fiber composite applications very limited. Studies on the effects of the number of layers of bio-fiber reinforcements on moisture absorption, thickness swelling, volume swelling, and density as a function of immersion time were done by Masoodi and Pillai [29]. It was observed that the moisture diffusion rate into composites increases with an increase in the jute fiber to epoxy ratio. The effect of moisture absorption on mechanical properties of short roselle and sisal fiber reinforced hybrid polyester composite were investigated and compared with the composites containing dried fibers [30]. Increasing fiber content and length at dry condition, the tensile and flexural strength increased. At wet conditions, the tensile and flexural strength significantly decreases. The impact strength was reduced with fiber content and length at dry and wet conditions. Exposure to moisture caused a significant drop in mechanical properties due to the degradation of the fiber–matrix interface [30]. Further, many researchers [30–32] have reported that moisture absorption by the composites containing natural fibers had several adverse effects on their properties and largely affected their long-term performance. This decrease in mechanical properties reported in most studies has been attributed to the effect of moisture coupled with poor fiber–matrix interface.

2.4.3   FIBER–MATRIX INTERFACE

The fiber–matrix interface is also an inherent challenge faced by developers of bio-fiber composites. The interface between fiber and the matrix decides the final mechanical and durability characteristics of the composite part. There is a limit beyond which the fiber–matrix interface between two constitutes of the composite cannot be increased. This limitation of interface arises due to multiple factors such as actual surface of the fiber (both form and chemistry), swelling due to absorbed moisture and incompatibility to the resin system itself. Hence, a proper strategic selection of the fiber and the resin treatment, suitable surface modification of the fiber both physical and chemical is essential to obtain the best bio-fiber composite product with the best fiber–matrix interface [26,27]. The study by Gassan concluded that the quality of jute and flax fiber–matrix interface has a significant effect on fatigue behavior of both brittle polyesters and ductile polypropylene [33]. For composites using both the resins it was also observed that the damage initiation and damage propagation were more rapid for fibers that were not pretreated or surface modified hence having poor interface properties. Some of the treatments for improvement of composite properties and overcoming the above-mentioned challenges/limitations are mentioned in Section 2.5 of this chapter.

2.5 FIBER MODIFICATION SUITING BIO-FIBER THERMOSET COMPOSITES

There has been increasing interest in the scientific community to analyze and understand the physical and chemical nature of the interface between the reinforcing entity and the matrix, which in the case of bio-fiber composites is very weak owing to the limitations of bio fibers mentioned in the previous section. This growing interest is precisely because the resulting reinforced composite properties depend significantly on the quality of the interface that is formed between both the reinforcement and the polymer matrix. The quality of the fiber matrix interface is significant for the application of natural fibers as reinforcement for plastics. Since the fibers and matrices are chemically different, strong adhesion at their interfaces is required for an effective transfer of stress throughout an interface. A good harmony and agreement between cellulose fibers and nonpolar matrices is achieved from polymeric chains that will favor entanglements and interdiffusion with the matrix. Many theories of adhesion have been proposed for better interface characteristics for composites both for reinforcements and for resin systems. These include those based on mechanical interlocking, electronic or electrostatic theory, theory of weak boundary layers, thermodynamic or adsorption model (also referring to wettability), diffusion or interdiffusion theory, and finally, chemical bonding theory. The following section gives information about some physical and chemical treatment methods that improve the fiber–matrix adhesion, further helping overcome the limitations of bio-fibers.

2.5.1 PHYSICAL TREATMENTS

Physical treatments for fiber modification are those treatments that change structural and surface properties of the fiber and thereby influence the mechanical bonding with the matrix. These include treatments such as surface fibrillation, electric discharge, corona, cold plasma, and so on. Surface modification by discharge treatment such as low temperature plasma, sputtering, and corona discharge is of great interest in relation to the improvement in functional properties of bio-fibers. Using plasma treatment, improvement of interlaminar shear strength and flexural strength in bio-fiber composites has been reported to be increased up to 35% and 30%, respectively [34]. Corona treatment is also an interesting technique for surface oxidation activation and has been explored by researchers. It changes the surface energy of cellulosic fibers, which in turn affects the melt viscosities compared to untreated materials. Gassan and Gutowski, in their research, have used corona plasma and UV to treat jute fibers and found both physical treatments increase the polarity of fibers but decreased fiber strength leading to reduced composite strength. Improvement of up to 30% flexural strength of epoxy matrix composites with UV treatment was noted in this study [35].

2.5.2 Chemical Treatments

Even though the physical treatments are explored and experimented with, the most widely used treatments are the chemical ones. These treatments can clean the fiber surface, modify the chemistry on the surface, lower the moisture uptake, and increase the surface roughness. Many chemical treatments such as mercerization, isocyanate treatment, acrylation, permanganate treatment, acetylation, silane treatment, and peroxide treatment with various coupling agents and other pretreatments for bio-fibers have been reported in the literature [7,9,33–37]. In the following paragraphs, some of the treatments are explained in detail.

Alkali treatment, also referred to as mercerization, is the most widely used chemical treatment for surface modification of bio-fibers. It is a treatment by solution of NAOH, that is, treatment of bio-fiber by alkalis. Alkaline treatment leads to fibrillation which causes the breaking down of a composite fiber bundle into smaller fibers. In terms of chemistry alkali sensitive hydroxyl (OH) groups present among the molecules of bio-fibers are broken down, which then react with water molecules (H-OH) and move out from the fiber structure. Further, the remaining molecules lead to the formation of an (–O–Na) end group on the fiber surface and in between the cellulose molecular chains. Thus, this reduction or removal of hydrophilic hydroxyl groups increases the moisture resistance property of the bio-fiber. It also takes out a certain portion of hemicelluloses, lignin, pectin, wax, and oil covering materials resulting in a cleaner fiber surface. Researchers also suggest an optimum concentration of the NAOH during the alkali treatment because a higher concentration can cause excess delignification of the bio-fiber which results in weakening or damaging the fibers. Treatment of jute fibers by 5% NaOH at room temperature for varying times showed an overall improvement in properties both as fibers as well as reinforced composites. For composites comprising of jute fibers and vinylester resins, it was observed that with 4 h-treated fibers, the flexural strength improved by 20%, modulus improved by 23%, and laminar shear strength increased by 19% [38].

Coupling agents usually improve the degree of cross-linking in the interface region and offer a perfect chemical bonding. Silane is one such multifunctional molecule which is used as a coupling agent to modify bio-fiber surfaces. The composition of silane forms a chemical link between the fiber surface and the matrix through a siloxane bridge. It undergoes several stages of hydrolysis, condensation, and bond formation during the treatment process of the fiber. Silanes have been found to increase hydrophobicity of bio-fibers and strength of bio-fiber composites predominantly because the covalent bonding occurs between silane and the matrix [39,40]. The efficiency for such silane treatment in most cases can be improved if the fibers are pretreated by alkaline solution such as NAOH as reported by Van de Weyenberg for flax fibers. Silane coupling agents operate by reducing the number of cellulose hydroxyl groups in the fiber–matrix interface, hence minimizing fiber sensitivity to humidity. Bio-fibers exhibit micro-pores on their surface and the other mechanism which takes place during silane coupling agents' treatment is of surface coating as the silane penetrates into the pores and develops mechanically interlocked coatings on the fiber surface. Hence, silane treated fiber composites provide better tensile strength properties than the alkali treated bio-fiber composites.

There is a range of other treatments tried and tested for different bio-fibers for their surface modification [41]. Grafting is an effective method for the modification of bio-fiber surface. Synthesis of graft copolymers by creation of an active site, a free radical, or a chemical group which may get involved in an ionic polymerization or in a condensation process, on the preexisting polymeric backbone is one of the common methods of surface modification of bio-fibers. The most widely used monomers for grafting process of bio-fibers are benzoyl chloride, maleated polypropylene, maelic anhydride, MAH/PP, acrylate, and titanate [24,27,37]. Acetylation, a surface treatment that was originally applied to wood cellulose, has been extended to bio-fibers which are to be utilized for composite manufacturing. Acetylation treatment protects the cell walls against moisture; it improves dimensional stability and environmental degradation of cellulosic fibers by esterification. In a study pertaining to acetylation treatment of coir, oil palm fiber, flax, and jute fibers using acetic anhydride, it was reported that at a reaction temperature of 120°C there was damage to the fiber structure of coir fibers which further resulted in poor mechanical properties, whereas at 100°C the modified fibers exhibited improved performance when compared with control samples [42]. Peroxide treatment of bio-fibers has been popular because of the ease of processing the treatments and the improvement in mechanical properties of bio-fiber. Organic peroxides are easily decomposed to give free radicals, which further react with the hydrogen group of the matrix and cellulose fibers. In peroxide treatment, fibers are treated with 6% benzoyl peroxide or dicumyl peroxide in acetone solution for about 30 min after alkali pretreatment [43,44] conducted at a temperature of 70°C to support the decomposition of peroxide. Permagnate treatment of bio-fibers enhances chemical interlocking at the interface and provides better adhesion with the matrix [44]. Other treatments such as acrylation [9,26], isocyanate [45], and sodium chloride treatment [8,43] are also found in the literature being used for chemical modification of bio-fibers. Fungal treatment has been recently considered as a promising alternative for surface modification of natural fibers. This biological treatment is environmentally friendly and efficient. Fungi treatment is used to remove noncellulosic components (such as wax) from the fiber surface by the action of specific enzymes. This treatment also causes removal of lignin and hemicellulose, along with creation of more micro-pores and hence roughening the surface for better interlocking. Li et al. studied fungal treatment using white rot fungi on hemp fiber and reported that a 28% higher composite strength was achieved in comparison to an untreated one [46].

2.6 COMPOSITE MANUFACTURING TECHNOLOGIES

Bio-fiber thermosets normally are processed by simple processing techniques such as hand layup and spraying, compression molding, resin transfer molding, injection molding, and pressure bag molding operations like conventional thermoset composites. A few other methods such as laminating, filament winding, pultrusion, rotational molding, and vacuum forming are being used for composites but these methods for natural fiber composites are rarely used and so are hardly reported. In thermoset polymers, fibers are used as unidirectional tapes or mats. The production of these composites is optimized in relation to raw material parameter such as type of resin, form of reinforcement, and so on, and the processing parameters as temperature, pressure, and molding time depending on type of bio-fiber. It is often necessary to preheat natural fibers to reduce moisture before processing composites. High temperatures degrade the cellulose; thus, negatively affecting mechanical properties of the composites. Inconsistent fiber dispersion in the matrix causes fiber agglomeration which is not suitable for some of the processes and yields very poor mechanical characteristics. Most of the previous research on bio-fiber composites focused on reinforcements such as flax, hemp, sisal and jute, and conventional thermoplastic and thermoset matrices. Recently composites have been produced using matrices made of derivatives from cellulose, starch, and lactic acid along with conventional and newly developed bio-fibers to develop fully biodegradable composites or biocomposites. The emerging diversity of applications of natural fiber composites has seen the production of sandwich structures based on natural-fiber composite skins. In some cases, these sandwich composites have been produced from paper honeycomb and natural fiber–reinforced thermoplastic or thermoset skins, depending on the applications. The main criteria for the selection of the appropriate process technology for bio-fiber thermoset composite manufacture include the desired product geometry, the performance needed, as well as cost and ease of manufacture. The fabrication methods for these bio-fiber composites are similar to those used for conventional glass fibers. The most commonly used manufacturing processes are presented in the following section. Although many variants on these techniques exist, this overview gives a good indication of production possibilities.

2.6.1 HAND LAMINATION

This is a low production process that involves manual placement of fibers in a mold which is of the shape of the final part, and then application of thermoset resin on the fibers also squeezing the resin manually via rollers. Curing of the applied resin is done using a vacuum bag, where the excess air is removed and atmospheric pressure is applied to compact the part. Sometimes an autoclave is also used for curing the resin infiltrated part. Simplicity, low cost of tooling, and flexibility of design are the main advantages of this process. Long production time, intensive labor, and low automation potential are some of the disadvantages of this process [9,32].

2.6.2 RESIN TRANSFER MOLDING (RTM)

Liquid composite molding processes include processes such as resin transfer molding (RTM), vacuum assisted resin transfer molding (VARTM), structural reaction injection molding (S-RIM), co-injection resin transfer molding (CIRTM), and other subsets with a common processing approach. In these processes, a liquid resin along with other constituents such as hardeners and release agents are injected in a bed of stationary preforms placed inside the closed mold cavity. The RTM process has become a popular composite manufacturing process due to its capability for high volume. The resin transfer molding technique requires the fibers to be placed inside a mold consisting of two solid parts. A tube connects the mold with a supply of liquid resin, which is injected at low pressure (or high pressure in the case of HP-RTM) through the mold, impregnating the fibers. The part is then cured inside the mold at specific temperature and sometimes at room temperature or above until the end of the curing reaction; postcuring is also sometime essential (see Figure 2.10). Injection pressure, fiber content, and mold temperature along with the profiles of pressure and temperatures play a vital role in optimizing the process and composite properties. In the RTM process, the resin injection pressure, temperature of the mold, permeability of the fiber mat, preform architecture and permeability, resin viscosity, gate location and configuration, vent control and preform placement techniques are the major processing variables. In general, higher injection pressure and mold temperature would shorten the manufacturing cycle time due to the low viscosity of resin.

The RTM process technique has the advantage of rapid manufacturing of large, complex, and high performance parts. Several types of resins (epoxy, polyester, phenolic, and acrylic) can be used for RTM as long as their viscosity is low enough to ensure a proper wetting of the fibers. The hemp fiber composites manufactured with this RTM process were found to have a very homogeneous structure with no noticeable defects. The mechanical properties of these materials were found to increase with increasing fiber content [47,48]. In this study, it was also shown that the processing time for a 35 vol% hemp fiber composite could be decreased to 40 min from almost 2 h presently without any major changes to the set-up. In order to achieve high fiber contents with hemp fibers in a process such as RTM, the need of a prepressing stage at 100°C was suggested. This additional step reduced greatly the spring-back behavior of the fibers, making the closure of the mold much easier. Oksmann [49] in her study on flax-based epoxy composite concluded that RTM is a suitable processing technique for natural fiber composites when high quality laminates are preferred. Idicula and colleagues [47] and Rodriguez and colleagues [50] reiterated such an observation in the case of hybrid natural fiber composites manufactured using RTM.

An alternative variant of this process is vacuum injection or vacuum-assisted resin transfer molding (VARTM), where a single solid mold and a foil (polymeric film) are used. This process has also been widely reported for bio-fiber thermoset composite manufacturing. One such study is reported by Rodriguez et al. [51], where they studied mechanical properties of composites based on different natural fibers and glass fibers using unsaturated polyester and modified acrylic as matrix manufactured using a vacuum infusion process. The VARTM process is a very clean and low cost manufacturing method: resin is processed into a dry reinforcement on a vacuum-bagged tool, using only the partial vacuum to drive the resin. As one of the tool faces is flexible, the molded laminate thickness depends partially on the compressibility of the fiber-resin composite before curing and the vacuum negative pressure.

2.6.3 COMPRESSION MOLDING

Compression molding is one other process among the most widely used processes for bio-fiber composite manufacturing. Many studies have been conducted on the feasibility of using bio-fibers as reinforcement mixed with polymers to form a new

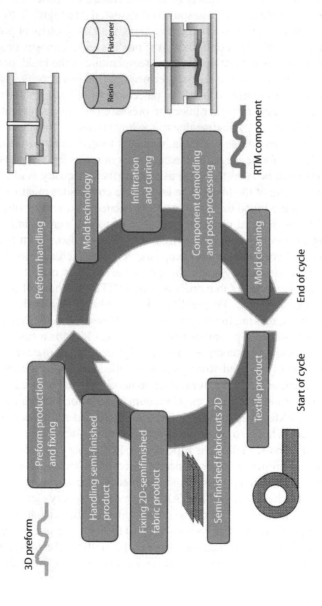

FIGURE 2.10 Process cycle for a typical **RTM** process. (From Fraunhofer Institute of Chemical Technology, Pfinztal, Germany.)

class of composites through compression molding process [51–56]. This process is a combination of a hot-press process and an autoclave process. Previously, this process was more popular for thermoplastic-based composite manufacturing. However, thermoset-based systems are now popularly tried in various compression molding experiments and product developments and have been reported for sisal/polyester and abaca-sisal hybrid thermoset composites using polyester resin [53–55]. Sheet molding compounds (SMCs) and bulk molding compounds (BMCs) are traditional initial charges for compression molding process. With the use of close-molds, the precut and weighed amount of fibers already infiltrated with resin (in the form of chopped, mat, or stitched) are stacked together and placed inside a preheated mold cavity. This amount of precut and weighed fibers usually covers 30%–70% of the female mold cavity surface. The mold is closed and then pressurized before temperature is applied where the compounds are flowing to form the shape of the cavity. After curing of the compounds, the mold is opened and the part is ejected (see Figure 2.11). In this process, the fibers are treated comparatively gently inside the mold and no shear stress and vigorous motion are applied, so the damage of fibers can be kept to a minimum. Long fiber can be used to produce composites with higher volume fracture by this process. Other advantages of compression molding are the very high volume production ability, the excellent part reproducibility, and short cycle times. Processing times of <2 min are reached during compression molding of three-dimensional components with a high forming degree. A big concern with compression molding that always needs to be considered is the maximum pressure and temperature before damage of fibers and structure especially in the case of bio-fibers and thermosets.

2.6.4 PULTRUSION

Pultrusion technology is traditionally used to manufacture composites using thermosetting resin systems. These profiles are produced by pulling a carefully specified mass of wetted reinforcement material through a heated metal die containing cavity of the desired cross-section. The pultrusion process has been successfully applied in fabrication of traditional composite profiles using fiber reinforcements like glass fiber and carbon fiber. There are a number of advantages for pultrusion over other composite making processes which include increased composite strength owing to the alignment of the tensioned fibers during the process and process capacity to achieve higher fiber volume fraction. Pultrusion is also a highly automated process with very little manual interface, so that it enables high volume production of constant cross-section parts with consistent quality and excellent structure. Moreover, pultrusion is relatively a cost-effective processing system having the potential of making composites with high fiber content. Recently, studies utilizing pultrusion technology for kenaf fiber [57], jute fiber [58], and hybrid jute/glass and kenaf/glass fiber [59] composites have also been reported. Based on its distinct advantages, pultrusion is considered a good choice for manufacturing bio-fiber composites to improve consistency of the composite quality and the overall properties by means of better impregnation, distribution, and alignment of the reinforcing fibers.

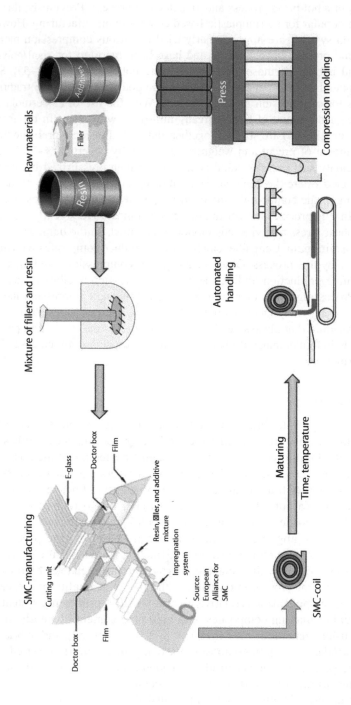

FIGURE 2.11 Process cycle for a typical SMC compression molding process. (From Fraunhofer Institute of Chemical Technology, Pfinztal, Germany.)

2.7 PROPERTIES OF BIO-FIBER THERMOSET COMPOSITES

2.7.1 MECHANICAL

Bio-fiber composite materials during their service life are subjected to different loads or forces. Thus, it is very important to understand the mechanical behavior of these materials so that the product made from it will not result in any failure during its life in service. An adequate knowledge of the mechanical properties of material helps in selection of specific material for suitable applications. As far as mechanical properties are concerned, bio-fibers are not stiffer then glass or carbon fibers, but the specific stiffness of these bio-fibers is higher than those of glass and carbon fibers. The high stiffness of reinforcing materials is often the basic requirement for their structural applications although for use as textile yarns and fabrics it may be an undesirable characteristic. Mishra [60] reported in his study in line with previous researchers that the tensile modulus of composites depends mainly on fiber volume fraction and not on the physical structure of the fibers. It was also observed that for jute-epoxy composites with 48 wt.% fiber loading, the highest tensile strength and modulus is exhibited compared to those with fiber loading% more or less than 48 wt.%. Similarly, he found the flexural properties of the manufactured composites were lowest for 12 wt.% loading and proportionally continue to increase after that [61]. In his study on tensile behavior of jute epoxy composites manufactured by varied preform stacking by VARI process, Hossain [62] reported that in longitudinal direction, the tensile strength and stiffness of 0-0 laminate composites have higher tensile properties compared to that of 0-45 or 0-90 laminate composites. This can be attributed to higher degree of fiber pull out in the 0-0 direction, which causes a relatively higher level of fracture surface. In the transverse direction, both the tensile and bending strengths of 0-0 laminate composites have been found to be lower compared to that of 0-45 or 0-90 laminate composites [59]. Ray et al. [38] subjected jute fibers to alkali treatment with 5% NaOH solution for different time intervals at 30°C. The study reported that the modulus of the jute fibers improved by 12%, 68%, and 79% after 4, 6, and 8 h of treatment, respectively. The tenacity of the fibers also showed an improvement by 46% after 6 and 8 h treatment and the percentage of breaking strain was reduced by 23% after 8 h treatment. For the composites manufactured using treated and nontreated jute fibers, the composites with jute fiber treated for 4 h showed an increase in flexural strength from 199.1 to 238.9 MPa by 20%, modulus improved from 11.89 to 14.69 GPa by 23%, and laminar shear strength increased from 0.238 to 0.283 MPa by 19%. The summary of the results from the above-mentioned study can be seen in Table 2.5.

Similar studies for epoxy resin and jute fiber were carried out by Doan [19]. The study reported the weight loss of fiber in NaOH treatment. This loss is more for 5% NAOH treatment when compared to NaOH 1% at the same period. It can be expected that the reaction between the NaOH and hemicellulose, which is thought to consist principally of xylan, polyuronide, and hexosan, and to be very sensitive to the action of NaOH, increased with increasing alkali concentration. NaOH 1% treatment for 4 h was reported to be the optimal condition of alkali treatment for the fiber tensile strength. The combination of alkali treatment and silane coupling agent (Y9669 from Momentive)

TABLE 2.5

Mechanical Properties of Untreated and Alkali-Treated Jute/Vinylester Composites

Jute (vol%)	Type of Fiber	Modulus (GPa) Mean Value	S.D.	Flexural Strength (MPa) Mean Value	S.D.	Breaking Energy (J) Mean Value	S.D.	Toughness [kJ/m2] Mean Value	S.D.
0	–	2.915	0.1	120.7	11.18	0.8227	0.33	34	13.63
8	Untreated	4.22	0.33	106.3	7.88	0.2948	0.02	11.75	0.79
	Treated for 2 h	3.446	0.53	96.27	3.99	0.2497	0.03	10.13	1.21
	Treated for 4 h	4.205	0.71	121.2	4.02	0.3634	0.03	14.75	1.21
	Treated for 6 h	3.967	0.66	101.8	6.14	0.227	0.04	9.28	1.63
	Treated for 8 h	3.13	0.43	93.97	4.95	0.2488	0.02	9.92	0.79
15	Untreated	5.544	0.55	128.6	3	0.3399	0.02	13.8	0.81
	Treated for 2 h	6.024	0.04	134.7	3.5	0.353	0.01	14.53	0.41
	Treated for 4 h	6.539	0.42	146.5	6.4	0.4016	0.05	16.3	2.02
	Treated for 6 h	5.546	0.21	121.5	7.93	0.2569	0.04	10.24	1.59
	Treated for 8 h	5.337	0.26	127.6	3.5	0.3351	0.03	13.6	1.21
23	Untreated	7.355	0.51	145.7	12.6	0.3531	0.08	14.08	3.19
	Treated for 2 h	8.065	0.79	157.7	1.1	0.4048	0.04	16.61	1.64
	Treated for 4 h	9.384	0.44	172.7	6.7	0.4198	0.05	16.74	1.99
	Treated for 6 h	8.542	0.41	155.4	10.7	0.3553	0.04	14.37	1.61
	Treated for 8 h	7.132	0.45	145.8	6.5	0.3762	0.04	15.27	1.62
30	Untreated	10.03	0.62	180.6	1.4	0.4799	0.02	19.14	0.79
	Treated for 2 h	10.99	0.27	189.4	15	0.4816	0.1	19.55	4.05
	Treated for 4 h	12.85	0.27	218.5	13.5	0.5061	0.02	20.69	0.81
	Treated for 6 h	12.49	0.94	195.9	10.04	0.4319	0.05	17.53	2.02
	Treated for 8 h	11.17	0.42	197.5	7.6	0.5042	0.05	20.32	2.01
35	Untreated	11.89	0.62	199.1	7.6	0.5543	0.07	22.1	2.79
	Treated for 2 h	12.7	0.37	205.2	6	0.457	0.04	18.55	1.62
	Treated for 4 h	14.69	0.85	238.9	17.6	0.5695	0.1	21.92	3.84
	Treated for 6 h	14.89	1.17	232	7.7	0.5678	0.02	23.05	0.81
	Treated for 8 h	12.32	0.35	204.2	1.2	0.5099	0.02	19.97	0.78

Source: Reproduced from *Composites Part A: Applied Science and Manufacturing* 32(1), Ray D, Sarkar B, Rana A, Bose N. The mechanical properties of vinylester resin matrix composites reinforced with alkali-treated jute fibres, 119–27, Copyright 2001, with permission from Elsevier.

or sizing in the form of silane + epoxy dispersion (APS+XB) lead to slightly improved interfacial interaction than only the NaOH treatment was reported. Therefore, the storage modulus (see Figure 2.12) of the NaOH/(APS+XB) and NaOH/Y9669 treated fiber composites was observed to be higher than both untreated and NaOH treated ones.

Deshpande et al. [20] investigated fiber treatment and filler addition on mechanical properties of jute polyester composites. The study concluded that the tensile, flexural, and impact strength properties of filled polyester decreases with increase

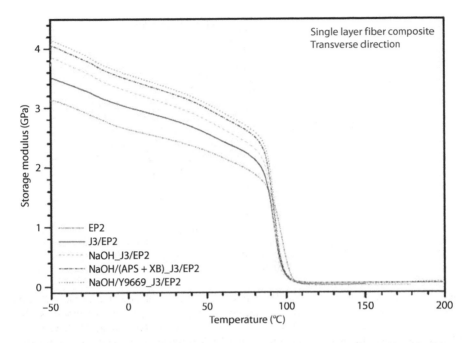

FIGURE 2.12 Storage modulus curves for epoxy, jute-epoxy composites with and without surface modification treatment as a function of temperature at 1 Hz frequency. (From Doan, T.T.L. 2006. Investigation on jute fiber and their composites based on polypropylene and epoxy matrices. PhD Thesis. Technical University of Dresden.)

in filler content. They also found that by 20% addition of filler, the change/decrease in the tensile was surprisingly not significant. Furthermore, alkali treatment to the jute fiber enhances mechanical properties of jute polyester composites. The study underlined the need to arrive at a detailed understanding and balance of the complex interaction among filler, fiber treatment, and resin for optimum bio-fiber composite manufacturing. Gassan and Bledzki [4] reported improvement of the mechanical properties of bio-fiber reinforced thermosets, as a result of optimization of the properties of tossa jute fibers by the use of an NaOH treatment process. Composite strength and stiffness increased as a consequence of the improved mechanical properties of the fibers by NaOH treatment. This increase in properties was reported to be highest for fibers with NAOH treatment carried under isometric conditions, that is, under zero shrinkage conditions. The increase in mechanical properties is diagrammatically reported in Figure 2.13.

Richardson and Zhang [63] studied the effects of nonwoven hemp on mechanical properties of phenolic resin and their microstructural features. They found a significant increase in flexural strength and modulus in phenolics with the introduction of nonwoven hemp. Rouison et al. [48,64] also studied the optimization of the resin transfer molding (RTM) process in manufacturing hemp fiber composites and their mechanical properties. Linear increase in tensile strength of hemp fiber/polyester composites was reported with increasing fiber content above 11% fiber volume fraction. A maximum tensile strength of 60 MPa was achieved for fiber volume fraction of 35%.

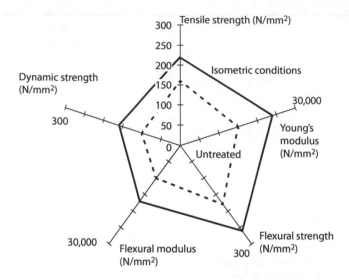

FIGURE 2.13 Influence of the NaOH treatment (26 wt.% NaOH, treatment time = 20 min, temp = 20°C, isometric conditions) on the mechanical properties of jute/epoxy composites (fiber content = 40 vol%). (Reproduced from *Composites Science and Technology*, 59(9), Gassan J, Bledzki AK, Possibilities for improving the mechanical properties of jute/epoxy composites by alkali treatment of fibres, 1303–9, Copyright 1999, with permission from Elsevier.)

The results for tensile modulus also showed a similar trend. The flexural strength and flexural modulus also increased linearly with increasing fiber volume fraction. Akil et al. [65] studied the mechanical properties of hemp and kenaf-fiber reinforced polyester composites, untreated and with alkali treatment. The alkali treated fibers of both types of composites showed superior flexural strength and flexural modulus values compared to untreated fibers. The flexural stiffness of these composites was found to be close to that of glass fiber composites. A considerable improvement was also observed in flexural strength. The relationship of surface modification and tensile properties of sisal fibers was studied by Li et al. [66]. Their modification methods include alkali treatment, H_2SO_4 treatment, conjoint H_2SO_4 and alkali treatment, benzol/alcohol, acetylated treatment, thermal treatment, alkali-thermal treatment, and thermal-alkali treatment. The results are summarized in Table 2.6.

In their research, Singh et al. studied sisal fiber reinforced polyester composites with different surface modification treatments as mentioned in Table 2.7 [67]. They reported by improving interfacial adhesion the moisture-induced degradation of composites can be reduced. Treated fiber composites absorb moisture at a slower rate than untreated counterparts, probably because of the formation of a relatively more hydrophobic matrix interface region by co-reacting organo functionality of the coupling agents with the resin matrix. Though significant reductions in tensile strength and flexural strength were reported for both untreated and surface-treated sisal composites in Table 2.7, the strength retention of surface treated composites is higher than that of composites containing untreated sisal fibers. Further, sisal epoxy composites manufactured from silane treated sisal fibers were studied by Bisanda and Ansell. They reported that the

TABLE 2.6
Effect of Treatment Methods on Tensile Properties of Sisal Fibers

Property	Tensile Strength (g/tex)	Tensile Modulus (×103 g/tex)	Elongation at Break (%)
Untreated	30.7	1.18	2.5
Benzol/alcohol	38.8	0.99	3.7
Acetic acid + alkali	9.3	0.39	2.6
Alkali	31.7	0.53	7.5
Acetylated	33.2	0.35	8.3
Thermal	42	1.22	3.5
Alkali–thermal	27.6	0.7	4.7
Thermal–alkali	25.7	0.71	4.4

Source: Reproduced from *Composites Science and Technology* 60(11), Li Y, Mai Y-W, Ye L. Sisal fibre and its composites: A review of recent developments, 35:53–60, Copyright 2000, with permission from Elsevier.

TABLE 2.7
Effect of Surface Treatments of Sisal Fibers on the Properties of Sisal/ Polyester Composites

Property	Untreated	N-Substituted Methacrylamide Treated	Silane Treated	Titanate Treated	Zirconate Treated
Density (g/cm³)	0.99	1.05	1.12	1.02	1.00
Void content (%)	16.11	5.88	3.01	12.55	13.30
Tensile strength (MPa)	29.66	39.48	34.14	36.26	34.69
Elongation (%)	9.52	9.75	5.71	8.00	9.51
Tensile modulus (GPa)	1.15	2.06	1.75	1.67	1.39
Energy to break (MJ/m²) × 10⁵	7.96	11.06	4.78	8.60	10.13
Flexural strength (MPa)	59.57	76.75	96.88	75.59	72.15
Flexural modulus (GPa)	11.94	15.35	19.42	15.13	14.46

Source: Singh B, Gupta M, Verma A: Influence of fibre surface treatment on the properties of sisal-polyester composites. *Polymer Composites.* 1996. 17. 910–918. Copyright Wiley–VCH Verlag GmbH & Co. KGaA. Reproduced with permission.

treatment of sisal fibers with silane, preceded by mercerization, provides improved wettability, mechanical properties, and water resistance of sisal-epoxy composites [68].

Oksman [49] studied the mechanical properties of traditionally retted unidirectional flax-epoxy composites and Arctic-flax-epoxy using the RTM technique. The results showed that the stiffness and strength values of the Arctic-flax fiber composites are promising compared to the earlier work on natural fiber composites. A study also concluded that RTM is a suitable processing technique for natural fiber composites when high quality laminates are preferred. Van de Weyenberg et al. [69] in their studies reported alkalization of flax fibers to be a simple and effective method to enhance the fiber/epoxy matrix bonding thus improving the flexural properties of UD flax-epoxy composites. Alkali treatment was reported to be beneficial to clean the fiber surface, modify the chemistry on the surface, lower the moisture uptake and increase the surface roughness. The treatment removed the impurities and waxy substances from the fiber surface and created a rougher topography which facilitated the mechanical interlocking. Also, the purified fiber surface further enhanced the chemical bonding between fiber and matrix and altogether enhanced the mechanical properties of the flax fiber thermoset composites as reported by Yan et al. [70].

Coir being an important lignocellulosic fiber has gained exposure in the bio-fiber thermoset composites recently. These fibers incorporated in polymers like unsaturated polyester in different ways are said to be achieving desired properties and texture suitable for composite applications. Rout et al. [71,72] studied the influence of fiber surface modification on mechanical properties of coir-polyester composites. The study reported that the composites containing 25 wt.% of fiber (untreated) improved tensile and flexural strength by 30% and 27%, respectively, in comparison to neat polyester. The work of fracture (impact strength) of the composite with 25 wt.% fiber content was found to be 967 J/m. The studies included surface modification by dewaxing, alkali (5%) treatment, aqueous graft copolymerization of methyl methacrylate (MMA) onto 5% alkali treated coir for different extents using $CuSO_4$–$NaIO_4$ combination as an initiator system and cyanoexhylation. It was reported that all the surface treatment techniques increased the mechanical performance of manufactured composites. Further significant improvement in mechanical strength was also observed for composites prepared from 5% PMMA grafted fiber. Alkali treatment showed reduction in moisture absorption of most of the natural fibers further leading to improved mechanical properties. This has been demonstrated for various natural fibers such as abaca [73], date palm tree fiber [74], and banana fiber [75] in various studies.

Hybrid composites behave like a unique material that does not exist in nature and weighs the sum of the individual components. The properties of hybrid composites are governed by the fiber length, orientation, fiber/matrix interfacial bonding, fiber content, extent of intermingling of fibers, and arrangement of both of the fibers. Several researchers developed hybrid composites by combining bio-fibers with polymeric matrices. Idicula [47] in his studies tested the tensile strength of short randomly oriented banana/sisal hybrid fiber-reinforced polyester composites fabricated by compression molding and RTM at different fiber loading by keeping the volume ratio of banana and sisal at 1:1. He found that resin transfer molded composites showed enhanced static and dynamic mechanical properties, compared with the compression molded samples. He attributed this phenomena to the lower water sorption and lower void content in the RTM composites. The effect of hybridization on mechanical properties of untreated woven jute and glass fabric-reinforced isothalic polyester composites was evaluated by Ahmed et al. experimentally [76]. The values of mechanical properties of hybrid composites plotted against glass fiber weight fraction show significant improvement after the inclusion of glass fiber. Addition of 16.5 wt.% glass fiber, in a total fiber weight fraction of 42%, enhances the tensile, flexural, and interlaminar shear strength (ILSS) by

37%, 31.23%, and 17.6%, respectively. This study also reported water absorption for different periods of immersion for jute-glass hybrid composites; it was found that the hybrid composites offer better resistance to water absorption. Further it was concluded in the study that the hybrid laminate of glass and jute fabric typically delaminates at the jute and glass interface when tested in tension [76]. Mohan et al. [77] also reported that with small amounts of surface layer reinforcements (t/d < 0.3) of jute with glass fibers, there were substantial improvements in flexural properties, moisture resistance, and toughness of the composite laminate. Raghvendra et al. [78] in their investigation revealed that, though the properties of jute reinforced composites properties are inferior to those of glass reinforced in terms of mechanical, when the tribological application of the jute fiber composites are considered, then they are far superior in properties than neat and glass-reinforced epoxy composites. The authors of this chapter also tried to study hybrid composites with jute and glass with epoxy resin in the RTM process. As a feasibility study the layup structure was altered in various trials of high pressure RTM composite manufacturing using fast cure epoxy resins and the mechanical properties were reported. It was found that a jute-glass epoxy laminate with glass layers (P representing plain woven, T representing twill woven glass fiber fabrics) on the top and bottom drastically improved the flexural strength by 3 times although the increase in weight was minimum (0.2 times) compared to neat jute composites. A summary of the experimental results can be seen in Figure 2.14.

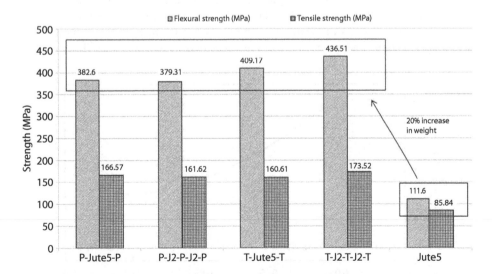

FIGURE 2.14 Mechanical properties of neat jute and hybrid jute-glass epoxy composites manufactured by RTM process (fiber content = 45 vol.%). (From Fraunhofer Institute of Chemical Technology, Pfinztal, Germany; Rajpurohit, A. et al., Centre of excellence for composites at ATIRA: overview of composites manufacturing technology and ongoing R & D projects, International Conferences on Composite Materials and Technology, ATIRA Ahmedabad, 2014.)

2.7.2 ENERGY ABSORPTION

The impact failure of composites occurs mainly due to factors such as fiber pullout, fiber and/or matrix fracture, and fiber/matrix debonding. Fiber pullout dissipates more energy compared to fiber fracture. The fiber fracture is common in composites with strong interfacial bonding, whereas occurrence of fiber pullout is a sign of weak bond. The load applied on composite transfers to the fibers by shear and may exceed the fiber–matrix interfacial bonding strength and debonding occurs. When the stress level surpasses the fiber strength, fiber fracture takes place. The fractured fibers may be pulled out of the matrix, resulting in energy dissipation. Mishra observed that the impact strength of jute epoxy composites increases with increase in fiber loading [60]. A similar trend of increase in impact strength with increase in needle-punched nonwoven fiber loading has also been reported by few researchers [61,79].

Structural properties of the fiber itself such as microfibrillar angle play an important role in deciding the impact properties of bio-fiber composite. Gassan and Bledzki [4] in their study reported that impact damping was distinctly affected by the shrinkage state of the fibers during the NaOH treatment because of its influence on yarn toughness. A good correlation was found between composite impact damping and yarn toughness for jute/epoxy composites as shown in Figure 2.15. Pavithran et al. [80] showed that for polyester composites reinforced with banana, pineapple, sisal, and coir fibers, the charpy impact work of fracture is significantly dependent on microfibrillar angle. The highest work of fracture was obtained for sisal fiber reinforcements with a microfibrillar angle of 20 and a higher fiber fracture toughness then for pineapple (microfibrillar angle = 15) and banana fibers (microfibrillar angle = 11). It was found that the microfibrillar angle and other structure parameters can be affected by using alkali treatment. These structural effects lead, among others, to changes in fiber and yarn toughness as measured in tensile mode. Taking fiber

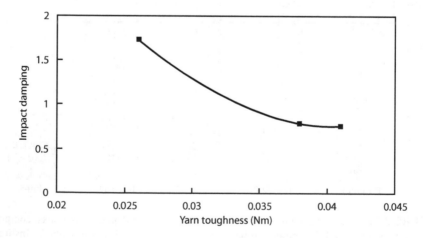

FIGURE 2.15   Influence of yarn toughness on the impact damping of UD jute/epoxy composites (fiber content = 30 vol.%). (Reproduced from *Composites Science and Technology*, 59(9), Gassan J, Bledzki AK, Possibilities for improving the mechanical properties of jute/epoxy composites by alkali treatment of fibres, 1303–9, Copyright 1999, with permission from Elsevier.)

toughness into account, impact damping is greatly dependent on this fiber property, where an increase in toughness leads to a decrease in impact damping, measured in nonpenetration impact tests.

Substantial improvement in toughness was reported by Richardson and Zhang for hemp fibers when utilized for phenolic composite applications. This improvement was attributed to two factors. First that hemp is tougher and stronger mechanically than phenolic matrix and leads to increase in mechanical properties. The second is that the presence of hemp has the capability of reducing the number of voids (defects) and their dimensions, which significantly contributes to improvement in mechanical properties [63]. In the case of kenaf–glass hybrid fibers, unsaturated polyester hybrid composite fabricated by sheet molding compound process displays higher tensile, flexural, and impact strength obtained from treated kenaf fiber. The kenaf with 15/15 v/v kenaf–glass fibers on treatment with 6% NAOH treatment (mercerization method) for 3 h yielded better mechanical strength to the composite. Researchers concluded that kenaf fiber alone (30% volume fraction) or higher percentage (22.5% volume fraction) cannot withstand higher impact load leading to brittleness and less toughness in hybrid composite [81]. Impact strength property of composites increases significantly as fiber is incorporated in the matrix. Rout et al. [71,72] report that even at 10% loading, the impact strength of polyester resin increases from 15 J/m (neat polyester) to 277 J/m. Furthermore, increase in fiber loading from 10% to 25% leads to a phenomenal increase in impact strength to 967 J/m, which then reduces to 568 J/m for 35% loading of coir fibers. This is attributed to poor wetting of the fiber at too high fiber fractions. This study concluded that for their experimental setup 17% to 25% of fiber weight% can lead to optimum mechanical properties including the impact properties of the composites. The effects of varying textile yarn linear density and textile weave configuration and their stacking sequence on the fracture behavior and fracture toughness of epoxy reinforced flax composites was investigated by Liu and Hughes [82]. It was reported that the addition of woven textile resulted in a 2–4 times improvement in toughness over that of the unreinforced epoxy, with a value of around 6 MPa $m^{1/2}$ for a cross-ply laminate. Both fracture behavior and toughness were found to be strongly dependent on the linear density of the weft yarn and the direction of crack propagation with respect to the orientation of the textile. However, the influence of weave type was not significant. It was concluded that toughness is dominated by fiber volume fraction, rather than the reinforcement architecture. The study concluded that the properties of the fiber and the fiber volume fraction actually dominate the toughness and the impact properties, rather than their microstructural arrangement. Further fracture toughness is strongly dependent on whether the composite is being tested in the weft or warp directions; there are strong anisotropic effects.

One of the interesting studies on energy absorption was conducted by Meredith and colleagues [83]. They studied the potential for natural fibers' ability to replace synthetic fibers for future environmentally friendly energy absorption structures. Conical test specimens of jute, flax, and hemp were manufactured using vacuum assisted resin transfer molding (VARTM) and subjected to impact testing.

The study reports that natural fiber cones exhibited high values of specific energy absorption: unwoven hemp 54.3 J/g, woven flax 48.5 J/g, and woven jute

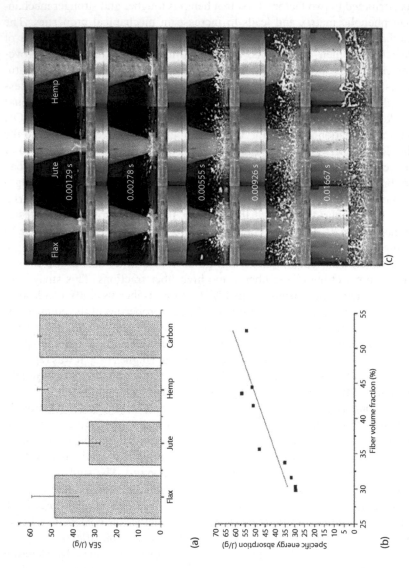

FIGURE 2.16 (a) Specific energy absorption versus fiber type, (b) correlation between specimen fiber volume fraction and degree of specific energy absorption, and (c) pictures from impact tests at different time points. (Reproduced from *Composites Science and Technology*, 72(2), Meredith J, Ebsworth R, Coles SR, Wood BM, Kirwan K, Natural fiber composite energy absorption structures, 211–7, Copyright 2012, with permission from Elsevier.)

32.6 J/g (see Figure 2.16a). These values were influenced primarily by fiber volume fraction of the composites where a high FVF leads to high specific energy absorption (see Figure 2.16b). Natural fiber composites have the potential to be widely applied as low cost, sustainable energy absorption structures. The woven flax and jute samples were said to exhibit characteristics of brittle fracture. The formation of multiple longitudinal cracks on impact is followed by the initiation of lamina bending. The jute displays a large longitudinal crack on impact, which suggests that the dominant energy dissipation feature is that of crack propagation through resinous areas. The longitudinal cracks propagate less effectively in flax signifying the importance of woven fiber density on composite toughness. The lamina bundles are observed to fracture under the influence of transverse bending forces, limiting energy absorption due to lamina bundle bending. In contrast, the unwoven hemp exhibits characteristics attributable to progressive collapse. There is no indication of terminal longitudinal crack formation on crush initiation. The initiation of and continued development of lamina bundles illustrates the dominant influence of crack growth limitation on fracture mode. The study concluded that nonwoven hemp with its low embodied energy is a promising candidate for sustainable energy absorption structures with properties comparable to carbon composites.

2.7.3 Fatigue

A fundamental problem concerning the engineering uses of bio-fiber reinforced thermosets is the determination of their resistance to combined states of cyclic stress. These composite materials exhibit very complex failure mechanisms under static and fatigue loading because of anisotropic characteristics in their strength and stiffness. Fatigue causes extensive damage throughout the specimen volume, leading to failure from general degradation of the material instead of a predominant single crack. A predominant single crack is the most common failure mechanism in static loading of isotropic, brittle materials such as metals. There are four basic failure mechanisms in composite materials as a result of fatigue: matrix cracking, delamination, fiber breakage, and interfacial debonding. Thus far very little work has been reported in the literature with respect to natural fiber composites. One of the most talked about works in the field of fatigue for bio-fiber composites is presented by Gassan and Bledzki. They reported in their studies that higher fiber volume content leads to a decrease in fatigue behavior for jute epoxy composites. It was also found that the alkali treated fiber (isometric conditions) utilized for composite manufacturing showed a lower fatigue performance compared to the untreated fiber [4]. It was reported that both the dynamic modulus and the fatigue behavior of composites with alkali-treated fibers are dependent on the shrinkage state of fibers (see Figure 2.17a). Further, Gassan and colleagues [4,33] concluded in their fatigue studies that decrease in fatigue behavior, that is, reduction in dynamic modulus with increasing number of cycles, for both types of composites, and the fatigue behavior of composites with alkali-treated fiber (in isometric conditions) are slightly poorer than that of composites with untreated jute fibers as observed in Figure 2.17b.

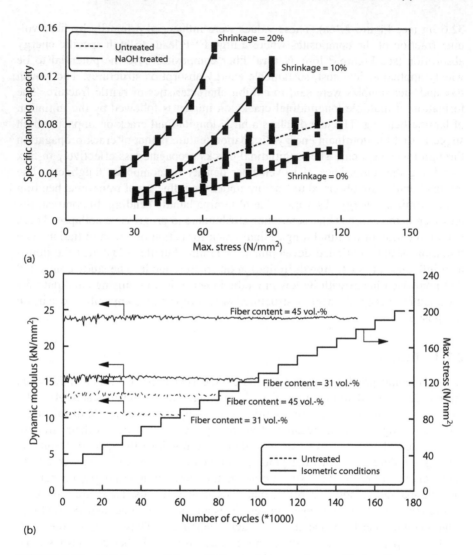

FIGURE 2.17 (a) Influence of the applied max. Stress on the specific damping capacity for UD jute/epoxy. (b) Influence of fiber content on the dynamic modulus of jute epoxy composites with untreated (fiber content = 36 vol.%) and NaOH treated fibers with different shrinkage states (26 wt.% NaOH, treatment time = 20 min, treatment temperature = 20°C; fiber content = 30 vol.%). (Reproduced from *Composites Science and Technology*, 59(9), Gassan J, Bledzki AK, Possibilities for improving the mechanical properties of jute/epoxy composites by alkali treatment of fibres, 1303–9, Copyright 1999, with permission from Elsevier.)

Though other extensive studies in the field of fatigue properties of bio-fiber composites are not reported, here efforts were made by some of the researchers to characterize and compare the results of fatigue experimental data with simple fatigue models. One such study was presented by Abdullah et al. [84]. This study reported on epoxy as well as epoxy and kenaf reinforced composites subjected to tension-tension fatigue loading. It concluded that the fatigue life is affected by the amount of fiber volume ratio but it may not show any significant improvement at very high number of cycles (as shown in Figure 2.18). Further investigation made by comparing experimental data with three simple fatigue models (Mandell, Manson-Coffin, and Hai-Tang) indicated that there are no applicable models close to mimicking the fatigue life of such bio-fiber composites.

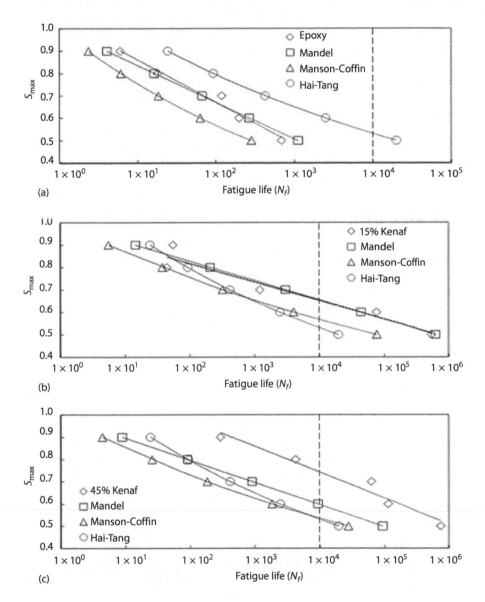

FIGURE 2.18 Comparison between experimental data and selected fatigue model. (a) Pure epoxy specimens, (b) 15% kenaf reinforced epoxy composite, and (c) 45% kenaf reinforced epoxy composite [84]. (From Abdullah AH et al. *Engineering Journal* 16(5):1–10, 2012. With permission.)

2.7.4 ABRASION

Wear of solid is generally considered to be a mechanical process. However, wear sometimes can also be attributed to corrosion, oxidation, and other chemical processes as exemptions. Among all the wear types, abrasive wear is one of the major problems encountered by composite products used in industrial application. Generally, abrasive wear occurs when hard asperities on one surface move across a softer surface under load, penetrate and remove material from the softer surface, leaving grooves. Mishra observed that the specific wear rate of composites increases with increase in normal load for jute bio-fiber epoxy composites similar to what is being reported for glass fiber thermoset composites [60]. At increased load, contact temperature between jute epoxy composite and rubber wheel increases; this leads to easy detachment of the softened layer from the surface of the material. The study also concluded that the composite with 36 wt.% fiber loading exhibits minimum specific wear rate compared to the other composites. This indicates a proper stress transfer and distribution by the matrix to the fibers which in turn increases the wear resistance of the composites. When the fiber content increases beyond optimum value (i.e., 36 wt.%), short jute fiber gets agglomerated, thus reducing the interaction between the fiber and matrix. This causes debonding of fiber from the epoxy matrix. Thus, at higher fiber loading, the interfacial adhesion between the matrix and fiber is not adequate enough to resist the sliding force resulting in higher wear.

Raghvendra et al. [78] in their studies on jute fiber epoxy composites and glass fiber epoxy composites concluded that jute composites give better erosion resistance properties when compared to neat epoxy and glass fiber reinforced epoxy composites. They attributed this to the rough surface of the jute fiber and to the presence of cellulose and lignin in the jute fibers. The morphologies of the eroded glass and jute fiber composites from the study are shown in Figure 2.20a and b.

Form Figure 2.19a, it is clearly observed that a large groove is formed due to blasting by silicon sand. It is also observed that the epoxy and glass fiber bonding is less; cracks and large material losses are also observed. From Figure 2.19b, it is

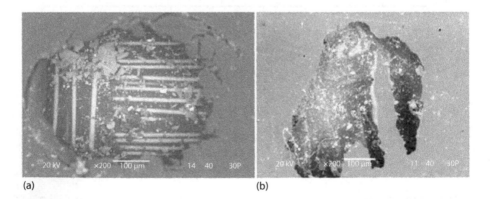

(a)   (b)

FIGURE 2.19 Pictographs of worn samples at 200×. (a) Glass fiber composites, (b) jute fiber composites. (From Raghavendra G et al. *Journal of Composite Materials* 48(20):2537–47, 2014. With permission.)

observed that no cracks are formed and also tight bonding of epoxy with jute fiber is observed, this is due to the lingo-cellulosic nature of jute fiber. The material loss is also less. Studies on abrasion behavior of bio-fiber thermoset composites are still lacking and need more attention from the researchers as little research experience shows better abrasive performance of bio-fiber thermosets compared to other high performance composites.

2.7.5 THERMAL

Thermal analysis is the concept that purely reflects the reactions that occur at the molecular level of materials. Figure 2.20 shows a curve of jute fiber and neat epoxy at different heating rates. The area under the TGA curve for raw jute fiber is less compared with the neat epoxy matrix, indicating lower thermal stability of raw jute fiber. From the figure it is clear that for jute fiber the initial decrease in weight below 100°C is due to moisture loss from bio-mass material at this temperature. The next step is thermal degradation of biomass. In thermal degradation, first at 155–169°C, lignin is the first component that decomposes followed by hemicellulose component, which starts decomposing from 230°C to 307°C. After hemicelluloses decomposition in the final step, major weight loss occurs within the temperature range of 323–392°C as a consequence of the cellulose decomposition.

Similar results were also reported by Yao et al. [85] on their study on thermal decomposition kinetics of natural fibers. Figure 2.21 presents the overall thermogravimetric decomposition process of natural fibers at a heating rate of 2°C/min from the above-mentioned study. It can be seen that thermal decomposition of cellulose was observed with an obvious "shoulder" (arrow) (Figure 2.21a), which was normally considered a result of thermal decomposition of hemicelluloses in an inert atmosphere. These low-temperature hemicelluloses' shoulder peaks, however, were overlapped in the cellulose main peaks and consequently they were not obvious in some cases (Figure 2.21b).

Despite previous research that has contributed to the understanding of thermal decomposition kinetics of bio-fibers and thermoplastic resins, a simplified prediction approach for thermal decomposition of these fibers and composites under normal

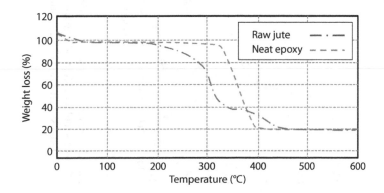

FIGURE 2.20   Thermogravimetric analysis (TGA) of jute and epoxy.

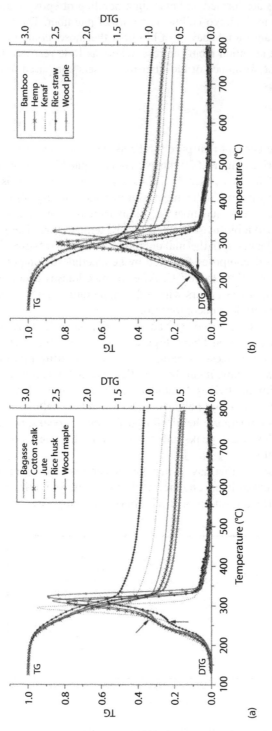

FIGURE 2.21 Overall thermogravimetric decomposition process of natural fibers at a heating rate of 2°C/min with (a) obvious and (b) not obvious hemicelluloses shoulders. (Reproduced from *Polymer Degradation and Stability*, 93(1), Yao F, Wu, Lei Y, Guo W, Xu Y, Thermal decomposition kinetics of natural fibers: Activation energy with dynamic thermogravimetric analysis, 90–8, Copyright 2008, with permission from Elsevier.)

processing temperatures of polymer composites has not been established thus far. For developing bio-fiber composites for specialized applications in automotive or civil engineering, it has now become essential for researchers to study the thermal behavior of the bio-fiber composite materials.

2.8 APPLICATIONS OF BIO-FIBER THERMOSET COMPOSITES

In the past decades, growing interest in bio-fiber composites has resulted in extensive research. The driving forces have been cost reduction, weight reduction, ecological benefit, and marketing advantage. Technical requirements were of less importance; hence, applications remained limited to non-structural parts for a long time. The reason for this was the traditional shortcomings of bio-fiber composites: the low impact resistance and moisture degradation. Recent research, however, showed that significant improvements of these properties are possible, resulting in applications such as automotive and civil and construction industry. Experiences from bio-fiber composite applications in western countries can be applied in developing new products, suitable for agro-industrial countries. In the case of bio-fiber thermoset composites, the bio-fiber (after specific treatments) serves as reinforcement by giving strength and stiffness to the structure while the polymer resin in this case thermoset serves as the adhesive to hold the fibers in place so that suitable structural components can be made. In recent years, a new class of fully biodegradable "green" composites has been made by combining natural fibers with biodegradable resins both thermoplastic and thermosets. The major attraction of these green composites is that they are fully degradable and sustainable, that is, they are truly "green." Green composites may be used effectively in many applications with short lifecycles or products intended for one-time or short-term use before disposal.

2.8.1 BIO-FIBER COMPOSITES IN AUTOMOTIVE

In the last two decades, bio-fiber thermoset composites have created substantial commercial markets for value-added products especially in the automotive sector. This share is expected to be steadily increasing owing to much research and development activities currently ongoing in terms of bio-fiber modification and newer bio resin development. The automotive industry is a prime driver of these green bio-fiber composites because the industry is facing multiple issues for which green materials offer a solution. It has been found that vegetable fibers perform favorably (in terms of specific performance; i.e., the ratio of mechanical performance to the density) than the glass fiber composites for all fiber architectures and also sometimes when compared to aluminum or steel. A practical translation of this parameter is that a panel with the same weight and surface area can be more than 10% stiffer in bio-fiber composites compared to glass. For consumers, bio-fiber composites in automobiles provide better thermal and acoustic insulation than fiber glass and reduce irritation to the skin and respiratory system. Also, lower density of plant fibers also reduces vehicle weight further leading to cuts in fuel consumption and carbon dioxide emissions.

Currently flax and hemp bio-fibers dominate the total bio-fibers used in the automotive field (see Figure 2.22) and their dominance is likely to be challenged by

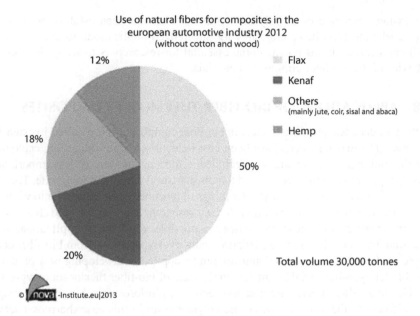

FIGURE 2.22   Use of bio-fibers in European automotive industry. (From Nova Institute's market study. Wood-Plastic Composites (WPC) and Natural Fiber Composites (NFC): European and Global Markets 2012 and Future Trends in Automotive and Construction, 2015. With permission.)

various other fibers which are constantly and extensively researched in the current time [86]. Bio-fibers have intrinsic properties like mechanical strength, low weight and low cost, ecological sustainability, low energy requirements for production, end of life disposal, and carbon dioxide neutrality—that have made them particularly attractive to the automobile industry. The growing interest in lignocellulosic fibers is mainly due to their economical production, with few requirements for equipment and low specific weight compared to glass fiber composites; for example, using bast fibers in the automotive industry leads to weight savings of between 10% and 30% and corresponding cost savings.

Many components for the automotive sector are now made from bio-fiber composite materials (Figure 2.23). As in developments with these materials, the German automotive manufacturers Audi/Volkswagen, BMW, and Mercedes-Benz have been considerably more proactive in specification of bio-fibers and have built up a considerable knowledge in this field. Similarly, the Canadian, North American, and Brazilian automotive industries have been following in these footsteps and are putting forward efforts for newer material and process combinations for bio-fiber thermoset composites [87]. In door panels, for example, weight reduction of 25% to 30% was achieved using bio-fiber/polypropylene (PP) composites by different automotive manufacturers. In consecutive research and development, the Mercedes-Benz E class replaced the wood fiber materials by flax/sisal fiber reinforced epoxy thermoset composites. These bio-fiber thermoset composites were 20% lighter and provided better mechanical performance in case of an accident. In addition, this combination of bio-fiber thermoset composite also extended the flexibility of three-dimensional shapes

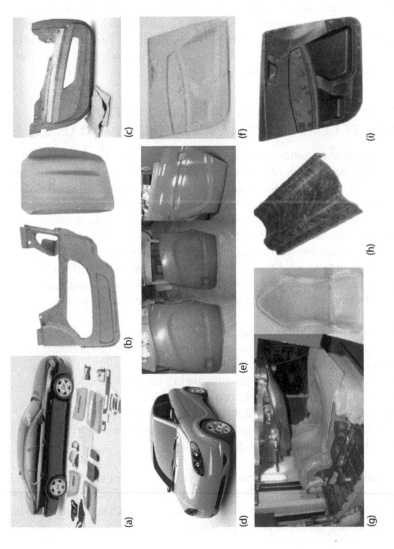

FIGURE 2.23 Examples of bio-fiber thermoset composites in automotive applications: (a) Composite parts fabricated with bio-fibers for BMW sedan (from BMW), (b) Natural fiber-reinforced acrylic resin composites, (c) BMW 7 Series door panel using bio-fiber and BASF Acrodur resin (From BASF SE), (d) Kestrel: Hemp bio composite car, (e) door panel made using hemp fiber and bio thermoset resin, (f) and (g) demonstrator parts manufactured using hemp and kenaf fibers and acrylic resins, (h) and (i) demonstrators manufactured from natural fibers and bio resins such as Furolite and BioRez. (From BASF and BMW via emails; other sources adequately cited.)

manufacturing. Mercedes has also been producing for specific classes a range of polyurethane-based bio fiber composites since the 1990s. As an overview, Mercedes A-Class, C-Class, and BMW employ 24 kg, 17 kg, and 20–24 kg, respectively, of bio-fibers for various applications including the thermoset resin based composites. The Flaxpreg project from France successfully developed flax fiber based composites for trunk load floor with extreme light weight and extreme high performance using Acrodur thermoset resin from BASF. New trends are apparently coming into the picture; these include the interest of automotive manufacturers to not only use bio-fiber composites but also to show them to their customers. Therefore, bio-fiber thermoset composites are now aiming to extensively serve the structural and performance needs of automotive components.

2.9 CONCLUSION AND FUTURE TRENDS IN BIO-FIBER THERMOSET COMPOSITES MANUFACTURING AND USAGE

Bio-fibers have excellent specific strength and modulus and have been used in various conventional applications in fiber tow and fabric form for centuries. There has been ever-increasing interest in the manufacture and utilization of ecologically friendly bio-fibers composite materials over the past two decades. Hence, it is no surprise that bio-fibers and composites made from these fibers have been extensively studied and their potential is explored. Research has shown that it is possible to manufacture, surface treat bio-fibers, and make effective reinforcements for a wide variety of thermoset resins and different composite manufacturing processes. The research in this field has been developed continuously over the years and comprises of activities specifically in the following areas covering the complete process chain:

1. Exploring newer plant/bio-materials for development of new bio-fibers
2. Efficient agricultural production of existing bio-fibers and development of improved processes for extraction of fibers
3. Fiber surface modification and manufacturing of reinforcement (semi-finished textile product) suitable for composite manufacturing
4. Development of new generation of thermoset resins based on bio-materials
5. Modification, development, and optimization of composite manufacturing processes ideal for industrial applications

With the continuous emergence of new and improved technologies and processes, bio-fiber thermoset composite materials are becoming more suitable than ever for use in a wide variety of commercial applications. Some future trends in the field of bio-fiber thermoset composites are proposed below:

1. Grading of fibers and utilizing the fibers of specific grade for composite manufacturing. Development of concepts for quality consistent bio-fiber manufacturing specifically for structural composite applications.
2. Reinforcements with specific orientation of fibers in specific directions, development of specific surface treatment suiting the resins and processing

techniques, and developing hybrid reinforcements containing one or more bio-fiber type or a combination of bio-fiber and high performance fiber.

3. Development of suitable resin system and specific composite manufacturing technology where manufacturing of the composites does not cause any harm to bio-fibers which in general have poor thermal properties and are degraded during the composite manufacturing process.

4. Improvement of performance of the bio-fiber composite shall be one of the most interesting topics for researchers as it is reported that increase in mechanical performance by a factor of 25% can lead to an increase in consumption volume by a factor of 50%.

5. Fully biodegradable composites will be an important field of research and development. The key factor here will be the development of new resin systems thermoplastics and thermosets, which are either recyclable or biodegradable.

It can be said that the natural fiber composites in the high performance applications have just arrived and they are here to stay.

ACKNOWLEDGMENTS

The authors are grateful to CRC Press and the editors of this book, Dr. O. Faruk, Dr. J. Tjong, and Prof. M. Sain, for inviting us to contribute to this book in the form of this chapter. The authors are also thankful to the publishers who granted permissions to use tables and figures from their journals.

REFERENCES

1. Witten, W. and Kraus, E. 2014. Report: Market developments, trends, challenges and opportunities. *Composites Market Report.*
2. Lucintel, Natural Fiber Composites Market Trend and Forecast 2011–2016: Trend, Forecast and Opportunity Analysis.
3. Karus, M., Kaup, M., and Ortmann, S. (Eds.). *Use of Natural Fibres in Composites in the German and Austrian Automotive Industry.* Nova-Institut GmbH, Hürth.
4. Gassan, J. and Bledzki, A.K. 1999. Possibilities for improving the mechanical properties of jute/epoxy composites by alkali treatment of fibres. *Composites Science and Technology.* 59(9):1303–9.
5. Bledzki, A.K., Faruk, O., and Sperber, V.E. 2006. Cars from bio-fibres. *Macromol. Mater. Eng.* 291(5):449–57.
6. Faruk, O., Bledzki, A.K., Fink, H., and Sain, M. 2014. Progress report on natural fiber reinforced composites. *Macromol. Mater. Eng.* 299(1):9–26.
7. Charlet, K. 2012. Natural fibres as composite reinforcement materials: Description and new sources. In: John, M.J. and Thomas, S., Eds. *Natural Polymers,* Vol. 1. Cambridge, England: Royal Society of Chemistry. pp. 37–62.
8. Saheb, D.N. and Jog, J.P. 1999. Natural fiber polymer composites: A review. *Advances in Polymer Technology.* 18(4):351–63.
9. Cristaldi, G., Latteri, A., Recca, G., and Cicala, G. 2012. Composites based on natural fiber fabrics. In: Dubrovski, P.D., Ed. *Woven Fabric Engineering* In-tech Open, November.

10. Shahzad, A. 2012. Hemp fiber and its composites—A review. *Journal of Composite Materials*. 46(8):973–86.
11. Yan, L., Chouw, N., and Jayaraman, K. 2014. Flax fibre and its composites—A review. *Composites Part B: Engineering*. 56:296–317.
12. Heijenrath, R. and Peijs, T. 1996. Natural fibre mat reinforced thermoplastic composites based on flax fibres and polypropylene, *Adv. Comp.* 5:81–5.
13. Berglund, L.A. and Ericson, M.L. 1995. Glass mat reinforced polypropylene. In: Karger-Kocsis, J., Ed. *Polypropylene: Structure, Blends and Composites*, Vol 3, Chapman & Hall, London, pp. 202–27.
14. Mohanty, A.K. and Misra, M. 1995. Studies on jute composites—A literature review. *Polymer-Plastics Technology and Engineering*. 34(5):729–92.
15. Sain, M. and Panthapulakkal, S. 2006. Bioprocess preparation of wheat straw fibers and their characterization. *Ind. Crops Prod.* 2:1–8.
16. Joshi, S., Drzal, L., Mohanty, A., and Arora, S. 2004. Are natural fiber composites environmentally superior to glass fiber reinforced composites? *Composites Part A: Applied Science and Manufacturing*. 35(3):371–6.
17. Sahari, J. and Sapuan, S.M. 2011. Natural fiber reinforced biodegradable polymer composites. *Reviews on Advanced Materials Science*. 30:166–74.
18. Furtado, S.C.R., Silva, A.J., Alves, C., Reis, L., Freitas, M., and Ferrão, P. 2012. Natural fibre composites: Automotive applications. In: John, M.J. and Thomas, S., Eds. *Natural Polymers*, Vol. 1. Cambridge, England: Royal Society of Chemistry. pp. 118–39.
19. Doan, T.T.L. 2006. Investigation on jute fiber and their composites based on polypropylene and epoxy matrices. PhD Thesis. Technical University of Dresden.
20. Deshpande, A.P., Kanakasabai, P., and Malhotra, K. (Eds.). *Effect of Fabric Treatment and Filler Content of Jute Polyetser Composites*. IIT Madras, Chennai.
21. Mussig, J. and Slootmaker, T. 2010. What are natural fibers? In: Mussig, J., Ed. *Industrial Applications of Natural Fibres: Structure, Properties and Technical Applications*, Wiley, New York, pp. 44–8.
22. Mohanty, A.K., Misra, M., and Drzal, L.T., Eds. 2005. *Natural Fibers, Biopolymers, and Biocomposites*, CRC Press, Boca Raton, FL.
23. Faruk, O., Bledzki, A.K., Fink, H., and Sain, M. 2012. Biocomposites reinforced with natural fibers: 2000–2010. *Progress in Polymer Science*. 37(11):1552–96.
24. Chollakup, R., Smitthipong, W., and Suwanruji, P. 2012. Environmentally friendly coupling agents for natural fibre composites. In: John, M.J. and Thomas, S., Eds. *Natural Polymers*, Vol. 1. Cambridge, England: Royal Society of Chemistry. pp. 161–82.
25. John, M.J.. and Thomas, S. 2012. *Natural Polymers: An Overview*. In: John, M.J. and Thomas, S., Eds. *Natural Polymers*, Vol. 1. Cambridge, England: Royal Society of Chemistry. pp. 1–7.
26. George, J., Sreekala, M.S., and Thomas, S. 2001. A review on interface modification and characterization of natural fiber reinforced plastic composites. *Polymer Engineering and Science*. 41(9):1471–85.
27. Ichhaporia, P. Composites from Natural Fibers. 2008. PhD Thesis submitted to North Carolina State University.
28. Goswami, D.N., Amsari, M.F., Day, A., Prasad, N., and Baboo, B. 2008. Jute-fiber glass-plywood/particle board composites. *Indian Journal of Chemical Technology*. 15:325–31.
29. Masoodi, R. and Pillai, K.M. 2012. A study on moisture absorption and swelling in bio-based jute-epoxy composites. *Journal of Reinforced Plastics and Composites*. 31(5):285–94.
30. Athijayamani, A., Thiruchitrambalam, M., Natarajan, U., and Pazhanivel, B. 2009. Effect of moisture absorption on the mechanical properties of randomly oriented natural fibers/polyester hybrid composite. *Materials Science and Engineering: A*. 517(1–2): 344–53.

31. Girischa, C., Sanjeevamurthy, Rangasrinivas, G., and Manu, S. 2010. Mechanical performance of natural fiber-reinforced epoxy-hybrid composites. *International Journal of Engineering Research and Applications.* 2(5):615–9.

32. Holbery, J. and Houston, D. 2006. Natural-fiber-reinforced polymer composites in automotive applications. *Journal of The Minerals, Metals & Materials Society.* 80–6.

33. Gassan, J. 2002. A study of fibre and interface parameters affecting the fatigue behaviour of natural fibre composites. *Composites Part A: Applied Science and Manufacturing.* 33(3):369–74.

34. Seki, Y., Sever, K., Sarikanat, M., Güleç, H.A., and Tavman, I.H. 2009. The influence of oxygen plasma treatment of jute fibers on mechanical properties of jute fiber reinforced thermoplastic composites. In: 5th International Advanced Technologies Symposium (IATS'09), May 13–15, 2009, Karabük, Turkey, pp. 1007–10.

35. Gassan, J. and Gutowski, V.S. 2000. Effects of corona discharge and UV treatment on the properties of jute-fibre epoxy composites. *Compos. Sci. Technol.* 60(15): 2857–63.

36. Pickering, K.L., Aruan Efendy, M.G., and Le, T.M. 2015. A review of recent developments in natural fibre composites and their mechanical performance. *Composites Part A: Applied Science and Manufacturing.* In press.

37. Kalia, S., Durfresne, A., Cherian, B.M., Kaith, B.S., Averous, L., Njuguna, J., and Nassiopoulos, E. 2011. Cellulose-based bio- and nanocomposites: A review. *International Journal of Polymer Science.* Volume 2011, Article ID 837875, 35 pages.

38. Ray, D., Sarkar, B., Rana, A., and Bose, N. 2001. The mechanical properties of vinylester resin matrix composites reinforced with alkali-treated jute fibres. *Composites Part A: Applied Science and Manufacturing.* 32(1):119–27.

39. Rachini, A., Le Troedec, M., Peyratout, C., and Smith, A. 2012. Chemical modification of hemp fibers by silane coupling agents. *J. Appl. Polym. Sci.* 123(1): 601–10.

40. Pickering, K.L., Abdalla, A., Ji, C., McDonald, A.G., and Franich, R.A. 2003. The effect of silane coupling agents on radiata pine fibre for use in thermoplastic matrix composites. *Composites Part A.* 34(10): 915–26.

41. Goda, K., Sreekala, M.S., Malhotra, S.K., Joseph, K., Thomas, S. (Eds.). *Advances in Polymer Composites: Biocomposites – State of the Art, New Challenges, and Opportunities.* John Wiley & Sons, New York.

42. Hill, C.A.S., Khalil, H.P.S., and Hale, M.D. 1998. A study of the potential of acetylation to improve the properties of plant fibres. *Ind. Crops Prod.* 8(1):53–63.

43. Sreekala, M.S., Kumaran, M.G., Joseph, S., Jacob, M., and Thomas, S. 2000. Oil palm fibre reinforced phenol formaldehyde composite: Influence of fibre surface modifications on the mechanical performance, *Appl. Compos. Mater.* 7:295–329.

44. Paul, A., Joseph, K., and Thomas, S. 1997. Effect of surface treatments on the electrical properties of low-density polyethylene composites reinforced with short sisal fibers. *Compos. Sci. Technol.* 57:67–79.

45. Kokta, B.V., Maldas, D., Daneault, C., and Beland, P. 1990. Composites of polyvinyl chloride 1: Effect of isocynate as a bonding agent. *Polymer Plast. Technol. Eng.* 29(1–2):87–118.

46. Li, Y., Pickering, K.L., and Farrell, R.L. 2009. Analysis of green hemp fibre reinforced composites using bag retting and white rot fungal treatments. *Industrial Crops and Products.* 29(2–3):420–6.

47. Idicula, M., Sreekumar, P.A., Joseph, K., and Thomas, S. 2009. Natural fiber hybrid composites—A comparison between compression molding and resin transfer molding. *Polym. Compos.* 30(10):1417–25.

48. Rouison, D., Sain, M., and Couturier, M. 2004. Resin transfer molding of natural fiber reinforced composites: Cure simulation. *Composites Science and Technology.* 64(5):629–44.

49. Oksman, K. 2001. High quality flax fibre composites manufactured by resin transfer molding process. *Journal of Reinforced Plastics and Composites.* 20(07):621–7.

50. Rodriguez, E., Giacomelli, F., and Vazquez, A. 2004. Permeability-porosity relationship in RTM for different fiberglass and natural reinforcements. *Journal of Composite Materials.* 38(3):259–68.
51. Rodriguez, E., Petrucci, R., Puglia, D., Kenny, J.M., and Vazquez, A. 2005. Characterization of composites based on natural and glass fibers obtained by vacuum infusion. *J. Compos. Mater.* 39(3):265–82.
52. van Voorn, B., Smit, H.H.G., Sinke, R.J., and de Klerk, B. 2001. Natural fibre reinforced sheet moulding compound. *Composites Part A: Applied Science and Manufacturing.* 32:1271–90.
53. Jacobs, W. 2006. Is NF always the best material choice for a product? Example: An automotive doorpanel. In: Proceedings of the 6th Global Wood and Natural Fibre Composites Symposium.
54. Idicula, M., Malhotra, S.K., Joseph, K., and Thomas, S. 2005. Dynamic mechanical analysis of randomly oriented intimately mixed short banana/sisal hybrid fibre reinforced polyester composites. *Composites Science and Technology.* 65:1077–87.
55. Idicula, M., Boudenne, A., Umadevi, L., Ibos, L., Candau, Y., and Thomas, S. 2006. Thermophysical properties of natural fibre reinforced polyester composites. *Composites Science and Technology.* 66:2719–25.
56. Rajpurohit, A., Henning, F., Jung, T., and Chaudhari, R. 2014. Centre of excellence for composites at ATIRA: Overview of composites manufacturing technology and ongoing R & D projects. International Conferences on Composite Materials and Technology, ATIRA Ahmedabad.
57. Nosbi, N., Akil, H.M.D., Ishak, Z.A.M., and Abu Bakar, A. 2009. Degradation of compressive properties of pultruded kenaf fiber reinforced composites after immersion in various solutions. *Materials and Design.* 31:4960–4.
58. Omar, M.F., Akil, H.M.D, Ahmad, Z.A., Mazuki, A.A.M., and Yokoyama, T. 2010. Dynamic properties of pultruded natural fibre reinforced composites using split hopkinson pressure bar technique. *Materials and Design.* 31:4209–18.
59. Akil, H.M.D., De Rosa, I.M., Santulli, C., and Sarasini, F. 2010. Flexural behavior of pultruded jute/glass and kenaf/glass hybrid composites monitored using acoustic emission. *Materials Science and Engineering A.* 527:2942–50.
60. Mishra, V. 2014. Physical, mechanical and abrasive wear behavior of jute fiber reinforced polymer composites. PhD Thesis. National Institute of Technology, Rourkela.
61. Patnaik, A. and Tejyan, S. 2014. Mechanical and visco-elastic analysis of viscose fiber based needle punched nonwoven fabric mat reinforced polymer composites: Part I. *Journal of Industrial Textiles.* 43: 440–57.
62. Hossain, M.R., Islam, M.A., van Vuurea, A., and Verpoest, I. 2013. Tensile behavior of environment friendly jute epoxy laminated composite. *Procedia Engineering.* 56:782–8.
63. Richardson, M. and Zhang, Z. 2001. Nonwoven hemp reinforced composites. *Reinforced Plastics.* 45:40–4.
64. Rouison, D., Sain, M., and Couturier M. 2006. Resin transfer molding of hemp fiber composites: Optimization of the process and mechanical properties of the materials. *Composites Science and Technology.* 66(7–8):895–906.
65. Akil, H.M., Omar, M.F., Mazuki, A., Safiee, S., Ishak, Z., and Abu Bakar, A. 2011. Kenaf fiber reinforced composites: A review. *Materials & Design* 32(8–9):4107–21.
66. Li, Y., Mai, Y.-W., and Ye, L. 2000. Sisal fibre and its composites: A review of recent developments. *Composites Science and Technology.* 60(11)35:53–60.
67. Singh, B., Gupta, M., and Verma, A. 1996. Influence of fibre surface treatment on the properties of sisal-polyester composites. *Polymer Composites.* 17:910–18.

68. Bisanda, E.T.N. and Ansell, M.P. 1992. Properties of sisal-CNSL composites. *Journal of Materials Science*. 27:1690–1700.
69. Van de Weyenberg, I., Chi Truong, T., Vangrimde, B., and Verpoest, I. 2006. Improving the properties of UD flax fibre reinforced composites by applying an alkaline fibre treatment. *Composites Part A*. 37(9):1368–76.
70. Yan, L.B., Chouw, N., and Yuan, X.W. 2012. Improving the mechanical properties of natural fibre fabric reinforced epoxy composites by alkali treatment. *J Reinf Plast Comp*. 31(6):425–37.
71. Rout, J., Tripathi, S.S., Misra, M., Mohanty, A.K., and Nayak, S.K. 2001. The influence of fiber surface modification on the mechanical properties of coir-polyester composites. *Polymer Composites*. 22(4):468–76
72. Rout, J., Tripathi, S.S., Misra, M., Mohanty, A.K., and Nayak, S.K. 2001. The influence of fiber surface modification on the mechanical properties of coir-polyester composites. *Composites Science Technology*. 61:1303–10.
73. Ramadevi, P., Sampathkumar, D., Srinivasa, C.V., and Bennehalli, B. 2012. Water absorption of abaca. *Bioresources*. 7(3):3515–24.
74. Alawar, A., Hamed, A.M., and Al-kaabi, K. 2009. Characterization of treated date palm tree fiber as composite reinforcement. *Composites Part B*. 40(7): 601–6.
75. Venkateshwaran, N. and Elayaperumal, A. 2010. Banana fiber reinforced composites— A review. *Journal of Reinforced Plastics and Composites*. 29(15):2389–96.
76. Ahmed, K.S., Vijayarangan, S., and Rajput, C. 2006. Mechanical behavior of isothalic polyester-based untreated woven jute and glass fabric hybrid composites. *Journal of Reinforced Plastics and Composites*. 25(15):1549–69.
77. Mohan, R. and Kishore. April 1985. Jute-glass sandwich composites. *Journal of Reinforced Plastics and Composites*. 4:186–94.
78. Raghavendra, G., Ojha, S., Acharya, S., and Pal, S. 2014. Jute fiber reinforced epoxy composites and comparison with the glass and neat epoxy composites. *Journal of Composite Materials*. 48(20):2537–47.
79. Tejyan, S., Patnaik, A., Rawal, A., and Satapathy, B. K. 2011. Structural and mechanical properties of needle-punched nonwoven reinforced composites in erosive environment, *Journal of Applied Polymer Science*. 123: 1608–707.
80. Pavithran, C., Mukherjee, P.S., Brahmakumar, M., and Damodaran, A.D. 1987. Impact properties of natural fibre composites. *J Mater Sci Lett*. 6:882.
81. Atiqah, A., Maleque, M.A., Jawaid, M., and Iqbal, M. 2004. Development of kenaf–glass reinforced unsaturated polyester hybrid composite for structural applications. *Compos Part B Eng*. 56:68–73.
82. Liu, Q. and Hughes, M. 2008. The fracture behaviour and toughness of woven flax fibre reinforced epoxy composites. *Composites Part A*. 39(10):1644–52.
83. Meredith, J., Ebsworth, R., Coles, S.R., Wood, B.M., and Kirwan, K. 2012. Natural fiber composite energy absorption structures. *Composites Science and Technology*. 72(2):211–17.
84. Abdullah, A.H., Alias, S.K., Jenal, N., Abdan, K., and Ali, A. 2012. Fatigue behavior of Kenaf fiber reinforced epoxy composites. *Engineering Journal*. 16(5):1–10.
85. Yao, F., Wu, L.Y., Guo, W., and Xu, Y. 2008. Thermal decomposition kinetics of natural fibers: Activation energy with dynamic thermogravimetric analysis. *Polymer Degradation and Stability*. 93(1):90–8.
86. Nova Institute's market study. 2015. Wood-Plastic Composites (WPC) and Natural Fiber Composites (NFC): European and Global Markets 2012 and Future Trends in Automotive and Construction.
87. Wilson A. 2015. Automotive Composites, the Make-or-Break Decade for Carbon and Natural Fibers. Textile Media Services, UK.

3 Wood Fiber Reinforced Thermoplastic and Thermosets Composites

Arne Schirp

CONTENTS

3.1 INTRODUCTION

Wood and natural fibers have been used in automotive applications for a long time. In 1942, Henry Ford developed the first prototype composite car made from hemp fibers (Suddell and Evans 2005). Due to economic limitations, the car did not make it into general production. Some eight years later, in Europe, the East German Trabant car was based on cotton fibers embedded in a polyester matrix and went into serial production until 1990. Since then, many automotive producers have introduced wood and natural fibers into their models (Bledzki et al. 2006), with the main components being side and back door panels, hat racks, spare tire lining, and instrumental panels. Recently, a racecar manufactured almost entirely with wood composite materials was designed and built by Joe Harmon ("Splinter"; Figures 3.1 and 3.2). The structure is made from a variety of bent and laminated wood veneers, with bodywork

FIGURE 3.1　Splinter car (www.joeharmondesign.com).

FIGURE 3.2　Splinter car (www.joeharmondesign.com).

produced from woven veneer. The car was completed in 2015 and featured at the Essen Motor Show in Germany from November 27 to December 6, 2015.

Wood and natural fibers represent a lightweight, sustainable, and recyclable option for car manufacturers. Policy frameworks such as the US Corporate Average Fuel Economy (CAFE) regulations and the EU end-of-life vehicle directive have contributed to increased interest in renewable fibers. Regarding recycling, good options exist by regrinding or incineration with energy recovery. Compared to glass fibers, there is less abrasion and occupational health hazard in assembly and handling with wood and natural fibers. The weight reduction due to the use of natural fibers is around 10%–30% in comparable parts (Bledzki et al. 2006). On the other hand, a challenge with natural fibers is their residual emission problem (odor and fogging).

In 2012, wood and natural fiber composites had a market share of 15% of the total European composites market, which includes glass, carbon, wood, and natural fibers (Carus 2015). This relates to a total production volume of 352,000 t for biocomposites of which 150,000 t are allocated to composites for automotive applications (Carus 2015). With natural fibers, compression molding is mostly used as a processing technique whereas with wood fibers, compression molding, sheet extrusion and thermoforming are almost equally used (Table 3.1). Injection-molding has not been used as extensively as other processing techniques in the past but is expected to increase due to many novel material and processing developments.

TABLE 3.1

Biocomposites with Natural Fibers, Wood Fibers, and Recycled Cotton in the European Automotive Production in 2012

Biocomposites	Volume Fibers in 2012 (t)	Volume Biocomposites in 2012 (t)	Processing Technologies	Matrices
Natural fiber composites	30,000	60,000	95% compression molding, 5% injection molding and others	55% thermoplastics, 45% thermosets
Wood-plastic composites	30,000	60,000	45% extrusion and thermoforming, 50% compression molding, 5% injection molding and others	Extrusion: 100% thermoplastics; compression-moulding: >90% thermoset
Recycled cotton reinforced plastics	20,000	30,000	Mainly compression molding	>90% thermoset
Total	80,000	150,000		

Source: Carus, M. (2015): Biocomposites in the automotive industry: Technology, markets and environment. Presentation at the Sixth WPC & NFC Conference, Cologne, December 16–17, 2015, Cologne, Germany.

A comprehensive list where wood and natural fibers are used in automotive interior components is contained in, for example, Du et al. (2014) and Huda et al. (2008). Detailed information on wood-plastic composites can be found in Klyosov (2007) and Oksman Niska and Sain (2008).

3.2   DEFINITIONS AND SCOPE

The term "wood fiber" is often used differently and may encompass wood flour, cellulose (pulp) fibers, or thermomechanical pulp. First of all, a fiber is a particle that is significantly longer than it is wide, in contrast to cubically shaped particles. Wood fiber is an anatomical term and refers to a single fiber. A wood fiber is an elongated, narrow cell with a lumen. By definition, this excludes vessels and parenchyma cells, and includes libriform fibers and fiber tracheids in the case of hardwoods, as well as tracheids in the case of softwoods. Cell types in softwoods and hardwoods are shown in Figures 3.3 and 3.4. In softwoods, tracheids perform the tasks of reinforcement as well as water conduction (Table 3.2). Ninety percent or more of softwood volume is composed of tracheids. In hardwoods, separate cell types are responsible for reinforcement (fiber tracheids and libriform fibers) and water conduction (vessels). Libriform fibers account for 50%–60% of the total mass of hardwoods (Wagenführ 1989), hence they have a major influence on density, strength properties, and workability of hardwoods.

The wood cell wall is mostly composed of cellulose with a percentage of up to 50% (Table 3.3).

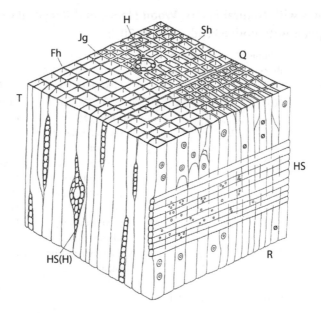

FIGURE 3.3   Softwood structure, microscopic view (schematic). Q: cross cut; T: tangential cut; R: radial cut; Fh: early wood tracheids; Sh: late wood tracheids; Jg: annual ring limit; HS: ray; H: resin channel; HS (H): ray with resin channel. (From Wagenführ, R. (1989): *Anatomie des Holzes* (Wood anatomy; in German). VEB Fachbuchverlag Leipzig. 4th Edition.)

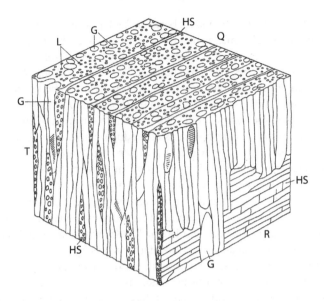

FIGURE 3.4 Hardwood structure, microscopic view (schematic). Q: cross cut; T: tangential cut; R: radial cut; G: vessels; L: libriform fibers; HS: ray. (From Wagenführ, R. (1989): *Anatomie des Holzes* (Wood anatomy; in German). VEB Fachbuchverlag Leipzig. 4th Edition.)

TABLE 3.2
Function of Cell Types in Softwoods and Hardwoods

Function	Softwood	Hardwood
Reinforcement	Tracheids	Fiber tracheids, libriform fibers
Water conduction	Tracheids	Vessels
Nutrient storage	Parenchyma cells	Parenchyma cells, libriform fibers (only while alive)

TABLE 3.3
Cell Wall Composition of Wood

Component	Proportion (%)
Holocellulose	60–85
Cellulose	40–50
Hemicellulose	20–35
Lignin	15–35

Source: Wagenführ, R. (1989): Anatomie des Holzes (Wood anatomy; in German). VEB Fachbuchverlag Leipzig. 4th Edition.

Tracheid (fiber) length of softwoods is between 2000 μm and 5000 μm with diameters between 20 and 40 μm. This means that the aspect ratio (ratio between fiber length and diameter) of softwood fibers is 1:100. Fibers are usually longer in softwoods than in hardwoods. Average hardwood fiber length is between 1000 and 1500 μm (Wagenführ 1989).

Wood flour comprises fiber bundles, rather than individual fibers, with aspect ratios typically only about 1–5 (Clemons 2008). An example is shown in Figure 3.5. For the production of wood-plastic composites (WPC), wood flour is most often used. By definition, wood-plastic composites are materials or products consisting of one or more natural fibers or flours and one or a mixture of polymers which are produced using plastic manufacturing technologies (EN 15534-1, 2014). In Europe, the term wood-polymer composite is often used instead of wood-plastic composite to avoid a possible negative connotation of plastics. In the past, the term wood-polymer composite has been used to address solid wood or wood-based materials that were impregnated with monomers or synthetic resin solutions (Schneider and Brebner 1985, Wittmann and Wolf 2005). The plastic components were then cured thermocatalytically or by radiation.

In addition to wood fibers and wood flour, natural fibers are also used to process thermoplastic and thermoset composites. Natural fibers may come from a variety of different plant sources, for example, hemp, flax, sisal, coconut, cotton, kenaf, jute, abaca, and bamboo. In addition, agricultural residues such as rice husks or wheat straw are used. A comprehensive review on the use of natural fibers for thermoplastic and thermoset matrices was recently provided by Faruk et al. (2012).

Wood and natural fiber reinforced composites are processed using various technologies. Extrusion is used for profiles and sheets. Injection molding or compression molding is used for more complex, three-dimensionally shaped parts, and flat-pressing or calendering for films and sheets. The contents of lignocellulosic particles or natural fibers and polymers depend on the application and the processing techniques. In commercial profile extrusion, up to 75% (by wt.) of lignocellulose is included while in injection-molding, typically, less than 50% (by wt.) of lignocellulose is used. In compression-molding with thermoplastics or thermosets, up to 90% of wood and natural fibers are included.

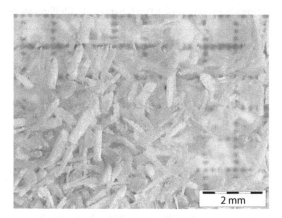

2 mm

FIGURE 3.5 Example of wood flour, shown here is "Lignocel BK 40-90" from J. Rettenmaier & Söhne GmbH + Co., Holzmühle, Germany. Particle size is mainly between 300 and 500 μm.

3.3   PULPING PROCESSES TO OBTAIN WOOD AND CELLULOSE FIBERS

In general, there are three types of pulping processes:

- Mechanical pulping, resulting in ground wood pulp which is used for news-papers, beer mats, display boards, and so on
- Thermomechanical pulping, resulting in wood fibers for the production of fiberboard
- Chemo-thermomechanical pulping, resulting in cellulose for paper production

The disintegration of larger wood elements into fibers consists of two steps. First, logs are reduced to wood chips, followed by the conversion of the chips to pulp fibers in a refiner or defibrator. Chips are produced in rotating disk chippers, equipped with specialized knives which determine the chip geometry. The mechanical pulping process works without any extra thermal or chemical treatment. The pulping step is essentially a mechanical process that takes place at elevated temperatures. The benefit of using high temperature is that power consumption may be significantly reduced. A laboratory-scale refiner plant, the process, and the resulting fibers and fiberboard panels are shown in Figures 3.6, 3.7, and 3.8.

FIGURE 3.6   Refiner plant at Fraunhofer WKI.

FIGURE 3.7   Medium-density fiberboard and thermomechanical pulp, the raw material for fiberboard.

During the pulping or refining process, wood chips are continuously added to a digester via a plug screw. Typical conditions are a pressure of 7–8 bars and a temperature of 150–170°C. After softening of the wood, the material is ground between two refiner discs to extract fibers and fiber bundles from the wood matrix. The objective is to disintegrate the chips into individual fibers. During refining, the structural integrity of the middle lamella, which contains the highest concentration of lignin across the cell wall and acts as glue for the wood matrix, is weakened and the lignin is softened and plasticized. Once the fibers have passed the refiner disks, the resin, for example, urea-formaldehyde, is applied to the fibers via the so-called blow-line. A small amount of wax is often used to impart hydrophobation to the fibers. After resin and wax application, the fibers are dried in a tube drier and collected for further processing. Size distribution of wood fibers after the thermomechanical refining process is heterogeneous, and in addition to fibers, very fine particles are obtained. Generally, wood fiber lengths after refining are between 20 μm and 4500 μm with widths between less than 1 μm and 80 μm (Table 3.4; Lohmann 2003).

Chemo-thermomechanical pulp (Figure 3.9) is produced by using the sulfite and sulfate (kraft) processes of which the latter predominates. This type of pulp is used to produce cellulose and paper. Chemical pulping is a process in which lignin is completely removed, in contrast to mechanical or thermomechanical pulping.

For the thermomechanical pulping process, softwoods are primarily used in Europe due to availability and high fiber aspect ratio, which result in good mechanical properties of fiberboards. With regard to costs, wood fibers are

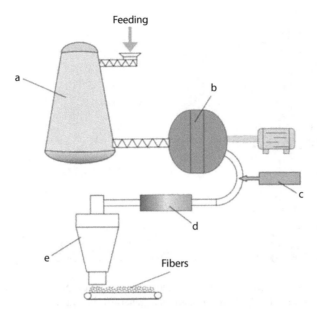

FIGURE 3.8 Fiber production process. (a) Digester, (b) refiner, (c) blow-line blending, (d) tube-dryer, (e) cyclone. (From Kasal, B., S. Friebel, J. Gunschera, T. Salthammer, A. Schirp, H. Schwab, V. Thole (2015): Wood-based materials. *Ullmann's Encyclopedia of Industrial Chemistry*, Wiley-VCH Verlag GmbH & Co. KGaA, Weinheim, Germany.)

TABLE 3.4
Components of Thermomechanical Pulp and Their Particle Sizes

Pulp Component	Length (μm)	Width (μm)
Splinter	Variable	Variable
Long fibers	800–4500	25–80
Short fibers	200–800	2.–2.8
Sludge	Up to 200	ca. 1
Flour	20–30	1–30
Dust		<1

Source: Lohmann, U. (2003): *Holz-Lexikon* (in German; Wood Encyclopaedia). DRW-Verlag Weinbrenner GmbH & Co., Leinfelden-Echterdingen, Germany.

---

significantly cheaper than hemp or cellulose fibers (e.g., sulfite or kraft fibers). Wood fibers produced in MDF plants are normally not available for sale as a product as they are usually processed into fiberboards on-site immediately following the refining step. One exception is the product "Woodforce" by Sonae Industria (Maia, Portugal). It consists of small dice cut out of a thin fiberboard

FIGURE 3.9   Unbleached (left) and bleached (right) cellulose pulp (http://www.knowpulp .com/english/demo/english/pulping/bleaching/1_general/frame.htm).

which are promoted for use as reinforcement for thermoplastics, for example, in automotive applications. "Woodforce" shall offer price-performance ratio between talc fillers and glass fibers, with lower weight than either. It is claimed that the material can offer properties equivalent to 20% glass fiber-filled thermo-plastics (Markarian 2015).

Spun cellulose fibers such as rayon tire cord yarn (e.g., produced by Cordenka GmbH, Obernburg, Germany), Tencel, viscose and Carbacell fibers have also been used in combination with thermoplastic polymers (Fink and Ganster 2006), resulting in excellent properties which are close to those of glass-fiber reinforced PP, thus permitting applications in the field of engineering thermoplastics such as polycarbonate-acrylonitrile butadiene styrene blends (PC-ABS).

3.4   WOOD FIBER AND WOOD FLOUR-BASED THERMOPLASTIC COMPOSITES

3.4.1   COMPOSITION, PROCESSING, AND APPLICATIONS

For thermoplastic composites, virgin or recycled polymers are used, most often polypropylene (PP), polyethylene (PE), and poly(vinyl chloride) (PVC). Biopolymers such as bio-PE or bio-PP (PE or PP made from sugarcane ethanol), polylactic acid (PLA), and poly(hydroxybutyrate) (PHB) can also be used to process thermoplastic composites.

As either filler or reinforcement, mechanical pulp (Mendez et al. 2007, Lerche et al. 2014), thermomechanical pulp (Stark and Rowlands 2003, Schirp and Stender 2010, Thumm and Dickson 2013, Schirp et al. 2014), and chemo-thermomechanical pulp (Klason et al. 1984, Raj et al. 1988, Maldas and Kokta 1989,

1990, Migneault et al. 2009) have been used; however, on an industrial scale, mostly wood flour (Figure 3.10, second from left) is used because pulp is difficult to feed using typical dosing equipment. Recently, some novel compounds based on chemo-thermomechanical pulp (cellulose fibers from kraft pulping processes) have entered the market also (see Section 3.7.2.2).

The primary lignocellulosic resource for the preparation of wood flour is soft-wood (spruce, fir, and pine) due to wide availability, color, and familiarity with pro-cessing. In the United States, which has a significant market share of wood-plastic composites, hardwoods such as maple and oak are also used (Clemons 2008). Wood species that contain a large amount of extractives which may be leached by water are avoided. This water can migrate to the surface and evaporate, leaving behind extractive stains. In addition, extractives can volatilize or darken during processing at high temperatures.

Grades of commercially manufactured wood flours for WPC are supplied in dif-ferent particle size ranges and typically fall within 180–840 μm. Wood flour is pro-duced by size reduction using various mills and classification by screening. Particle size distribution is mostly determined using traditional shaking sieve or air jet sieve analysis. Aspect ratio of commercial pine wood flour (mesh size between 20 and 140, i.e., between 0.85 mm and 0.106 mm) is between 3.4 and 4.5 (Stark and Rowlands 2003). An important requirement for feeding of wood particles into dosing equip-ment used in compounding is sufficiently high bulk density of the filler. Wood flour bulk density is typically between 0.19 and 0.22 g/m^3 (Clemons 2008).

To process thermomechanical pulp or long natural fibers in conventional com-pounding and extrusion equipment, bulk density of the fibers needs to be reduced by either pelletizing (Bengtsson et al. 2007) or by using pressed and cut panels (small wood dice, "Woodforce" by Sonae Industria). Alternatively, wood fibers may be mixed with thermoplastic fibers and processed using a heating-cooling mixer (Schirp and Stender 2010). The agglomerate can then be added to profile extrud-ers using a crammer feeder. If kneaders are used, fiber feeding is not a challenge. Kneaders are used not only on a laboratory scale for formulation development but

FIGURE 3.10 Different types of wood particles. From left to right: Wood particles for par-ticleboard, wood flour, thermomechanical pulp for medium-density fiberboard produced with refiner disc distance of 1 mm, 0.7 mm, 0.4 mm, and 0.15 mm. The wood species is beech (*Fagus sylvatica*).

also for processing lignocellulosic raw materials on a large scale. German company Advanced Compounding Rudolstadt (ACR) in Rudolstadt, Germany, has been using such a kneader to produce compounds based on long natural fibers from hemp, kenaf, flax, sisal, and cellulose. Filling levels are up to 50%, and polymers that can be used include technical polymers such as polyamides. Applications are, for example, toys, automotive interior parts, crates, boxes, packaging, garden furniture, and pressed panels.

Controversial results regarding the benefits of using wood fibers as compared to wood flour in thermoplastic matrices have been reported. This is to a large degree because the term "wood fiber" is interpreted differently and may encompass wood flour, cellulose fibers, or thermomechanical pulp. Stark and Rowlands (2003) determined that the use of wood fibers (here: sourced from recycled hardwood pallets, converted into wood chips and then into thermomechanical pulp) resulted in higher tensile and flexural strengths at 20% and 40% filler levels, and higher moduli at the 40% filler level compared to wood flour. Aspect ratios of the refiner fibers and wood flour (40-mesh, equivalent to 0.425 mm) were calculated to be 16–26 and 3, respectively. It was claimed that the higher aspect ratio of the fibers enhanced stress transfer from the matrix to the fiber. However, the use of wood fiber had little effect on impact energy. Migneault et al. (2009) also found that increasing fiber length/ diameter ratio had a beneficial effect on the mechanical properties of WPC while the effect on water absorption was negative. They determined that aligned fibers in injection-molded samples resulted in better stress transfer between the wood fibers and HDPE matrix, and therefore improved mechanical performance compared to extrusion.

Beg and Pickering (2008a) investigated the influence of fiber length, fiber beating, and hygrothermal aging on properties of Kraft fiber reinforced PP composites. They found that tensile strength, modulus of elasticity, and impact strength of composites decreased with decreasing fiber length.

Using an internal kneading mixer, beech TMP with increasing fiber length and coarseness was processed, and fiber changes before kneading and after injection-molding were determined (Schirp et al. 2014). Overall, the flexural strength of the WPC was not improved when TMP fibers were used instead of wood flour. However, increasing the refiner gap width from 0.15 mm to 1 mm resulted in an improvement in flexural strength of the WPC despite the fact that fiber length and L/W ratio were strongly reduced during compounding. This was mainly attributed to the smaller surface area of the smaller fibers compared to the larger ones, as indicated by convexity. In addition, fiber surface composition, topography, and changes in L/D frequency distributions before and after processing may be responsible for this observation.

Thumm and Dickson (2013) determined the critical fiber length for radiata pine thermomechanical pulp used in a PP matrix. The critical fiber length l_c is defined as (Bowyer and Bader 1972):

$$l_c = \frac{d\sigma_\tau}{2\tau}$$

To calculate the critical fiber length, the diameter d of the fibers, the tensile strength σ_t of the fiber, and the interfacial shear strength τ need to be known. These parameters and properties have been determined for larger, natural fibers such as flax in a PP matrix (Van den Oever and Bos 1998) but are difficult to determine for short wood fibers. Thumm and Dickson used a method in which an initially long fiber fraction was screened from a TMP. The fraction was then reduced in fiber length in a controlled and predictable manner by manual cutting. A wet forming technique and compression molding were used for composite production to minimize any further length reduction and damage that is known to occur during compounding. Using this method, the critical fiber length was determined to be 0.8 μm for radiata pine TMP. Critical fiber lengths for various natural fibers (not in the scope of this chapter) are summarized in Pickering et al. (2016).

The main applications for extruded wood flour-filled thermoplastic composites (wood-plastic composites in the sense of EN 15534-1, 2014) are decking, cladding (siding), panels, fencing, building profiles, and furniture. Windows, window sills, door sills, and indoor flooring have also been produced with WPC as a material. Recently, interest for indoor applications of such composites has increased. This may be partially due to the fact that replacements for wood-based composites bonded with formaldehyde-based resins are sought. Some interesting new applications based on wood or cellulose fiber reinforced thermoplastic materials during the recent past include, for example, a co-extruded pencil called "Wopex" by German company Staedler, injection-molded kitchen furniture by Finnish company Puustelli based on material from UPM Kymmene, and an office chair produced by Werzalit using back-injection molding of wood veneer. In addition, material combinations of wood or bamboo particles and PLA have also been used for 3D printing filaments (e.g., "Woodfill" by company ColorFabb).

Density of extruded or injection-molded wood flour or fiber-reinforced thermoplastics is typically relatively high compared to traditional wood-based composites, with values between 1.1. and 1.4 g/cm^3, although the densities of the matrix and filler are much lower. Wood density, including pores, is in the range between 0.32 and 0.72 g/cm^3 (Simpson and TenWolde 1999) while density of polyolefins (PP, PE) is between 0.90 and 0.96 g/cm^3 (Osswald and Menges 1996). High composite density is attributed to the fact that the wood particles are heavily compressed during processing and that pores may be filled with the thermoplastic matrix. Density values for the composites reach values for the density of the wood cell wall (1.4 g/cm^3). Considering that commonly used inorganic fillers have densities of up to 2.8 g/cm^3, density of compressed wood is still considerably lower. Wood fiber/flour and natural fiber-based composites are therefore considered a lightweight option in automotive applications. Foaming can be used to further reduce the density of injection-molded or extruded composites.

Additives for WPC are added in small quantities but play an important role in achieving product performance. To achieve bonding of the matrix and filler, coupling agents are employed. The most popular coupling agents are maleic anhydride-grafted polyethylene (MAPE) or polypropylene (MAPP). Maleic anhydride is believed to form covalent ester linkages and hydrogen bonds when reacting with hydroxyl groups at the cellulose interface (Felix and

Gatenholm 1991) whereas the polymer backbone of the coupling agent entangles mechanically with the polymer matrix. This leads to a reduction of WPC water uptake (Bledzki et al. 1998) and higher flexural strength compared to uncoupled formulations (Woodhams et al. 1984, Dalväg et al. 1985). Various factors influence the effectiveness of the coupling agent, among them graft content, molecular weight, acid number, applied concentration, wood species and natural fiber type, fiber processing, and processing. Isocyanates, silanes, and other chemicals have also been used as coupling agents (Lu et al. 2000). When PVC is used as a matrix, the coupling agent may be omitted due to the polar nature of both matrix and wood filler.

In addition to coupling agents, lubricants, antioxidants, acid scavengers, UV protecting agents such as hindered amine light stabilizers (HALS), mineral fillers (e.g., talc and calcium carbonate), biocides, foaming agents, pigments, and flame retardants may be included in WPC formulations, depending on the intended processing technology and application. Oksman Niska and Sain (2008) provide an overview of additive types and amounts used to prepare WPC formulations based on PE, PP, or PVC as matrix.

3.4.2 MECHANICAL AND PHYSICAL PROPERTIES

Although the mechanical properties of wood and natural fibers are much lower than those of glass fibers (Table 3.5; Wambua et al. 2003), their specific properties, especially modulus of elasticity, are comparable to the values reported for glass fibers. Moreover, wood and natural fibers are about 50% lighter than glass fibers, renewable, recyclable, biodegradable, and cheaper. These attributes provide an incentive for the use of lignocellulosics in composites. The mechanical properties of wood fibers are difficult to determine due to inhomogeneity, sample preparation, and shortness of fibers compared to natural fibers. Groom et al. (2002) determined the mechanical properties of loblolly pine individual fibers. They found that the latewood fiber modulus of elasticity and strength is relatively constant with respect to vertical location within an individual tree but increased with increased distance from the pith. Average latewood fiber MOE values ranged from 6.6 GPa at the pith to as high as 27.5 GPa in the mature portion of the tree. Average fiber ultimate tensile stress values ranged from 410 MPa at the pith to 1420 MPa in the mature portion of the tree.

In general, incorporation of wood fibers or flour into a thermoplastic matrix tends to increase the modulus of elasticity, thermal stability, UV resistance, and workability compared to unfilled thermoplastics (Wolcott 2001). On the other hand, thermoplastics impart moisture resistance, resistance to decay, and thermoforming characteristics. Composite strength is highly dependent on several factors, for example, wood particle size and origin, plastic type, processing method, and bonding quality of wood and thermoplastic. Decreased flexural and tensile strength are generally cited with increasing wood content (Wolcott 2001). Impact resistance suffers when wood particles are included as fillers; however, the use of elastomeric coupling agents may reduce this effect (Oksman and Clemons 1998).

TABLE 3.5
Properties of Natural Fibers and E-Glass Fibers

Properties	E-Glass	Hemp	Jute	Ramie	Coir	Sisal	Flax	Cotton
				Fibers				
Density, g/cm^3	2.55	1.48	1.46	1.5	1.25	1.33	1.4	1.51
Tensile strength, MPa	2400	550–900	400–800	500	220	600–700	800–1500	400
E-modulus, GPa	73	70	10–30	44	6	38	60–80	12
Specific, E/d[a]	29	47	7–21	29	5	29	26–46	8
Elongation at failure, %	3	1.6	1.8	2	15–25	2–3	1.2–1.6	3–10
Moisture absorption, %	–	8	12	12–17	10	11	7	8–25

Source: Wambua, P., J. Ivens, I. Verpoest (2003): *Composites Science and Technology* 63: 1259–1264.
[a] Specific E-modulus = E-modulus/density.

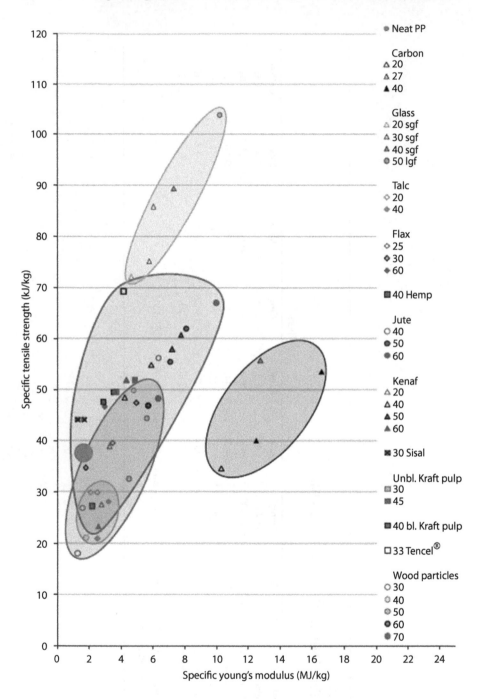

FIGURE 3.11 Ashby plots showing the specific tensile strength versus specific Young's modulus (property divided by density). (From Sobczak, L., R. W. Lang, A. Haider (2012): *Composites Science and Technology* 72: 550–557.)

Sobczak et al. (2012) compiled available data regarding the mechanical properties of polypropylene-based composites reinforced or filled with natural fibers and wood flour and illustrated the results using Ashby plots (Figures 3.11 and 3.12). A comparison with glass and carbon fibers as well as talcum was included. Only data generated using injection molding were used. The authors found that WPC superseded PP-talc composites both in terms of modulus and strength while NFC largely overlapped with PP-glass fiber-composites in terms of modulus and approach their lower end regarding strength. While NFC partially overlaps with the PP-talc composites in terms of impact strength, WPC exhibits inferior impact strength.

Material performance is significantly influenced by the processing method used. Consequently, reported property values for wood and natural fiber reinforced composites need to be carefully evaluated. Strength and modulus values may be improved by design (profile geometry) of the products, choice of matrix polymers, additives, and processing method. Manufacturers have proprietary formulations; hence, each manufacturer has to develop its own design values, similar to many engineered wood products, such as laminated veneer lumber and I-joists.

Knowledge regarding the long-term creep performance of wood fiber reinforced polymers is essential to widen the applications for such composites. To predict the deflection of flexural members subjected to both quasi-static ramp loading and long-term creep, Hamel et al. (2014) developed a finite-element material model. Predictions were made for six different WPC products, encompassing a variety of polymers and cross-sections. These predictions were compared with experimental

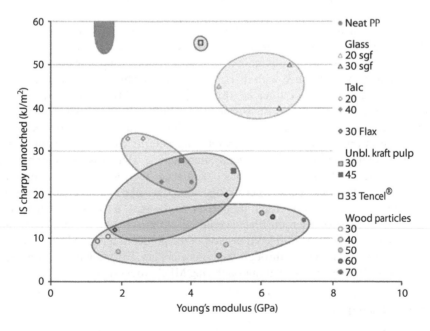

FIGURE 3.12 Ashby plots showing unnotched impact strength versus Youngs's modulus of various PP compounds. (From Sobczak, L., R. W. Lang, A. Haider (2012): *Composites Science and Technology* 72: 550–557.)

testing. Creep predictions were more accurate for solid polyethylene-based materials than polypropylene-based hollow box sections. The mechanical and time-dependent behavior of WPC subjected to tension and compression was also investigated (Hamel et al. 2012). Effects of fiber size on short-term creep behavior of HDPE-based WPC were recently reported by Wang et al. (2015). They determined that the largest wood fibers used (10–20 mesh, equivalent to 0.841 to 2.0 mm) provided the best composite creep resistance.

3.4.3 EFFECTS OF ADDITIONAL REINFORCEMENT BY GLASS, CARBON, AND NANO-SIZED FIBERS

To improve composite properties, in the past decade, many attempts have been made by including reinforcing fillers such as manufactured glass and carbon fibers, as well as natural mineral materials such as kaolin, talc, nanoclays, and others. Turku and Kärki (2014) studied the effects of micro-sized fillers, carbon fiber (CF) and glass fiber (GF), and nano-sized filler, MMT organoclay, on the mechanical properties, wettability, and morphology of a wood flour/PP composite. The tensile properties and hardness of the hybrids was improved with the GF and CF incorporation. It was found that GF has a better reinforcing effect than CF. Adding nanometric MMT improved the tensile modulus of the WPC, whereas the other tested mechanical properties, tensile and impact strength, as well as the hardness of the composite were reduced. The main reason for the decreased mechanical properties was attributed to poor dispersion of the nanoclay particles throughout the composite.

Thermoplastic polymers have also been reinforced with microfibrillated cellulose (MFC). Boldizar et al. (1987) were among the first to reinforce polyolefins with MFC and reported values for the modulus of elasticity of the composites which greatly exceeded the values obtained with inorganic fillers and which almost equaled those recorded for glass fiber or mica reinforced polymers. Microfibrillated cellulose (MFC) is produced by forcing suspensions of cellulose fibers through mechanical devices, such as high-pressure homogenizers (Klemm et al. 2011). This mechanical treatment delaminates the fibers and liberates microfibrils with diameters of approximately 5–60 nm and a length of several micrometers, which display a high aspect ratio, resulting in fibers with enormous potential for reinforcement of composites, coatings, and adhesives. These microfibrils exhibit gel-like characteristics in water, with pseudoplastic and thixotropic properties. In the past, a major impediment for commercial success of microfibrillated cellulose was the very high energy consumption amounting to over 25,000 kWh per ton in the production of MFC as a result of the required multiple passes through the homogenizers. Efforts have been made to overcome the high energy consumption during MFC production.

3.4.4 DURABILITY

The term durability comprises a variety of aspects, such as weathering, which covers the effects of moisture, UV light, heat, frost, plus combinations of these effects. In addition, durability refers to degradation by biological agents, including fungi and bacteria. If materials including wood and natural fibers are used in automotive interior components, which is mostly the case, biological durability should usually not be an issue although resistance against mold fungi should be tested.

The hydroxyl groups in wood and natural fibers are primarily responsible for the absorption of water which causes swelling of the fibers. This can lead to stress in the wood particles and the matrix and to the formation of microcracks. After drying of the composite, cracks form in the plastic and interfacial gaps contribute to penetration of water into the composite. This in turn will lead to loss in mechanical properties (Stark 2001). The diffusion of water into the composite is generally slow, which means it can take a long time to reach equilibrium moisture content. There is a strong gradient regarding the distribution of absorbed water in the material with higher moisture content in the outer surface layer than in the core. Cumulative moisture exposure can cause some irreversible damage to wood and fiber-reinforced composites. Freeze-thaw cycles may also lead to loss in mechanical performance. Test methods usually include moisture resistance under cyclic conditions (e.g., immersion in water at room temperature, freezing, drying, multiple cycles).

In thermoplastic composites, usually, the coupling agent used to bond wood or natural fibers and thermoplastic reduces water uptake of the composite. Maleated thermoplastics (maleic anhydride-modified polypropylene and polyethylene; MAPP and MAPE) represent the most important coupling agents in industrial processing technologies due to their effectiveness, applicability, and affordability. Usually, they are added in concentrations between 1% and 3% (by weight). Maleic anhydride grafted on synthetic polymers such as polypropylene is believed to form covalent ester linkages and hydrogen bonds when reacting with hydroxyl groups at the cellulose interface (Felix and Gatenholm 1991) which leads to a reduction of water uptake (Bledzki et al. 1998).

Thermal changes, which consist of thermal expansion, mechanical creep, and thermo-oxidative degradation, can also affect composites. An important parameter is the linear thermal expansion coefficient, which can be determined using thermomechanical analysis (ISO 11359-2) with a temperature range between $-20°C$ and $80°C$. Thermo-oxidative degradation can occur when composite formulations do not contain sufficient amounts of antioxidants. This is especially problematic in climates where products are exposed to high temperatures.

Accelerated weathering tests are used to develop material formulations and to evaluate their performance, in addition, depending on the application, outdoor (field) tests should always be run in parallel as they provide the most valuable insight into weathering behavior under real conditions. During accelerated weathering, test samples are exposed to cycles of ultraviolet or xenon radiation as well as water spray over a set period of time. During weathering, color changes of the composite are mostly due to lightening of the wood fibers or flour. The wood components are susceptible to degradation by UV radiation with the lignin being primarily responsible for UV absorption. Of the total amount of UV light absorbed by wood, lignin absorbs 80%–95% (Fengel and Wegener 1983). UV light degrades lignin into water-soluble compounds that are washed from the wood with rain, leaving a cellulose-rich surface with a fibrous appearance.

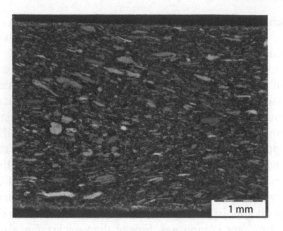

FIGURE 3.13   Microscopic view of WPC including thermally treated beech wood particles (cross cut).

For mass coloration of thermoplastic composites, several options are available: (1) use of pigments or dyes (colorants) which are added to the wood substrate prior to compounding or extrusion; (2) addition of pigments during compounding; and (3) use of master batches during compounding or extrusion in which the pigment is predispersed in the polymer matrix. Usually, methods (2) or (3) are used, and the wood substrate is not stained or colored prior to processing. However, if the wood particles are treated prior to compounding, this can improve the color stability of wood fiber-based composites (Schirp et al. 2015).

In addition to color change, mechanical property loss can be expected due to weathering. Mechanical performance is suffering due to composite surface oxidation, matrix crystallinity changes, and interfacial degradation. If information regarding mechanical performance of the composite is sought, weathering exposure should be for periods of at least 4000 h.

Fourier transform infrared (FTIR) spectroscopy has been intensively used to determine changes in surface chemistry of thermoplastic composites during weathering (Stark and Matuana 2004a,b). Fiber swelling due to moisture absorption is the primary cause for loss in mechanical properties after weathering. Cracks formed in the matrix due to fiber swelling as well as synergism between UV radiation and water exposure contribute to the loss in mechanical properties.

Regarding biological degradation, the two major components of wood fiber-based thermoplastic composites, polyolefins and lignocellulosic fibers or particles, display different behavior. Polyolefins are highly resistant to biodegradation, especially without prior abiotic oxidation because their backbone is solely built of carbon atoms. Other formulation components such as plasticizers, lubricants, stabilizers, and colorants may cause fungal colonization and decay on plastic materials. Nondurable wood is also prone to fungal decay, however, if the wood particles are well encapsulated by the thermoplastic matrix (i.e., at filler levels ≤ 50% by wt.) and if moisture uptake can be minimized, fungal decay should not be an issue (Clemons and Ibach 2004, Schirp and Wolcott 2005, Schirp et al. 2008). It is important to remember

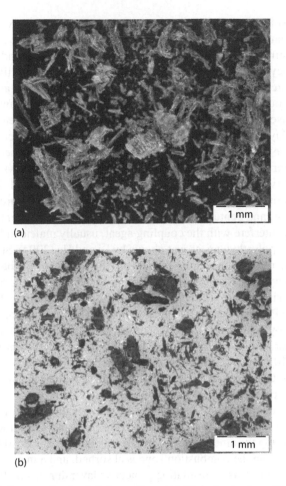

(a)

(b)

FIGURE 3.14 Microscopic view of thermally treated beech wood particles used in WPC before (a) and after (b) processing. Particles were extracted from the matrix using xylene as solvent.

that the availability of moisture is a prerequisite for biological decay. Once water has entered into the material, it will leave only very slowly since the plastic in WPC provides a barrier to gas evaporation. One way to reduce moisture uptake of the composites is to use thermally treated wood particles. In Figure 3.13, WPC based on thermally treated beech wood flour is shown. Particle size reduction occurred during processing (Figure 3.14).

3.4.5 Fire Retardancy

Good reaction to fire performance is important for many applications in the residential construction (decking, siding, roof tiles, window frames), automotive, transportation, and furniture industries. Wood fibers per se are flammable materials, as are the thermoplastic matrices. If PVC is used as matrix, this provides the benefit that the polymer is self-extinguishing. Although there is comprehensive information in the literature regarding the fire retardancy of wood and plastics, the fire performance of wood and natural fiber reinforced composites is still not fully understood.

Fire retardancy is expected to be improved if both matrix and filler or reinforcement are adequately protected. However, at present, fire retardants are added mostly during the compounding step without filler pretreatment. Ammonium polyphosphate (APP) is often used as FR (Stark et al. 2010, Winandy 2013). When APP is heated, an intumescent layer is created on the polymer surface which protects the underlying material from the action of the heat flux or the flame. The proposed mechanism is based on the charred layer acting as a physical barrier which slows down heat and mass transfer between gas and condensed phases (Bourbigot et al. 2004). In traditional intumescent FR systems, APP has been used as the acid source, melamine as the blowing agent, and pentaerythritol (PER) as the carbonific agent. When used in WPC, PER may interfere with the coupling agent, usually maleic-anhydride-grafted polypropylene or polyethylene, as was shown by Li and He (2004). Stark et al. (2010) reported positive results concerning the use of APP as FR for PE-based WPC compared to alternative FRs, for example, a bromine-based FR and zinc borate. APP was also used by Ayrilmis et al. (2012) in flat-pressed WPC based on PP as matrix. In total, four FR (APP, zinc borate, decabromodiphenyl oxide, and magnesium hydroxide) were tested by Ayrilmis et al. (2012) of which APP performed best.

Another phosphate-based FR, melamine phosphate (MP), was investigated by Stark et al. (2010). During decomposition of MP, nitrogen and ammonia are produced which dilute fuel gases while melamine aids in char formation. The authors reported inferior performance of this FR compared to APP.

Metal hydroxides, typically aluminum- or magnesium-based, are another category of common FRs. Metal hydroxides act by releasing a significant amount of water, hence diluting the amount of fuel required to sustain combustion during a fire. In addition, heat from the combustion zone is absorbed, and a metal oxide coating is generated which can act as an insulating protective layer during combustion. Positive effects of magnesium hydroxide (MH) to improve the fire resistance of WPC were reported by Stark et al. (2010) and Sain et al. (2004) while no benefit regarding the use of aluminum hydroxide was indicated by Arao et al. (2014).

Combinations of APP, red phosphorus (RP), and expandable graphite (EG) for use in WPC were examined by Seefeldt et al. (2012). They showed that EG has the highest potential for flame retardancy of WPC; however, due to its expansion it cracked the formed residue layer. Combinations of EG with RP or high amounts of APP were able to suppress this cracking. Naumann et al. (2012) showed that expandable graphite in WPC was effective not only as a flame retardant but also against fungal decay.

A recent review covers some aspects of fire retardancy of WPC (Nikolaeva and Kärki 2011). The effects of profile geometry, material composition, and moisture content of WPC were investigated by Seefeldt and Braun (2012a) using cone calorimetric measurements. Using hollow-shaped geometries, the fire load was reduced but at the same time, fire propagation was increased. Best results were achieved using high amounts of wood filler. Guo et al. (2007) reported that using a small amount of nanoclay can significantly improve the flame retarding properties of WPC based on HDPE. With increasing amounts of clay, the flame retarding properties were enhanced.

Another way to increase fire retardancy is to use co-extrusion (Turku et al. 2014). Co-extrusion may be an effective, economic solution because fire-retardants need to be added only in the co-extruded layer but it needs to be considered that where profiles are cut, protection is needed also. Alternatively, intumescent coatings containing fire-retardants could be a solution but this requires an additional processing step, and coatings adhesion may be difficult without pretreatment of the WPC surface. Most often, mass protection is sought to enhance fire retardancy of WPC. In this case, it may be assumed that fire retardancy of WPC can be increased if both polymer matrix and wood filler are protected with a FR. However, while there is information available on the burning behavior and fire-retardant treatments of wood and polymers, respectively, studies on the flammability of WPC where both wood flour and polymer matrix are treated by flame retardants are scarce. Seefeldt and Braun (2012b) impregnated wood particles with the salt of a phosphoric-acid derivative (Disflamoll TP LXS 51064, Lanxess) and used the dried material to produce WPC. Their investigations were focused on the decomposition and fire behavior of the flame-retarded WPC while mechanical and physical properties of the composites were not determined. Hämäläinen and Kärki (2014) modified WF using two phosphate-based FR solutions and melamine formaldehyde resin. Their results showed that fire performance of WPC was enhanced with the treated wood flour compared to WPC based on untreated wood flour. However, it was not investigated if additional treatment of the polymer matrix (PP) leads to a further improvement in fire-retardancy. Schirp and Su (2016) used hardly inflammable particleboard converted into wood flour in combination with several halogen-free FRs. The fire performance and thermal properties of the WPC (matrix: recycled polypropylene) were investigated using limiting oxygen index analysis (LOI), a modified single-flame source test, fire shaft test, and thermogravimetric analysis (TGA). The combination of B1-particleboard flour, expandable graphite, and red phosphorous led to the highest LOI (38.3). However, at the same time, relatively high water absorption of this formulation was observed which was attributed to the reaction of ammonium polyphosphate in the treated particleboard flour and red phosphorous which was added during the compounding step, leading to increased hydrophilicity of the composite. The combination of pretreated particleboard flour and ammonium polyphosphate (APP) or phosphorized expandable graphite to protect the polymer matrix offered the best overall performance when taking into account LOI (here: 34–35), tensile strength, modulus of elasticity, and water absorption values.

An alternative to pretreated particleboard is the use of thermomechanical pulp, which can be equipped with fire-retardants using a continuous process which can be easily scaled-up on an industrial level. The feasibility of this approach was shown in the EU-project HIFIVENT (http://www.hifiventproject.eu/). Here, the main application is the development and construction of a WPC siding system; however, the compounds based on TMP as well as on rice husks and wood flour can also be used for other applications such as automotive. It was shown that a high level of flame retardancy of WPC based on HDPE may be achieved if both wood particles and polymer matrix are protected with flame retardants. Using the single-burning item

(SBI) test according to EN 13823, B-s2d0-classification (EN 13501-1) was achieved, which means that the product is hardly flammable (Schirp et al. 2016).

The effects of fire retardants on outdoor durability (color change, mechanical properties, water uptake, and swelling) of WPC have not been investigated much thus far. Garcia et al. (2009) found that fire-retardants worsened the outdoor durability. Stabilized composites with aluminum hydroxide as fire retardants showed the best overall results with a color fading degree lower than stabilized, non-fire-retarded composites. However, color change (ΔE) of the WPC including $AlOH_3$ and stabilizer after 600 h in a QUV weathering chamber was still at a high value of 14.5.

Weathering of fire-retarded, co-extruded WPC based on recycled pulp cellulose and PP was investigated by Turku and Kärki (2016). Interestingly, fire-retardant loaded samples displayed smaller color change compared to non-fire-retardant treated samples. Fire performance after weathering was not evaluated.

For automotive components and systems, there are currently no standards or laws that are calling for flame retardance, except for the US Federal Motor Vehicle Safety Standard (FMVSS) 302 for interior components (Hörold 2016). The FMVSS is a flame spread test, and foams, films, and textiles need flame retardants to pass it whereas injection-molded parts usually do not. However, the industry is moving toward increased fire safety, as evidenced by the publication of the US National Fire Protection Association (NFPA) guide to fire and hazard (NFPA 556).

3.4.6 EMISSIONS AND ODOR

Because extruded WPC were initially developed for outdoor applications, little research has addressed the emission of volatile organic compounds (VOC) and odor from WPC. Recently, Félix et al. (2013) characterized and quantified VOC of WPC produced with low density polyethylene (LDPE) and polyethylene/ethylene vinyl acetate (PE/EVA) films and sawdust, in each stage of production, by solid phase microextraction in headspace mode (HS-SPME) and gas chromatography–mass spectrometry (GC–MS). An odor profile was also obtained by HS-SPME and GC–MS coupled with olfactometry analysis. More than 140 compounds were observed in the raw materials and WPC samples. Hexanoic acid, acetic acid, 2-methoxyphenol, acetylfuran, diacetyl, and aldehydes were identified as the most important odorants.

Reduction of odor and emissions of wood and natural-fiber reinforced composites for automotive applications is particularly important because these parts are used in the car interior. Odors may originate from the polymer matrix, wood and natural fibers, from a combination of both, or from additives. The matrix can emit unwanted odors, caused by residual monomers or by decomposition products that occur during processing due to non-optimized temperature profiles. Additives such as compatibilizers, flame-retardants, and so on may add to the odor spectrum. Natural fibers often display a specific odor which may stem from fiber treatment prior to addition to the matrix. Mold and fiber degradation can occur during harvest or storage of the fibers. Cellulose is not affected as much as the pectin which forms odor-active substances (Bledzki et al. 2003). A major factor influencing the odor is moisture. To reduce odors, all production steps, beginning with the raw materials selection and processing (polymer synthesis; fiber processing) need to be considered. Odors may be reduced, for example, by the use of additives, degassing during processing, compounding conditions (temperature profile), and tempering.

Sensorial methods to gauge odors are olfactometry and electronic noses. With olfactometry, odor concentration, strength of odor reception, and hedonic impression (judged as pleasant or unpleasant) are determined (Bledzki et al. 2003). The olfactometer is a dilution apparatus that mixes odorous air in specific ratios with odor-free air for presentation to a panel of assessing persons. Electronic noses are highly developed sensors, gas sensor arrays, which produce digital fingerprints of scents. These sensors detect gases and classify them by comparison with stored data.

Rüppel et al. (2016) determined olfactory properties of injection-molded composites based on PP and PET fibers and filled with wood flour (mixture of spruce and oak, medium fiber length 75 µm) intended for automotive applications. Human sensory methods did not indicate any significant differences between pure materials, those reinforced with fibers, and aged materials. Partially very unpleasant odors of individual fractions of aged samples were perceived which for the most part could not be detected by mass spectrometry. For unaged samples, no emissions that produced odor were perceived.

3.5 WOOD FIBER AND WOOD FLOUR-BASED THERMOSET COMPOSITES

3.5.1 COMPOSITION AND PROPERTIES

Thermosetting resins used in combination with wood and natural fibers are predominantly melamine, polyester, acrylic, phenolic, polyurethane, and epoxy resins. In contrast to thermoplastic polymers, they cannot be reshaped after processing. However, they exhibit, for example, excellent high-temperature, solvent and creep resistance. In the past, high weight percentages of wood flour have been used with thermosets such as phenolic or urea-formaldehyde resins to produce molded products (Clemons and Caulfield 2005). The wood flour was added to improve toughness and reduce shrinkage on curing. Such types of composite were very prevalent in the twentieth century, often under trade names such a "Bakelite." Nowadays, polyester, polyurethane and epoxy resins are used more often. Aranguren et al. (1999) prepared and tested composites based on a styrene cross-linked unsaturated polyester (UP) resin and oak sawdust as well as hazelnut shells. Samples were prepared with 20% by weight of the two filler types. Fillers were modified by a simple alkaline treatment or by further addition of acrylic acid. It was found that the addition of any of the fillers facilitated the manufacturing of the samples due to reduced shrinkage during the curing step. Regarding mechanical performance, flexural strength of the UP resin was 132 MPa while flexural strength of the best filled variation (UP and acrylic acid treated oak sawdust) was only 60 MPa. However, the flexural modulus of the pure matrix was improved in all cases when the oak sawdust was pretreated. This was attributed to improved compatibility between the matrix and filler. Incorporation of untreated milled hazelnut shells in the UP matrix resulted in composites with higher MOE compared to the pure resin but very low flexural strength which was explained with the lack of fibrous structure in hazelnut particles. Marcovich et al. (1999) investigated the moisture diffusion in polyester-wood flour composites also.

To increase the dispersion and adhesion to the UP matrix, the wood filler may be pretreated with NaOH or modified with maleic anhydride (Marcovich et al. 2001). By modifying the wood flour with maleic anhydride, up to 75% of filler could be incorporated in the matrix. For comparison, nontreated wood flour could only be used by up to 50%. In addition, increased hydrophobicity of the modified wood flour was obtained resulting in lower water absorption of the composite as compared to the composite based on untreated wood filler. An improvement in ultimate deformation and toughness was also observed for the maleic-anhydride treated composites.

To enhance the toughness of UP matrix-based wood flour-composites, incorporation of long natural fibers could be a solution. Nunez et al. (2006) added nonwoven mats of sisal fibers (length ranging from 10 to 18.5 cm with diameters in the range of 0.11–0.23 mm) to pine wood flour (200 μm) dispersed in UP resin. Hybrid composites were prepared by using one or two layers of mats of sisal fibers, using a hand-lay-up technique followed by compression molding. Addition of 7% (by wt.) of sisal fibers to wood flour-filled UP resin resulted in an almost 10-fold increase in the work of fracture of notched specimens. Crack growth resistance in the hybrid composites was effectively achieved by fiber bridging and pull-out.

During the last 10 years, a lot of efforts have been made to combine bio-based thermoset polymers and natural or wood fibers and wood flour (Quirino et al. 2012, Mosiewicki and Aranguren 2013). A review on the synthesis, properties, and potential applications of novel thermosetting biopolymers from soybean and other natural oils was provided by Li and Larock (2005). Quirino et al. (2012) prepared soybean and linseed oil-based thermosets reinforced with pine, oak, and maple wood flours and a mixture of hardwood fibers. Soybean oil is one of the most prevalent vegetable oils worldwide. It consists of a triglyceride with a fatty acid composition of 11% palmitic acid, 3% stearic acid, 22% oleic acid, 55% linoleic acid, and 9% linolenic acid. Quirino et al. (2012) determined that a filler load of 80% (by wt.) was the most practical for composite preparation which was done by manual mixing of fillers and resin, transferring into a mold and compression molding at 180°C and 600 psi. The composites were then removed from the mold and postcured in a convection oven for 2 h at 200°C at ambient pressure. Parameters, such as cure time, filler load, filler particle size, origin, and resin composition were varied and analyzed. It was shown that maleic anhydride serves as a good filler-resin compatibilizer and that the composites reinforced with fibers show significantly higher mechanical properties than those filled with wood flour.

With a similar fatty acid composition but a higher degree of unsaturation, linseed oil is also a potential bio-based monomer for the preparation of "green" composites (Henna et al. 2007). Mosiewicki et al. (2005) used a polyester resin synthesized from linseed oil which was further crosslinked with styrene. Composite materials were then made from the thermoset resin in combination with various percentages of pine wood flour and tested. The wood flour used had a tremendous effect on the glass transition temperature of the polymer (upward shift of 40°C in the sample with 30% [by wt.] of wood flour), accompanied by increased glassy and rubbery storage modulus. These observations pointed to strong interfacial interactions. The use of concentrations of wood filler higher than 30% (by wt.) led to voids in the composite.

Polyurethane-pine wood flour composites prepared from a modified tung oil were studied by Mosiewicki et al. (2009). Tung oil, also called china wood oil, is obtained from the seeds or nuts of the tung tree and is used in the paint and varnish industry as drying oil because of its fast polymerization in the presence of oxygen. Pine wood flour and microcrystalline cellulose were used as reinforcements, with the best performance corresponding to the wood flour-based composites which was attributed to excellent dispersion of the wood flour in the polyurethane matrix. Good fiber-matrix adhesion could be expected due to the chemical reaction between cellulosic hydroxyl groups and isocyanate groups. In addition, it was found that incorporation of the filler into the polymer improved the thermal stability of the composites at high temperatures. The pine wood flour-PU composites were further studied to determine the effect of the wood flour concentration on the material properties (Casado et al. 2009). Excellent compatibility of the main phases led to higher tensile strength and modulus of elasticity but also higher deformability and fracture resistance than the neat polymer.

Bio-based epoxy resins have also been used in combination with wood flour. Shibata et al. (2010) used two types of glycerol-based epoxy resins with different viscosities in combination with tannic acid and wood flour made from Japanese Sanbu cedar. Optimum tensile properties (51 MPa for strength and 5.1 GPa for modulus of elasticity) were obtained using a wood filler content of 60% by wt. The wood filler was tightly incorporated into the cross-linked epoxy matrix, without the use of a coupling agent or surface modification treatment.

In addition to the industrially often used polyester, polyurethane, and epoxy resins, starch has been investigated as a potential matrix for wood and natural fibers. Starch is a thermoplastic material but can be chemically modified by cross-linking such that thermosetting behavior is obtained. Duanmu et al. (2010) used enzymatically degraded, modified potato starch as thermoset matrix for bleached kraft softwood pulp fibers. When 40% (by wt.) of pulp fibers were incorporated, compression-molded specimens displayed high flexural strength of 128 MPa and a modulus of elasticity of 4500 MPa. Instead of pulp fibers, microfibrillated cellulose has also been used as reinforcement by the same reseachers (Duanmu et al. 2012).

3.5.2 PROCESSING

For automotive applications, needled, resin-impregnated mats are usually compression-molded in a hot tool, followed by a cold tool. Generally, compression molding is the most widely used method to process thermoset composites, whether long or short fibers are used. In addition, resin transfer molding (RTM) has been used, mostly for long natural fibers such as hemp and flax. For example, O'Donnell et al. (2004) used a vacuum-assisted resin transfer molding (VARTM) process to produce composite panels out of plant oil-based resin (in this case, acrylated epoxidized soybean oil) and natural fiber mats made of flax, cellulose, pulp, and hemp. The VARTM process is a variation of vacuum-infusion RTM in which one of the solid tool faces is replaced by a flexible polymeric film. It is a very clean and economical manufacturing method; only the partial vacuum is used to drive the resin which reduces worker contact with the liquid resin and decreases volatile emissions. In addition, the process increases component mechanical properties and fiber content by reducing void percentage compared to other large-part manufacturing processes, such as hand lay-up. Various cellulose-based reinforcements were successfully used in this process, for example, air-laid cellulose fiber, chemo-thermomechanical pulp, fluff pulp, and recycled paper.

Thermoset resins may also be used to process lightweight honeycomb core sandwich panels. In this case, two pieces of thin and stiff skin layers are bonded to a lightweight core. Polymer foams, metal foams, balsa wood, and various honeycomb structures are commonly used as core materials. As skin materials, natural fiber-reinforced thermoset polymer composites can be used., A sandwich construction using natural fiber mat and polyurethane-based polymer skins plus paper honeycomb core is marketed under the trademark Baypreg®. This panel has been used by automotive manufacturers, such as BMW and Audi, in sunshade, spare tire cover, and load floor applications (Du et al. 2012). Du et al. (2012) used paper-reinforced polymer (PRP) composites as skin materials for lightweight sandwich panels. These PRP composites displayed comparable bending rigidity and flexural load bearing capability but lower areal weights compared to commercial products suggesting their use as an alternative to glass fiber-reinforced polymer composites as skin materials in sandwich panels. An advantage of the thermoset-based sandwich composites compared to thermoplastic-based composites is their resistance to creep (Du et al. 2013a). The same research group also investigated a variety of different pulp fibers (hardwood and softwood chemi-thermomechanical pulp, bleached Kraft softwood, and cellulose pulp) for their suitability in combination with two thermoset matrices, UP resin and vinyl ester resin (Du et al. 2013b). In this case, fiber sheets were impregnated with the resins, and the resulting prepregs were pressed followed by a postcuring step. The absence of lignin on the fiber surfaces of the bleached kraft and cellulose pulps meant that in this case, vinyl ester was a more suitable matrix than the UP resin. Lignin on the two chemi-themomechanical pulp types acted as a natural coupling agent for these pulps and the more hydrophobic UP resin. These results were reflected in the tensile properties of the composites.

Wood-plastic composites are usually processed using thermoplastic matrices; however, efforts have been made in the recent past to accommodate thermosets also (Haider et al. 2012, Englund and Chen 2014). Because thermosets generally require different processing techniques which accommodate low viscosity resins and the cure kinetics of the cross-linking reaction, some important modifications needed to be considered and developed. A modified melamine resin was developed by Borealis (previously, Agrolinz Melamine GmbH) which contained both a thermoplastic and thermoset phase (HIPE®ESIN). The thermosetting reaction relies on an acid catalyst and is accelerated with elevated temperature. A thermoplastic melamine was developed by incorporating polymer/oligomer chains between the melamine resin molecules (Englund and Chen 2014, and references therein). The polymer chains should allow the resin to become moldable prior to the beginning of the cross-linking reaction, allowing thermoplastic processing equipment such as extrusion and injection-molding to be used. It was determined that a temperature profile which allows for the softening of the resin without inducing a curing reaction during extrusion is useful to maximize the mechanical properties of the WPC. A post-heating step at 175°C for 6 h ensured curing of the resin.

Thermosets such as various isocyanates and phenol-formaldehyde resin have also been used as coupling agents for WPC (Lu et al. 2000 and references therein). Small amounts of thermoset resins are sometimes also incorporated into WPC by way of MDF sawdust, which is used as filler instead of wood flour.

3.6 RECYCLING AND LIFE-CYCLE ASSESSMENT (LCA) OF COMPOSITES

In general, recycling of wood and natural fiber reinforced thermoplastic composites is well possible due to the thermoplastic polymer matrix. Regarding processing, internal and end-of-life recycling need to be distinguished. Internal recycling refers to reuse of materials during the production whereas end-of-life recycling addresses the reuse of materials and products after their service life. Many producers of wood-plastic composites are performing internal recycling because of the achievable level of material performance and cost savings. A lot of publications deal with internal recycling of WPC and NFC (e.g., Augier et al. 2007, Bourmaud and Baley 2007, 2009, Englund and Villechevrolle 2011, Beg and Pickering 2008b,c). Most of these publications have shown that after re-extrusion, the mechanical properties of the composites are improved and water uptake is reduced, due to improved dispersion of wood particles in the matrix, increased thermal exposure of the wood particles (leading to a reduction in hydrophilicity), and cross-linking. Only after a number of reprocessing cycles are mechanical properties affected, which is mostly due to thermal and mechanical degradation of the polymer matrix.

Dickson et al. (2014) compared the effect of reprocessing (six injection molding and extrusion cycles) on the mechanical properties of polypropylene reinforced with either thermomechanical pulp, flax fibers, or glass fibers. PP reinforced with thermomechanical pulp showed the best mean property retention after reprocessing (87%) compared to flax (72%) and glass (59%) fibers. Property reductions were attributed to reduced fiber length. Thermomechanical pulp displayed the lowest reduction in fiber length. Flax fibers showed larger damage (cell wall dislocations) with reprocessing than thermomechanical pulp.

Solutions for end-of-life recycling of WPC are not available yet in Europe mostly because the amounts of recycled materials and products available at present are still too small and because of efforts required to collect and separate WPC based on different matrices, filler, and reinforcement types. Most manufacturers recommend that small quantities of WPC are disposed of via the household waste system, and that larger quantities are disposed of either as recyclable waste wood material or as heat source ("energy recycling"). A specific waste legislation for WPC does not exist at present.

Regarding life-cycle assessment (LCA) of wood and natural fiber reinforced composites as well as WPC, limited information is available at present. Xu et al. (2008) based their study on wood fiber-reinforced PP preforms produced by compression molding. They introduced a term called "material service density," which is defined as the volume of material satisfying a specific strength requirement. When material service density was used as a basis, the composite demonstrated superior environmental friendliness compared to pure PP.

Bolin and Smith (2011) compared alkaline copper quaternary (ACQ)-treated lumber used for decking with WPC decking. They determined that under the assumptions of their LCA, the use of ACQ-treated lumber for decking offers lower fossil fuel use and environmental impacts than WPC.

Mahalle et al. (2014) addressed wood fiber-reinforced PLA and PLA/ thermoplastic starch (TPS) biocomposites, including a comparison to neat PP. PLA was found to cause more environmental burden than TPS. Environmental performance of the biocomposite can be improved by substituting locally available TPS for PLA. The composite can outperform neat PP in all impact categories except eutrophication effects.

Recycling of thermoset composites is more difficult than recycling of thermoplastic composites because simple remelting and reprocessing into shape is not possible. Only grinding of thermoset composites and addition of ground particles as filler to virgin materials is feasible.

3.7 AUTOMOTIVE APPLICATIONS

3.7.1 OVERVIEW

The processing options to obtain wood and natural fiber reinforced composites for automotive applications are shown in Table 3.6 while commercial wood and natural fiber-based composite products are listed in Du et al. (2014). Three types of processing technologies are of primary interest: compression molding (predominant technology), injection-molding, and sheet extrusion. Compression-molding is performed with both thermoplastics as well as thermosets while injection-molding and sheet extrusion are performed with thermoplastics only. Sheet extrusion is followed by thermoforming ("Renolit" Wood-Stock, used by Yanfeng Automotive Interiors). Maximum temperatures during the individual processing steps are shown in Table 3.6. The most important requirements from the viewpoint of the part producer are consistent quality, low emissions and fogging, appropriate mechanical properties, stability, and durability as well as low water uptake and swelling.

3.7.2 THERMOPLASTIC COMPOSITES

3.7.2.1 Hot-Pressing Followed by Compression-Molding

The first step in the production of a thermoplastic compression-molded part is the preparation of a needle felt with a mixture of wood or natural fibers and thermoplastic (e.g., polypropylene) fibers (Figure 3.15). This felt is produced by carding or airlaying. After hot-pressing in a three-dimensionally shaped tool, the resulting part is transferred to a cold tool. One example using this process is the product EcoCor (Figure 3.16) by Yanfeng Automotive Interiors, which is a joint venture of Yanfeng Automotive Trim Systems and Johnson Controls. EcoCor consists of 50% flax, hemp, or kenaf fibers blended with 50% PP fibers (Klusmeier 2015). The EcoCor

TABLE 3.6

Classification of Wood and Natural Fiber Reinforced Composites for Automotive Applications

Lignocellulosic Substrate	Polymer	Process	Application
Fibers: Wood, flax, hemp, rape seed, jute, kenaf	Thermoplastic: Polypropylene, (co-) polyester	Hot-pressing (190–220°C) followed by compression-molding in a cold tool (<50°C)	Door panels, pillars, seat back panel
Short fibers or flour: Wood, flax, hemp, kenaf, ground coconut shells, rice husks, wheat straw	Thermoplastic: Polypropylene, co-polyester, polybutylene succinate (PBS), acrylonitrile butadiene styrene (ABS), polyamides	Injection-molding (<200°C)	Door panels, armrest, storage bin
Wood flour	Thermoplastic: Polypropylene	Sheet extrusion and thermoforming	Door panels, parcel shelves, sides
Fibers: Wood, flax, hemp, rape seed, jute, kenaf, sisal	Thermosetting: melamine, polyester resin, acrylic resin, PUR, epoxy	Impregnation or spraying of fiber felts with resin and compression-molding in a hot tool (140–220°C, depending on the resin)	Instrument panel, door panel, seat back panel

mat is compression-molded and the cover stock can be applied in a one-step manufacturing process. The mat is preheated to approximately 210°C and then pressed in a low-temperature tool (Figure 3.17).

The EcoCor-mats can also be used in combination with injection-molding (back injection-molding, BIM). This means that parts can be directly attached to the panel in a single step. Challenges of the process are a reduction of odor and emissions as well as improvement of adhesion with polyurethanes, which may be achievable, for example, using plasma treatment.

Instead of PP fibers, PET fibers have also been used in combination with wood fibers to prepare mats for compression molding. This product is marketed under the trade name "LignoFlex" by Faurecia. Final application is for interior-trim components, and the composition is 70% wood fibers, 20% PET fibers, and 10% phenol-formaldehyde or acrylic resin. The PET fibers hold the wood fibers and PF resin together during initial prepreg production and help prevent tearing of the fiber mat during molding, allowing designs with undercuts and deeper ribs/sharper radii than could be produced with earlier products (http://www.compositesworld.com/articles /interior-innovation-the-value-proposition; accessed on April 27, 2016). The PF resin cross-links at temperatures of 200°C, leaving the PET fibers, which melt at

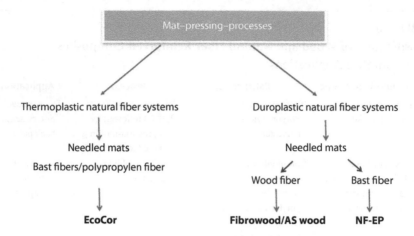

FIGURE 3.15 Different mat pressing processes. (From Klusmeier, W. (2015): Natural fiber reinforced components for vehicle interiors, status and development. Presentation at the Sixth WPC & NFC Conference, Cologne, December 16–17, 2015, Cologne, Germany; copyright: Yanfeng Automotive Interiors.)

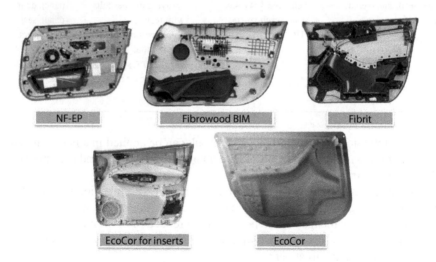

FIGURE 3.16 Examples of lightweight door panel solutions incorporating wood and natural fibers. (From Klusmeier, W. (2015): Natural fiber reinforced components for vehicle interiors, status and development. Presentation at the Sixth WPC & NFC Conference, Cologne, December 16–17, 2015, Cologne, Germany; copyright: Yanfeng Automotive Interiors.)

220°C, intact to contribute to better elongation in molded substrates. "LignoFlex" can be bonded to hardwood veneers during compression molding to form trim panels which are lighter than conventional composites and offer a real-wood surface. "Lignolight" is a lower density version of the same product (1400 g/m^2 compared to 1800–2000 g/m^2).

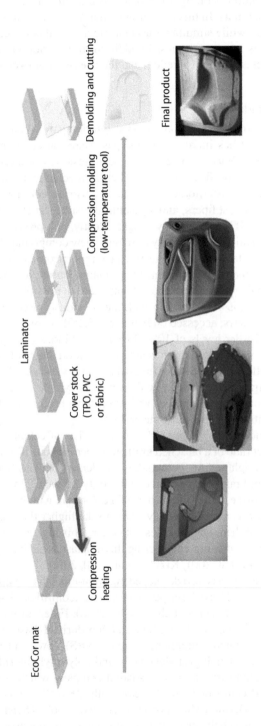

FIGURE 3.17 EcoCor process used by Yanfeng Automotive Interiors. (From Klusmeier, W. (2015): Natural fiber reinforced components for vehicle interiors, status and development. Presentation at the Sixth WPC & NFC Conference, Cologne, December 16–17, 2015, Cologne, Germany; copyright: Yanfeng Automotive Interiors.)

Last, but not least, compression hybrid molding is a technique available for processing consolidated fiber mats. In this case, heated mats are placed into the injection molding tool and pressed while simultaneously plastic material is injected wherever needed which eliminates mat tear issues. In addition to a reduction of processing steps, further advantages are design freedom and the use of a lightweight and easily workable material.

3.7.2.2 Injection-Molding

Injection-molding is less common for automotive interior parts than compression-molding; however, interest has increased in the past years based on new material and process developments. Various types of lignocellulosic raw materials are used in this case: wood flour, wood fibers from the thermo-mechanical pulping (MDF) process (example: product "Woodforce" by Sonae Industria), cellulose fibers (e.g., bleached kraft pulp), short bast fibers, and agricultural residues such as wheat straw. Filler levels are between 20% and 40% by weight. Mostly, polypropylene is used as the matrix. Several companies have recently developed new compounds for injection-molding of automotive parts. Inno-Comp in Hungary reports on its web page that it created a consortium together with Milliken, a filler manufacturer, and MecaPlast, an automotive supplier, and developed a compound that makes it possible to manufacture plastic parts whose weight is 7% less than with the previously used basic material (https://inno-comp.hu/en/news; accessed on February 16, 2016). Inno-Comp products are available with wood filler levels of 10%–30% in a PP block-copolymer matrix. The wood fiber source is not disclosed so it is not clear if wood flour or fibers are used.

The product "Woodforce" by Sonae Industria consists of small dice of flat-pressed thermomechanical pulp bonded with polymers which is currently being tested by various compounding companies and automotive producers. The material has passed the specification of Ford (Klusmeier 2015).

Mondi offers a product called "Fibromer" which consists of cellulose fibers combined with PP. The aspect ratio of a bleached kraft pulp fiber is about 100 which offers a high reinforcement and weight savings potential. While density of glass-fiber reinforced PP is 2.5 g/cm³, density of kraft-fiber reinforced PP is only 1.5 g/cm³. In addition, since cellulose fibers are lignin-free, there is less change in color than with lignin-containing fibers, and fiber coloration is easier to achieve. Due to the higher thermal stability of cellulose, it is claimed that there are no issues with odor. This higher thermal stability also leads to the possibility of using engineering thermoplastics such as polyamide 6 (Jacobson et al. 2001, Sears et al. 2001, Kiziltas et al. 2010, 2014).

US company Weyerhaeuser has also developed injection-moldable compounds based on cellulose fibers named "Thrive" (http://weyerhaeuser.com/files/8114/2506/2924/THRIVE_brochure.pdf; accessed on February 16, 2016). Filler levels are between 10% and 40% by wt. A range of fossil-fuel and bio-derived base polymers are available, including acrylonitrile butadiene styrene (ABS), low- and high-density polyethylene (LDPE, HDPE), polypropylene (PP), and polyvinyl chloride (PVC).

UPM is another large forestry and biomass-based company which is offering cellulose fiber-reinforced PP compounds, in this case with 20%–50% filler level (UPM Formi). In 2014, UPM and Helsinki Metropolia University of Applied Sciences launched the Biofore concept car in which the majority of parts traditionally made of plastics were replaced with UPM Formi and UPM Grada (thermoformable wood). UPM Formi was used in the front mask, side skirts, dashboard, door panels, and interior panels.

Italian compounding company So.F.Ter introduced its Polifor NF product line in September 2015 (http://www.softergroup.com/en/natural_fibers_improve _performance_and_lightness). The Polifor NF product line is based on polypropylene reinforced with natural fibers, and according to the company´s website, "offers sustainability, lightness and improved performance for a wide range of industries, from automotive to design." Various (undisclosed) vegetable fillers derived from renewable sources are used in the compounds. Polifor compounds are claimed to offer the same rigidity as polypropylene compounds with the same percentage of talc, but are much lighter because of lower density (–8%), provide better resistance to high temperatures (+17%), and offer a significantly higher level of resistance to impacts (up to +67% in the notched Izod test). Possible applications in the automotive industry include interior and exterior parts, both structural and aesthetic, such as the cowl vent grille, internal consoles and pillars, structural dashboard carrier, and various parts of the luggage compartment. The Polifor NF composites are said to withstand operating temperatures up to 110–120°C and therefore can also be used for under-the-hood applications such as the air filter box. The high impact resistance of Polifor NF products makes them also suitable for the production of external casings, or for covers and protective housings of technical parts. More applications in the field of furniture and design are under development.

In addition to wood fibers, short bast fibers from hemp, flax, kenaf, jute, and ramie have also been used in injection-molding; however, with limited success due to problems with odor emission. An exception is the Nafilean (Natural Fiber for Lean Injected Design) compound, which is sold by automotive supplier Faurecia (http://na.faurecia.com/en/faurecia-naias/interior-systems; accessed on February 17, 2016). The compound is based on hemp fiber and polypropylene. It was introduced to the market in the 2013 Peugeot 308 where it is used for instrument panels, center consoles, and door panels. Nafilean was recognized with the 2014 Innovation award in the Green Category from the European Automotive Association (CLEPA). According to Faurecia, Nafilean cannot only be shaped into complex forms and architectures, but also it reduces the weight of injected parts by up to 25% compared with standard injection-molded plastic parts and provides a 40% better fit and finish.

Faurecia is partnering with Interval, an agricultural cooperative in France, through a joint venture called Automotive Performance Materials (APM), to process the compound. Production in France is underway, and the business is set to expand to the North American market in 2016, with plans to produce locally in Asia by 2018 (http://na.faurecia.com/en/faurecia-naias/interior-systems). APM will also soon begin industrial production of BioMat, a 100% bio-based compound made from 25% hemp-fiber reinforced polybutylene succinate (PBS), developed in a partnership with Mitsubishi Chemical.

Biobased solutions are also offered by German company Röchling, which has partnered with Corbion Purac to develop PLA-based compounds called "Plantura," which are 95% biobased. Plantura is available in different grades, for example, with 30% wood fibers or various levels of glass fibers. Interior trim parts and an air filter box have been developed and made with Plantura.

Pellets based on long hemp fibers and polypropylene are used by German automotive parts supplier HIB Trim Parts Solutions for injection-molding of decorative trim parts (http://www.hib-solutions.com/de/index.php). Up to 50% of hemp fibers which are up to 2 cm long prior to pelletization are used in combination with thermoplastic. After injection-molding, fiber length is retained to a large degree, resulting in an interesting design. Impact bending strength is not as high as with glass fiber reinforced plastics; however, it is sufficiently high for decorative trim applications. In addition, the material is 20% lighter than alternative material options. HIB is working closely with German company Badische Faserveredelung (BAVE), which is providing hemp fibers in a pelletized form. BAVE is also involved in the Naturtruck-EU-project (www.naturtruck.eu), which is coordinated by Aimplas, a Spanish plastics research institute. In this project, injected plastic parts are developed for the commercial vehicles industry (mainly cabin truck parts) made with thermoplastic composite materials from renewable resources (modified polylactic acid and natural fibers), with improved thermal and flame retardancy properties and high quality surface finishing to be used in car internal parts. These biocomposites are aimed to be an alternative to standard ABS grades at a competitive cost. Volvo as car manufacturer is participating in this project.

In 2015, PolyOne introduced "reSound NF," a series of natural fiber reinforced plastics, to the market. According to the company's website, it is a highly engineered, strong and sustainable alternative to glass fiber reinforced formulations, with densities 5% to 10% lower than comparable glass fiber formulations. Tensile and flexural properties are claimed to be improved by 20%, impact strength by 50%, compared to other natural fiber reinforced solutions. Heat deflection temperature is said to be 10°C to 20°C higher. The superior thermo-mechanical properties of reSound NF solutions are claimed to be a result of the novel process used to manufacture them. Customers can process these specialty polymers with standard machinery and tooling at low injection molding temperatures.

US-based company RheTech, which was acquired by Hexpol compounding in January 2015 (http://rhetech.com), is offering compounds reinforced with ground coconut shells, rice hulls, agave, flax, or wood fibers. The wood fiber reinforced options include pine and maple wood. RheTech stresses that both maple and pine wood come from secondary processes and that no trees are cut down to produce their materials. Wood fiber is sourced from the production of cabinetry and cut lumber. Wood shavings are hammer-milled to a coarseness that confers "maximum physical properties" to the compound. The following compound characteristics are listed by RheTech:

• High stiffness and dimensional stability
• Lower specific gravity than mineral-filled compounds
• Excellent chemical and mold resistance
• Colorability and uniqueness

Prototypes based on wheat straw were produced by Schulman with its "AgriPlas" compounds which contain up to 30% of wheat straw in a PP matrix. According to the company's website, the commercialization resulted in the first ever injection moldable automotive application utilizing wheat straw fiber in a polypropylene compound.

The development process to commercialize wheat straw began through a consortium set up by the Ontario BioCar Initiative. This consortium was started and led by the University of Waterloo and included the Ford Motor Company, A. Schulman, Inc., and Omtec Inc. The consortium commercialized the utilization of the AgriPlas™ BF20H-31 product in the Ford Flex third row quarter trim bin and inner lid. Some advantages of the material are summarized as follows:

* Compared to similar filled materials, AgriPlas™ provides ~10% weight savings.
* Wheat straw does not compete with food sources or alternative fuel sources.
* All physical property values and application testing requirements were equivalent or improved.

Currently, Schulman is investigating tomato pumice in PP compounds, in conjunction with Heinz and Ford Motor Company. In addition, a technology patented by Biobent Polymers is explored to exfoliate bio-waste soy meal into a PP compound (Markarian 2015). Biobent is also working with other polyolefins and various plastics from renewable resources (Mapleston 2016). The company claims that by replacing 30%–40% of a petroleum-based resin with a bio-feedstock costing as little as 10% of the polymer that it is replacing, Biobent can offer savings that range from 2%–18% or more (Mapleston 2016).

Summarizing the above-mentioned developments, it becomes clear that at present, PP is still the number one choice as polymer matrix in automotive applications. However, there are a number of polymers that may be of interest in combination with wood and other natural fibers. The primary criterion in selecting polymers is the thermal degradation temperature of wood and natural fibers (beyond approximately 250°C). Above this temperature, unmodified lignocellulosic materials undergo rapid thermal degradation which limits the choice of polymers to polyolefins. However, efforts have been made to use engineering plastics also. In the case of WPC decking, polymethylmethacrylate (PMMA) has been used based on a direct extrusion process. The decking product is marketed under the trade name Plexiglas® Wood (http://www.plexiglas.de/product/plexiglas/de/produkte/plexiglas-wood/pages /default.aspx). Attempts with limited success to use nylon (polyamide) 6, which displays a melting point of approximately 220°C, were reported as early as 1984 by Klason et al. Severe discoloration and thermal degradation of the cellulose fibers were observed. Sears et al. (2001) and Jacobson et al. (2001) later showed that it is possible to extrude cellulose fibers and PA 6 (as well as PA 6.6) by using either preblending or side-feeding methods, which utilize control of temperature resulting from viscosity shear-heating. Kiziltas et al. (2010, 2014) investigated microcrystalline cellulose filled PA 6 and poly(ethylene terephthalate) (PET)-poly(trimethylene terephthalate) (PTT) blend composites. Blends of high-density polyethylene and PET were prepared and successfully mixed with wood flour in a second compounding step (Lei and Wu 2012). It was reported that the HDPE-PET-wood-based composites showed 65% higher tensile strength, 95% higher tensile modulus, 42% higher flexural strength, and 64% higher flexural modulus, respectively, compared with conventional HDPE-wood-based composites.

Additional polymers used to process wood fiber-based thermoplastic composites include, for example, ABS (Yeh et al. 2009), PS (Klason et al. 1984), and PA 12 (Lu et al. 2007).

3.7.2.3   Sheet Extrusion and Thermoforming

Extruded PP sheets with wood powder have been used for a long time, especially in southern Europe, as a very cost-effective trim material. These sheets are limited, however, in their ability to stretch to create contoured trim parts. In addition, mechanical performance is limited (tensile strength, brittleness). To improve these properties, a particular amount of natural fibers may be added to the formulation. One of the best-known products where a combination of wood flour and thermoplastic resin is used is "Wood-Stock" by Renolit, which is used for trim parts such as door panels or inserts. In one step, the substrate is formed, compressed together with the surface material and attached to the carrier. The extruded sheets are heated up with infrared energy to 210–220°C while surface materials (textiles, PVC, etc.) are also preheated. Both sheets and surface materials are then combined and pressed at a mold temperature of less than 40°C.

3.7.3   Thermoset Composites

3.7.3.1   Fibrowood

For automotive applications, needled mats are usually processed in combination with thermoset resins. In this case, either wood fibers ("Fibrowood," Yanfeng Automotive Interiors) or bast fibers ("Natural Fiber Epoxy Resin" or "NF-EP") are used. Again, the first step is the preparation of a needle felt. Next, the felt is sprayed or impregnated with resin and pressed in a hot tool, followed by a cold tool. This results in, for example, door panel substrates which may be further laminated and equipped with additional parts.

In the case of "Fibrowood," the wood fiber felt is impregnated with acrylic resin and mixed with synthetic fibers. Different fibers such as polyester, co-polyester or bicomponent fibers can be used, with melting points of approximately 120°C. The material was developed to replace the "Fibrit" process, a wet process similar to paper production. The advantage of "Fibrowood" is that the impregnated felts can be easily transported and further processed where required. In addition, highly contoured trim parts can be manufactured during compression-molding. The material is used as a carrier for door panels, door inserts, and seat back panels. The mats are pressed in a high-temperature tool at approximately 220°C. "Fibrowood" can also be used in combination with injection molding (back injection molding or BIM, Figure 3.18).

3.8   OUTLOOK

On a global level, the availability of wood and manufactured cellulose fibers surpasses natural fibers (hemp, flax, sisal, etc.). Hence, it can be expected that the market share of wood fibers and flour in the automotive industry will increase in the future.

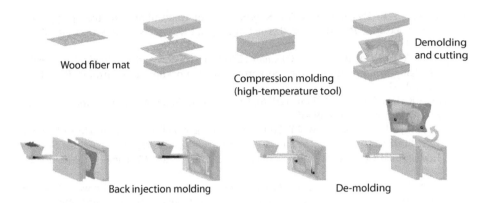

Wood fiber mat

Compression molding
(high-temperature tool)

Demolding
and cutting

Back injection molding

De-molding

FIGURE 3.18 Fibrowood-back-injection molding process. (From Klusmeier, W. (2015): Natural fiber reinforced components for vehicle interiors, status and development. Presentation at the Sixth WPC & NFC Conference, Cologne, December 16–17, 2015, Cologne, Germany; copyright: Yanfeng Automotive Interiors.)

This development is driven by political frameworks as well as advancements in the processing of parts and an increased supply of wood and cellulose fiber reinforced compounds for injection-molding. Further process improvements and combinations of processing steps as well as integration of functional elements are expected.

Recently, new processing techniques such as 3D printing have shown the potential to produce lightweight, individualized parts and composites. Wood and natural fibers may play a part in this development which is indicated by the availability of wood-filled filaments for fused deposition modeling. In addition, combinations of wood and carbon fibers have the potential to create new lightweight structural and design opportunities.

REFERENCES

Aranguren, M.I., M.M. Reboredo, G. Demma, J. Kenny (1999): Oak sawdust and hazelnut shells as fillers for a polyester thermoset. *Holz als Roh- und Werkstoff* 57: 325–330.

Arao, Y., S. Nakamura, Y. Tomita, K. Takakuwa, T. Umemura, T. Tanaka (2014): Improvement on fire retardancy of wood flour/polypropylene composites using various fire retardants. *Polymer Degradation and Stability* 100: 79–85.

Augier, L., G. Sperone, C. Vaca-Garcia, M.-E. Borredon (2007): Influence of the wood fibre filler on the internal recycling of poly(vinyl chloride)-based composites. *Polymer Degradation and Stability* 92: 1169–1176.

Ayrilmis, N., J.T. Benthien, H. Thoemen, R.H. White (2012): Effects of fire retardants on physical, mechanical, and fire properties of flat-pressed WPCs. *European Journal of Wood and Wood Products* 70: 215–224.

Beg, M.D.H., K.L. Pickering (2008a): Mechanical performance of Kraft fibre reinforced polypropylene composites: Influence of fibre length, fibre beating and hygrothermal ageing. *Composites: Part A* 39: 1748–1755.

Beg, M.D.H., K.L. Pickering (2008b): Reprocessing of wood fibre reinforced polypropylene composites. Part I: Effects on physical and mechanical properties. *Composites: Part A* 39: 1091–1100.

Beg, M.D.H., K.L. Pickering (2008c): Reprocessing of wood fibre reinforced polypropylene composites. Part II: Hygrothermal ageing and its effects. *Composites Part A* 39: 1565–1571.

Bengtsson, M., M. Le Baillif, K. Oksman (2007): Extrusion and mechanical properties of highly filled cellulose fibre-polypropylene composites. *Composites: Part A* 38: 1922–1931.

Bledzki, A.K., V.E. Sperber, O. Faruk (2006): Cars from bio-fibres. *Macromolecular Materials and Engineering* 291: 449–457.

Bledzki, A.K., V.E. Sperber, S. Wolff (2003): Measurement and reduction of odors of natural fiber filled materials employed in the automotive industry. Presentation at the Seventh International Conference on Woodfiber-Plastic Composites, Madison, WI, May 19–20, 2003.

Bledzki, A.K., S. Reihmane, J. Gassan (1998): Thermoplastics reinforced with wood fillers: A literature review. *Polymer Plastics Technology and Engineering* 37(4): 451–468.

Boldizar, A., C. Klason, J. Kubát, P. Näslund, P. Sáha (1987): Prehydrolyzed cellulose as reinforcing filler for thermoplastics. *International Journal of Polymeric Materials* 11(4): 229–262.

Bolin, C.A., S. Smith (2011): Life cycle assessment of ACQ-treated lumber with comparison to wood plastic composite decking. *Journal of Cleaner Production* 19: 620–629.

Bourbigot, S., M. Le Bras, S. Duquesne, M. Rochery (2004): Recent advances for intumescent polymers. *Macromolecular Materials and Engineering* 289: 499–511.

Bourmaud, A., C. Baley (2007): Investigations on the recycling of hemp and sisal fibre reinforced polypropylene composites. *Polymer Degradation and Stability* 92: 1034–1045.

Bourmaud, A., C. Baley (2009): Rigidity analysis of polypropylene/vegetal fibre composites after recycling. *Polymer Degradation and Stability* 94: 297–305.

Bowyer, W.H., M.G. Bader. On the re-inforcement of thermoplastics by imperfectly aligned discontinous fibres. *J Mater Sci* 1972;7:1315–1321.

Casado, U., N.E. Marcovich, M.I. Araguren, M.A. Mosiewicki (2009): High-strength composites based on tung oil polyurethane and wood flour: Effect of the filler concentration on the mechanical properties. *Polymer Engineering and Science* 49: 713–721.

Carus, M. (2015): Biocomposites in the automotive industry: Technology, markets and environment. Presentation at the Sixth WPC & NFC Conference, Cologne, December 16–17, 2015, Cologne, Germany.

Clemons, C., D. Caulfield (2005): Wood flour. Chapter 15 in: *Functional Fillers for Plastics*. Edited by M. Xanthos. Wiley-VCH Verlag GmbH & Co. KGaA.

Clemons, C.M., R.E. Ibach (2004): The effects of processing method and moisture history on the laboratory fungal resistance of wood-HDPE composites. *Forest Products Journal* 54(4): 50–57.

Clemons, C. (2008): Raw materials for wood-polymer composites. Pages 1–22 in: *Wood-Polymer Composites*. Edited by K. Oksman Niska and M. Sain. Woodhead Publishing Limited, Cambridge, England.

Dalväg, H., C. Klason, H.-E. Strömvall (1985): The efficiency of cellulosic fillers in common thermoplastics. Part II. Filling with processing aids and coupling agents. *International Journal of Polymeric Materials* 11: 9–38.

Dickson, A.R., D. Even, J.M. Warnes, A. Fernyhough (2014): The effect of reprocessing on the mechanical properties of polypropylene reinforced with wood pulp, flax or glass fibre. *Composites: Part A* 61: 258–267.

Duanmu, J., E.K. Gamstedt, A. Pranovich, A. Rosling (2010): Studies on mechanical properties of wood fiber reinforced cross-linked starch composites made from enzymatically degraded allylglycidyl ether-modified starch. *Composites: Part A* 41: 1409–1418.

Duanmu, J., E.K. Gamstedt, A. Rosling (2012): Bulk composites from microfibrillated cellulose-reinforced thermoset starch made from enzymatically degraded allyl glycidyl ether-modified starch. *Journal of Composite Materials* 46(25): 3201–3209.

Du, Y., N. Yan, M.T. Kortschot (2012): Light-weight honeycomb core sandwich panels containing biofiber-reinforced thermoset polymer composite skins: Fabrication and evaluation. *Composites: Part B* 43: 2875–2882.

Du, Y., N. Yan, M.T. Kortschot (2013a): An experimental study of creep behavior of light-weight natural fiber-reinforced polymer composite/honeycomb core sandwich panels. *Composite Structures* 106: 160–166.

Du, Y., N. Yan, M.T. Kortschot (2014): A simplified fabrication process for biofiber-reinforced polymer composites for automotive interior trim applications. *Journal of Materials Science* 49: 2630–2639.

Du, Y., T. Wu, N. Yan, M.T. Kortschot, R. Farnood (2013b): Pulp fiber-reinforced thermoset polymer composites: Effects of the pulp fibers and polymer. *Composites: Part B* 48: 10–17.

EN 13501-1 (2010): Fire classification of construction products and building elements—Part 1: Classification using data from reaction to fire tests.

EN 13823 (2002): Reaction to fire tests for building products: Building products excluding floorings exposed to the thermal attack by a single burning item.

EN 15534-1 (2014): Composites made from cellulose-based materials and thermoplastics (usually called wood-polymer composites (WPC) or natural fibre composites (NFC))—Part 1: Test methods for characterisation of compounds and products.

Englund, K., L.-W. Chen (2014): The rheology and extrusion processing performance of wood/melamine composites. *Journal of Applied Polymer Science* 131: 39858–39864.

Englund, K., V. Villechevrolle (2011): Flexure and water sorption properties of wood thermoplastic composites made with polymer blends. *Journal of Applied Polymer Science* 120: 1034–1039.

Faruk, O., A.K. Bledzki, H.-P. Fink, M. Sain (2012): Biocomposites reinforced with natural fibers: 2000–2010. *Progress in Polymer Science* 37: 1552–1596.

Felix, J.M., P. Gatenholm (1991): The nature of adhesion in composites of modified cellulose fibers and PP. *Journal of Applied Polymer Science* 42: 609–620.

Félix, J.S., C. Domeño, C. Nerín (2013): Characterization of wood plastic composites made from landfill-derived plastic and sawdust: Volatile compounds and olfactometric analysis. *Waste Management* 33; 645–655.

Fengel, D., G. Wegener (1983): *Wood: Chemistry, Ultrastructure, Reactions.* De Gruyter, Berlin.

Fink, H.-P., J. Ganster (2006): Novel thermoplastic composites from commodity polymers and man-made cellulose fibers. *Macromolecular Symposia* 244: 107–118.

Garcia, M., J. Hidalgo, I. Garmendia, J. Garcia-Jaca (2009): Wood-plastics composites with better fire retardancy and durability performance. *Compos: Part A* 40: 1772–1776.

Groom, L., L. Mott, S. Shaler (2002): Mechanical properties of individual Southern pine fibers. Part I. Determination and variability of stress-strain curves with respect to tree height and juvenility. *Wood and Fiber Science* 34(1): 14–27.

Guo, G., C.B. Park, Y.H. Lee, Y.S. Kim, M. Sain (2007): Flame retarding effects of nanoclay on wood-fiber composites. *Polymer Engineering and Science* 47: 330–336.

Hämäläinen, K., T. Kärki (2014): Effects of wood flour modification on the fire retardancy of wood-plastic composites. *European Journal of Wood and Wood Products* 72: 703–711.

Haider, A., U. Müller, U. Panzer (2012): Melamine-resin based WPC—Control of the processing and the properties (in German; Melaminharzbasierende duromere WPC—Steuerung der Verarbeitung und der Eigenschaften). *European Journal of Wood and Wood Products* 70: 579–585.

Hamel, S.E., J.C. Hermanson, S.M. Cramer (2014): Predicting the flexure response of wood-plastic composites from uni-axial and shear data using a finite-element model. *Journal of Materials in Civil Engineering* 26(12): 04014098.

Hamel, S.E., J.C. Hermanson, S.M. Cramer (2012): Mechanical and time-dependent behavior of wood-plastic composites subjected to tension and compression. *Journal of Thermoplastic Composite Materials* 26(7): 968–987.

Henna, P.H., D.D. Andjelkovic, P.P. Kundu, R.C. Larock (2007): Biobased thermosets from the free-radical copolymerization of conjugated linseed oil. *Journal of Applied Polymer Science* 104: 979–985.

Hörold, S. (2016): Formulating fire retardant PAs. *Compounding World*, December 2015, www.compoundingworld.com.

Huda, M.S., L.T. Drzal, D. Ray, A.K. Mohanty, M. Mishra (2008): Natural-fiber composites in the automotive sector. Pages 221–268 in: *Properties and Performance of Natural-Fibre Composites*, edited by K. L. Pickering, CRC Press, Boca Raton, FL.

ISO 11359-2 (1999): Plastics—Thermomechanical analysis (TMA)—Part 2: Determination of coefficient of linear thermal expansion and glass transition temperature.

Jacobson, R., D. Caulfield, K. Sears, J. Underwood (2001): Low temperature processing of ultra-pure cellulose fibers into nylon 6 and other thermoplastics. Pages 127–133 in: The Sixth International Conference on Woodfiber-Plastic Composites. Forest Products Society, Madison, WI.

Kasal, B., S. Friebel, J. Gunschera, T. Salthammer, A. Schirp, H. Schwab, V. Thole (2015): Wood-based materials. *Ullmann's Encyclopedia of Industrial Chemistry,* Wiley-VCH Verlag GmbH & Co. KGaA, Weinheim, Germany.

Kiziltas, A., B. Nazari, D.J. Gardner, D.W. Bousfield (2014): Polyamide 6-cellulose composites: Effect of cellulose composition on melt rheology and crystallization behavior. *Polymer Engineering and Science* 54(4): 739–746.

Kiziltas, A., D.J. Gardner, Y. Han, H.-S. Yang (2010): Determining the mechanical properties of microcrystalline cellulose (MCC)-filled PET-PTT blend composites. *Wood and Fiber Science* 42(2): 165–176.

Klason, C., J. Kubát, H.-E. Strömvall (1984): The efficiency of cellulosic fillers in common thermoplastics. Part 1. Filling without processing aids or coupling agents. *International Journal of Polymeric Materials* 10: 159–187.

Klemm, D., F. Kramer, S. Moritz, T. Lindström, M. Ankerfors, D. Gray, A. Dorris (2011): Nanocelluloses: A new family of nature-based materials. *Angewandte Chemie International Edition* 50: 5438–5466.

Klusmeier, W. (2015): Natural fiber reinforced components for vehicle interiors, status and development. Presentation at the Sixth WPC & NFC Conference, Cologne, December 16–17, 2015, Cologne, Germany.

Klyosov, A.K. (2007): *Wood-Plastic Composites*. John Wiley & Sons, Inc., Hoboken, NJ.

Lei, Y., Q. Wu (2012): High density polyethylene and poly(ethylene terephthalate) *in situ* sub-micro-fibril blends as a matrix for wood plastic composites. *Composites: Part A* 43: 73–78.

Lerche, H., J.T. Benthien, K.U. Schwarz, M. Ohlmeyer (2014): Effects of defibration conditions on mechanical and physical properties of wood fiber/high-density polyethylene composites. *Journal of Wood Chemistry and Technology* 34: 98–110.

Li, B., He J. (2004): Investigation of mechanical property, flame retardancy and thermal degradation of LLDPE-wood-fibre composites. *Polymer Degradation and Stability* 83: 241–246.

Li, F., R.C. Larock (2005): Synthesis, properties, and potential applications of novel thermosetting biopolymers from soybean and other natural oils. Pages 727–750 in: *Natural Fibers, Biopolymers, and Biocomposites*, edited by A. K. Mohanty, M. Misra, and L. T Drzal. CRC Press, Taylor & Francis Group, Boca Raton, FL.

Lohmann, U. (2003): *Holz-Lexikon* (in German; Wood Encyclopaedia). DRW-Verlag Weinbrenner GmbH & Co., Leinfelden-Echterdingen, Germany.

Lu, J. Z., Q. Wu, H.S. McNabb (2000): Chemical coupling in wood fiber and polymer composites: A review of coupling agents and treatments. *Wood and Fiber Science* 32(1): 88–104.

Lu, J.Z., T.W. Doyle, K. Li (2007): Preparation and characterization of wood-(nylon 12) composites. *Journal of Applied Polymer Science* 103: 270–276.

Maldas, D., B.V. Kokta (1989): Improving adhesion of wood fiber with polystyrene by the chemical treatment of fiber with a coupling agent and the influence on the mechanical properties of composites. *Adhesion Science and Technology* 3(1): 529–539.

Maldas, D., B.V. Kokta (1990): Effects of coating treatments on the mechanical behavior of wood-fiber-filled polystyrene composites. I. Use of polyethylene and isocyanate as coating components. *Journal of Applied Polymer Science* 40: 917–928.

Mahalle, L., A. Alemdar, M. Mihai, N. Legros (2014): A cradle-to-gate life cycle assessment of wood fibre-reinforced polylactic acid (PLA) and polylactic acid/thermoplastic starch (PLA/TPS) biocomposites. *International Journal of Life Cycle Assessment* 19: 1305–1315.

Mapleston, P. (2016): Natural fibres aim to compete. *Compounding World*, March 2016, www.compoundingworld.com.

Marcovich, N.E., M.I. Aranguren, M.M. Reboredo (2001): Modified wood flour as thermoset fillers Part I. Effect of the chemical modification and percentage of filler on the mechanical properties. *Polymer* 42: 815–825.

Marcovich, N.E., M.M. Reboredo, M.I. Aranguren (1999): Moisture diffusion in polyester-woodflour composites. *Polymer* 40: 7313–7320.

Markarian, J. (2015): Renewable reinforcements hit the road. *Compounding World*, March 2015, www.compoundingworld.com.

Méndez, J.A., F. Vilaseca, M.A. Pèlach, J.P. López, L. Barberà, X. Turon, J. Gironès, P. Mutjé (2007): Evaluation of the reinforcing effect of ground wood pulp in the preparation of polypropylene-based composites coupled with maleic anhydride grafted polypropylene. *Journal of Applied Polymer Science* 105: 3588–3596.

Migneault, S., A. Koubaa, F. Erchiqui, A. Chaala, K. Englund, M. P. Wolcott (2009): Effects of processing method and fiber size on the structure and properties of wood-plastic composites. *Composites: Part A* 40: 80–85.

Mosiewicki, M.A., M.I. Aranguren (2013): A short review on novel biocomposites based on plant oil precursors. *European Polymer Journal* 49: 1243–1256.

Mosiewicki, M.A., U. Casado, N. E. Marcovich, M.I. Aranguren (2009): Polyurethanes from tung oil: Polymer characterization and composites. *Polymer Engineering and Science* 49: 685–692.

Mosiewicki, M., J. Borrajo, M.I. Aranguren (2005): Mechanical properties of woodflour/linseed oil resin composites. *Polymer International* 54: 829–836.

Naumann, A., H. Seefeldt, I. Stephan, U. Braun, M. Noll (2012): Material resistance of flame retarded wood-plastic composites against fire and fungal decay. *Polymer Degradation and Stability* 97: 1189–1196.

Nikolaeva, M., T. Kärki (2011): A review of fire retardant processes and chemistry, with discussion of the case of wood-plastic composites. *Baltic Forestry* 17(2): 314–326.

Nunez, A.J., M.I. Aranguren, L.A. Berglund (2006): Toughening of wood particle composites—Effects of sisal fibers. *Journal of Applied Polymer Science* 101: 1982–1987.

O'Donnell, A., M.A. Dweib, R.P. Wool (2004): Natural fiber composites with plant oil-based resin. *Composites Science and Technology* 64: 1135–1145.

Oksman, K., C. Clemons (1998): Mechanical properties and morphology of impact modified polypropylene–wood flour composites. *Journal of Applied Polymer Science* 67: 1503–1513.

Oksman Niska, K., M. Sain (2008): *Wood-Polymer Composites*. Woodhead Publishing Limited, Cambridge, England.

Osswald, T.A., G. Menges (1996): *Materials Science of Polymers for Engineers*. Carl Hanser Verlag, New York.

Pickering, K.L., M.G. Aruan Efendy, T.M. Le (2016): A review of recent developments in natural fibre composites and their mechanical performance. *Composites: Part A* 83: 98–112.

Raj, R.G., B.V. Kokta, D. Maldas, C. Daneault (1988): Use of wood fibers in thermoplastic composites: VI. Isocyanate as a bonding agent for polyethylene-wood fiber composites. *Polymer Composites* 9(6): 404–411.

Rüppel, A., H.-P. Heim, C. Aßmann, R.-U. Giesen (2016): Investigation of the mechanical and olfactory properties of PP with wood and PET fibers for automotive application fields. Presentation, 18th Conference on Odour and Emissions of Plastic Materials, March 7–8, Kassel, Germany.

Quirino, R.L., J. Woodford, R.C. Larock (2012): Soybean and linseed oil-based composites reinforced with wood flour and wood fibers. *Journal of Applied Polymer Science* 124: 1520–1528.

Sain, M., S.H. Park, F. Suhara, S. Law (2004): Flame retardant and mechanical properties of natural fibre-PP composites containing magnesium hydroxide. *Polymer Degradation and Stability* 83: 363–367.

Schirp, A., A. Hellmann, A. Barrio, J. Hidalgo (2016): Development of flame-retarded wood-plastic composites. Presentation at the Applied Market Information (AMI) conference "Wood-Plastic Composites 2016," Vienna, March 7–9.

Schirp, A., S. Su (2016): Effectiveness of pre-treated wood particles and halogen-free flame retardants used in wood-plastic composites. *Polymer Degradation and Stability* 126: 81–92.

Schirp, A., Plinke, B., Napolow, D. (2015): Effectiveness of organic and inorganic pigments for mass colouration of thermo-mechanical pulp used in wood-plastic composites. *European Journal of Wood and Wood Products* 73: 5–16.

Schirp, A., Mannheim, M., Plinke, B. (2014): Influence of refiner fibre quality and fibre modification treatments on properties of injection-moulded beech wood-plastic composites. *Composites: Part A* 61: 245–257.

Schirp, A., Stender, J. (2010): Properties of extruded wood-plastic composites based on refiner wood fibres (TMP fibres) and hemp fibers. *European Journal of Wood and Wood Products* 68: 219–231.

Schirp, A., R.E. Ibach, D.E. Pendleton, M.P. Wolcott (2008): Biological degradation of wood-plastic composites (WPC) and strategies for improving the resistance of WPC against biological decay. In: Schultz, T.P., Militz, H., Freeman, M.H., Goodell, B., Nicholas, D.D. (Eds.): *Development of Commercial Wood Preservatives: Efficacy, Environmental, and Health Issues*. ACS Symposium Series 982, American Chemical Society (ACS), Washington, DC.

Schirp, A., M.P. Wolcott (2005): Influence of fungal decay and moisture absorption on mechanical properties of extruded wood-plastic composites. *Wood and Fiber Science* 37(4): 643–652.

Schneider, M.H., K.I. Brebner (1985): Wood-polymer combinations: The chemical modification of wood by alkoxysilane coupling agents. *Wood Science and Technology* 19: 67–73.

Sears, K., R. Jacobson, D. Caulfield, J. Underwood (2001): Reinforcement of engineering thermoplastics with high purity wood cellulose fibers. Pages 27–34 in: The Sixth International Conference on Woodfiber-Plastic Composites. Forest Products Society, Madison, WI.

Seefeldt, H., U. Braun (2012a): Burning behavior of wood-plastic composite decking boards in end-use conditions: The effects of geometry, material composition, and moisture. *Journal of Fire Sciences* 30(1): 41–54.

Seefeldt, H., U. Braun (2012b): A new flame retardant for wood materials tested in wood-plastic composites. *Macromolecular Materials and Engineering* 297: 814–820.

Seefeldt, H., U. Braun, M. Wagner (2012): Residue stabilization in the fire retardancy of wood-plastic composites: Combination of ammonium polyphosphate, expandable graphite, and red phosphorus. *Macromolecular Chemistry Physics* 213: 2370–2377.

Shibata, M., N. Teramoto, Y. Takada, S. Yoshihara (2010): Preparation and properties of bio-composites composed of glycerol-based epoxy resins, tannic acid, and wood flour. *Journal of Applied Polymer Science* 118: 2998–3004.

Simpson, W., A. TenWolde (1999): Physical properties and moisture relations of wood. Pages 3-1 to 3-24 in *Wood Handbook: Wood as an Engineering Material*. Forest Products Society, Madison, WI.

Sobczak, L., R.W. Lang, A. Haider (2012): Polypropylene composites with natural fibers and wood—General mechanical property profiles. *Composites Science and Technology* 72: 550–557.

Stark, N. (2001): Influence of moisture absorption on mechanical properties of wood flour-polypropylene composites. *Journal of Thermoplastic Composite Materials* 14(5): 421–432.

Stark, N.M., L.M. Matuana (2004a): Surface chemistry and mechanical property changes of of wood-flour/high-density-polyethylene composites after accelerated weathering. *Journal of Applied Polymer Science* 94(6): 2263–2273.

Stark, N.M., L.M. Matuana (2004b): Surface chemistry changes of weathered HDPE/wood-flour composites studied by XPS and FTIR spectroscopy. *Polymer Degradation and Stability* 86(1): 1–9.

Stark, N.M., R.E. Rowlands (2003): Effects of wood fiber characteristics on mechanical properties of wood/polypropylene composites. *Wood and Fiber Science* 35(2): 167–174.

Stark, N.M., R.H. White, S.A. Mueller, T.A. Osswald (2010): Evaluation of various fire retardants for use in wood flour-polyethylene composites. *Polymer Degradation and Stability* 95: 1903–1910.

Suddell, C., W.J. Evans (2005): Natural fiber composites in automotive applications. Pages 231–260 in: *Natural Fibers, Biopolymers, and Biocomposites*, edited by A. K. Mohanty, M. Misra, and L. T. Drzal. CRC Press, Taylor & Francis Group, Boca Raton, FL.

Thumm, A., A.R. Dickson (2013): The influence of fibre length and damage on the mechanical performance of polypropylene/wood pulp composites. *Composites: Part A* 46: 45–52.

Turku, I., T. Kärki (2016): Accelerated weathering of fire-retarded wood-polypropylene composites. *Composites: Part A* 305–312.

Turku, I., M. Nikolaeva, T. Kärki (2014): The effect of fire retardants on the flammability, mechanical properties, and wettability of co-extruded PP-based wood-plastic composites. *BioResources* 9(1): 1539–1551.

Turku, I., T. Kärki (2014): The effect of carbon fibers, glass fibers and nanoclay on wood flour-polypropylene composite properties. *European Journal of Wood and Wood Products* 72: 73–79.

Van den Oever, M.J.A, H.L. Bos (1998). Critical fibre length and apparent interfacial shear strength of single flax fibre polypropylene composites. *Adv Compos Lett* 7(3): 81–85.

Wagenführ, R. (1989): *Anatomie des Holzes* (Wood anatomy; in German). VEB Fachbuchverlag Leipzig. 4th Edition.

Wambua, P., J. Ivens, I. Verpoest (2003): Natural fibres: Can they replace glass in fibre reinforced plastics? *Composites Science and Technology* 63: 1259–1264.

Wang, W., H. Huang, H. Du, H. Wang (2015): Effects of fiber size on short-term creep behavior of wood fiber/HDPE composites. *Polymer Engineering and Science* 55: 693–700.

Winandy, J.E. (2013): State of the art paper: Effects of fire-retardant treatments on chemistry and engineering properties of wood. *Wood and Fiber Science* 45(2): 131–148.

Wittmann, O., F. Wolf (2005): Wood-based materials. *Ullmann's Encyclopedia of Industrial Chemistry.*

Wolcott, M.P. (2001): Wood-plastic composites. Pages 9759–9763 in K. H. J. Buschow, R. W. Cahn, M. C. Flemings, B. Ilschner, E. J. Kramer, and S. Mahajan, Eds. *Encyclopedia of Materials: Science and Technology*. Elsevier, Amsterdam.

Woodhams, R.T., G. Thomas, D.K. Rodgers (1984): Wood fibers as reinforcing fillers for polyolefins. *Polymer Engineering and Science* 24(15): 1166–1171.

Xu, X., K. Jayaraman, C. Morin, N. Pecqueux (2008): Life cycle assessment of wood-fibre-reinforced polypropylene composites. *Journal of Materials Processing Technology* 198: 168–177.

Yeh, S.-K., S. Agarwal, R.K. Gupta (2009): Wood-plastic composites formulated with virgin and recycled ABS. *Composites Science and Technology* 69: 2225–2230.

4 Bio-Based Thermoplastic and Thermosets Polymer

Hans-Josef Endres

CONTENTS

4.1 WORDING

Bioplastics are not a completely new kind of material, but rather a rediscovered class of materials within the big familiar group of materials known as plastics. The first polymer materials synthesized by humans, for example, caseins, gelatine, shellac, celluloid, cellophane, linoleum, rubber, and so on, were all bio-based, that is, based on renewable materials or on transformed natural materials. At that time there were simply no petrochemical materials available. Nowadays these first bio-based plastics have been almost completely displaced from the middle of the last century onward by petrochemical polymer materials apart from a few exceptions (cellulose and rubber-based materials).

Due, in particular, to an increasing awareness among the public, politicians, the industry, and research of ecological aspects as well as the limitations of petrochemical resources and, in part, to successful research and development activities leading to new innovative property profiles, bioplastics are now experiencing a renaissance.

Unfortunately, at the same time, a range of different terms are also in use for bioplastics, such as biopolymers, green plastics, ecoplastics, and so on, and at the same time, there is varying and confusing discussion regarding the correct usage of these terms. The differentiation between bioplastics as useful materials and biopolymers as macromolecular substances is very important in order to achieve a clearly defined terminology and a unified nomenclature in the field of bioplastics.

First, while biopolymers are indeed the basis for the creation of bioplastics, the term "biopolymer" does not describe the resulting material itself. In the vast majority of cases, biopolymers have to be "refined" to varying degrees, that is, modified or through additives (stabilizers, plasticizers, colors, processing agents, fillers, etc.) and blended in order to make "ready-to-use materials" with satisfactory processing and performance properties (see Figure 4.1) like conventional polymers.

Therefore, in this chapter, the term "biopolymer" describes the macromolecules, while the term "bioplastic" represents the materials used in engineering. Also as with conventional plastics, in bioplastics there are now a variety of materials based on the same basic polymer types. For the sake of a unified nomenclature, it is therefore suggested that this differentiation between the initial polymer and the "finished" material also be applied with appropriate consistency when speaking about bioplastics.

The best general definition for the term "bioplastic" is a polymer material which possesses *at least one* of the following properties:

1. Is made from bio-based (renewable) raw materials
    and/or
2. Is biologically degradable

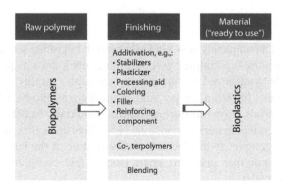

FIGURE 4.1   From raw polymer to bioplastic.

If this definition is adhered to, the following three fundamental bioplastic/biopolymer groups exist:

1. Degradable petrochemical-based bioplastics
2. Degradable (primarily) bio-based bioplastics
3. Nondegradable bio-based bioplastics

Biologically degradable plastics can be based on petrochemical raw materials as well as on renewable raw materials. The degradability of the biopolymer materials is influenced exclusively by the chemical and physical microstructure and not by the origin of the raw materials or carbon used as feedstock. This means that biopolymers need not necessarily be made exclusively from renewable materials. Biologically degradable biopolymers can also be produced on the basis of petrochemical ingredients such as various polyesters, some polyvinyl alcohols, polycaprolactone, polyesteramides, and so on (Figure 4.2, bottom right). Conversely, not all biopolymers based on renewable ingredients are necessarily biologically degradable, for example, highly substituted vulcanized rubber, cellulose acetates, bio-based polyamides or bio-based resins, (Figure 4.2, top left). Typical examples for the group of bio-based and biologically degradable bioplastics (Figure 4.2, top right) are starch-based plastic blends, polylactic acid (PLA) or polyhydroxyalkanoates (PHA).

When speaking of biopolymers or bioplastics, it is imperative that the most precise nomenclature possible is used, that is, it is advisable to speak specifically of degradable or bio-based bioplastics in order to avoid misunderstandings. Degradability here means a functional property or disposal option at the end of the material's lifecycle, irrespective of the origin of the raw materials, while conversely, bio-based describes exclusively the origin of the raw ingredients of the polymer and provides no statement whatsoever regarding its degradability. These two different approaches are still being pursued and form the technical basis for a variety of bioplastics (Figure 4.3).

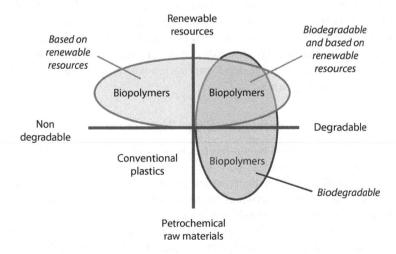

FIGURE 4.2 Bioplastics and the three fundamentally different biopolymer groups. (From Endres, H.-J. and Siebert-Raths, A. *Engineering Biopolymers*. Carl Hanser Verlag, Munich, 2011.)

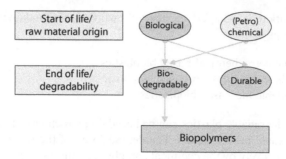

FIGURE 4.3   Raw material basis and degradability of bioplastics. (From Endres, H.-J. and Siebert-Raths, A. *Engineering Biopolymers*. Carl Hanser Verlag, Munich, 2011.)

4.1.1   DEGRADABLE PETROCHEMICAL-BASED BIOPLASTICS

Biopolymers on the basis of petrochemical raw materials are, like conventional plastics, based on the various hydrocarbon monomers and oligomers produced from crude oil, natural gas, or coal through fractionated distillation and targeted cracking processes as well as their derivatives (e.g., polyols, carboxylic acids). Just as in the past, the properties profile of conventional polymers could be varied and adjusted to suit an enormous range of applications through the use of a wide variety of starting monomers, polymerization reactions, process parameters, resulting polymer micro-structures, and additives. The property profile of polymer materials can be further expanded in particular by the inclusion of various heteroatoms in the molecule (primarily oxygen and nitrogen). A significant property that can be influenced by this is degradability. While with conventional plastics the focus in the past has mostly been on durability, that is, a high level of resistance to chemical, microbiological, or other environmentally defined influences, with the degradable, petrochemical-based bio-polymers an appropriate molecule and material design is pursued with the objective of creating a polymer material that is not very resistant to environmental influence. The purpose of this is to achieve a material that, as a result of environmental influences or as part of a targeted (industrial) composting process, can be broken down and achieve the simplest possible depolymerization with a further ultimate degradation of the molecule fragments.

4.1.2   DEGRADABLE (PRIMARILY) BIO-BASED BIOPLASTICS

The renaissance of bio-based bioplastics has been initiated over the last few decades by the second subgroup of these polymer materials, that is, by polymer materials based on renewable raw ingredients which are at the same time compostable (see Figure 4.2, top right). These activities led to the recognition and naming of bioplastics as innovative materials around 30 years ago.

Renewable raw materials that can be used to create degradable polymers and bioplastics include, in particular, oligo and polysaccharides such as cellulose, starch, sugar, and vegetable oils as well as some lignins and proteins and chemical and bio-technological derivatives based on them (e.g., acids and alcohols).

4.1.3 NONDEGRADABLE BIO-BASED BIOPLASTICS

These biopolymer materials are in part materials produced on the basis of renewable raw materials. The final polymer structures are not biodegradable, even though they are based on a bio-based degradable feedstock. In particular in this context there is currently (as yet) no minimum share of bio-based material components for polymer blends and copolymers or terpolymers to be declared as bio-based bioplastics, although in recent years suitable methods have been developed for determining the share of bio-based carbon in bioplastics.

In addition, there also exist—similarly to conventional plastics—many copolymers and terpolymers as well as polymer blends in which a combination of various monomers or a mixture of the various aforementioned biopolymer groups are created in order to optimize the resultant properties.

4.1.4 OLD AND NEW ECONOMY BIOPLASTICS

Around 100 years ago, when petrochemical raw materials were not yet available, bio-based polymers constituted the first polymers and thereby, from today's perspective, biopolymers. These so-called *old economy bioplastics* were based on renewable plant-based raw materials like cellulose, vegetable oils, and natural latex or on animal proteins and fats. Of these old economy bioplastics, the only ones still of economic significance on the plastics market are natural rubber and different cellulose based plastics (regenerated cellulose and cellulose derivatives [cellophane, viscose, celluloid, cellulose acetate, etc.]) as well as linoleum in smaller volumes (see Figure 4.4).

But there are also some *new economy bioplastics* available, which consist of modified renewable feedstock. Among those, the starch-based biopolymers and blends made out of them, which have been researched for around 30 years, occupy a leading role due to the low commodity prices, good availability, and very good degradability.

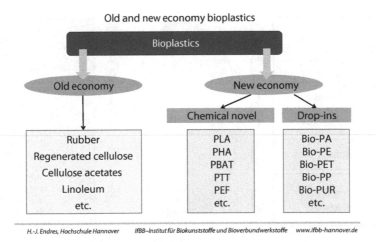

FIGURE 4.4 Traditional (old economy) and novel bioplastics (new economy). (IfBB. Available at www.ifbb-hannover.de.)

Regardless of the source or the provision of feedstock new economy bioplastics are made up of two basic groups: the chemically novel biopolymers, that is, unknown in the field of plastics from a chemical point of view until a few years ago (e.g., starch-based polymers, novel bio-based polyesters such as polylactic acid [PLA] or polybutylene adipate terephthalate [PBAT]), and so-called *"drop-ins"* which are identical in chemical structure but partially or completely bio-based plastics. Alongside these, work is currently being carried out on further drop-ins including in the field of thermoset (e.g., bio-based EP resins) or elastomer polymer materials (e.g., bio-based EPDM or bio-based polyurethanes).

In the current developments in bioplastics for technical utilization like automotive applications, bio-based durable materials, that is, the long-term availability of raw materials, are becoming of concern. Within this group of nondegradable, bio-based bioplastics, one development that has made strong and technically very successful advances in the last 5 to 10 years in particular is what is known as drop-in solutions. These are, simply put, an effort to retain the established final chemical structures as well as the methods of synthesis based on petrochemical raw materials while substituting as completely as possible petrochemical feedstock by biogenous raw materials. Due to the identical chemical structure, with the same additivities the drop-ins have completely the same properties profiles as their petrochemical equivalents. This means that when conventional plastics are substituted by the respective drop-ins, no changes are to be anticipated in the areas of processing, usage, and recovery or recycling. Examples of these are fully bio-based polyethylene (bio-PE), partly bio-based polyethylene terephthalate (bio-PET), fully or partly bio-based polyamides (bio-PA), polyurethanes (bio-PUR), and some thermosets based on a variety of renewable raw materials or their bio-based derivatives.

Depending on the perspective, this means that there are a number of different types of bioplastics at the end (see Figure 4.5). In order to avoid misunderstandings,

Bioplastics

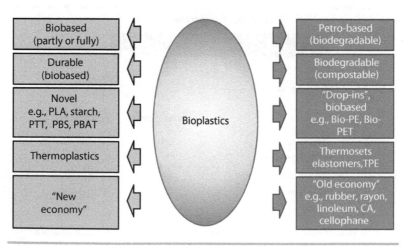

Hochschule Hannover   *IfBB–Institute for Bioplastics and Biocomposites–www.ifbb-hannover.de*

FIGURE 4.5   Various types of bioplastics. (IfBB. Available at www.ifbb-hannover.de.)

bioplastics should therefore generally not be mentioned without further specifying, through additional information, which group is meant.

Within the next section, only the new economy bioplastics are considered in more detail because the old economy bioplastics are already sufficiently described in the literature. Under these new economy bioplastics in particular the long-term resistant materials are considered for technical applications.

4.2 PROCESS TECHNOLOGIES

As already demonstrated, biopolymers can, in principle, be based on both biogenic and petrochemical feedstocks. For the production of biopolymers, the following fundamental production methods therefore apply (Figure 4.6):

1. Chemical synthesis of petrochemical raw materials
2. Chemical synthesis of bio-technologically manufactured polymer feedstock
3. Direct biosynthesis of polymers
4. Modification and/or polymerization of native or modified molecular, renewable feedstock
5. Production of blends and co-/terpolymers from these groups

In Table 4.1, the most important associated bioplastics are assigned to these various process routes.

Further possibilities for "new" materials are co- and terpolymers or blends using the mono- or polymers described in Table 4.1. This is shown in Table 4.2.

That means that there are two main pathways to produce durable bio-based plastics, chemical synthesis and polymerization of bio-technologically manufactured

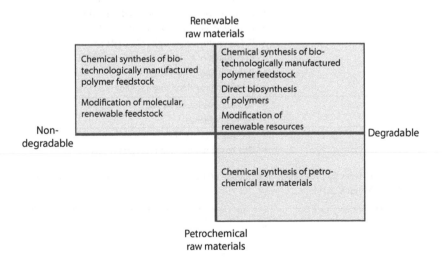

FIGURE 4.6 Synthesis routes in biopolymers. (From Endres, H.-J. and Siebert-Raths, A. *Engineering Biopolymers*. Carl Hanser Verlag, Munich, 2011.)

TABLE 4.1
Production Routes for Various Biopolymers

Synthesis Process	Examples for Biopolymers/Bioplastics
Chemical synthesis and polymerization of petrochemical raw materials	• Polyesters • Polyester amides • Polyester urethanes • Polyvinyl alcohols (PVOH) • Polycaprolactone (PCL)
Chemical synthesis and polymerization of bio-technologically manufactured polymer feedstock	• Polyethylene (Bio-PE) • Polyester (e.g., PLA, Bio-PET) • Polyamide (e.g., bio-based PA 11)
Direct biosynthesis of polymers	• Polyhydroxyalkanoates (e.g., PHB)
Modification of molecular, renewable feedstock	• Cellulose regenerates • Starch derivates • Cellulose derivates (e.g., CA)

Source: Modified from Endres, H.-J. and Siebert-Raths, A. *Engineering Biopolymers.* Carl Hanser Verlag, Munich, 2011.

TABLE 4.2
Examples for Co- and Terpolymers and Blends

Co-/terpolymers	• Polybutylene terephthalate (Bio-PBT) • Polybutylene succinate (PBS) • Polybutylene adipate terephthalate (PBAT) • Polybutylene succinate terephthalate (PBST) • Polyethylene terephthalate (Bio-PET) • Polytrimethylene terephthalate (PTT) • Polyamide (e.g., Bio-PA 4.10, 6.10, 10.10) • Polyurethane (Bio-PUR) • Ethylene propylene diene rubber (EPDM)
Blends	• Starch or cellulose blends • Polyester blends
Thermosets	• Bio-based epoxy resins • Bio-based unsaturated polyester resins

Source: Modified from Endres, H.-J. and Siebert-Raths, A. *Engineering Biopolymers.* Carl Hanser Verlag, Munich, 2011.

mono- or oligomers as polymer feedstock and modification or polymerization of native or modified molecular, renewable feedstock.

4.3 MATERIALS

4.3.1 BIO-BASED POLYESTERS

In most cases, bio-based polyesters are manufactured from a bio-based diol and a dicarboxylic acid or from an ester generated from the diacids. There are a lot of different possibilities to produce bio-based alcohols or acids (Li Shen et al. 2009). Therefore, there are as many different (co-) or (ter-) polyesters with a wide range of properties. If terephthalic acid or dimethyl terephthalate are used as acid components, the resulting polyalkylene terephthalates are aliphatic-aromatic polyesters. By contrast, the polyesters made from aliphatic, petro-, or bio-based dicarboxylic acids and diols are entirely aliphatic bio-polyesters.

4.3.1.1 Bio-Based Polyethylene Terephthalate (Bio-PET)

For the group of biopolymers and for all future new economy bioplastics as well as for the bio-based polyesters, partially bio-based polyethylene terephthalate (Bio-PET) is the most important bioplastic material economically. Its bio-based basis is supplied by bioethanol, which is thus far mainly produced from sugarcane or corn starch (see Figure 4.7). In a series of chemical reactions, it is converted to bio-based monoethylene glycol (MEG) as alcohol component. The next step is to start the transesterification (also known from conventional PET) with the petro-based terephthalate acid (PTA) to produce Bio-PET.

In this case, the percentage of bio-based feedstock is 30 wt-% (therefore, named as Bio-PET 30). In the final product Bio-PET 30, however, only about 23% of the carbon is bio-based, due to the differing portions of carbon in the two polymer components as feedstock.

Research is currently underway to develop entirely, that is, 100% bio-based PET (Bio-PET 100). This means that the alcohol component as well as the aromatic acid component are fully bio-based (see Figure 4.8).

Because bio-based PET is a chemical identical material as the conventional PET in Bio-PET, the same characteristics can be expected. Therefore, Bio-PET can also be used to produce fibers therefrom.

4.3.1.2 Polytrimethylene Terephthalate (PTT)

In addition to Bio-PET, aliphatic-aromatic bio-copolyesters and bio-terpolyesters with a structure similar to PET have been developed, in particular to further optimize mechanical and thermal properties for engineering applications. If, for example, potentially bio-based 1,3-propanediol (bio-PDO) is combined with the petrochemical based aromatic components terephthalic acid or terephthalic acid dimethylester, the result is the partially bio-based copolyester polypropylene terephthalate (PPT), often designated as polytrimethylene terephthalate (PTT) (see Figure 4.9), PTMT (also for polytrimethylene terephthalate), PPT (polypropylene terephthalate), 3GT, or Sorona™. In the presence of a PDO excess, methanol and DMT or water and PTA of low molecular weight are

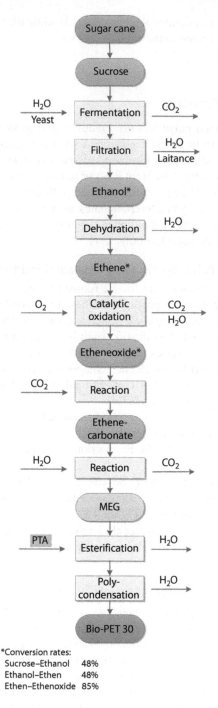

FIGURE 4.7   Process routes and material flows in Bio-PET 30 (i.e., 30% bio-based) production. (From IfBB 2015. *Biopolymers—Facts and Statistics*. Edition 2. Available at www.downloads.ifbb-hannover.de. Last updated on September 9, 2013.)

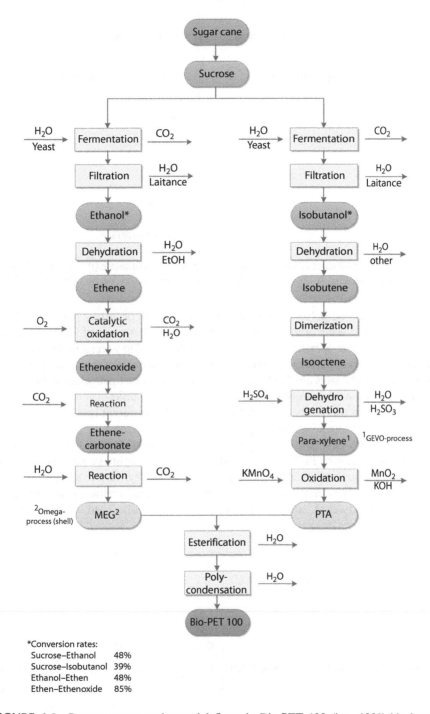

FIGURE 4.8 Process routes and material flows in Bio-PET 100 (i.e., 100% bio-based) production. (From IfBB 2015. *Biopolymers—Facts and Statistics*. Edition 2. Available at www.downloads.ifbb-hannover.de. Last updated on September 9, 2013.)

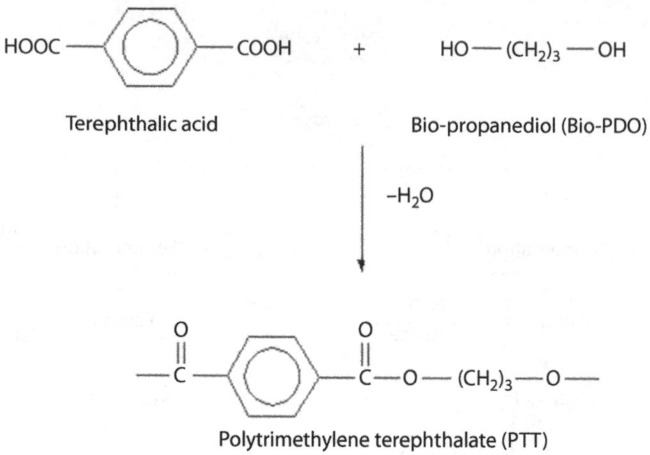

HOOC —⟨O⟩— COOH   +   HO — (CH$_2$)$_3$ — OH

Terephthalic acid   Bio-propanediol (Bio-PDO)

−H$_2$O

$$\begin{array}{c} O \\ \| \\ \end{array}$$
—C—⟨O⟩—C—O — (CH$_2$)$_3$—O —

Polytrimethylene terephthalate (PTT)

FIGURE 4.9 Copolyester synthesis of polytrimethylene terephthalate (PTT). (From Endres, H.-J. and Siebert-Raths, A. *Engineering Biopolymers*. Carl Hanser Verlag, Munich, 2011.)

removed. In a second polycondensation step, chain growth occurs by removal of PDO and the remaining water or methanol. The removal of the last traces of PDO takes place in a series of reactors operating under high temperatures and low pressure. The bio-PDO also enables the manufacturing of other biopolyetser. The processing and use properties of these bio-copolyesters are fundamentally similar to those of petrochemical PET and PBT, depending on the monomers used (Albertsson 1995, Guillet et al. 1995).

Figure 4.10 provides a comparative overview of the different chemical structures of these two aliphatic aromatic bio-polyesters.

PTT has a very desirable property profile as an engineering thermoplastic. It combines the properties of PET, that is, rigidity, strength, toughness, heat, static resistance, and chemical resistance with the good processability, surface appearance, and gloss of the poly(butylene terephthalate) (PBT) or the good resiliency and wearability of nylon (Brady et al. 2008, Shen et al. 2009). PTT is also used to produce fibers

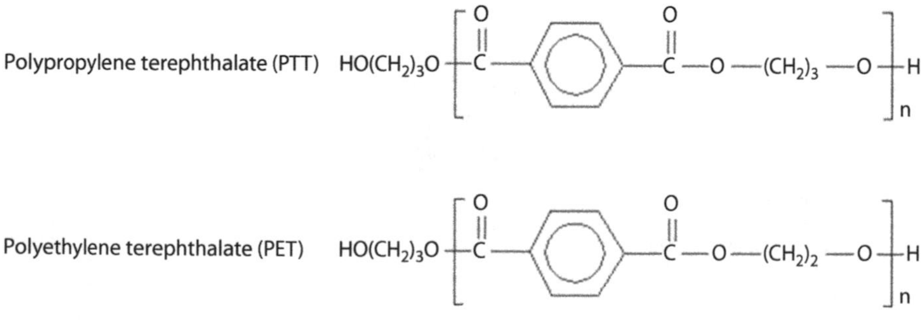

Polypropylene terephthalate (PTT)   HO(CH$_2$)$_3$O —[C—⟨O⟩—C —O —(CH$_2$)$_3$ —O]$_n$ H

Polyethylene terephthalate (PET)   HO(CH$_2$)$_3$O —[C—⟨O⟩—C —O —(CH$_2$)$_2$ —O]$_n$ H

FIGURE 4.10 Polypropylene terephthalate (PPT) or polytrimethylene terephthalate structure (PTT) in comparison to PET. (From Endres, H.-J. and Siebert-Raths, A. *Engineering Biopolymers*. Carl Hanser Verlag, Munich, 2011.)

which offer the softness and stain resistance suitable for automotive carpet, fabrics, door trims, seat covers, or other plastic parts.

4.3.1.3 Polylactide (PLA)

Beside aromatic polyesters such as PET or PTT, there also exist aliphatic polyesters within the group of biopolymers that are synthesized in part or completely from materials produced by means of biotechnology. These esters are generally not as durable and are therefore preferably used in the packaging industry or for no long-life and degradable products. In terms of volume, currently one of the most important biopolymers in this group is polylactide (PLA) based on lactic acid. Lactic acid (2-hydroxypropinic acid) is a ubiquitous, natural acid that occurs in two optically active forms, L(+) and D(−) lactic acid. In addition to its use as a building block for biopolymers, it is used especially as an acidifier, as flavoring, and as a preservative in the food, textiles, leather, and pharmaceutical industries, and as a basic material for synthesizing a number of additional chemicals, such as acetaldehyde. Worldwide approximately 70%–90% of lactic acid is manufactured by fermentation (Chen et al. 2005, Ehrenstein et al. 2015, Ramakrishna et al. 2004).

The production of PLA is connected to multiple process steps. Figure 4.11 shows the main process or conversation steps from the bio-based feedstock to PLA.

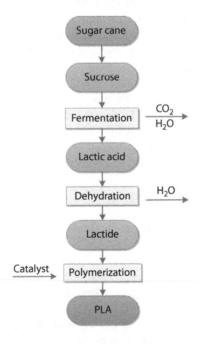

*Conversion rates:
Sucrose–Lactic acid 85%

FIGURE 4.11 Process route of PLA based on sugar cane. (From IfBB 2015. *Biopolymers— Facts and Statistics*. Edition 2. Available at www.downloads.ifbb-hannover.de. Last updated on April 9, 2013.)

By optimizing processing technology and increasing output (scaling effects), the price of PLA was reduced from what was originally considerably more than 10 €/kg to between 1.5 and 2.0 €/kg in the last 20 years (Endres et al. 2011). Further significant reductions in the manufacturing cost seem possible in the future, especially when raw materials costs are reduced, that is, by the use of biogenic residues or wastes, such as whey, molasses, or wastes containing ligno-cellulose. However, there has been less documented experience with these materials compared to the use of glucose or starch containing substrates.

There are a number of microorganisms capable of generating lactic acid. Especially Gram-positive, non-spore-forming, facultative anaerobic homo- and heterofermentative lactic acid bacteria are used for the industrial production of lactic acid. Fermentative lactic acid production creates specific optically active forms of lactic acid. The less productive homo-fermentative lactic acid bacteria generate only L(+) lactic acid as a fermentation product. By contrast, heterofermentative lactobacteria generate a racemic mixture of L and D lactic acid in which the D content dominates. The ratio of L to D lactic acid essentially depends on the bacterial culture itself, its age, as well as on the pH value (Bastioli 2005, Fritz et al 1994, Jacobsen 2000, Kaplan 1998). The generated PLA microstructure (conformation) and the resulting product quality (crystallinity, mechanical characteristics, T_g) as well as the final properties can be influenced by the expensive generation of pure monomers and/or dimers or by purification of the racemic mixtures as base monomers (Siebert-Raths et al. 2012). As with conventional polymers, increasing degrees of polymerization and increasing crystallinity of PLAs always lead to increased strength, internal and elastic deformation resistance, and glass transition and melting temperature.

The properties of PLA also show a wide range and can be tailored by the ratio of lactic acid isomers (L and D lactic acid) used, by blend components and additives compounded into PLA, and by molecular weight and crystallinity. PLA is therefore manufactured with differenced processes like injection molding, foaming, film or fiber production and it is suitable for both durable and degradable applications.

Essential advantageous properties of PLA are (Endres et al. 2011):

- Property ranges depending on the ratio of isomers used and variable molecular weight
- High modulus of elasticity
- High scratch resistance
- High transparency (low degree of cristallinity), low haze, and high gloss
- Good dyeability
- High surface energy, i.e., very good printability and easy to metallize
- Good odor and flavor barrier properties
- Oil, fat, water, and alcohol resistance
- UV resistance
- Good contour accuracy

However, some properties of PLA are disadvantageous, especially for engineering applications:

- Relatively strong hydrophilic and water vapor permeable
- Poor carbon dioxide barrier
- Moderate oxygen barrier
- Requires sophisticated engineering for injection molding processing
  - Slow crystallization when injection molded (relatively long cycle times)
  - Hot-runner advisable
  - Good purging required (no mixing with other polymers, such as PET)
  - Tends to hydrolyze during processing
  - Good predrying required
  - Machine dwell times as short as possible
- Brittle without additives (glass transition temperature above 50–55°C)
- Low heat resistance, that is, low softening temperature
- Low resistance to solvents, acids, and bases
- Only degradable at elevated temperatures (above 60°C), that is, not home compostable

Currently, intensive research effort is spent to optimize PLA properties and to overcome the obstacles to engineering applications. PLA is being blended typically with other polyesters to increase its impact strength and PLA-based copolymers are being developed. Also, different companies offer modifiers for PLA.

PLA materials feature especially high elastic distortion resistance under tensile or bending load. They have an inherently high modulus of elasticity that can be increased even further by increasing crystallinity (e.g., due to a higher racemic purity in monomers or the use of talc as a nucleation agent), drawing, and fiber reinforcement. Currently, there are first natural fiber-reinforced PLA grades for increasing heat resistance temperature available. Here, the greater chemical compatibility of polar PLA with cellulose-based natural fibers (compared with nonpolar polypropylene or polyethylene) is utilized.

There are more fully or partly bio-based polyesters, but these do not have any great significance in the field of biocomposite materials.

4.3.2 BIO-BASED POLYETHYLENE (BIO-PE)

Bio-based PE is a drop-in; as with conventional PE, for bio-PE, the synthesis conditions of the polymer formation reactions (temperature, pressure, monomer concentration, catalysts, inhibitors, etc.) also ultimately determine the resulting microstructure and thereby the macroscopic properties. As expected, the properties of the bio-PE can also, through further measures such as co-monomers, additives, blending, and cross-linking, be configured in exactly the same way as is known from conventional PE. The only significant difference between conventional and bio-based polyethylene lies in the feedstock or the process route. More specifically, the process routes differ from one another only as far as the source of the bio-based ethylene compared to the petrochemical variant.

Depending on the selected raw material, for one tonne of bio-PE, between 0.48 and 3.1 hectares are required (see Figure 4.12). Due to the higher starch and sugar yields for sugar cane in the sugar plants and corn starch in the starches, these demonstrate also the highest land-use efficiency. In principle, the values for the land requirements for bio-PE are slightly higher than for PLA because with bio-PE, the oxygen as an integral part of the polysaccharide starting product is not present in the molecular structure.

During the development and market introduction of Bio-PE was successful, other polyolefins such as a bio-based PP are still in an early development stage.

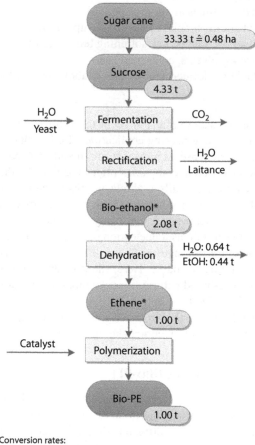

*Conversion rates:
Sucrose–Ethanol 48%
Ethanol–Ethene 48%
(Conventional technology)

FIGURE 4.12 Process routes of bio-based polyethylene (Bio-PE). (From Endres, H.-J., Siebert-Rath, A., Behnsen, H., and Schulz, C. *Biopolymers—Facts and Statistics*, Hannover 2016, ISSN: 2363-8559; IfBB. Available at www.downloads.ifbb-hannover.de. Last updated on April 9, 2013.) *(Continued)*

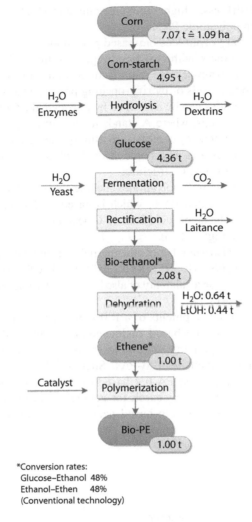

*Conversion rates:
Glucose–Ethanol 48%
Ethanol–Ethen   48%
(Conventional technology)

FIGURE 4.12 (CONTINUED)   Process routes of bio-based polyethylene (Bio-PE). (From IfBB 2015. *Biopolymers—Facts and Statistics*. Edition 2. Available at www.downloads.ifbb -hannover.de. Last updated on August 26, 2013.)

4.3.3   Bio-Based Polyamides (Bio-PA)

There are a number of different bio-polyamides known, both drop-ins and new poly-amides. Bio-based polyamides offer a wide range of properties and with regard to the processing and utilization properties they are fully comparable to conventional polyamides. Exactly as in conventional polyamides also in bio-based polyamides, intermolecular interactions (that determine properties) decrease with an increasing number of methyl groups or with an increase in the repeating $CH_2/CONH$ ratio, due to the accompanying decreasing polarity. As with conventional polyamides, bio-polyamide PA6 therefore has a higher density, higher water absorption, significantly higher strength and stiffness, a higher melting point, and a higher continuous operating temperature than polyamide PA11.

Also as conventional polyamides bio-based polyamides can be further classified into homo- and copolymers with two basic repeating units (diamines and dicarboxylic acids). Homopolyamides are manufactured either by polycondensation of bio-based aminocarbonic acids or by ring-opening polymerization of cyclic amides (lactams). Homopolyamides can be described by a so-called ACAC structure on the basis of their reactive groups, where A stands for the amino groups and C for the carboxyl groups. Copolymers, by contrast, are usually manufactured by polycondensation of various diamines and dicarbonic acids. Looking at their reactive groups, the copolyamides have an AACC structure.

The fully bio-based (homo-) polyamides include PA 11, which is based on castor oil or undecanoic acid, and PA 6, which is based on fermentatively produced ε-caprolactam (6-amino hexanoic acid lactam, 6-hexan lactam, azepan-2-one) as an initial raw material (see Figure 4.13).

In addition, there are also a number of other partially or fully bio-based (co-) polyamides such as PA 4/10, 6/6, 6/9, 6/10, 6/12, 10/10, and 10/12. The PA 10/10 is available as a 100% bio-based PA due to the availability of also bio-based 1,10-decamethylenediamine (DMDA) derived from castor oil

Another example of developments of a 100% bio-based PA is PA 10/12, based on bio-based DMDA and bio-based dodecanedioic acid or PA 5/10 based on DMDA and 1,5-diaminopentane via lysine from glucose fermentation. These polyamides are rather expensive Bio-PAs. Successful research has also been done on generating PA 4/4 and PA 6/4 based on fermentatively generated succinic acid. Further progress in research for generating biopolyamide was made on PA 6/9, where HMDA and bio-based azelaic acid represent the two basic components.

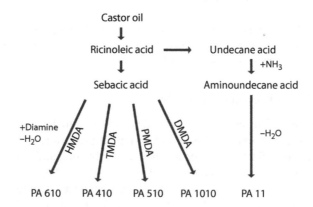

FIGURE 4.13 Bio-based polyamides derived from castor oil. (From Endres, H.-J. and Siebert-Raths, A. *Engineering Biopolymers*. Carl Hanser Verlag, Munich, 2011.)

4.3.4   Bio-Based Polyurethanes (Bio-PUR)

Polyurethanes with their typical repeating urethane linkages [-NH-CO-O-] in their backbone have been known since the early 1950s. They are generally produced by polyaddition of bi- or multivalent alcohols with di- or polyfunctional aromatic or aliphatic isocyanates resulting in the formation of linear, branched, or cross-linked polymers. The resulting microstructure, and with it the macroscopic property profile, can be varied over a wide range by selection of functional groups, the particular stoichiometric proportions of precursors, alcohol valence, as well as the targeted use of catalysts, chain extenders, blowing agents, surfactants, and fillers.

Typically, both monomer components are based on petrochemical feedstock. The most important isocyanates, especially for PUR foams, are petro-based aromatic toluene diisocyanate (TDI) and methylene diphenyl isocyanate (MDI). Most commonly used aliphatic isocyanates are petro-based hexamethylene diisocyanate (HDI) and polymeric isocyanates (PMDI).

The development of bio-based polyurethanes dealt strongly with developments in "natural-oil polyols," or NOPs, which are produced from renewable raw materials such as soybean, palm, sunflower, rapeseed, euphorbia, or castor oil. These NOPs can be used to decrease the petrochemical content of PUR formulations. While the diols or polyols of bio-PUR are bio-based, thus far the di- or polyfunctional aromatic (e.g., MDI or TDI) or aliphatic isocyanates (HDI) are still petro-based (see Figure 4.14). To produce polyols from bio-based resources, there are three approaches: producing polyether (e.g., sucrose or sorbitol) or polyester polyols (e.g., ethylene glycol or glycerol) or producing oleochemical polyols from vegetable oils (Härkönen et al. 1995). To prepare polyols from vegetable oils there are four methods: oxidation and epoxidation, esterification, hydroformylation, and ozonolysis (Endres and Siebert-Raths 2011).

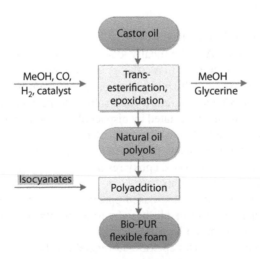

FIGURE 4.14 Bio-based polyurethanes derived from castor oil. (From Endres, H.-J., Siebert-Raths, A., Behnsen, H., and Schulz C. *Biopolymers—Facts and Statistics*, Hannover 2016, ISSN: 2363-8559.; IfBB. Available at www.downloads.ifbb-hannover.de. Last updated on April 9, 2013.)

As conventional PUR resins, bio-PURs offer a large range of chemical linear or cross-linked structures and they are extremely versatile plastics ranging from elastomers, thermosets, and thermoplastic materials, for example, for rigid or flexible foams, coatings, glues, fibers, or insulation. Nowadays first bio-based PURs are commercially available. Based on bio-polyols they have a biocontent of 40–95 wt-%.

4.3.5 Bio-Based Thermosetting Resins

Approximately 15% of all polymers produced are thermosets. This is one reason why R&D on bioplastics has primarily addressed thermoplastics. Beside Bio-PUR, epoxy and unsaturated resins are the most significant partly bio-based materials in the field of thermosettings (Chattopadhyay et al. 2007, Akesson et al. 2010, Raquez et al. 2010). Another reason is probably because resins consist of various reaction components that need to be created first bio-based and the material is finally formed during product manufacture.

4.3.5.1 Bio-Based Epoxy Resins

Today, approximately 75% of all "epoxies" are derived from diglycidyl ether (DGEBA). The so-called DGEBA is based on the two monomers epoxide epichloro-hydrin and bisphenol A. Traditionally these two monomers have been derived from petrochemical sources. The conventional petrochemical process of producing epi-chlorohydrin is the chlorohydrination of allyl chloride. The petro-based allyl chloride in turn is made by chlorination of propylene. But now, there are three general opportunities for obtaining partly bio-based epoxy resins or to increase the bio-based content in the resins:

1. Production of bio-based DGEBA

   Bio-based DGEBA is chemically identical to the petrochemical one and there is hence no difference in product properties, but it is derived from bio-based epichlorohydrin, which in turn is produced from bio-based glycerol. Bio-based DGEBA can therefore fully substitute petrochemical DGEBA. Due to DGEBA being dominated by bisphenol A (epichlorohydrin accounts only for approximately 20% of the molecular weight), the biocontent in bio-DGEBA is not high and using bio-based DGEBA results only in a low bio-based content in the bio-based epoxy resins.

2. Using bio-based curing agents

   Another way to obtain bio-based epoxy resins or to increase the bio-based content is the use of bio-based curing agents in combination with conventional or bio-based DGEBA.

3. Blending epoxy resins with epoxidized vegetable oils

   The third way to increase the biocontent in epoxy resins is to blend epoxidized vegetable oils with (bio-based) epoxy resins.

Especially when bio-based epoxy resins are used as a matrix in composites reinforced with bio-based fibers like flax or hemp fabrics, the bio-based content can be further increased. The development, the production, and the characterization of these composites are part of biocomposite research and are described in Miyagawa et al. (2004), Mehta et al. (2004), Mohanty et al. (2005), and Campaner et al. (2010).

4.3.5.2 Bio-Based Unsaturated Polyester Resins

Unsaturated polyester resins (UPRs) are produced by polycondensation of unsaturated and saturated dicarboxylic acids with polyols. Traditionally as unsaturated acid components, maleic anhydride and fumaric acid are primarily used. As saturated dicarboxylic acid, phthalic acid is used in all standard unsaturated resins (adipic acid and other compounds for special grades). The most widely used diol for standard unsaturated polyester resins is petro-based propylene glycol (also referred to as 1,2-propanediol). To form the final thermoset resin, these UPR components are dissolved in a vinyl monomer (usually styrene), which reacts with the unsaturated double bonds of the polyester and provides the cross-linked thermoset network.

While all these compounds are of petrochemical origin thus far, the development of bio-based polyols also enables bio-based UPRs. Examples of bio-based UPR diols are propylene glycol from glycerol or bio-based 1,3-propanediol (so called Bio-PDO and also used for manufacturing PTT) replacing petro-based 1,2-propanedial (Frattini 2008).

4.4 PROPERTIES

During the course of these development stages, various biopolymers as well as bioplastic material types based on these biopolymers have been developed which exhibit widely differing property profiles. The range of bioplastics thereby ranges—similar to conventional plastics—from less-expensive bioplastics which are produced in large quantities through to high-quality and higher-priced materials for technical applications (see Figure 4.15). Pyramid compares the material performance of the most important bio-based and biodegradable bioplastics and shows the range of materials now available.

Similar to conventional bulk plastics, economically priced bioplastics are now available with prices of around 2 €/kg. Bio-based polyamides, however, are higher-priced materials. The prices for bio-based polyamides are currently still 20%–50% higher than the prices for conventional polyamides. These bio-based polyamide materials, however, partially offer innovative or specifically better property profiles. As regards drop-ins, such as bio-based PE or PET, the technical properties of the bioplastics are identical with those of their petrochemical equivalents. In terms of price they are, due to the currently modest production scales, approximately 20%–30% more expensive than their petrochemical equivalents. Drop-ins can currently therefore only be marketed on the strength of their sustainability and partly or fully bio-based feedstock.

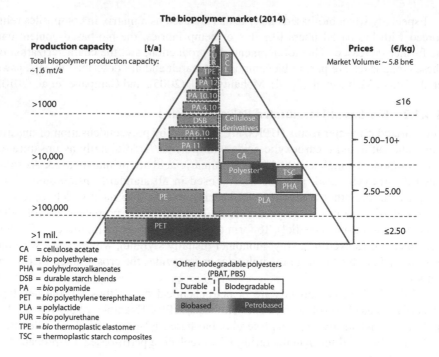

FIGURE 4.15 Production capacities and prices for various bioplastics. (From Endres, H.-J., Jürgens, F., Habermann, C., Spierling, S., Behnsen, H., and Schulz, C. *Eine nachhaltige Alternative? Über Sinn und Unsinn des Einsatzes von Biokunststoffen.* Kunststoffe 7/2015, S. 22–27.; IfBB. Available at www.downloads.ifbb-hannover.de.)

As regards the thermal properties (see Figure 4.16), biopolyamides are durable bioplastics with a higher thermal resistance; there are, however, currently no bioplastics that achieve the level of high-temperature-resistant conventional plastics such as PEEK (HDT/B = 240°C) or PPS (HDT/B = 215°C). However, it must be thereby noted that the developments in bioplastics until now have not been targeted at their use in high-temperature ranges.

As regards the mechanical properties (see Figure 4.17), bioplastics now also cover a greater range, but there is still a need for further optimization. To further optimize the thermal and mechanical property profiles of bioplastics with regard to technical applications, the known methods used in conventional plastics, such as the production of specific blends, reinforcement with fibers, or cross-linking, can and should be used. Drop-ins offer in most cases with the same additives also the same processing, utilization, and disposal features as their petro-chemical counterparts.

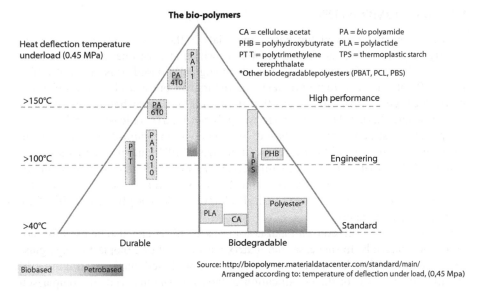

FIGURE 4.16   Heat distortion temperatures (HDT/B) of various bioplastics. (From Endres, H.-J., Jürgens, F., Habermann, C., Spierling, S., Behnsen, H., and Schulz, C. *Eine nachhaltige Alternative? Über Sinn und Unsinn des Einsatzes von Biokunststoffen.* Kunststoffe 7/2015, S. 22–27.)

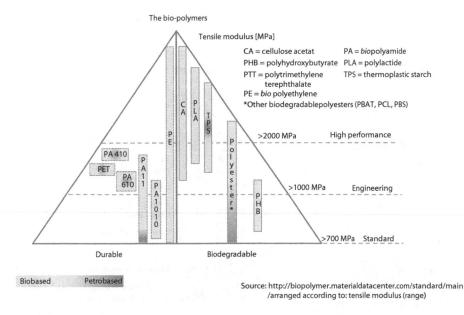

FIGURE 4.17   Tensile modulus of various bioplastics. (From Endres, H.-J., Jürgens, F., Habermann, C., Spierling, S., Behnsen, H., and Schulz, C. *Eine nachhaltige Alternative? Über Sinn und Unsinn des Einsatzes von Biokunststoffen.*)

4.5 BIOCOMPOSITES

Another aspiring group of bio-based materials is the biocomposites. This term usually describes fiber-reinforced plastics in which at least one material component (matrix or reinforcing component) is biologically based or is made of a bioplastic. Bioplastics can therefore serve as the matrix but can also be reinforcing fibers (see Figure 4.18). This means that in the case of a petrochemical-based non-biodegradable thermoplastic or thermoset polymer matrix, at least the reinforcing component must be biologically based. Well-known biocomposite materials from this group are the natural fiber reinforced plastics (NFRP) and wood plastics composites (WPC), that is, polyolefins reinforced or filled with natural plant fibers or wood fibers and wood dust. Superior bio-based synthetically produced reinforcing fibers, that is, bioplastic fibers (e.g., PLA [polylactic acid] Bio-PA or Bio-PET fibers), could also be used.

Conversely, however, conventional non-bio-based fibers can also be used for biocomposite materials. In this case, the matrix must consist of a bioplastic (e.g., glass fiber-reinforced Bio-PA or carbon fiber-reinforced bio-based duromers). Figure 4.19 provides an overview of the classification of biocomposite materials in comparison with conventional composite materials.

Alternatively, both components can, of course, have a bio-based origin, such as wood fiber-reinforced polylactide, viscose fiber-reinforced Bio-PA, or components made from natural-fiber weaves and bio-based resins. Faruk et al. (2012) give an overview on the development of natural fiber reinforced composites.

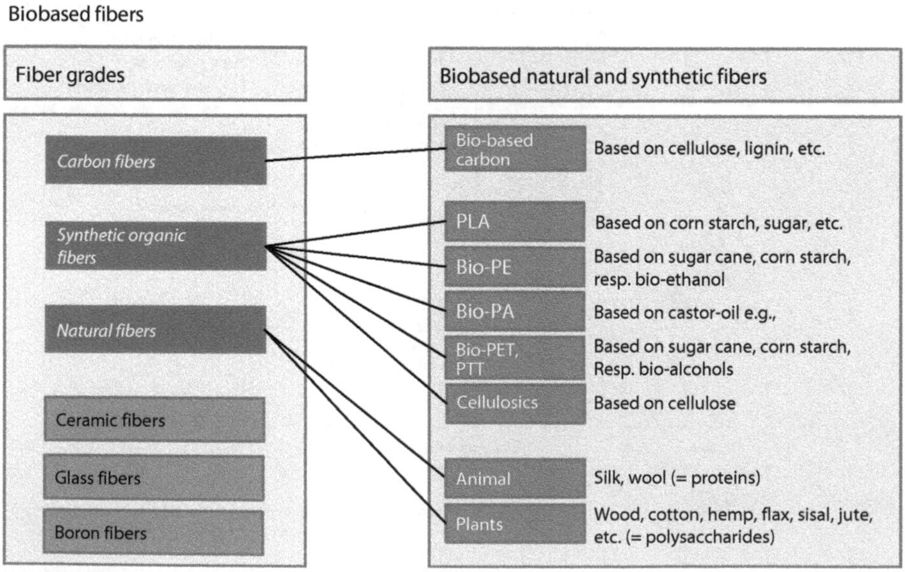

FIGURE 4.18 Bioplastics and bio-based fibers. (From Endres, H.-J., Koplin, T., and Habermann, C. *Technology and Nature Combined*. Kunststoffe International, 2012.)

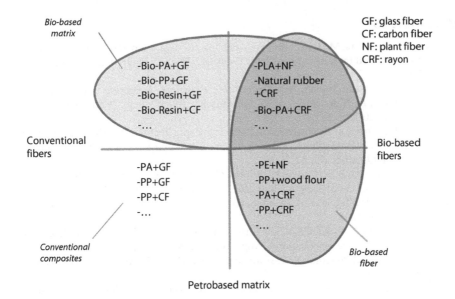

FIGURE 4.19 Biocomposite materials. (From Endres, H.-J., Koplin, T., and Habermann, C. *Technology and Nature Combined.* Kunststoffe International, 2012.)

4.6 CONCLUSIONS

The biopolymer group of materials is not an entirely new type of material. Instead they are innovative polymer materials within the well-known class of plastics materials.

Bioplastics are now experiencing a renaissance: this is due, in particular, to ecological aspects as well as the limitations of petrochemical resources and, in part, to innovative property profiles. This is combined with an increasing awareness among the public, politicians, the industry and, in particular, research and development.

In the early 1980s, the newly developed biopolymers went through a euphoric phase as the future polymer materials independent of crude oil. However, since the materials properties were still unproven and the price to performance ratio of this first generation of biopolymers was sobering, the euphoria soon cooled off and was followed by the further development and/or optimization of the innovative biopolymer materials. In recent years, what is now the further developed biopolymer has meanwhile experienced dynamic, annual double-digit growth.

In terms of materials development, biopolymers are still in their early phase. Future materials developments will, as they did with conventional plastics, not only concentrate on new monomers or innovative polymers, but also increasingly on the further development of existing polymers by generative co- and terpolymers, blending, reinforcement, additivizing, cross-linking, and so on. To this end, the extensive existing experience in the field of conventional plastics can and should definitely be reverted. The relationships between microstructural composition and macroscopic processing, application and disposal properties apply in exactly the same way for biopolymers as they have always applied for conventional plastics.

REFERENCES

Akesson, D., Skrifvars, M., Lv, S., Shi, W., Adekunle, K. et al. Preperation of nanocomposites from biobased thermoset resins by UV-curing, *Progress in Organic Coatings*, 67, 281, 2010.

Albertsson, A.-C. and Huang, S.J. *Degradable Polymers, Recycling and Plastic Waste Management*, New York, 1995. Technische Biopolymere. Carl Hanser Verlag, München, 2009.

Bastioli, C. *Handbook of Biodegradable Polymers*. Rapra Technology, Shrewsbury, 2005.

Brady, M. and Brady P. Handbook of Bioplastics and Biocomposites Engineering Applications, *Reinforced Plastics*, 52, 37, 2008.

Campaner, P., Amico, D.D., Longo, L., Stifani, C., and Tarzia, A. Innovative bio-based composites for automotive applications. *JEC Composite Magazine*, (56), 46, 2010.

Chen, P. Molecular Interfacial Phenomena of Polymers and Biopolymers. Woodhead, Cambridge, 2005.

Chattopadhyay, D.K. and Raju, K.V.S.N. Structural Engineering of Polyurethane Coatings for High Performance Applications, *Progress in Polymer Sciences*, 32, 352, 2007.

Ehrenstein, G.W. and Pongratz, S. *Resistance and Stability of Polymers*, Carl Hanser Verlag, München, 2015.

Endres, H.-J. and Siebert-Raths, A. *Engineering Biopolymers*. Carl Hanser Verlag, Munich, 2011.

Endres, H.-J., Koplin, T., and Habermann, C. *Technology and Nature Combined*. Kunststoffe International, 2012.

Endres, H.-J., Jürgens, F., Habermann, C., Spierling, S., Behnsen, H., and Schulz, C. *Eine nachhaltige Alternative? Über Sinn und Unsinn des Einsatzes von Biokunststoffen.* Kunststoffe 7/2015, S. 22–27.

Endres, H.-J., Siebert-Raths, A., Behnsen, H., and Schulz, C. *Biopolymers—Facts and Statistics*, Hannover 2016, ISSN: 2363-8559.

Faruk, O., Bledzki, A., Fink H.-P., and Sain, M. Biocomposites reinforced with natural fibers: 2000–2010, *Progress in Polymer Science*, 37(11), 1552–1596, 2012.

Frattini, S. Demand is increasing for renewable resourced resins, *JEC Composite Magazine*, (38), 32, 2008.

Fritz, H.-G., Seidenstücker, T., Endres, H.-J. et al. *Production of Thermo-Bioplastics and Fibres based Mainly on Biological Materials*. European Commission, Directorate XII, Brussels, 1994.

Guillet, J. et al. Synthesis and applications of photodegradable poly(ethyleneterephthalate). In: A. Albertsson and S.J. Huang (Eds.). *Degradable Polymers, Recycling, and Plastics Waste Management*. Marcel Dekker, New York, 1995, pp. 231–241.

Härkönen, M. et al. Properties and polymerization of biodegradable thermoplastic poly(esterurethane), *J. Macromol-Sci. Pure Appl. Chem.*, 32, 857–862, 1995.

IfBB 2015. *Biopolymers—Facts and Statistics*. Edition 2.

Jacobsen, S. Darstellung von Polylactiden mittels reaktiver Extrusion (Dis.). [Hrsg.] Institut für Kunststofftechnologie Universität Stuttgart, Stuttgart, s.n., 2000.

Kaplan D.L. *Biopolymers from Renewable Resources*. Springer-Verlag, Berlin, 1998.

Mehta, G., Mohanty, A.K., Misra, M., and Drzal L.T. Biobased resin as a toughening agent for biocomposites, *Green Chemistry*, 6, 254, 2004.

Miyagawa, H., Mohanty, A.K., Misra, M., and Drzal, L.T. Thermo-physical and impact properties of epoxy containing epoxidized linseed oil, 1: Anhydride-cured epoxy, *Macromolecular Materials and Engineering*, 289, 629, 2004.

Mohanty, A.K., Misra, M., Drzal, L.T., Selke, S.E., Harte, B.R., and Hinrichsen, G. Natural fibers, biopolymers and biocomposites: An introduction. In: Mohanty, A.K., Misra, M., and Drzal, L.T. Eds., *Natural Fibers, Biopolymers and Biocomposites*, CRC Press, Boca Raton, FL, pp. 1–35, 2005.

Ramakrishna, S., Huang, Z.-M. et al. *An Introduction to Biocomposites*. Imperial College Press, London, 2004.

Raquez, J.M., Deleglise, M., Lacrampe, M.F., and Krawczak, P. Thermosettings (bio) materials from renewable resources: A critical review, *Progress in Polymer Science*, 35, 487, 2010.

Shen, L., Haufe, J., and Patel, M.K. Product overview and market projection of emerging bio-based plastics, Copernicus Institute for Sustainable Development and Innovation Utrecht University, commissioned by European Polysaccharide Network of Excellence and European Bioplastics, 2009.

Siebert-Raths, A. and Endres, H.-J. *Modification of Polylactide for Technical Applications*. Kunststoffe International 5/2011, pp. 61–65.

5 Bio-Based EPDM Rubber and Sustainable EPDM Compounding

Martin van Duin, Philip Hough,
Joyce Kersjes, Marjan van Urk,
M. Montserrat Alvarez Grima,
and Niels van der Aar

CONTENTS

5.1 GENERAL INTRODUCTION

One of the greatest challenges facing our industries today is the need to design and develop sustainable technical solutions to address mega-trends in society, such as mobility and urbanization. Key issues that have to be dealt with are emissions and fuel efficiency. Another challenge is the need to offer solutions to support a lower dependence on fossil fuels. To this end, ARLANXEO Performance Elastomers has made a pioneering move toward exploring a future based on renewable resources, by developing the world's first bio-based EPDM rubber commercialized under the tradename Keltan® Eco. In addition, the potential of sustainable alternatives for traditional plasticizer oils and (reinforcing) fillers was evaluated in an effort to develop a Keltan Eco based rubber compound with the highest sustainable content without compromising the technical performance.

5.2   KELTAN EPDM RUBBER

Polymers of ethylene, propylene, and a nonconjugated diene (EPDM) represent an important class of elastomers.[1,2,3] The fully saturated, flexible polymer backbone provides excellent resistance against oxygen, ozone, heat, and irradiation and, thus, is the product of choice for outdoor and elevated-temperature applications. Figure 5.1 shows the application segments for EPDM rubber.* Clearly the main segment for EPDM is automotive, certainly when it is also considered that automotive is the main outlet for plastic modification and oil additives.

Typical automotive applications for EPDM rubber include weather-stripping, seals, hoses (radiator, brake, and air conditioning hoses), belts, wipers, and mounts. About 50% of the automotive and the building and construction segments consist of sealing systems. Typical examples of EPDM seals are

- Automotive solid and sponge seals for the doors, trunk, sun roof, front hood, and so on
- Window seals for busses, trucks, trains, and metro
- Window profiles for buildings and houses
- Other seals like tunnel seals, brake seals, seals for air conditioning systems, valve seals, seals for sewage systems and potable water seals, and so on

Although EPDM has been used for sealing purposes since its birth in the 1960s, a lot of developments are still taking place:

1. Focus on costs: faster extrusion, lower scrap rates, and low cost rubber compounds. ARLANXEO has high ENB grades for fast vulcanization, has a strong focus on product quality and consistency to limit scrap rates at its customers, and has several EPDM grades for low cost compounding, which can be loaded to high total compound levels of 600–700 phr.
2. Focus on performance: improved aesthetics, more complex designs to improve isolation performance, and improved flame resistance. ARLANXEO has developed special grades to give extruded sponge with improved surface smoothness. These grades combine easy mixing, excellent filler dispersion, and fast extrusion with a high level of physical properties. The patented ARLANXEO Controlled Long Chain Branching technology (CLCB) assures good collapse resistance, facilitating more complex seal designs. Also, the inherent ability of EPDM to be filled with high loading levels of fillers gives the opportunity to compound using hydrated fillers for improved

---

* Company estimates based on analysis, forecasts and trends reporting by companies like markets and markets, mrs research group, research and market, a.o.

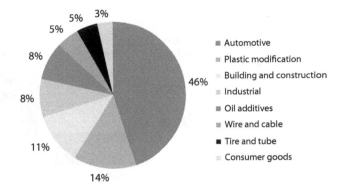

FIGURE 5.1   Application segments for EPDM rubber.

flame resistance. The addition of maleic anhydride grafted Keltan EPM gives improved coupling between hydrated fillers and the polymer matrix leading to significantly improved tensile properties.

3. Focus on sustainability: use of green material, weight reduction, and recyclability. ARLANXEO has recently launched the world's first bio based EPDM. This "green" Keltan EPDM is based on sugar cane produced in Brazil and has found its first commercial application in passenger bus window seals. Weight reduction can be achieved by applying micro dense profiles or by using thermoplastic vulcanizates (TPV) seals instead of conventional rubber profiles. TPVs can be processed like plastic and can also be recycled. ARLANXEO has best-in-class EPDM grades for the production of TPVs. Consumption of TPV is gradually growing with increased use in new passenger cars. Glass-run channel seals and waist belts are more and more based on TPVs.

In conclusion, the broad portfolio of Keltan EPDM grades is well equipped to meet the current and future demands of the rubber industry for the production of high performance applications.

5.3 KELTAN ECO BIO-BASED EPDM RUBBER[4,5]

5.3.1 INTRODUCTION

Regular EPDM products, including ARLANXEO Performance Elastomers' Keltan EPDM products, are produced from fossil raw materials, that is, the ethylene and propylene are produced via cracking of natural oil. Keltan Eco EPDM is a recent ARLANXEO Performance Elastomers' development, produced from bio-based ethylene. Keltan Eco is produced in the ARLANXEO's EPDM plant in Triunfo, Brazil, by means of a solution polymerization process using Ziegler Natta catalyst technology. Keltan Eco is produced from bio-based ethylene supplied by Braskem S.A., which originates from sugar cane (Figure 5.2). The sugar from sugar cane is converted to ethanol, which is then dehydrated to ethylene by Braskem in their Triunfo plant. This bio-based ethylene is transported via a pipeline to the neighboring

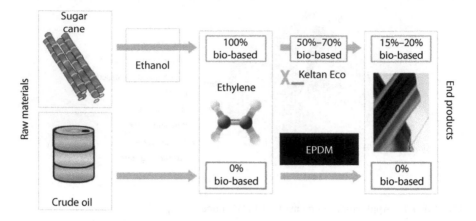

FIGURE 5.2   Route to bio-based EPDM.

ARLANXEO EPDM polymerization plant. Depending on the ethylene content of the particular grade, the bio-based content of Keltan EPDM rubber ranges between 50 and 70 wt%. Translating this to final articles produced from mixed EPDM compounds, a bio-based content of 15–20 wt% can be achieved if Keltan Eco is the only bio-based ingredient of the rubber compound (Figure 5.2).

Keltan Eco gives the following benefits:

- Reduced dependence on fossil resources.
- Reduced carbon footprint due to use of sugar cane.
- Truly sustainable as validated by a life cycle assessment performed by PE International.
- Bio-based content can be measured and traced back by ASTM D6866 carbon-14 test performed by Beta Analytic Inc. (Figure 5.3).
- No compromise on quality.

Figure 5.4 shows that green polyethylene (PE) offers a significant carbon footprint improvement compared to petrochemical PE. The difference is more than 4 ton of emitted CO_2 per ton of product.[6]

Figure 5.5 shows global warming potential of Keltan Eco. This represents an opportunity to reduce the EPDM carbon footprint by up to 82%, depending on the ethylene content of the polymer.

In essence, the Keltan Eco grades look, feel, and behave like conventional oil-sourced EPDM. Thus, Keltan Eco EPDM is an elastomer, which has exceptional elasticity, flexibility, and durability. It can be molded, extruded, and calendared to produce functional articles with excellent aesthetics. ARLENXEO has five Keltan Eco grades commercially available in its portfolio (Table 5.1).

Keltan Eco has received a positive response in the market and commercialization at customers is on-going in different application segments, like automotive sealing systems, construction seals, O-rings, petroleum, and plastics additives. Some examples are displayed below.

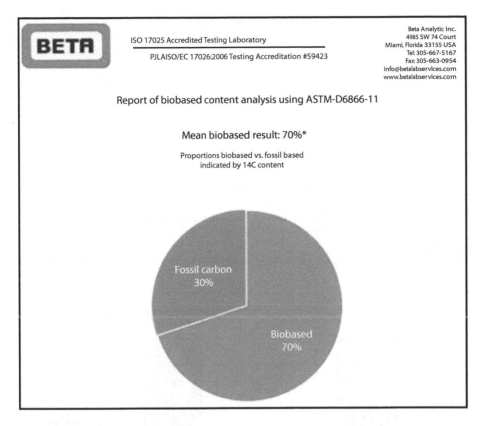

FIGURE 5.3 Results from the C-14 measurement of Keltan Eco 5470.

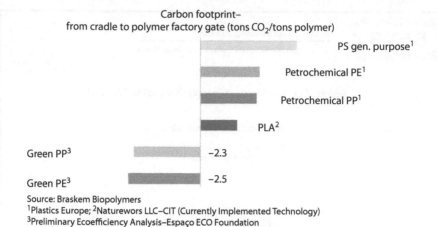

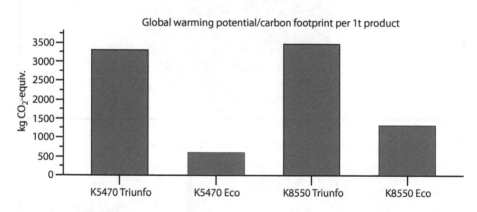

FIGURE 5.4 Carbon footprint from cradle to polymer factory gage (ton CO_2/ton polymer).

FIGURE 5.5 Global warming potential of Keltan Eco 5470 and 8550 (kg CO_2-equivalent per ton polymer) (PE International data).

TABLE 5.1
Keltan Eco EPDM Portfolio and Key Properties

Grades	Viscosity ML(1+4) (@ Shown °C) [MU]	C_2 [wt%]	ENB [wt%]
Keltan Eco 0500R	11 g/10 min. (MFI)	49	–
Keltan Eco 3050	51 (@ 100°C)	49	–
Keltan Eco 5470	55 (@ 125°C)	70	4.6
Keltan Eco 8550	80 (@ 125°C)	55	5.5
Keltan Eco 6950	65 (@ 125°C)	48	9.0
Keltan Eco 9950	60 (@ 150°C)	48	9.0

TABLE 5.2

Overview of Polymer Characteristics of Keltan (Eco) 5470

Characteristic	Unit	Keltan 5470	Keltan Eco 5470
ML (1+4) 125 °C	MU	55	57
Δδ	–	22	23
Composition (FTIR)			
Ethylene	wt%	70	70
ENB	wt%	4.6	4.6
Crystallinity (DSC)			
Tc	°C	18	18
ΔHc	J/g	39	39
MWD	–	2.7	2.6
Bio-based Ethylene content	%	n.a.	70

5.3.2 EXPERIMENTAL

5.3.2.1 Polymer Evaluation

Keltan Eco 5470 was produced in the Triunfo plant with Ziegler Natta catalysis, using the same production conditions as those used for the production of regular Keltan 5470. The aim was to obtain a polymer with exactly the same polymer characteristics as the existing Keltan 5470, with the only difference being the source of the ethylene feed (bio-based vs. oil-based). The polymer characteristics of both polymers have been measured and compared to prove their similarity (Table 5.2). The Mooney viscosities of the two polymers are very similar, both falling within the specification limits for Keltan 5470 (55 ± 4 MU). The Δδ values of the polymers are almost identical, showing that they have equivalent levels of long-chain branching.[7] Furthermore, the chemical composition of the polymers, the crystallinity characteristics (Figure 5.6), and the molecular weight distribution (MWD; Figure 5.7) are identical. The only difference between the polymers is the amount of bio-based carbon, where, as expected, ^{14}C analysis shows that the Keltan Eco 5470 polymer contains 70% bio-based carbon which corresponds exactly with the 70% ethylene content of the polymer as shown with infrared spectroscopy.

5.3.2.2 Compound Evaluation

Keltan 5470 and Keltan Eco 5470 have been compared in a typical sulfur-vulcanized extrusion compound (Table 5.3), which is a relatively low filled (polymer-rich) formulation and, therefore, quite sensitive to small variations in the polymer characteristics. The compounds were mixed, vulcanized, and evaluated at the ARLANXEO Performance Elastomers laboratory in Geleen, Netherlands. Mixing was carried out in a Shaw K-1 Intermixer with a load factor of 54%. The mixing characteristics for each compound will be shown in the results section. After mixing, the batches were transferred to a Troester WNU-5 two roll mill, where the curing ingredients were added to the compounds at 60°C.

The Mooney viscosity of the compounds [ML (1+4) at 100°C] was measured, and the extrusion behavior of the test compounds was studied by means of a Garvey die

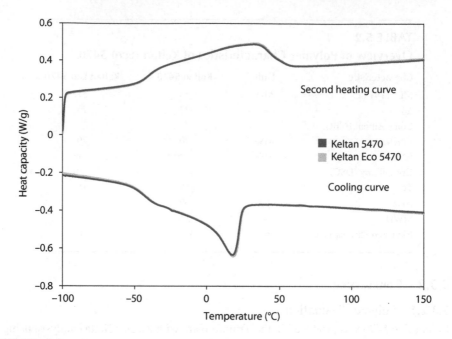

FIGURE 5.6　DSC analysis of temperature behavior.

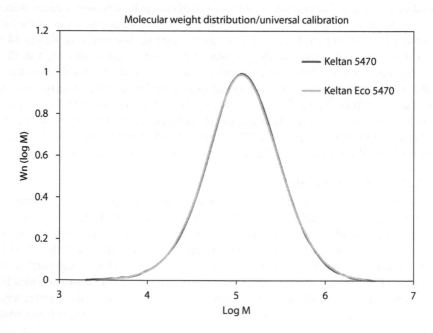

FIGURE 5.7　GPC analysis of molecular weight distribution.

TABLE 5.3
Keltan (Eco) 5470 Compound Formulations (in phr)

	Keltan 5470	Keltan Eco 5470
EPDM	100	100
ZnO-active	5	5
Stearic acid	1	1
Carbon black N-550	70	70
Carbon black N-772	40	40
Paraffinic oil	70	70
MBTS-80	1.31	1.31
ZBEC-70	0.7	0.7
ZDBP-50	3.5	3.5
Vulkalent E/C	0.5	0.5
S-80	1.25	1.25
Total phr	293.26	293.26

extrusion test using a cold-feed extruder with a screw speed of 50 rpm. The Mooney scorch of the test compounds was measured at 125°C. The vulcanization behavior of the compounds was evaluated using a Monsanto MDR 2000 rheometer, according to ISO 6502, for 20 min at 180°C. With this measurement, the scorch time (ts2), vulcanization time (t90), and maximum rheometer torque difference (MH-ML, as a measure for cross-link density) were determined.

In order to measure the physical properties of the compounds, press plates of 2 mm and 6 mm thickness were compression molded at 40 bar and 180°C. The 2 mm press plates were vulcanized for a time equivalent to t90 + 10% and the 6 mm press plates for t90 + 25%. The hardness (IRHD) was measured according to DIN 53505 before and after hot air aging for 168 h at 100°C. The tensile properties (tensile strength, elongation at break, and modulus at 100 and 300% elongation) were measured using dumbbell #2 according to ISO 37 before and after hot air aging for 168 h at 100°C. The Delft tear strength was measured according to ISO 188 before and after aging for 168 h at 100°C The compression set was measured for 24 h at 70°C and 24 h at 100°C according to ISO type B.

5.3.3 RESULTS AND DISCUSSION

The mixing curves from the test compounds are shown in Figure 5.8, which shows that both polymers show identical mixing behavior. Table 5.4 summarizes the mixing characteristics and shows the Mooney viscosities of the compounds as well as the results of Garvey die extrusion tests. Both compounds have very similar Mooney viscosities and almost identical Garvey die extrusion behavior. It can therefore be concluded that the mixing and processing behavior of these two polymers is equivalent.

The scorch and vulcanization behavior of the compounds are shown in Figure 5.9 and Table 5.5. It can clearly be concluded that the scorch and vulcanization behavior of the two polymers are identical. Since most physical properties are strongly

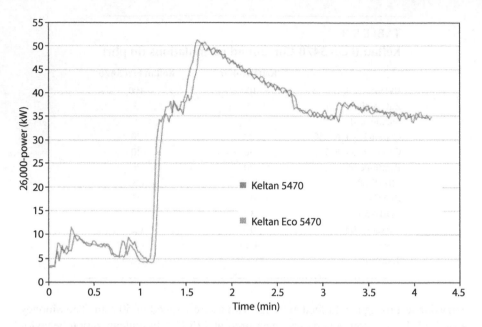

FIGURE 5.8 Mixing curves of Keltan (Eco) 5470 in low-filled compounds.

TABLE 5.4

Mixing Characteristics of Keltan (Eco) 5470 in Low-Filled Compounds

Property	Unit	Keltan 5470	Keltan Eco 5470
Mixing Characteristics			
Mixing time	S	249	251
Power maximum	kW	51	51
Dump temperature	°C	117	118
ML (1+4) 100 °C	MU	55	54
Garvey Die Extrusion			
Output	g/min	503	505
Die swell	%	37	38
Head pressure	bar	53	51
Rating swell and porosity	–	4	4
Rating 30° edge	–	3	3
Rating surface	–	4	4
Rating corners	–	4	4

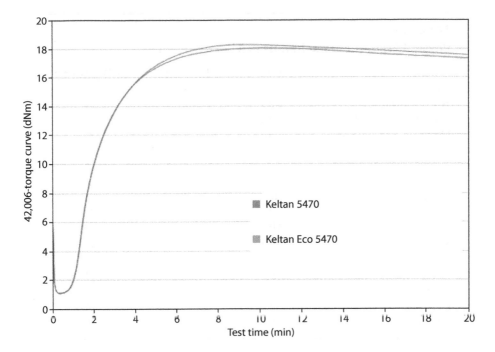

FIGURE 5.9   MDR cure curves of Keltan (Eco) 5470 compounds.

correlated with the cross-link density, it can be expected that compounds showing the same level of cross-link density (and having the same level of crystallinity) will have very similar physical properties. Table 5.6 presents an overview of the physical properties measured for these two compounds and shows that the properties are very similar indeed. All the vulcanizate properties, measured both before and after aging, are equivalent for the two vulcanized compounds. Any small differences observed in the values are within the statistical limits of the testing methods and typical batch-to-batch variations.

TABLE 5.5
Vulcanization Data for Keltan (Eco) 5470 Compounds

Property	Unit	Keltan 5470	Keltan Eco 5470
	Vulcanization Behavior (MDR)		
MH-ML	dNm	17	17
ts2	min	1.2	1.2
t90	min	4.7	4.7
	Mooney Scorch		
Initial	MU	55	53
t2	min	23	24
t5	min	31	31
t35	min	45	46

TABLE 5.6

Physical Properties of Keltan (Eco) 5470 Compounds

Property	Unit	Keltan 5470	Keltan Eco 5470
Hardness	IRHD	68	68
Tensile strength (TS)	MPa	15	15
Modulus at 100% (M100%)	MPa	2.8	2.6
Modulus at 300% (M300%)	MPa	10	9
Elongation at break (EB)	%	486	507
Tear strength Delft (TSD)	N	51	49
After aging 168 h @ 100°C			
Hardness	IRHD	70	71
TS	MPa	15	15
M100%	MPa	3.7	3.5
M300%	MPa	12	12
EB	%	386	409
TSD	N	51	51
Compression set (CS)			
CS 24 h @ 70°C	%	9	9
CS 24 h @ 100°C	%	25	26

The fact that the two polymers show identical behavior in this compound evaluation confirms that Keltan Eco 5470 is technically equivalent to Keltan 5470, as concluded also from the polymer evaluation, and therefore the only difference between the two polymers is the source of the ethylene monomer (bio-based vs. oil-based).

5.4   SUSTAINABLE EPDM COMPOUNDING[8]

5.4.1   INTRODUCTION

Typically, rubber products not only consist of elastomers, but also of (reinforcing) fillers, plasticizers, cross-linking agents, and other additives. EPDM products may contain up to 400 phr of compounding ingredients incorporated into 100 phr of EPDM polymer. Carbon black is produced via incomplete combustion of a hydrocarbon feed with natural gas. Silica is produced via precipitation from a silicate salt solution. Inert white fillers, such as clay, talc, and chalk are extracted from the ground in open mines and milled to fine powders. Traditional extender oils for EPDM are refinery fractions of crude oil. All of these ingredients, typically used for EPDM compounding, lack sustainability.

Numerous studies have been performed on more sustainable rubber compound ingredients.[9] With respect to fillers, one may consider all sorts of natural fibers (jute, palm, sisal, hemp, etc.) and natural flours and powders (wood, cork, soy, etc.)[10–13] as well as recycled carbon black, produced via pyrolysis of waste tires.[14] The former should be viewed as inert, mineral filler alternatives that only dilute the compound but do not contribute to its performance, whereas the latter has a promising performance close to that of traditional carbon black. Natural oils, such as palm, rice bran, ground nut, soybean, mustard, and sunflower oils, have been explored as sustainable plasticizers in rubber products but with only limited success due to their high polarity and/or high levels of unsaturation.[15–22]

In further efforts to increase the sustainability of EPDM rubber products based on Keltan Eco, we have explored the potential of using sustainable alternatives for traditional plasticizer oils and (reinforcing) fillers leading to application compounds for automotive seals with up to 90% sustainability (Figure 5.10).

The emphasis of this sustainable compounding evaluation was on technical aspects, such as compound mixing, processing, vulcanization, and properties (before and after aging) of the final vulcanizates. There will be no further discussions on the level of sustainability nor on the costs of the alternative ingredients and the final compounds.

In a first screening study, a variety of bio-based oils were studied as possible replacements for traditional mineral oil. The natural oils studied, for example, linseed oil, tung oil, coconut oil, and olive oil, are all triglycerides with fatty acid chains varying in length and unsaturation. These four natural oils were selected because they provide a nice spread in their levels of unsaturation (as witnessed by the iodine number) and their melting points. Butter fat is a triglyceride obtained from cow's milk and was also included because it has a rather low iodine number, though a relatively high melting point. Modified natural oils, such as hydrogenated coconut oil (almost no unsaturation) and mono-esters produced via trans-esterification of natural oils, that is, ethylhexyl oleate (reaction product of high-oleate sunflower oil with ethylhexyl alcohol) and isotridecyl stearate (saturated mono-ester), were included in a second phase of the oil screening study to overcome some of the compatibility and vulcanization issues experienced with the natural oils and butter fat.

Factice, which is a vulcanized vegetable oil, was originally introduced to the rubber industry around 1847 as a partial economic substitute for natural rubber,

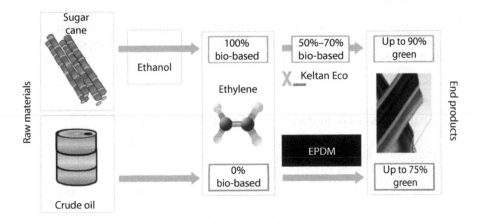

FIGURE 5.10 Green compounding of Keltan Eco EPDM, allowing EPDM products with up to 90% sustainable ingredients.

but is used today as a processing agent.[1] Factice was tested as a full oil replacement because it consists of highly cross-linked natural oil and thus could overcome some of the issues encountered with the other natural oils. Factice was also tested in combination with other natural oils with the idea that it could act as a sort of sponge to permit higher loadings of the natural oil. Finally, 2,6,10,15,19,23-hexamethyltetracosane (squalane) was evaluated. Squalane is a fully saturated C30 hydrocarbon, which resembles an EPM hexamer and is sometimes used as a high-boiling point solvent for EPDM in academic studies. Squalane is traditionally obtained via hydrogenation of the triterpene squalene from shark liver, but more recently a process was developed where plant sugar is converted via genetically engineered yeast into trans-ß-farnesene,[23] which is dimerized and subsequently hydrogenated to squalane.[24]

In a second study, sustainable fillers have been investigated as replacements for standard carbon black and inert mineral fillers. Pyrolysis black was evaluated versus standard furnace black. The tire pyrolysis product combines both the original carbon black used in the production of tires and additional black formed on pyrolysis of the rubber and plasticizer in the end-of-life tire waste.[14] While one can argue that pyrolysis black is not derived from a bio-based source, a proportion is in fact derived from natural rubber. In addition, (tire) rubber end-of-life waste is a major environmental issue and pyrolysis seems to be one of the preferred recycling technologies, which in the end reduces the CO_2 production by 5 tons per ton of rubber compound. In the context of this study, we chose to use the term sustainable ingredient.

Rice husk ash is basically silica, recovered via burning off the organic fraction of rice husk which is obtained during the rice cleaning process.[25] Micro-cellulose is a natural fiber, produced through a process of chemical disintegration of different woods. In the final study we explored EPDM compounds with the highest level of sustainable ingredients, while still maintaining the performance of high-quality EPDM compounds.

5.4.2 EXPERIMENTAL

For a brief description and the characteristics of the EPDM polymers, the sustainable oils, and the fillers used in this study, see Tables 5.7 through 5.9, respectively. The details have been provided by the suppliers. The solubility parameters of the oils

TABLE 5.7
Characteristics of Keltan EPDM (Eco) Polymers Used

Keltan	Bio-Based	Ethylene (wt%)	ENB (wt%)	ML 1+4 @ 125°C	Oil (phr)
5470	–	70	4.6	55	0
6471	–	67	4.7	65	15
8550	–	55	5.5	80	0
Eco 5470	+	70	4.6	55	0
Eco 8550	+	55	5.5	80	0

TABLE 5.8

Characteristics of Oil Plasticizers Used

Chemical Composition	Product	Supplier	Molar Mass (g/mol)	Density (g/ml)	Melting Point (°C)	Solubility Parameter (cal/cm³)$^{0.5}$	Iodine Number (g/100 g)
Paraphinic mineral oil	Sunpar 2280	Sunoco	~700	0.89		7.8	0
Refined linseed oil		Rutteman	880	0.93	−20	8.4	193
Chinese tung oil		Rutteman	870	0.94	4	8.4	168
Olive oil		Albert Heyn	880	0.92	−1	8.4	84
Butter fat		Friesland Campina	840	0.91	36	8.3	34
Refined coconut oil		Rutteman	720	0.92	25	8.4	11
Hydrogenated coconut oil	Agri-pure™ AP-620	Cargill	722	0.92	32	8.4	2
Isotridecyl stearate	Loxiol G40	EmeryOleo	466	0.86	−5	8.1	1
Ethylhexyl oleate		Hansen+Rosenthal	394	0.87	0	8.1	0.64
Squalane		Amyris	422	0.81	−38	7.5	0.13
Factice	Rhenopren EPS	Rhein Chemie					
EPDM	K8550	LANXESS	300.000	0.86		7.7	11.6

TABLE 5.9

Characteristics of Fillers Used

Chemical Composition	Product	Supplier	Particle/Fiber (μm)	Density (g/ml)	Surface Area	Comments
Carbon black	Corax N550	Orion engineered carbons	0.039–0.055	1.8	40[a]	Fast extrusion furnace (FEF) black
Carbon black	BBC 500	Black bear	0.005–0.1	1.75	77[a]	Residue of pyrolysis of waste tires; also contains 5% ZnO/S and 1%–27% silica
Aluminosilicate	Polestar 200R	Imerys performance minerals	~2	2.6	8.5[b]	Soft calcined clay
Silica	Ultrasil VN 3	Evonik industries	~0.015	2	180[a]	Precipitated silica
Calcium carbonate	Superfine S whiting		~2	2.7	2.8[b]	Superfine calcium carbonate
Amorphous silica	Rice husk ash	Geradora de energia eléctrica alegrete	<8	2.2	1.4[b]	Residue of burning rice husk
Micro-cellulose	Arbocel UFC M8	J. Rettenmaier & Söhne	14 × 5	1.5		Chemical disintegration of wood
Micro-cellulose	Arbocel FD 600-30	J. Rettenmaier & Söhne	45 × 25	1.5		Idem

a BET measurement using nitrogen adsorption (m^2/g).

b Measured by oil (DBP) absorption (m^2/g).

have been calculated using a group contribution method. First, the plasticizer oils were screened in sulfur-vulcanized EPDM compounds with varying oil and carbon black contents (33/50, 33/150, 67/100, 100/50, and 100/150 phr/phr) The compounds were based on an amorphous, high Mooney EPDM (Keltan 8550) (Table 5.10). In the next study, pyrolysis black was evaluated in a sulfur-vulcanized, automotive solid seal compound based on a crystalline EPDM (Keltan Eco 5470), as can be seen in Table 5.11. Finally, rice husk ash and micro-cellulose was evaluated in a non-black bull's-eye compound based on a 15 phr oil-containing, crystalline EPDM (Keltan 6471) (Table 5.12). As a concluding study, the best options for bio-based plasticizers and sustainable fillers were combined in two highly filled, automotive, solid seal compounds (Table 5.13). The first has good low-temperature flexibility for dynamic sealing applications and was, therefore, based on an amorphous, high Mooney EPDM (Keltan Eco 8550).The second compound is typical of a static seal, where low-temperature flexibility is less important and allowed the use of a crystalline, medium Mooney EPDM (Keltan Eco 5470). All sulfur accelerators were from the Rhenogran product range supplied by LANXESS RheinChemie Additives and the TMQ and ZMMBI heat stabilizers were from the Vulkanox family supplied by LANXESS Advanced Industrial Intermediates.

The rubber compounds were prepared in an internal mixer (1.5 liter capacity) from Harburg Freudenberger. The mixer was equipped with PES5 inter-meshing rotors with a thermostatically temperature-controlled body using circulating water. Mixing was carried out according to ISO 2393 following an upside-down mixing protocol with 72% fill factor, 45°C mixer body temperature, 8 bar ram pressure and 50 rpm rotor speed. During the first 30 seconds of the mixing cycle, the EPDM polymer was crumbled; next the compounding ingredients with the exception of the cure system were added and mixed for 210 seconds, giving a total mixing time of 240 seconds The compounds were then transferred to a two-roll mill having

TABLE 5.10
EPDM Compound Compositions with Varying Oil and Black Contents Used for Screening the Bio-Based Oils

K8550 EPDM	100
N550 carbon black	50/50/100/150/150
Bio-based or mineral oil	33/100/67/33/100
TMQ	1
ZMMBI	1
ZnO active	5
Stearic acid	1
Sulfur-80	1.25
MBTS-80	1.31
ZBEC-70	0.7
ZDBP-50	3.5
Vulkalent E/C	0.5
Total	198/265/282/298/365

TABLE 5.11

Automotive, Solid Seal EPDM Compound Composition Used for Evaluating Pyrolysis Carbon Black Versus Standard Furnace Black

Compound Formulation	FEF Black	Pyrolysis Black
Keltan Eco 5470 EPDM	100	100
N550 carbon black	120	
BBC 500 carbon black		135
Superfine S whiting	85	85
Flexon 876 paraffinic oil	80	80
PEG 4000	2	2
CaO-80	10	10
ZnO active	5	5
Stearic acid	1	1
Sulfur-80	1.8	1.8
CBS-80	2.1	2.1
TMTD-80	0.5	0.5
ZDMC-80	1.2	1.2
ZDBC-80	2.5	2.5
Total	411.1	426.1
MDR rheometry @ 180°C		
ML (dNm)	1.1	1.6
MH-ML = ΔS (dNm)	18	17
ts2 (min)	0.6	0.7
tc90 (min)	1.7	3.7
Vulcanisate properties		
Hardness (IRHD)	68	68
Modulus @ 100% (MPa)	2.8	2
Modulus @ 300% (MPa)	8	6
Tensile strength (MPa)	11	10
Elongation at break (%)	440	501
Rebound resilience (Schob) (%)	35	34
Compression set 24 h @ 100°C (%)	63	59
After aging for 168 h @ 100°C		
Hardness (IRHD)	78	78
Modulus @ 100% (MPa)	6.8	5.4
Tensile strength (MPa)	11	11
Elongation at break (%)	183	207

tempered rollers (20 cm diameter, 50°C and 20 rpm speed) where the cure system was added and final dispersion was accomplished by cutting, rolling up, and rotating the rolled rubber sheet by 90° through the mill nip three times. For the study to compare the alternative white fillers with silica, the starting mixer body temperature was 70 or 130°C. After mixing for 240 seconds the rotor speed was increased to achieve a batch temperature of 150°C and then mixing was continued for 180 seconds at 150°C to complete silanization.

TABLE 5.12

Bull's-Eye EPDM Compound Composition Used for Evaluating Rice Husk Ash and Micro-Cellulose Versus Silica

Compound Formulation	Silica	Silica	Rice Husk Ash	Micro-Cellulose 1	Micro-Cellulose 1 + Maleated EPM	Micro-Cellulose 2
Starting Mixer Temperature (C)	70	130	130	130	130	130
Keltan 6471 EPDM	115	115	115	115	115	115
Keltan 1519R maleated EPM					5	
PoleStar 200R clay	110	110	110	110	110	110
Ultrasil VN 3 silica	30	30				
Rice husk ash			30			
Arbocel UFC M8 micro-cellulose				30	30	
Arbocel FD 600-30 micro-cellulose						30
Titanium dioxide	9	9	9	9	9	9
Sunpar 2280 mineral oil	70	70	70	70	70	70
Si 69 coupling agent	2	2	2	2	2	2
TEA	2	2	2	2	2	2
ZnO	5	5	5	5	5	5
Stearic acid	1	1	1	1	1	1
S-80	0.6	0.6	0.6	0.6	0.6	0.6
DPTT-70	1.14	1.14	1.14	1.14	1.14	1.14
DTDM-80	1.2	1.2	1.2	1.2	1.2	1.2
MBT-80	1.2	1.2	1.2	1.2	1.2	1.2
TMTD-70	1.14	1.14	1.14	1.14	1.14	1.14
ZDBC-80	2.5	2.5	2.5	2.5	2.5	2.5
Total	351.8	351.8	351.8	351.8	356.8	351.8
Compound properties						

(Continued)

TABLE 5.12 (CONTINUED)
Bull's-Eye EPDM Compound Composition Used for Evaluating Rice Husk Ash and Micro-Cellulose Versus Silica

Compound Formulation	Silica	Silica	Rice Husk Ash	Micro-Cellulose 1	Micro-Cellulose 1 + Maleated EPM	Micro-Cellulose 2
ML 1+4 @ 100°C (MU)	18.8	28.9	17.6	24.0	28.3	23.8
ts5 (min)	14.2	13.6	10.4	13.8	14.8	14.6
ML (dNm)	0.28	0.52	0.23	0.34	0.40	0.32
MH − ML = ΔS (dNm)	5.7	8.2	5.7	6.3	6.7	6.7
ts2 (min)	1.4	1.2	1.3	1.4	1.4	1.5
tc90 (min)	3.5	3.1	2.6	3.2	3.3	3.2
Vulcanisate properties						
Hardness (Sh A)	42.1	49.3	41.5	46.7	47.1	48.1
Modulus @ 100% (MPa)	1.5	1.8	1.4	1.9	2	1.6
Modulus @ 300% (MPa)	3.9	4.7	3.9	4.6	4	4
Tensile strength (MPa)	7.2	9.4	5.4	6.1	6.3	5.7
Elongation at break (%)	683	647	574	487	718	551
Tear resistance Delft (N/mm)	27.1	31	23.3	26.4	29.2	22.5
Compression set 72 h @ 23°C (%)	12.1	10.5	11.1	12.8	13.5	12.2
Compression set 24 hr @ 70°C (%)	19.9	18.2	18.1	19.7	22.2	20.3
Compression set 4 hr @ 100°C (%)	43.2	36.4	44.7	37.5	41.5	40.7
After aging for 168 hr @ 100 °C						
Hardness (Sh A)	46.6	53.8	45.6	50.5	50.7	51.4
Modulus @ 100% (MPa)	1.9	2.7	1.9	2.5	2.7	2
Modulus @ 300% (MPa)	5.1	6.7	4.4	5.5	5.1	4.8
Tensile strength (MPa)	6.6	8.7	4.8	6.6	5.8	5.3
Elongation at break (%)	468	459	381	407	466	365
Tear resistance Delft (N/mm)	–	31.1	20.3	24.6	28.7	21.3

TABLE 5.13

Composition of Highly Filled, Automotive, Solid Seal Compounds Based on Amorphous or Crystalline EPDM, Used to Maximize Content of Sustainable Ingredients

Compound Formulation	Dynamic, Automotive Seal		Static, Automotive Seal	
	Traditional	Sustainable	Traditional	Sustainable
Keltan 8550 EPDM	100			
Keltan Eco 8550 EPDM		100		
Keltan 5470 EPDM			100	
Keltan Eco 5470 EPDM				100
N550 carbon black	155		147	
BBC 500 carbon black		174		165
Superfine S whiting	77		110	
Rise husk ash		77		110
Sunpar 2280 mineral oil	98		100	
Squalane		98		100
CaO	5	5	5	5
PEG 1000	1.5	1.5	1.5	1.5
ZnO active	3	3	3	3
Stearic acid	1	1	1	1
Sulfur-80	0.8	0.8	0.8	0.8
DPG-80	0.5	0.5	0.5	0.5
TBBS-80	0.625	0.625	0.625	0.625
CBS-80	1.25	1.25	1.25	1.25
ZBEC-70	1.7	1.7	1.7	1.7
TP-50	2	2	2	2
CLD-80	1	1	1	1
Vulcalent E/C	0.8	0.8	0.8	0.8
Total	449.2	468.2	476.2	494.2
Percentage bio-based/sustainable (%)	0	86	0	90
Compound properties				
Extruder throughput (g/min.)	32.6	32.8	35.2	34.8
Extruder head pressure (bar)	27	21	26	21
Die swell (%)	11.9	18.9	8.0	9.0
Garvey die ranking (Total score)	12	12	12	12
ML 1+4 @ 100°C (MU)	59.5	47.8	55.0	44.5
ts5 @ 125°C (min)	18.0	26.4	19.1	28.0
ML (dNm)	1.82	2.63	1.33	2.31
MH − ML = ΔS (dNm)	13.2	13.6	14.1	14.4
ts2 (min)	1.0	1.1	1.2	1.1
tc90 (min)	2.8	3.2	3.6	3.3
Vulcanisate properties				
Hardness (Sh A)	67.3	66.9	67.9	68.9
Modulus @ 100% (MPa)	3.5	2.3	3.2	2.4

(Continued)

TABLE 5.13 (CONTINUED)

Composition of Highly Filled, Automotive, Solid Seal Compounds Based on Amorphous or Crystalline EPDM, Used to Maximize Content of Sustainable Ingredients

Compound Formulation	Dynamic, Automotive Seal		Static, Automotive Seal	
	Traditional	Sustainable	Traditional	Sustainable
Modulus @ 300% (MPa)	9.5	–	8.3	7.3
Tensile strength (MPa)	9.5	7.5	9.5	8.3
Elongation at break (%)	304	291	368	349
Tear resistance Delft (N/mm)	28.4	20.6	31.6	25.6
Compression set 24 h @ –25°C (%)	70.8	42.3	97.6	97.8
Compression set 72 h @ –23°C (%)	9.2	12.7	27.2	25.4
Compression set 24 h @ 70°C (%)	12.1	13.1	13.7	17.3
Compression set 24 h @ 100°C (%)	27.2	36.7	29.1	44.9
After aging for 168 h @ 135°C				
Hardness (Sh A)	72.8	83.3	75.8	84.8
Modulus @ 100% (MPa)	6.1	–	5.2	–
Tensile strength (MPa)	9.7	7.7	9.7	10.1
Elongation at break (%)	179	7	226	46
Mass change upon aging	–0.57	–20.18	–0.69	–19.2
After aging for 336 h @ 135°C				
Hardness (Sh A)	74.77	93.43	76.73	93.63
Modulus @ 100% (MPa)	7.3	–	6.2	–
Tensile strength (MPa)	10.2	11.1	10.3	13.7
Elongation at break (%)	161	14	207	26

The compound Mooney viscosity (1 + 4) was measured at 100°C (ML) (DIN 53523 part 3) and the Mooney scorch characteristics, such as t2 at 125°C (DIN 53523 part 4) were measured on a Mooney Viscometer. The cure characteristics of the compounds, such as scorch time ts2, vulcanization time tc90, and maximum torque difference MH – ML = ΔS were determined with a Monsanto MDR 2000E Rheometer according to DIN 53529 part 3 at 180°C.

Test plates (2 and 6 mm thick) were compression molded at 180 bar and 180°C. The 2 mm test plates used for tensile and tear measurements were obtained after curing for tc90 plus 10%, whereas the 6 mm press plates, used for hardness and compression set measurements, were cured for tc90 plus 25%. Evaluation of the cured compounds focused on the following properties: hardness Shore A or IRHD (DIN 53505), tensile properties such as tensile strength (TS), elongation at break (eab) and modulus at 100 (M100%) and 300% elongation (M300) using a dumbbell #2 (DIN 53504), tear resistance Delft (tear) (ISO 34) and compression set (CS) for 24 or 72 h at –25°C, 23°C, 70°C, 100°C, and/or 125°C (DIN ISO 7743). Hardness, tensile properties, and tear resistance were also measured at room temperature after hot-air aging for 7 days at 70, 100, and/or 125 or for 7 and 14 days at 135°C (DIN 53508). Oil volume swell was determined for IRM901 and 903 oils for 48 h at 70°C (DIN 53521).

All test specimens were prepared according to DIN ISO 23529 and data evaluation was done in accordance with DIN 53598.

The level of oil bleeding out of the uncured and cured samples was estimated by manually touching the surface of the compounds and subjectively ranking them according to surface feel. The unvulcanized milled compounds were stored for 1 week before being evaluated and the cured test sheets were stored for 4 weeks at room temperature in aluminum trays. Samples were ranked on a scale of 5 (no oil bleeding), 4 (surface slightly slippery), 3 (greasy surface), 2 (grease transfers to fingertips), and 1 (oil deposited in trays).

5.4.3 Results and Discussion

5.4.3.1 Bio-Based Oils

The results of the screening study for replacing mineral oil by bio-based oils in EPDM compounds (Table 5.10) will not be presented or discussed in detail, but will be presented in terms of general trends and issues that were encountered, focusing on any lack of compatibility and reduced state of vulcanization. A type of go/no-go elimination approach was followed, eventually resulting in a selection of bio-based oils that seemed to offer a feasible, technical alternative for mineral oil.

5.4.3.1.1 Compatibility

In the screening study of bio-based oils to replace mineral oil, compounds with varying oil/black ratios were evaluated (Table 5.10). Mixing of these EPDM based compounds did not pose any major obstacles. All green oil based compounds with 33/50, 33/150, 67/100, and 100/150 (phr/phr) oil/black compositions could be mixed well with mixing characteristics quite similar to the mineral oil references. It was noted that the compounds based on butter fat smelled like French fries, which could be considered objectionable. However, mixing of the 100/50 (phr/phr) natural oil compounds, that is, based on linseed oil, tung oil, coconut oil, olive oil, and butter fat, showed low power development and required very long mixing times. The resulting compounds appeared to be of relatively poor quality with varying degrees of stickiness/greasiness and a lack of coherence, probably because the amount of black filler was too low compared to the high level of relatively polar natural oils. The more polar natural oils and butter fat have a lower compatibility with the apolar EPDM, which makes them quite sensitive for these kinds of mixing issues. These incoherent/sticky 100/50 (phr/phr) oil/filler compounds will, therefore, receive no further consideration in this section.

The Mooney viscosities (ML) of the compounds with natural oils and butter fat were quite similar to those with equivalent levels of mineral oil, ranging between 35 and 75 MUs, although mineral oil usually gave the highest compound ML for any given oil/black ratio. The compound ML correlated quite nicely with the reciprocal of the oil/black ratio, as expected. The Mooney viscosity (ML) of all the compounds, including those based on mineral oil, having 33 phr oil and 150 phr black was extremely high (>150 MU), probably due to the extremely low plasticizer/filler ratio, preventing any practical application. The ML values of compounds with factice fully replacing mineral oil were also very high (>100 MU). In particular, when factice levels of 67 and 100 phr were used in combination with 100 and 150 phr carbon black, the ML was so high it could no longer be measured. This is because factice behaves like an elastic solid rather than a liquid plasticizer. Using factice as a 50/50 (w/w) mixture with butter fat resulted in a reduction of the compound ML, but only when used in combination with 50 phr carbon black were acceptable values of compound ML (<100 MU) obtained. The ML values of compounds with factice/butter fat having 67 and 100 phr carbon black were still too high for practical use. The ML results of the compounds where isotridecyl oleate, ethylhexyl stearate, and squalane were used were typically 20 MU below that of compounds containing equivalent levels of mineral oil because these three bio-based oils have much lower molecular weight than the mineral oil used (Table 5.8: ~400 vs. ~700 g/mol). The 33/150 (phr/phr) oil/filler compounds and those based on high factice levels all with very high compound ML were excluded from further evaluation.

The surfaces of compression molded plaques of the EPDM compounds with 33/50, 67/100, and 100/150 (phr/phr) oil/filler ratios based on ethylhexyl oleate, isotridecyl stearate, and squalane felt completely dry to the touch, showing no signs of bleeding after storage for 1 week at room temperature (5 on the scale of 1 to 5), and were comparable to their respective reference compounds based on mineral oil (Figure 5.11a). The compounds based on linseed oil, coconut oil, and hydrogenated

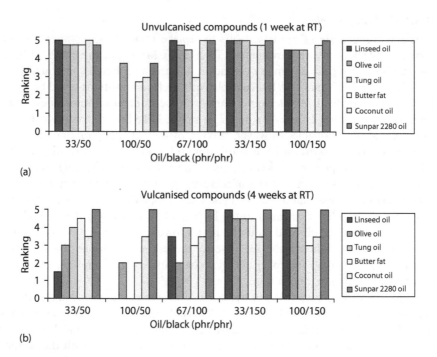

(a)

(b)

FIGURE 5.11 Bleeding of oil from samples with varying (bio-based) oil types and oil/black compositions: (a) milled sheets of unvulcanized compounds and (b) vulcanized compounds assessed after one and four weeks, respectively, of storage at room temperature (subjective ranking on a scale of 5: no oil bleeding; 4: surface slightly slippery; 3: greasy surface; 2: grease transfers to finger tips; 1: oil deposited in trays).

coconut oil (not shown) with 33/50 and 67/100 (phr/phr) compositions also remained dry, but the 100/150 (phr/phr) compositions showed a slight greasiness to the touch, achieving a rating of 4.5 on the scale of 1 to 5. For butter fat, only the 33/50 (phr/phr) compound remained dry with the other butter fat compounds becoming greasy (2 and 3 on the 1 to 5 scale). Obviously, the occurrence of bleeding of the natural oils and butter fat out of the compounds, especially at high oil levels (100 phr), is the result of a lack of compatibility of the polar oils for the apolar EPDM. Combinations of factice with butter fat (50/50 w/w) and with hydrogenated coconut oil (33/67 w/w) did not prevent the bleeding of the oil from the compounds.

In summary, the more polar nature of the natural oils and butter fat compared to the apolar EPDM, as witnessed by the relatively high solubility parameters (Table 5.8: 8.3–8.4 vs. 7.7 {cal/cm^3}$^{0.5}$), leads to reduced compatibility, resulting in poor compound quality and bleeding of the oil out of the compounds. This is especially the case in those compounds with high oil/black ratios. For butter fat there seems to be an extra effect related to its melting point being above room temperature. The good compatibility of isotridecyl oleate, ethylhexyl stearate, and squalane is because these oils have relatively lower polarities more closely matching that of EPDM, as witnessed by their solubility parameters (Table 5.8: 8.1, 8.1, and 7.5 vs. 7.7 {cal/cm^3}$^{0.5}$). The use of factice results in very high compound viscosity (ML) and when used in combination with natural oils does not reduce the bleeding.

5.4.3.1.2   Vulcanization

Because of the compatibility issues discussed in the previous sections, only the 33/50, 67/100, and 100/150 (phr/phr) oil/filler compounds will be further discussed in detail. Rheometry of the EPDM compounds showed that the natural oils, butter fat, and isotridecyl oleate had a detrimental effect on vulcanization compared to the mineral oil reference compounds (Figure 5.12). For a given compound composition, the rheometer torque difference ΔS (= MH – ML) decreases in the order: mineral oil > coconut oil > butter fat > tung oil > olive oil > linseed oil. ΔS also decreases with increasing levels of natural oil or butter fat. The compounds with hydrogenated coconut oil, ethylhexyl stearate, and squalane gave quite similar ΔS values to the mineral oil reference compounds. These results can be easily explained in terms of the level of unsaturation of the bio-based oils. Natural oils and fats are triglycerides of C12 – C18 fatty acids, both saturated (stearic, palmitic, myristic, and lauric acid) and unsaturated (mono-unsaturated: oleic and palmitoleic acid; di-unsaturated: linoleic acid; tri-unsaturated: linolenic and eleostearic acid). The unsaturation in these oils will compete with EPDM for sulfur vulcanization, resulting in a decrease of rubber cross-linking efficiency. Indeed, the ΔS sequence parallels the level of unsaturation of the natural oils, as witnessed by their iodine numbers (Table 5.8). Therefore, higher levels of natural oil will also result in more competition for sulfur vulcanization. As with mineral oil, hydrogenated coconut oil, ethylhexyl stearate, and squalane are virtually without unsaturation (Table 5.8) and for this reason will not compete with EPDM for sulfur vulcanization. Probably for the same reason, the rheometer scorch times ts2 follow the same trend as ΔS, both as a function of oil type and oil content. In view of the previous discussion, it is difficult to explain why

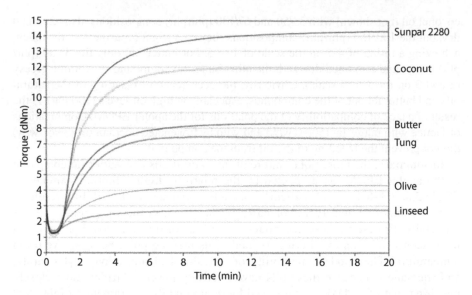

FIGURE 5.12 Overlay of MDR 2000E rheometer curves at 180°C for sulfur vulcanization of 33/50 (phr/phr) oil/black EPDM compounds.

the rheometer vulcanization time tc90 decreases in the series olive oil > butter fat > mineral oil > linseed oil > tung oil ~ coconut oil.

Figure 5.13 shows a plot comparing the normalized rheometer ΔS versus the calculated vulcanization efficiency. The normalized ΔS is defined as the ΔS of a particular EPDM compound with natural oil divided by ΔS of the corresponding mineral oil compound and ranges from zero (no vulcanization of the compound with natural oil) to unity (vulcanization identical to the mineral oil reference). The vulcanization efficiency is calculated as the molar ratio of the EPDM unsaturation and the total amount of unsaturation in both the EPDM and the oil, and therefore combines the effects of the level of unsaturation in the oil and the amount of oil used in the compound. This parameter also ranges from zero (infinite unsaturation in oil and/or infinite amount of oil) to unity (no unsaturation in oil and/or no oil). The data in Figure 5.13 show an excellent, linear correlation with a slope close to unity with the exception of the tung oil data. This shows that the decrease in rheometer ΔS for the natural oil based compounds is fully due to the competition for sulfur vulcanization with EPDM and that the unsaturation in the natural oils has the same molar reactivity toward sulfur vulcanization as ENB in EPDM rubber. The tung oil exception is probably related to the fact that it consists of ~80% of oleostearic acid, which is a conjugated triene, whereas the unsaturated fatty acids in the other natural oils and butter fat are monoenes (oleic and palmitoleic acid) or a nonconjugated diene (linolenic acid). By assuming a 30% efficiency for tung oil, the tung oil data in Figure 5.13 fits to the same line as the other (bio-based) oils.

Unfortunately, vulcanized plaques of the compounds with natural oils, hydrogenated coconut oil, and mixtures of factice with natural oils showed much more bleeding after storage for 4 weeks at room temperature than the corresponding

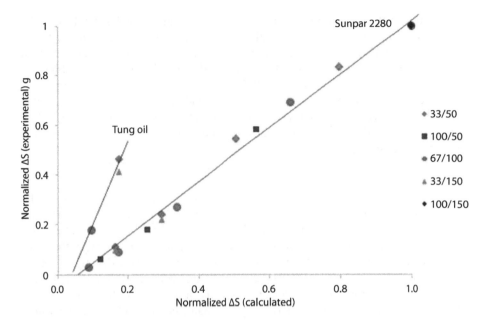

FIGURE 5.13   Normalized rheometer torque difference of EPDM compounds with varying oil types and varying oil/black levels versus calculated vulcanization efficiency.

unvulcanized compounds (Figure 5.11b). Of the 33/50 (phr/phr) natural oil/black compounds only the butter fat sample was considered to be just about acceptable (4.5 on scale from 1 to 5). All 67/100 (phr/phr) vulcanizates with natural oils were (very) greasy (2–4) and of the 100/150 (phr/phr) vulcanizates only the linseed oil and tung oil samples showed no bleeding. The vulcanizates with mineral oil and squalane were completely dry (5 on the 1 to 5 scale). For ethylhexyl oleate, a dry sample was obtained for the 33/50 (phr/phr) composition and slightly greasy samples for the 67/100 and 100/150 (phr/phr) compositions (5 and 4.5 ratings, respectively). For isotridecyl stearate dry vulcanizates were obtained for the 33/50 and 67/100 (phr/phr) compounds (5 ratings). As for the unvulcanized compounds, this shows that for the vulcanizates there is a lack of compatibility between the polar oils and apolar EPDM. Bleeding was especially high for the low black filled compounds because of insufficient porous black to absorb all the natural oil. There seems to be no clear explanation for the differences between the various oils; in addition, there seems to be no correlation with crosslink density.

In summary, the unsaturation present in natural oils competes with EPDM for sulfur vulcanization. Higher levels of unsaturation and higher levels of oil result in lower cross-link densities. Because squalane, isotridecyl stearate, ethylhexyl oleate, and hydrogenated coconut oil have very low levels of unsaturation, the compounds based on these oils achieve cross-link densities very close to the mineral oil references. It could be argued that sulfur vulcanization would reduce bleeding of polar oils out of the EPDM compounds by linking unsaturated oil molecules to the rubber network. However, calculations on a molar basis show that, although the level of unsaturation in the natural oils is sufficiently high to compete with EPDM for sulfur vulcanization, the fraction of the relatively small oil molecules thus linked to the network is negligible. Indeed, extractions with tetrahydrofuran for 2 days at room temperature performed on the 33/50 (phr/phr) oil/black vulcanizates yielded residue weights for the natural oil compounds (80%–83%), which were similar within experimental error to the mineral oil reference (80%) and also to the theoretical residue weight (81%). Only for linseed oil was a much lower residue weight collected (63%), which shows that the EPDM rubber in this compound was so poorly cross-linked that it partly dissolved.

5.4.3.1.3   Properties (After Aging)

In general, an excellent correlation was found between the physical properties of the vulcanized compounds before aging and cross-link density (rheometer ΔS) (Figure 5.14) as is commonly observed in rubber chemistry and technology studies. For compounds with a given oil/carbon black composition but of varying oil type, the hardness and moduli increases with cross-link density, whereas the elongation at break (EB), compression sets (CS) at 70, 100, and 125°C, and volume swell in IRM 901 and 903 oils decrease with cross-link density, typically in sequence of linseed oil < olive oil < tung oil < butter fat < coconut oil < hydrogenated coconut oil ~ isotridecyl stearate ~ ethylhexyl oleate < squalane < mineral oil. For the 33/150 (phr/phr) oil/black compounds, the tensile strength (TS) increases with ΔS. For the 33/50 (phr/phr) oil/black compounds, TS goes through a maximum versus ΔS. For the other compound compositions, TS shows signs of peaking at an optimum cross-link density. The tear resistance (tear) shows an optimum versus cross-link density for all compositions. CS at –25°C is close to 100% for the compounds based on the natural oils and only deviates for the reference mineral oil compounds to values below 70%. In summary, all physical properties correlate with the cross-link density in a well-known and expected way, and thus correlate with the levels of unsaturation in the (bio-based) oils as discussed in the previous section.

For a given oil the physical properties are usually found to correlate with the carbon black and oil levels (compare various fitted lines in Figure 5.14), again in a way that is commonly observed in rubber technology studies. Higher oil levels result in compounds of lower hardness, moduli, and TS before and after aging and in higher EB, CS, and oil swells. It is noted that for the unsaturated bio-based oils, the oil not only acts as a plasticizer, but also reduces the cross-link density. Therefore, for a given oil type, the level used affects properties in two ways, by coincidence in a parallel fashion. Higher levels of carbon black result in higher hardness, moduli, TS, and tear, in somewhat higher CS's at various test temperatures, and in lower oil volume swells. For some reason, the effects of carbon black content on EB and oil content on tear strength are not so straightforward.

In a small side study, an attempt was made to compensate for the loss of crosslink density due to competition for sulfur between EPDM and unsaturated, bio-based oils for a 33/50 (phr/phr) butter fat/black compound. Options included increasing the ENB content of the EPDM from 5.5 to 9.0 wt%, increasing the sulfur content in the compound from 1.25 to 2.5 phr, and increasing the sulfur curative package, that is, doubling the amount of sulfur (and accelerators). Each option resulted in an

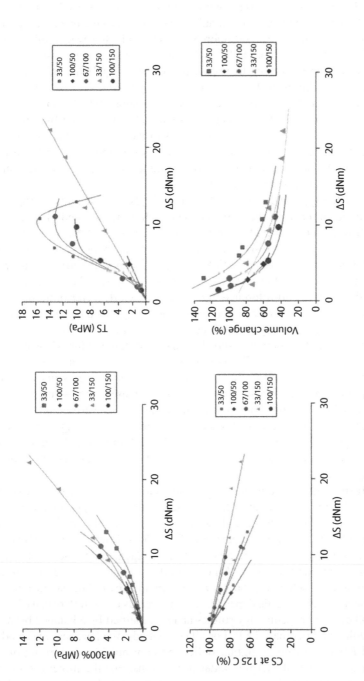

FIGURE 5.14 Plots of physical properties (top left: modulus at 300% elongation; top right: tensile strength; bottom left: compression set at 125°C and bottom right: oil swell in IRM 901 oil at 70°C) versus cross-link density (rheometer ΔS) for compounds with varying (bio-based) oil/carbon black compositions (33/50, 33/150, 67/100, 100/50, and 100/150 phr/phr).

increased cross-link density, as witnessed by the rheometer torque difference ΔS, and corresponding improvements of physical properties, with higher TS, lower high-temperature CS's, and lower oil swells. Although these measures resulted in desired changes, the absolute effects were too small, since the final vulcanizate properties did not reach those of the mineral oil reference compound. However, combining an increase of ENB content from 5.5 to 9.0 wt% with either increasing just the sulfur content or doubling the sulfur curative package did result in a butter fat compound with properties quite similar to those of the mineral oil reference. Only the CS at –25°C could not be repaired by these (combined) measures, which shows that the low-temperature CS of these EPDM compounds is not only limited by cross-link density, but also by solidification of the plasticizer in the case of butter fat (and the [modified] natural oils). Indeed, a DSC experiment showed that the 33/50 (phr/phr) butter fat compound displayed an extra melting point at +6°C as well as the typical EPDM glass transition temperature at –59°C.

The effects of the type and amount of bio-based oils on vulcanizate properties are fully explained in terms of their effects on cross-link density and, thus, are not especially exciting. Even the beneficial effects of increasing the ENB content of the EPDM and/or adding more sulfur (and accelerators) on the vulcanizate properties are simply the result of increased crosslink density. The properties after aging show a different behavior though. On first sight, the data are actually quite confusing. The properties of the mineral oil reference compounds typically deteriorate on aging and deteriorate more with harsher aging conditions (original unaged $\rightarrow$ 70°C $\rightarrow$ 100°C $\rightarrow$ 125°C), leading to higher hardness, lower TS, and tear. The compounds based on (hydrogenated) coconut oil, isotridecyl stearate, and ethylhexyl oleate show similar trends, but not as strong. The compounds based on squalane have proper-ties after aging that are closest to the mineral oil reference. Interestingly, for the squalane compounds, the hardness change on aging decreases with increasing aging temperature, showing virtually no hardness change at 125°C. Surprisingly, TS and tear of the aged olive oil vulcanizates actually increased on aging, whereas TS and tear of the aged vulcanizates based on butter fat and tung oil only show scatter around the starting values of the nonaged samples, almost suggesting some sort of heat stabilizing effect.

These confusing aging results can be rationalized again in terms of cross-link density. Obviously, for aged samples there is no rheometer torque difference that can be used as a measure for the cross-link density; therefore, the modulus at 100% elongation (M100%) is used for that purpose. Figure 5.15 shows a plot of TS versus the corresponding M100% for the 33/50 (phr/phr) oil/black compounds both before and after aging at various conditions. All data overlap on one curve with the excep-tion of the tung oil data, which is probably again explained by its special unsaturated structure (cf. Section 5.4.2.). This dependence of TS versus cross-link density is well known from other rubber technology studies and is explained by a balance between increased modulus and decreased EB. For all compounds, the hardness and moduli increased and EB decreased on aging, indicating continued cross-linking of the EPDM rubber as a result of thermo-oxidative reactions. For the mineral oil com-pound, an increase in cross-link density due to aging resulted in a decrease of TS (right flank of TS curve). For the coconut oil and butter fat compounds, TS increased

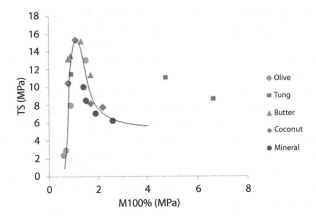

FIGURE 5.15 Plot of tensile strength versus modulus at 100% elongation as a measure for cross-link density of 33/50 (phr/phr) oil/black compounds before and after aging for 1 week at 70, 100, and 125°C.

and then decreased on aging (around the maximum of the TS curve). Finally, for the olive oil compound, increased cross-linking due to aging resulted in a TS increase (left flank of TS curve). Similar plots to Figure 5.15 have been constructed for tear strength of the 33/50 (phr/phr) oil/black compositions and for TS and tear of the other compound compositions. It is finally noted that these findings suggest a new way of producing (EPDM) rubber products with optimum heat aging resistance. By deliberately giving a compound a slight under-cure, vulcanizates with a somewhat suboptimum TS are obtained (just on the left side of TS optimum in Figure 5.15). On heat aging further cross-linking occurs, which results in a small increase of TS and on further aging in a small decrease of TS (passing TS maximum). Overall, such a suboptimum vulcanized EPDM product will show rather good TS retention on aging. Obviously, the suboptimum vulcanization of the original sample is not beneficial for the unaged elasticity and oil resistance. As usual, one has to find the best balance of properties for a particular application.

In summary, the properties of the EPDM vulcanizates with (bio-based) oils simply follow the rubber textbook correlations with crosslink density, which in its turn is determined by the competition for sulfur vulcanization between EPDM and the unsaturated oils. As a result, increasing the EPDM ENB content and/or the amount of sulfur (and accelerators) provides an easy way to compensate for the loss in cross-link density and vulcanizate properties. Plots are constructed showing that changes of TS and tear on aging are related to changes in cross-link density as a result of continued cross-linking.

5.4.3.2 Sustainable Fillers

5.4.3.2.1 Black Fillers

Our first study with respect to alternative fillers was to replace the standard FEF N550 carbon black by pyrolysis black in an automotive, solid seal formulation. Some previous compound optimization related to confirming differences in reinforcement behavior resulted in formulations as shown in Table 5.11 with the pyrolysis black level being 12.5% higher than the N550 (fast extrusion furnace black) level. The rheometer data showed that the scorch time ts2 and final state of cure ΔS are quite comparable, but that the compound with pyrolysis black showed a longer vulcanization time to achieve tc90. This is a known phenomenon of the use of pyrolysis black, probably due to the fact that it consists not only of carbon black, but also contains around 5 wt% of zinc oxide and zinc sulfide (Table 5.9). The presence of zinc oxide/sulfide is obviously related to the use of zinc oxide and possibly other zinc salts, like zinc soaps and/or accelerators, in the original tire composition. For the same reason the pyrolysis black also contains silica from the original tire formulation, with the level varying depending on the tire type used as the feedstock for the pyrolysis process. Compared to other commercially available pyrolysis blacks, the particular grade used in this study actually showed a rather limited increase of tc90. As an aside, it is worth noting that the polyaromatic hydrocarbon (PAH) content of the pyrolysis black used is below the detection limit, whereas for furnace blacks PAH levels are detectable. The data in Table 5.11 indicate that all the physical properties before and after heat aging of the vulcanizates are similar when the pyrolysis black content is increased by 12.5%. This is in agreement with the somewhat lower surface area and surface structure of the pyrolysis black compared to N550. The overall conclusion is that pyrolysis black can be considered a technical alternative for certain medium reinforcing furnace blacks.

5.4.3.2.2 White Fillers

A typical nonblack EPDM bull's-eye compound based on a silica and clay white filler system was used to evaluate the performance of rice husk ash and micro-cellulose as potential bio-based fillers as possible replacements of the reinforcing silica (Table 5.12). The compound Mooney viscosity (ML) of the first silica control compound, mixed with a starting mixer body temperature of 70°C, was rather low. This is probably due to polymer degradation, since it took an excessively long time (around 17 min) to reach a batch temperature of 150°C, which was required to reach completion of the silanization reaction. Therefore, all additional experiments were performed with a starting mixer body temperature of 130°C, which gave reasonably reproducible results for silica control compounds. Nevertheless, for both the rice husk ash and micro-cellulose compounds, lower compound ML were measured, suggesting a lack of coupling with the rubber. The addition of maleated EPM to the micro-cellulose compound resulted in a compound ML which is comparable to that of the silica reference, suggesting an improved coupling of the cellulose fibers to the EPDM rubber matrix. The rheometer data showed that the vulcanization kinetics of the compounds with the bio-based, white fillers were comparable to that of the silica reference compound (similar ts2 and tc90 values). However, the ΔS values of the bio-based filler compounds were significantly smaller than those of the silica compound. This probably does not reflect a lower state of cure for the former, but a lack of reinforcement, since the hardness, TS, EB, and tear of all the compounds with bio-based, white fillers are (much) lower, compared to the silica reference. Comparing the physical properties of the compounds with the bio-based white fillers shows that rice husk ash actually yields the poorest performance. The two micro-cellulose grades show similar properties. The addition of maleated EPM to the micro-cellulose compound resulted in improved EB and tear, which may be due to better dispersion/coupling of the cellulose fibers. The properties after aging are quite similar for all white fillers, when expressed as relative changes (data not shown), with micro-cellulose Arbocel UFC M8 being a positive exception (lowest relative change in tensile properties on aging). The CS values at 23 and 70°C are comparable for all fillers, but CS at 100°C is the lowest for the silica control and one of the micro-cellulose samples. The overall conclusion of this limited screening of bio-based white fillers is that they do not have the reinforcing properties of silica. Generally, similar performance has been observed in other studies on bio-based fillers in rubber compounds. This disappointing result is probably due to a lower surface area/structure of the bio-based fillers combined with a lack of reactivity to silane coupling. Despite this, these bio-based white fillers can be used as green alternatives for inert white fillers, as will be shown in the final study on compounds with maximized sustainability content in the next section.

5.4.3.2.3   Maximizing the Content of Sustainable Ingredients

In this final study, the best options from the screening studies described in the previous sections were combined in a final effort to maximize the content of sustainable ingredients in EPDM compounds. Two highly filled, automotive, solid seal EPDM compounds were used for this purpose. The first compound having low-temperature flexibility for dynamic sealing applications was based on an amorphous EPDM. The second compound is more typical of a static seal, where low-temperature flexibility is not required. This compound was based on a crystalline EPDM polymer (Table 5.13). It should be noted that these two automotive sealing compounds are more highly filled than would be typically practiced in industry to be able to maximize the sustainable ingredient content of the compounds. As mentioned in the earlier section, the amount of pyrolysis carbon black was 12.5% higher than the corresponding furnace N550 black content. Rice husk ash was used as an inert filler to replace calcium carbonate (whiting). Based on the compound formulations in Table 5.13, the total amount of green ingredients, that is, the sum of the bio-based ethylene of the Keltan Eco EPDM rubber, the green oil, and white filler, plus the sustainable black filler, amounted to 86% for the dynamic automotive seal and as high as 90% for the static automotive seal, due to the higher weight content of sugar cane derived ethylene in the crystalline Keltan Eco EPDM used.

Table 5.13 shows that the compound Mooney viscosities of both sustainable compounds are significantly below those of the reference compounds by about 10 MU. The Garvey strip extrusion results showed that the extruder throughputs of the sustainable compounds were comparable to those of the reference compounds, but the head pressures were much lower. As a result, the extruder throughputs normalized to the pressures were significantly lower for the sustainable compounds, which is in agreement with the lower compound ML. The total score for edge, corner, and surface quality according to the Garvey die ranking was the maximum of 12 for all extruded strips, showing excellent processing of the sustainable compounds based on squalane, pyrolysis black, and rice husk ash. The scorch sensitivity of the sustainable compounds is substantially less than of the reference compounds, as witnessed by much longer Mooney scorch times (ts5). The rheometer ts2 values of both sustainable compounds were similar to those of the reference compounds, but the tc90 times were significantly longer, which is probably related to the retarding effect of the pyrolysis black. The rheometer torque difference ΔS of the sustainable compounds was comparable to those of the reference compounds, which explains why hardness, elongation at break, and compression sets (CS) at 23 and 70°C are also comparable. The tensile and tear strengths and CS at 100°C of the sustainable compounds were somewhat inferior. The test results after aging of the sustainable compounds are clearly inferior to those of the reference compounds. This is fully accountable to the use of squalane as the bio-based oil, since the weight loss of the sustainable compounds on aging for 168 h at 135°C is similar to the original squalane content (19.2 vs. 20.2 wt% and 20.2 vs. 20.9 wt% for the dynamic and static seal compounds, respectively). There is hardly any weight loss for the reference compounds using Sunpar 2280 mineral oil on aging. It seems that despite an atmospheric boiling point of ~350°C, squalane is still too volatile compared to regular mineral oil. This seems to be no issue for aging at temperatures up to 125°C but becomes critical at 135°C and, therefore, is clearly a topic for further studies.

To have a fully sustainable EPDM compound with 100% sustainable ingredients, the residual 10%–15% nonsustainable content of the EPDM compounds in Table 5.13 should be further addressed. A major step would be to develop a second generation Keltan Eco EPDM not only based on green ethylene, but also using green propylene, which would bring the total bio-based content of the EPDM rubber to ~95% and of the EPDM compounds also to ~95%. Currently, production of propylene based on green resources is being explored,[26] among others via (1) production of methanol from wood, followed by conversion of methanol to propylene, (2) sugar-based routes either via ethanol to ethylene and then via metathesis to propylene or via isopropanol to propylene, and (3) direct fermentation of glucose using genetically engineered microorganisms to a mixture of olefins, including propylene. The finishing touch will be a green diene for EPDM. As an example to stimulate interest, it can be mentioned that first experiments with an amorphous EPDM with 6 wt% 2,4-dimethyl-2,7-octadiene (natural terpene supplied by Dérivés Résiniques & Terpénique) as the diene, showed reasonable sulfur vulcanization characteristics and corresponding vulcanizate properties, similar to a medium ENB-EPDM. The final step toward a fully sustainable EPDM rubber compound will require the development of bio-based rubber additives, especially the curatives, which considering their chemical structure will be a much longer and more challenging development.

5.5 CONCLUSIONS

Keltan Eco 5470, produced from bio-based ethylene, has been compared with regular Keltan 5470. The chemical composition, the rheology, and other molecular characteristics of the two polymers are identical. As a result, the properties of low-filled compounds and the corresponding vulcanizates are also the same within experimental error. The only obvious and expected difference was identified by ^{14}C analysis, which confirmed that Keltan Eco 5470 contains 70% bio-based carbon, exactly matching the amount of bio-sourced ethylene in the polymer as determined with infrared spectroscopy.

In an effort to develop compounds based on Keltan Eco EPDM with the highest level of sustainable ingredients, a series of bio-based/sustainable oils and fillers were screened. Typical issues encountered when exploring relatively polar and unsaturated natural oils and fats in EPDM compounds are a lack of compatibility (mixing issues and oil bleeding) and competition for sulfur vulcanization (reduced cross-link density and corresponding inferior vulcanizate properties). Modified natural oils, such as hydrogenated coconut oil or trans-esterified mono-esters, have improved compatibility and/or vulcanization performance. Squalane (EPM hexamer) provides the best bio-based alternative for mineral oil plasticizer in this study, since it is as apolar as EPDM and is fully saturated. As far as sustainable fillers are concerned, pyrolysis black was shown to have a reinforcing efficiency 90% of that of furnace N550 black. Rice husk ash and micro-cellulose do not show reinforcing properties, but can still be used as inert, white fillers, substituting certain traditional, mineral white fillers. Combining these leads has resulted in automotive solid seal EPDM compounds based on Keltan Eco with more than 85% sustainable content and properties reasonably comparable to the reference EPDM compounds, including heat aging resistance up to 125°C.

ACKNOWLEDGMENTS

We would like to acknowledge Dr. N. Kumar Singha of the Rubber Technology Centre of the Indian Institute of Technology in Kharagpur for a useful literature survey on green rubber compounding ingredients. Next, we would like to thank Rutteman, Friesland Campina, Amyris, Cargill, Hansen+Rosenthal, and EmeryOlea for supplying us with samples of bio-based oils and fat and Black Bear and J. Rettenmaier & Söhne for alternative filler samples used in this study. We also express our gratitude to the ARLANXEO Polymer Testing group in K10 Leverkusen, Germany and the former Keltan Rubber Processing & Testing Laboratory in Geleen, Netherlands for performing all the mixing and testing.

REFERENCES

1. W. Hofmann, Ethylene propylene rubber (EPM and EPDM), *Rubber Technology Handbook*, Hanser Publishers, Munich, 1989.
2. J.A. Riedel and R. Vander Laan, Ethylene propylene rubbers, in *The Vanderbilt Rubber Handbook*, 13th ed. R.T. Vanderbilt Co., New York, 1973, p. 123.
3. J.W.M. Noordermeer, Ethylene-propylene copolymers, in *Encyclopedia of Polymer Science and Engineering*, John Wiley & Sons, New York, 1986, Vol. 6, p. 178–196.
4. M. Alvarez Grima, P. Hough, D. Taylor, and M. van Urk, Bio-based EPDM selected for bus window seals in Brazil, *Eur. Rubber J.* (Mar–Apr 2013).
5. M. Alvarez Grima et al. Keltan® Eco: First Bio-based EPDM during International Rubber Expo, 182nd Technical Meeting of ACS Rubber Division in Cincinnati (2012), Sustainable Rubbers Meeting in London (2013), and Green Polymer Chemistry Meeting in Cologne (2014).
6. http://www.braskem.com.br/plasticoverde/eng/Produto.html

7. H.C. Booij, Long-chain branching and viscoelasticity of ethylene-propylene elastomers, *Kaustch. Gummi Kunstst.* 44(1991), 128.
8. M. van Duin, P. Hough and N. van der Aar, Green EPDM Compounding during International Rubber Conference/Deutsche Kautschuk Gesellschaft (2015) Meeting in Nürnberg, ACS Rubber Division (2015) Fall Meeting in Cleveland, and Plastics and Rubber Institute of Malaysia Conference (2016) in Kuala Lumpar.
9. K.A. Job, Trends in green tire manufacturing, *Rubber World* (March 2014), 32.
10. E. M. Fernandes et al., Cork based composites using polyolefins as matrix: Morphology and mechnical performance, *Comp. Sci. Technol.* 70 (2010), 2310.
11. J. Wang, W. Wu, W. Wang, and J. Zhang, Effect of a coupling agent on the properties of hemp-hurd-powder filled styrene-butadiene rubber, *J. Appl. Polym. Sci.* 121 (2011), 681.
12. A. Hassan, A.A. Salema, F.N. Ani, and A.A. Bakar, A review on oil palm empty fruit bunch fiber-reinforced polymer composite materials, *Polym. Comp.* (2010), 2079.
13. T. Vladkova, S. Vassileva, and M. Natov, Wood flour: A new filler for the rubber processing industry. I. Cure characteristics and mechanical properties of wood flour-filled NBR and NBR/PVC compounds, *J. Appl. Polym. Sci.* 90 (2003), 2734.
14. C. Twigg, R. Verberne, and J. Jonkman, Method for obtaining a carbon black powder by pyrolizing scrap rubber, *Rubber Fibres Plastics* 9 (2014), 34.
15. S. Dasgupta et al., Characterization of eco-friendly processing aids for rubber compound, *Polymer Testing* 26 (2007), 489.
16. W.G.D. Jayewardhana, G.M. Perera, D.G. Edirisinghe, and L. Karunanayake, Study on natural oils as alternative processing aids and activators in carbon black filled natural rubber. *J. Nat. Sci. Foundation Sri Lanka* 37 (2009), 187.
17. H. Ismail and H. Anuar, Palm oil fatty acid as an activator in carbon black filled natural rubber compounds: Dynamic properties, curing characteristics, reversion and fatigue studies, *Polymer Testing* 19 (2000), 349.
18. J. Clarke et al., Vegetable Oils as a Replacement for Petroleum Oils in Elastomer Compounds during Sustainable Rubbers Meeting in London (2013).
19. L.D. Beyer, C.M. Flanigan, D. Klekamp, and D. Rohweder, Comparative study of silica, carbon black and novel fillers in tread compounds, *Rubber Fibres Plastics* 8 (2013), 246.
20. A.A. Gujel et al., Development of bus body rubber profiles with additives from renewable sources: Part II—Chemical, physical–mechanical and aging characterization of elastomeric compositions, *Mat. Design* 53 (2014), 1112.
21. Z.S. Petrovic, M. Ionescu, M. Milic, and J.R. Halladay, Soybean oil plasticizers as replacement of petroleum oil in rubber, *Rubber Fibres Plastics* 9 (2014), 218.
22. C. Bergmann, J. Trimbach, and Z. Saleem, Replacement of phthalates by vegetable oil derivative plasticizer in NBR compounds, *Rubber Fibres Plastics* 9 (2014), 225.
23. S. Schofer et al., Biofene, a renewable monomer for elastomer materials with novel properties: Polymer development, characterization and use in elastomer formulation, *Rubber Fibres Plastics* 9 (2014), 235.
24. D. McPhee, A. Pin, L. Kizer, and L. Perelman, Deriving renewable squalane from sugarcane, *Cosm. Toil* (Jul/Aug 2014), 129.
25. M.R. Beaulieu et al., Rice Husk Ash Silica: Agricultural Byproduct to Silica Filler during Fall 186th Technical Meeting of ACS Rubber Division in Nashville (2014).
26. R. Rocle, Production of Bio-Propylene and other Light Olefins via Direct Fermentation during Future of Polyolefins in London (2015).

6 Carbon Fiber Composite Materials

Abdullah Al Mamun,
Moyeenuddin Ahmad Sawpan,
Mohammad Ali Nikousaleh,
Maik Feldmann, and Hans-Peter Heim

CONTENTS

6.1 INTRODUCTION

Carbon fibers assign to fibers that consist of at least 92 wt.% carbon in composites [1]. They can be long (continuous) or short depending on the fiber production process. Carbon fibers, structure can be amorphous, semi-crystalline, or crystalline. The crystal structure of graphite is an example of a crystalline fiber, which consists of sp^2 hybridized carbon atoms arranged two-dimensionally in a honeycomb structure in the x-y plane. Carbon atoms are bonded within a layer by covalent bonds provided by the overlap of the sp^2 hybridized orbitals and metallic bonding provided by the delocalization of the p_z orbitals. The bonding between the layers is van der Waals bonding, so the carbon layers can easily slide with respect to one another [2].

The difference between the in-plane and out-of-plane bonding provides different properties. For instance, graphite has a high modulus of elasticity parallel to the plane and a low modulus perpendicular to the plane. Thus, graphite is highly anisotropic. The carbon layers in graphite are stacked in an AB layer packing sequence, such that half of the carbon atoms have atoms directly above and below them in adjacent layers (Figure 6.1) [3].

In carbon fiber, there can be a graphite region of size L_c perpendicular to the layer and L_a parallel to the layers. The greater the degree of alignment of the carbon layers parallel to the fiber axis; that is, the stronger the fire texture, the greater the c-axis crystallite size, the carbon content, fiber's tensile modulus, electrical conductivity, and thermal expansion and as well as the smaller the fiber's coefficient of thermal expansion and internal shear strength [4,5].

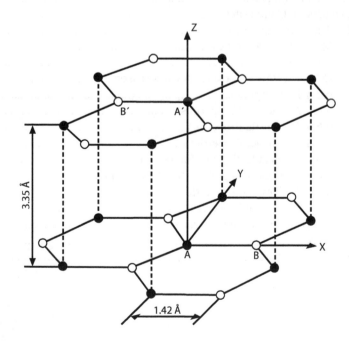

FIGURE 6.1 The crystal structure of graphite. (From Chung, D.D.L., *Carbon Fibre Composites*, Butterworth-Heinemann, Boston, pp. 3–4.)

The stratic carbon refers to a very specific structure in which adjacent aromatic sheets overlap with one carbon atom at the center of each hexagon. The turbostratic refers to aromatic sheets randomly oriented to each other. At the optimal interlayer separations, the corresponding binding energies (per carbon atom) of graphene sheet model dimers increase with increasing polycyclic aromatic hydrocarbon sizes for all three different stacking orders. It can be anticipated that the binding energy per carbon atom will monotonously decrease until the graphene sheet model is large enough to reproduce the experimental exfoliation energy of graphite [5].

In recent times, the most important raw material for most commercial production of carbon fiber is polyacrylonitrile (PAN), but due to high costs, these carbon fibers remain a value added product for use in high-end industrial applications, aerospace, sporting equipment, and different luxury automotive products [6]. Rayon-based carbon fibers are no longer in production. Pitch-based fibers satisfy the needs of niche markets, and show promise of reducing prices to make mass markets possible. Vapor-grown fibers are entering commercial production, and carbon nanotubes are full of promise for the future [7].

However, the interest in lignin for its use as an inextensible raw material for the production of carbon fibers of reasonable qualities is rapidly growing [8,9]. A renewed interest in the beginning of the 1990s occurred when researchers [10,11] prepared carbon fibers from steam-exploded lignin, modified through hydrogenolysis and phenolysis, which had a very good spinnability in the melt state. Advantages of using lignins as precursors to carbon fibers are related to the high carbon content and high carbon yield after carbonization and the lack of exposure to toxic elimination products during carbonization [11]. Furthermore, the accessibility of lignin in large quantities could push the development of inexpensive, carbon-fiber, lightweight materials. Carbon nano fibers with diameters as small as 200 nm have been successfully prepared from lignin using electro-spinning [12]. The key parameters in the production of high-strength lignin-based carbon fibers are as the glass-transition temperature, molecular weight, and molecular weight distribution [10].

Composite materials based on carbon fiber have been developed for many industrial applications as well as during the last couple of years the demand is increasing in the field of automobile applications.

BASF has developed its first carbon fiber reinforced PBT portfolio (polybutylene terephthalate) composites called Ultradur®. The main feature of Ultradur B4300 C3 is low electrostatic charge along with good conductivity. This makes it particularly suitable for components in sensitive areas of measurement and control technology for machines and automotive electronics. The antistatic behavior of PBT, less dust or dirt adheres to the component: this allows working reliably and permanently even in unfavorable usage conditions. The use of the conductive Ultradur grade also reduces the risk of electrostatic loading and possible sparking in areas with explosion hazards. Thus, it meets the increasing requirements on material and parts especially in automotive electronics. Parts made of the carbon fiber reinforced engineering plastic retain their antistatic property permanently and even after contact with media (e.g., fuels) and at high temperatures.

The other grade of Ultradur (B4300 C3 LS bk15126) is available in commercial quantities. The material, which is reinforced with 15% carbon fiber, has a low volume and surface resistance. It does not absorb moisture and its good mechanical properties are similar to those of a standard PBT with 30% glass fibers.

The other possible applications of carbon fiber reinforced PBT are parts in cars or machines with gases or fluids flowing through, fast-moving components in textile machines, or conveyor belt elements that are subject to static charge due to friction. Other fields of usage include machines in paper processing, printers, and transportation packaging for sensitive electronic goods that require electrostatic discharge (ESD) protection [13].

The Carbon Company SGL Group and the chemical company BASF have concluded the joint research of a new composite material system as an important development step of their collaboration. The system aims at enhancing the cost-effectiveness of manufacturing thermoplastic carbon-fiber composites, for example, in injection procedures (T-RTM: thermoplastic resin transfer molding) and reaction injection molding. The composite is based on a reactive polyamide system and compatible carbon fibers. Carbon fiber surface or sizing is specially designed for the matrix system as well as tailored thermoplastic reactive systems. It means that lightweight structural components for the automotive industry can now be manufactured quickly and easily [14].

Mitsui, Toray, and Hexagon Lincoln have entered into a joint development agreement to conduct a viability study into the joint manufacture and supply of carbon fiber reinforced high-pressure hydrogen cylinders for vehicles in Japan. The long-term aim of the venture is to manufacture high-pressure hydrogen carbon fiber cylinders for fuel cell vehicles in the Japanese market. The development of lighter parts and materials to reduce energy consumption by transportation equipment, especially in motor vehicles, is seen as an important approach to the solution of this problem. The market for lightweight materials, such as carbon fiber, is expected to expand rapidly [15].

The BMW i3 is constructed with carbon fiber reinforced plastic (CFRP) for electric drive. CFRP provides the weight reduction that effectively neutralizes the heft of the car's battery pack. The result is a four-passenger car that can go 100 miles on a charge or accelerate to 60 mph in less than 7 seconds due to weight reduction of 250–350 kilos from carbon fiber [16].

Significant amounts of carbon fiber are used in a vehicle production with no significant cost increment. Car production is 10,000 units per year. BMW has made roof or hood panels from carbon fiber, mostly for limited production performance models. There are also million-dollar McLarens and Lamborghinis with CFRP bodies [17].

One of BMW's goals was to make lifecycle energy costs be less than for a traditional vehicle [18]. BMW and SGL Automotive Carbon Fibers set up a new factory in Moses Lake, WA. The Moses Lake plant draws from utilities making heavy use of hydro power. The factory takes a polyacrylonitrile (PAN) precursor produced in Otake, Japan, by joint venture involving Mitsubishi Rayon (MRC) and SGL Group. Moses Lake unit turns the PAN fibers into carbon fibers with proper control of carbonization ovens. The next step turns them into lightweight carbon fiber fabrics in Wackersdorf, Germany, and, finally, they wind up in Leipzig, Germany, where the CFRP parts are finished and the i3 is assembled along with BMW's 1 Series sedan and X1 SUV [16,19].

BMW armored the cockpit of its next-generation 7 Series with carbon fiber in a move to improve safety and distance its big sport sedan from Mercedes-Benz and Audi as the world's finest, safest high-end car. BMW uses the term carbon core to describe the safety cell surrounding the passenger compartment that uses CFRP along with aluminum and high-strength steel. BMW must have confidence in the CFRP process and demand almost 1 million carbon fiber pieces to build the 7 Series between now and the end of 2016. Additionally, in the 5 Series and the 3 Series, and their SUV equivalents, BMW would need 10–20 million CFRP components a year [16,20].

6.2 PROPERTIES OF CARBON FIBER

There are many varieties of carbon fibers, depending on their production process, available in the economic market. Thus, the physical properties vary over a broad domain. Table 6.1 and Figure 6.2 show a relationship between fiber grades and tensile properties. The general purpose fibers are made from isotropic pitch, have modest levels of tensile strength and modulus. Furthermore, the strain to failure is about 2%. However, they are the least expensive pitch-based fiber and are useful in enhancing modulus or conductivity in many industrial applications [7].

Mesophase pitch fibers are appropriate for high temperature uses. They may be heat treated to very high modulus values, approaching the in-plane modulus of graphite at 1TPa. On the other hand, the strain to failure property is as low as 0.2%. PAN-based fibers are the strongest and are used for very special applications. However, when they are heat treated the modulus increases to up to 600 GPa and at the same time, the strength property decreases to 2.5 GPa. Due to the heat treatment, the strain to failure property reduced from 2% to 0.7%. Therefore, the application field of PAN based carbon fiber trended to narrow [21].

TABLE 6.1
Tensile Properties of Selected Carbon Fibers

Type	Manufacturer	Product Name	Tensile Strength (GPa)	Young's Modulus (GPa)	Strain to Failure (%)
PAN	Toray	T300	3.53	230	1.5
		T1000	7.06	294	2.0
		M55J	3.92	540	0.7
GP-Pitch	Hercules	IM7	5.30	276	1.8
HP-Pitch	Kureha	KCF200	0.85	42	2.1
	BP-Amoco	Thornel P25	1.40	140	1.0
		Thornel P75	2.00	500	0.4
		Thornel P120	2.20	820	0.2

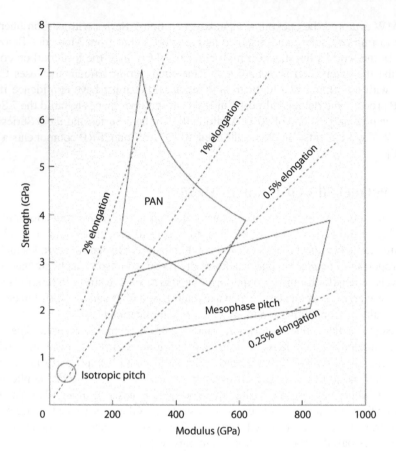

FIGURE 6.2 Tensile properties of various carbon fibers. (From Lavin, J.G., Carbon fibre, in *High Performance Fibres*, edited by Harle, J.W.S., Woodhead Publishing Limited, Cambridge, England, 2001, pp. 156–188.)

Electrical and thermal conductivity of carbon fibers are important in many applications, and thermal conductivity is illustrated in Figure 6.3. PAN-based carbon fiber shows very high electrical resistivity and low to moderate thermal conductivity. Isotropic pitch fiber shows moderate electrical resistivity and thermal conductivity. Mesophase pitch fiber has the lowest electrical resistivity and the highest thermal conductivity [22].

Finally, there is a property of high-performance carbon fibers, both PAN and mesophase pitch-based, which sets them apart from other materials. They are not subject to creep or fatigue failure. These are important characteristics for critical applications. In a comparison of materials for tension members of tension leg platforms for deep sea oil production, carbon fiber strand survived 2,000,000 stress cycles between 296 and 861 MPa. In comparison, steel pipe stressed between 21 and 220 MPa failed after 300,000 cycles [21].

Creep studies on PAN and pitch-based carbon fibers were conducted at 2300°C and stresses of the order of 800 MPa [22,23].

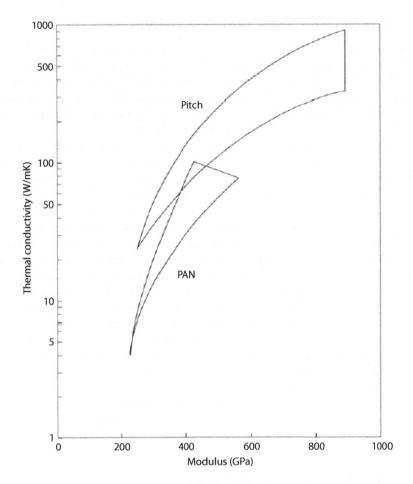

FIGURE 6.3 Thermal conductivity of carbon fibers. (From Lavin, J.G., Carbon fibre, in *High Performance Fibres*, edited by Harle, J.W.S., Woodhead Publishing Limited, Cambridge, England, 2001, pp. 156–188.)

The data obtained from projections at ambient temperatures indicate that creep deformations will be infinitesimally small. There is a surprising degree of cross-correlation in the physical properties of carbon fibers. The mechanisms for con-duction of heat and electricity are different in carbon fibers: heat is transmitted by lattice vibrations and electricity by diffusion of electrons and holes. However, there is a strong correlation between the electrical resistivity and thermal conductivity which allows thermal conductivity to be estimated by measurement of the electrical resistance, a much simpler measurement [24,25]. In the case of pitch based carbon fibers, the thermal conductivity decreased with increasing of electrical resistivity. At the stage of from 1000 W/mK to 500 W/mK, the thermal conductivity decreased sharply with increase of minimum electrical resistivity. Afterward, it decreased slowly.

6.3 FIBER PRODUCTION AND STRUCTURE

For high-strength carbon fibers, it is important to avoid the formation of voids within the fiber which come from raw material footprint. So the raw material source and raw material production process are important for carbon fiber quality, conversion rate, and density. The preferred process for high-strength fiber today is wet-spinning. Processes for melt-spinning PAN plasticized with water or polyethylene glycol have been developed, but are not practiced commercially [25].

The mechanism is presumed to be the removal of small impurities which can act as crack initiators. The molecular weight of raw PAN fibers and introducing co-monomers could assist the processing. The chemistry of conversion of PAN to carbon is quite complex. The first critical step in making carbon fiber from PAN fiber is causing the pendant nitrile groups to cyclize, as illustrated in Figure 6.4a. This process is thermally activated and is highly exothermic. The activation temperature is influenced by the type and amount of co-monomer used [26]. It is also important to keep the fiber under tension in this process, and, indeed, during the whole conversion process. The next step is to make the fiber infusible: this is accomplished by adding oxygen atoms to the polymer, again by heating in air. The reaction is diffusion limited, requiring exposure times of tens of minutes. When about 8% oxygen by weight has been added, the fiber can be heated above 600°C without melting. At such temperatures, the processes of decyanization and dehydrogenation take place, and above 1000°C large aromatic sheets start to form, as illustrated in Figure 6.4b. The weight loss experienced in the production of carbon fibers from PAN precursor is approximately 50% [25].

This leads to a structure containing many longitudinal voids and a density of 1.8 g/cm^3, compared with 2.28 g/cm^3 for pure graphite, and 2.1 g/m^3 for pitch-based carbon fibers [27].

6.3.1 HEAT TREATMENT AND PHASE INVERSION

One of most important steps of carbon fiber production is heat treatment (graphitization) which alters the phase arrangement. The phase inversion of carbon fiber is seen in Figure 6.5. It has direct effect on mechanical, thermal, and electrical properties of fiber [28].

6.4 SURFACE TREATMENT OF CARBON FIBERS

Carbon fibers have been extensively used for improving mechanical properties of polymer composite materials due to their excellent mechanical properties and light weight. An epoxy resin is often preferred for the host matrix due to excellent electrical properties, high mechanical strength, high resistance against aging/hydrolysis, and high bond strength to many other polymer materials [29]. In order to achieve high mechanical strength of the composites, fiber distribution, alignment, fiber damage, and interface between fiber surfaces and a polymer matrix need to be considered. In particular, strong adhesion between the fiber surfaces and the polymer matrix is one of the key issues for improving the longitudinal tensile strength of

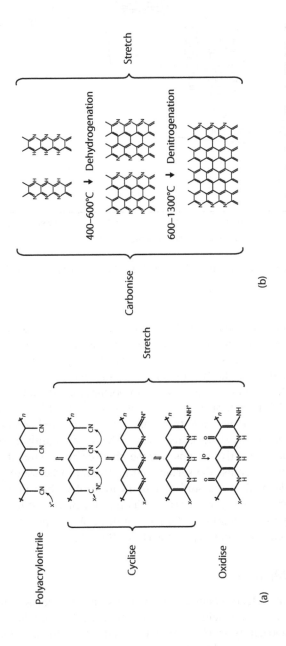

FIGURE 6.4 Carbon fiber chemistry: (a) cyclization and oxidation and (b) carbonization. (From Lavin, J.G., Carbon fibre, in *High Performance Fibres*, edited by Harle, J.W.S., Woodhead Publishing Limited, Cambridge, England, 2001, pp. 156–188.)

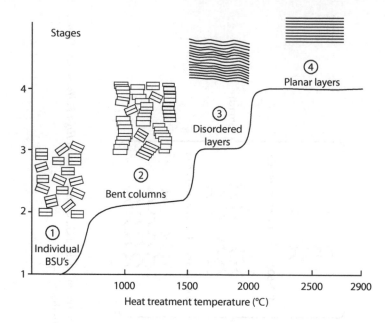

FIGURE 6.5   Stages of graphitization of carbon fiber. (From Oberlin, A., *Carbon*, 1984, 22, 521–540.)

carbon fiber reinforced polymers [30]. However, due to the non-polar nature of carbon fibers they are difficult to wet and almost impossible to chemically bond to general polymer matrices. It is noted that for adhesion improvement the chemical effect of oxygen containing polar functional groups such as -OH, =O, and -COOH at the carbon fiber surfaces is known to be more important than the mechanical effect of rough surfaces [31]. Proper surface modification thus should be chosen so that carbon fiber surfaces can be wettable by the polymer matrix and bond to it tightly. Adhesion can be improved by surface treatment of the fibers, mainly by oxidation of the surfaces, introducing reactive groups onto the fiber surfaces so that they can react with matrices as well as increase the surface energy for improved wetting. Extensive research has been devoted to the surface modification of carbon fires in order to improve their bonding to the resin matrix, including dry or wet oxidation, electrochemical methods, polymer coatings, plasma treatment, plasma polymerization, and plasma enhanced chemical vapor deposition (PECVD) [29,31].

6.4.1 Chemical Method

The wet oxidation and electrochemical methods use nitric acid, KMnO$_4$, H$_2$SO$_4$, sodium hypochlorite, chromic acid, and electrolytic NaOH, while the dry methods use oxygen, ozone, and catalysis. However, these kinds of chemical methods may be least preferable. For example, when the carbon fibers are oxidized in concentrated nitric acid, the equipment used must have good corrosion resistance and the acid absorbed on the fiber surfaces must be properly removed by subsequent washing, which is time-consuming and inevitably damages and tangles the carbon fibers [30]. These methods can also produce environmental pollution. On the other hand, plasma surface modification techniques are attractive for this application because they can be operated at room temperature, they do not require any use of solvents and toxic chemicals (environmentally friendly process), the bulk properties are retained, oxygen and/or nitrogen containing functional groups are easily introduced into the surfaces, which is often required for the application of adhesion improvement, a surface can be cleaned and weak layers can be eliminated simultaneously with the surface modification, and physical and chemical micro-etching is expected, improving the mechanical interlocking with adhesives [32].

6.4.2 PHYSICAL METHOD

Plasma surface modification can usually be divided into two categories with opposite effects, depending mainly on the process gas used. The first one mainly ablates the surfaces, and is usually called "plasma treatment," "plasma surface modification," "plasma ablation," or "nonpolymer-forming plasma." The second one is usually called "plasma polymerization," "polymer-forming plasma," or "plasma enhanced chemical vapor deposition." In the following, "plasma surface modification" is meant to cover both types while "plasma treatment" is used for the first one. If the used gas has high proportions of carbon and hydrogen atoms, double- or triple-bonds in its composition such as methane, ethylene, acetylene, and ethanol, or if they are precursors such as metal-organic (organometallic) gas, the plasma often results in plasma polymerization or PECVD. Here, metal-organic gases are those that contain a metal, particularly compounds in which the metal atom has a direct bond with a carbon atom. Otherwise, the plasma will have a tendency of ablation (plasma treatment). These techniques have been studied for adhesion improvement of carbon fibers [30].

This kind of plasma is generally obtained at low pressure. These plasma surface modifications at low pressures, however, suffer from the drawbacks that they require expensive vacuum systems, and the methods are only well-developed for batch or semi-batch treatments. To overcome these drawbacks, plasma surface modification at atmospheric pressure has been developed, avoiding expensive vacuum equipment, it permits the treatment of large objects and continuous treatment on production lines can readily be designed. Atmospheric pressure plasma has already been used to treat glassy carbon plates, which are thought to be ideal model specimens for fundamental studies of adhesive properties of carbon fibers due to the structural similarity and easier handling than carbon fibers [32,33].

6.5 PROCESSING OF COMPOSITES

In the case of carbon fiber composites, the fiber grade, sizing, surface modification and form of fire as well as the nature of the matrix material play an important role for the selection of the processing technique. Recently, the short and long carbon fibers are available in the market. Therefore, they could be compounded using usual industrial techniques: heating-cooling mixer, extrusion and pultrusion as well as compression and injection moulding.

Carbon fiber reinforced polypropylene and polyamide composites were produced using twin screw extruder and test specimens were produced using injection moulding. The main problems in the processing of carbon fibers are the dosing and distribution of the fibers. This problem could be minimized by means of a surface modification fiber and optimized process parameter and the screw geometry. Most of these have screws and barrels made up of smaller segments (mixing, conveying, venting, and additive feeding) so that the design can be changed to meet the production and product needs. Single-screw extruders can also be used for compounding with an appropriate screw design and static mixers after the screw. The selection of the components to be mixed (viscosities, additive carriers) is as important as the equipment.

6.6 PERFORMANCE OF COMPOSITES

Recycling carbon fibers (RCF) are collected from aircraft industry and compounded with matrix material using twin screw extruders. Carbon fibers were dried at 80°C in an air circulating oven for 6 h before compounding. Fibers at 30 wt% proportion were compounded in pellet form and pellets were cut into small pieces, dried and test samples were produced following by injection molding.

6.6.1 THERMAL PROPERTIES

The features of thermal analysis as a result of control PP, PP-RCF, and modified PP-RCF are shown in Figure 6.6. The starting decomposition temperatures at 1%

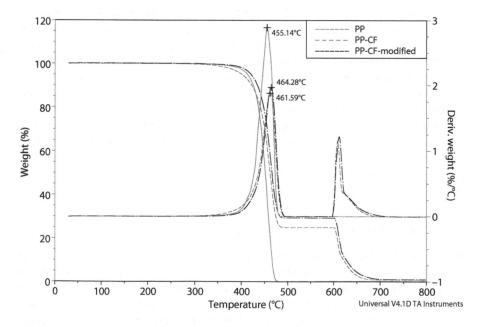

FIGURE 6.6 TGAs of PP, PP-RCF, and modified PP-RCF.

weight loss of PP, PP-RCF, and modified PP-RCF were observed at 267°C, 264°C, and 275°C, respectively. It was observed that the peak decomposition temperature for PP was as low as 455°C, and for PP-RCF was as low as 461°C. On the other hand, the peak decomposition temperature of modified PP-RCF is 464°C. The highest starting decomposition temperature and peak decomposition temperature were observed for modified PP-RCF. This could be reason that the MA/PP was found to improve fiber-matrix adhesion and alter the molecular architecture of PP, which could have an effect on rheological and thermal properties.

The thermo gravimetric analyses of recycling carbon fiber based polyamide composites are display in Figure 6.7. The peak decomposition temperatures of PA-RCF and modified PA-RCF are 438°C and 440°C.

The differential scanning calorimetry (DSC) investigations were carried out by a modulated DSC (TA Instruments), under a nitrogen atmosphere, at temperatures ranging from 0–260°C. The calibration of the temperature and heat flow scales at the same heating rate was performed with In, Zn, and Sn. A determination of the crystalline forms of polymers or composites was performed with identical conditions of the heating-cooling-heating process, at a temperature scanning speed of 10°C/min within the temperature range 0–260°C, and isothermal conditions for 5 min at 260°C.

DSC thermograms of recycling carbon fiber based PP and modified PP composites are seen in Figure 6.8. The crystalline peak temperatures and the melt temperatures found to shift slightly to the lower range due to the modification. The melt enthalpy remained the same for both cases.

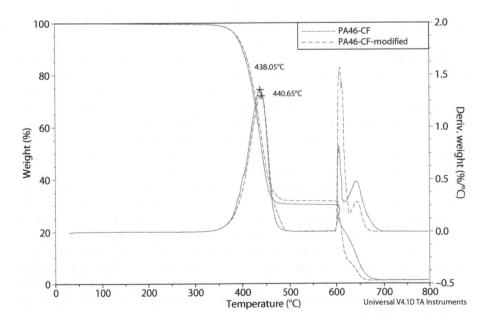

FIGURE 6.7   TGAs of PA-RCF and modified PA-RCF.

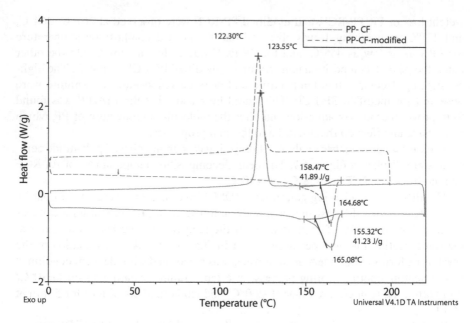

FIGURE 6.8   DSCs of PP-RCF composites and modified PP-RCF composites.

Besides an amorphous phase, semi-crystalline polyamide-6 can exhibit three main crystalline forms, the stable monoclinic α form, the metastable pseudohexagonal β form, and the unstable monoclinic γ form. The β and γ forms may reorganize into the α form during the DSC scan. The DSC thermograms for PA 6 in control PA and composites exhibit different kinds of melting peaks. It can be seen in Figure 6.9. The melt peaks including shoulder peak were observed for both types of composites.

These peaks may be due to the presence of the different crystalline forms of PA 6, the recrystallization during the DSC scan, the different order of crystal perfection, or the different thickness of the lamellae [34]. The higher perfection of the crystals and a higher content of fibrillar crystals lead to a higher-temperature endotherm [35]. The melt enthalpy reduced from 46 J/g to 36 J/g due to addition of carbon fiber. This property was improved a bit due to the addition of an impact modifier. The crystalline peak temperature shifted about 20°C due to addition of carbon fiber.

The viscoelastic properties of the polypropylene and polyamide CF composites were studied by dynamic mechanical analysis where a sinusoidal force (stress σ) is applied to a material and the resulting displacement (strain) is measured. For a perfectly elastic solid, the resulting strain and the stress will be perfectly in phase. For a purely viscous fluid, there will be a 90° phase lag of strain with respect to stress. Viscoelastic polymers have their characteristics in between where some phase lag will occur during DMA tests [36]. The storage modulus and mechanical loss factor (tan delta) of recycling carbon fiber reinforced PP and PA (modified and unmodified) composites were studied as a function of temperature from 25°C to 200°C and it was shown in Figure 6.10.

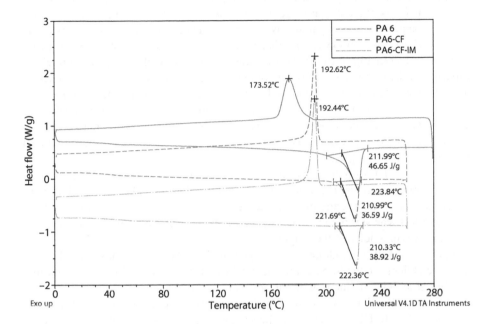

FIGURE 6.9   DSCs of PA, PA-RCF composites, and modified PA-RCF composites.

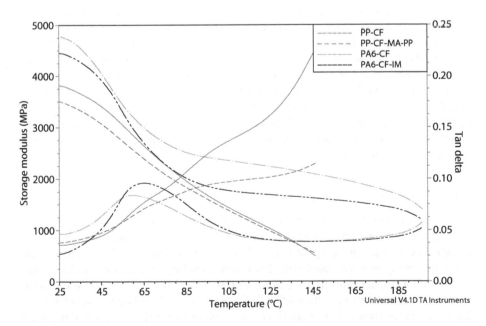

FIGURE 6.10   DMAs of PP-RCF composites and modified PP-RCF composites.

The variation of storage modulus as a function of temperature is graphically enumerated and it is evident that there was a notable increase in the storage modulus of RCF composites in comparison with virgin matrix (not seen in the figure). The same trend was also observed for other grain by-products composites. This is probably due to increasing of the stiffness toward the matrix with the reinforcing effect imparted by the fibers that allowed a greater degree of stress transfer at the interface. For PP composites, the storage modulus was significantly reduced in the regions of 0°C to 150°C. A similar trend was observed for modified PP composites. The lowering of the storage modulus is associated with the softening of the matrix at high temperature and the thermal expansion occurring in the matrix, resulting in reduced intermolecular forces [36,37].

In the case of PA composites, the storage modulus was decreased with increasing temperature until 100°C, afterwhich it decreased slowly. Due to the addition of impact modifier, the storage modulus decreased and after glass transition temperature the effect was significant. This is because the impact modifier reduces stiffness and improves elasticity of the polymer.

The rate of fall of modulus was compensated by the interactions caused in the presence of fibers in the filled composites, which also leads to an increase in thermal stability of the virgin matrix with the addition of fibers.

The heat deflection temperature (HDT) analysis was conducted on a DMA analyzer according to DIN EN ISO 75. The samples were analyzed with 1.8 MPa bending force, a heating rate of 2°K/min, and the HDT was measured at a fixed elongation of 0.21 mm.

The deflection temperature is a measure of a polymer's resistance to distortion under a given load at elevated temperatures. The deflection temperature results are useful as a means to measure the relative service temperature for a polymer when used in load-bearing parts. However, the deflection temperature test is a short-term test and should not be employed when designing a product. The value obtained for a specific polymer grade will depend on the base resin and on the presence of reinforcing agents. The heat deflection temperatures of PA-RCF composites and modified PA-RCF composites are presented in Figure 6.11. It was observed that the HDT of PA-RCF composite was 200°C and 5°C higher than those of modified PA-RCF composite. Due to the modification, the HDT reduced minimally (5°C). This could be due to softening of the matrix materials.

6.6.2 MECHANICAL PROPERTIES

The tensile, flexural, and impact properties of carbon fiber reinforced PP and PA6 composites are described below. MA/PP was used with PP based carbon fiber composites as a coupling agent and maleic anhydride functionalized elastomeric ethylene copolymer was used with PA based recycling carbon fiber (RCF) composites as an impact modifier. It can be seen in Figure 6.12 that the tensile moduli increased 6-fold due to addition of RCF for both cases. This is because RCF has a higher modulus than matrix materials, which incorporate the increase of modulus property. In the case of PP-RCF composites, the property reduced slightly by the addition of MA/PP. On the other hand, the property reduced about 10% due to

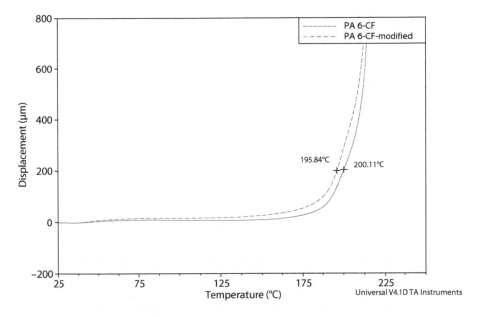

FIGURE 6.11 IIDTs of PA-RCF composites and modified PA-RCF composites.

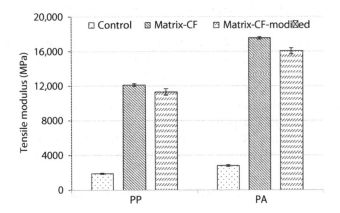

FIGURE 6.12 Tensile moduli of PP and PA based RCF composites.

the addition of impact modifier. An impact modifier could improve flexibility by forming secondary flexible phase in between two hard phases. Thus, the modulus property reduced.

The tensile strengths of carbon fiber reinforced PP and PA composites are depicted in Figure 6.13. Due to the incorporation of carbon fiber, the tensile strength was found to be increased 95% in the case of PP composites and 110% in the case of PA composites.

This is because carbon fiber shares its properties and facilitates the stress transfer ability in composites for both types of composites. The conventional coupling

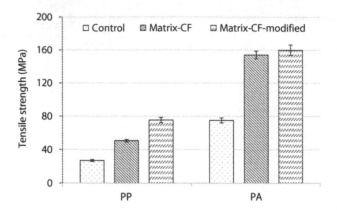

FIGURE 6.13 Tensile strengths of PP and PA based RCF composites.

agent (MA/PP) has a positive effect on the tensile strength. Carbon fiber based PP composites modified with MA/PP showed further 50% improvement in comparison to unmodified PP-RCF composites. It can be explained by an ester bond forming between the fiber and the matrix. Due to the addition of impact modifier in PA-RCF composites, the tensile strengths of composites was found to increase slightly.

The tensile elongation at break is shown in Figure 6.14. The elongation at break or strain is expressed as the ratio of total deformation to the initial dimension of the material body in which the forces are being applied. Elongation at break is also directly proportional to the molecular weight of polymer in a certain range. In the case of PP-RCF composites, the tensile elongation at break improved 140% due to the addition of coupling agent compared to unmodified composites. Whereas the elongation at break of PA-RCF composites increased about 20% in the case of using impact modifier. The ductility of PP-RCF composites was found to improve a large extent and for the PA-RCF improve to some extent.

The flexural strength of RCF composites is displayed in Figure 6.15. The flexural strength of PP based composites was found to improve 105% than control PP. On

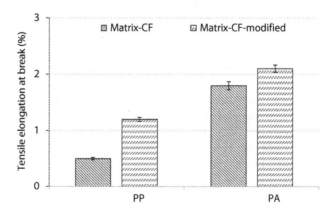

FIGURE 6.14 Tensile elongations at breaks of PP and PA based RCF composites.

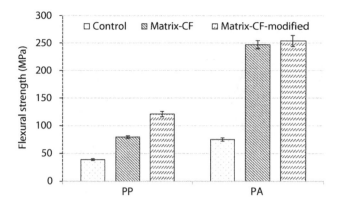

FIGURE 6.15 Flexural strengths of PP and PA based RCF composites.

the other hand, this property increased 230% for PA based composites. Due to the addition of MA/PP, the flexural strength increased an additional 50%. This is due to the formation of a strong bond between the fiber and matrix. With the addition of an impact modifier in PA based composites, the flexural strength of PA-RCF composites remains unchanged.

The falling weight impact test describes the force-displacement relation which could help to assess the multidimensional load-mechanical properties of material. The method provides a stiffness parameter for the material. Furthermore, the response of a material to a given impact energy can be split into a dissipation and a storage energy contribution. The dissipation energy is defined as the energy absorbed by material by means of damage initiation, deformation, development of deliminations, and fracture [38]. The storage energy describes the energy given back to the impactor. In this case, 35.77 Joules impact energy was chosen and the energy was high enough to puncture the material. So there was no storage energy left. The energy absorption and deformation of PP and PA based RCF composites are represented in Figures 6.16 and 6.17.

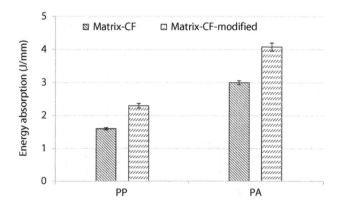

FIGURE 6.16 Energy absorption of PP and PA based RCF composites.

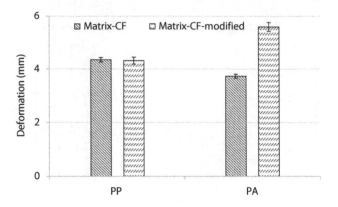

FIGURE 6.17 Deformation of PP and PA based RCF composites.

The energy absorption PA based RCF composites were found to be 90% higher than that of PP based RCF composites. Due to the modification, the energy absorption improved 37% for PP based composites and 30% for PA based composites. The modifier creates a secondary phase and subsequently absorbs a high amount of energy [38].

Owing to modification, the deformation property of PP based RCF composites remained unchanged and in the case of PA composites, the deformation property increased about 40% (in Figure 6.17). The increase of deformation is due to the increase of the stress distribution ability [39].

Figure 6.18 depicts the Charpy impact strength of PP and PA based carbon fiber composites. The Charpy impact strength of PP and PA based RCF composites showed about 6 kJ/m^2 and 45 kJ/m^2, respectively. This property was found to improve 150% by PP based composites and 21% by PA based composites owing to modification. The strong adhesion between carbon fiber and matrix with the addition of malic anhydride based coupling agent to PP based composites could be the main reason. The increment of this property for PA based composites is due to the formation of a secondary phase by the addition of an impact modifier.

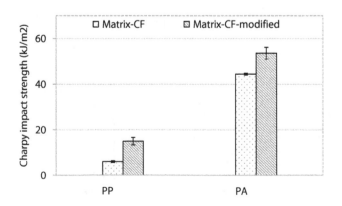

FIGURE 6.18 Charpy impact strengths of PP and PA based RCF composites.

6.6.3 MORPHOLOGY

The morphology of carbon fiber reinforced PP and PA 6 composites was investigated using scanning electron microscope (SEM) MV2300, CamScan Electron optics. Flexural samples were fractured after being submerged in liquid nitrogen and test specimens were prepared sputter coated with gold.

The scanning electronic micrograms of carbon fiber reinforced PP composites are seen in Figure 6.19. In the case of carbon fiber PP composites, the fibers are not well embedded in the matrix and there is a weak adhesion. There are also many fiber fractures and random distribution of fiber was observed. In the case of modified PP composites, the fibers are well embedded in the matrix and observed strong fiber matrix adhesion. Random distribution of fiber was also observed. A very few fiber fracture and pull-outs occurred.

The scanning electronic micrograms of carbon fiber reinforced PA 6 composites are displayed in Figure 6.20. Random distribution of fiber, a few pull-outs of fiber, and a few fractures of fiber were observed. Fibers were well embedded in the matrix and showed strong adhesion between fiber and matrix.

6.6.4 DIFFERENT PA COMPOSITES: A COMPARISON

Different types of polyamides were compounded with carbon fibers (RCF) using a twin screw extruder. Carbon fibers were dried at 80°C in an air circulating oven for 6 h before compounding. Fibers at 30 wt% proportion were compounded in pellet form and pellets were cut into small pieces, dried, and test samples were produced following injection molding. Polyamide 6, polyamide 66, polyamide 612, and polyamide 46 are considered for this research work. Polyamide 612 is a bio-based polyamide among them.

6.6.4.1 Morphology

The SEM of PA 66 based carbon fiber composites with modification are seen in Figure 6.21. Uneven breaks of matrix material and no phase interaction were observed in Figure 6.21a. On the other hand, the impact modifier was found to form a secondary phase (Figure 6.21b) with matrix phase which acts as a bridge. In both cases, fibers were distributed randomly and fracture and pull out of fiber occurred. Some small holes or pores on the surface of both composites were also seen. It could form by a vigorous mixture of differing PA chain lengths and chain architectures, owing to the addition of the impact modifier. This mixture is thermodynamically immiscible and forms an unstable blend with internal micro- and meso-cracks [39].

6.6.4.2 Flexural Properties

The flexural moduli of recycling different carbon fiber reinforced different polyamide composites are seen in Figure 6.22. Polyamide 46 based showed the best and polyamide 612 showed the lowest flexural modulus. On the other hand, the flexural modulus of polyamide 6 and polyamide 66 showed a similar level. Polyamide 46 showed about 15% better flexural modulus than polyamide 66. The better stiffness of polyamide 46 is due to its molecular architecture, rheology, and thermal

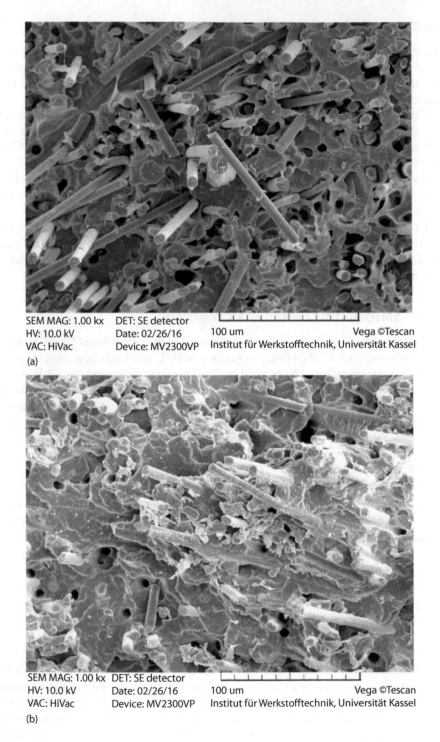

SEM MAG: 1.00 kx DET: SE detector
HV: 10.0 kV   Date: 02/26/16   100 um   Vega ©Tescan
VAC: HiVac   Device: MV2300VP   Institut für Werkstofftechnik, Universität Kassel
(a)

SEM MAG: 1.00 kx DET: SE detector
HV: 10.0 kV   Date: 02/26/16   100 um   Vega ©Tescan
VAC: HiVac   Device: MV2300VP   Institut für Werkstofftechnik, Universität Kassel
(b)

FIGURE 6.19   SEMs of PP based RCF composites; (a) Matrix-CF and (b) matrix-CF-modified.

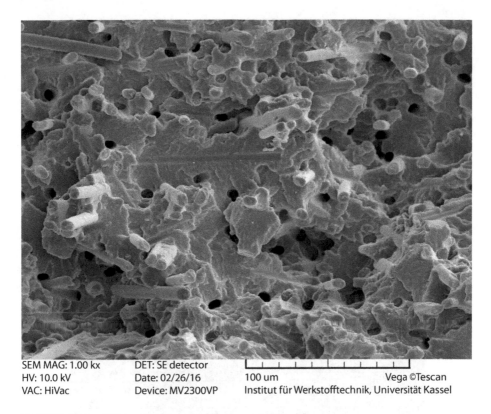

SEM MAG: 1.00 kx   DET: SE detector
HV: 10.0 kV   Date: 02/26/16   100 um   Vega ©Tescan
VAC: HiVac   Device: MV2300VP   Institut für Werkstofftechnik, Universität Kassel

FIGURE 6.20   SEMs of PA based RCF composites.

behavior. Owing to the modification, the flexural modulus reduced minimally but in the case of polyamide 46, the property reduced about 20%. This could be the thermal stability of the impact modifier. The average compounding and molding temperature was about 320°C.

Polyamide 46 based composites showed the best flexural strength in all polyamide composites (in Figure 6.23) and about 5% and 14% better than that of polyamide 66 and polyamide 6 composites, respectively. On the other hand, the flexural strength of polyamide 66 composites showed 10% improved than that of polyamide 6. Polyamide 612 composites showed the lowest flexural strength and about 10% lower than that of polyamide 6. The flexural strength increased some extent owing to the addition of impact modifier except polyamide 46 composites. The malic hydride based impact modifier primarily acts on improvement of adhesion between fiber and matrix and secondarily builds up a local soft phase which supports energy absorption property.

The flexural elongations at breaks of different PA based RCF composites is displayed in Figure 6.24. Polyamide 612 composites showed the highest and polyamide 4.6 showed the lowest flexural elongation at break. Polyamide 66 composites showed 17% lower flexural elongation at break than that of polyamide 6 composites. The impact modifier has a positive effect on flexural elongation at break. This property increased 18% for polyamide 6 composites, 12% for polyamide 66 composites, and

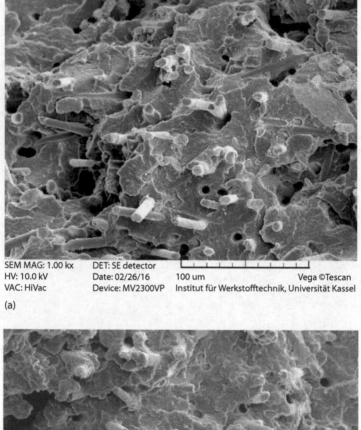

SEM MAG: 1.00 kx   DET: SE detector
HV: 10.0 kV   Date: 02/26/16   100 um   Vega ©Tescan
VAC: HiVac   Device: MV2300VP   Institut für Werkstofftechnik, Universität Kassel
(a)

SEM MAG: 1.00 kx   DET: SE detector
HV: 10.0 kV   Date: 02/26/16   100 um   Vega ©Tescan
VAC: HiVac   Device: MV2300VP   Institut für Werkstofftechnik, Universität Kassel
(b)

FIGURE 6.21 SEMs of PA 66 based carbon fiber composites; (a) matrix-CF and (b) matrix-CF-modified.

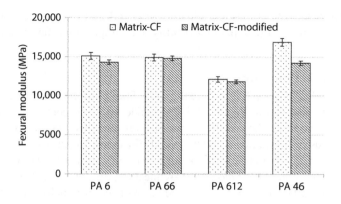

FIGURE 6.22 Flexural moduli of different PA based RCF composites.

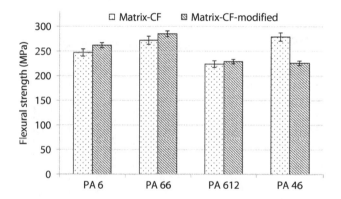

FIGURE 6.23 Flexural strengths of different PA based RCF composites.

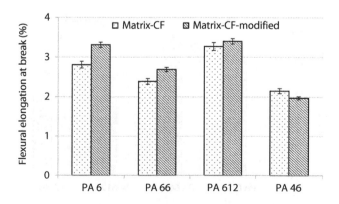

FIGURE 6.24 Flexural elongations at breaks of different PA based RCF composites.

4% for polyamide 612 composites. This property reduced in the case of polyamide 46 composites. This could be in compatibility of an impact modifier with polyamide 46 composites.

6.6.4.3 Tensile Properties

The tensile moduli of recycling different carbon fiber reinforced different polyamide composites are observed in Figure 6.25. Polyamide 46 based composites showed the best and polyamide 612 showed the lowest flexural modulus. On the other hand, the flexural modulus of polyamide 6 and polyamide 66 showed a similar level. Polyamide 46 showed about 15% better flexural modulus than polyamide 66. The better stiffness of polyamide 46 is due to their molecular architecture, rheology, and thermal behavior. Owing to the modification, the flexural modulus reduced minimally but in the case of polyamide 46, the property reduced about 20%. This could be the thermal stability of impact modifier. The average compounding and molding temperature was about 320°C.

The tensile strengths of recycling different carbon fiber reinforced polyamide composites are observed in Figure 6.26. Polyamide 46 based composites showed the

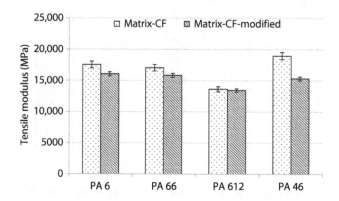

FIGURE 6.25 Tensile modulus of different PA based RCF composites.

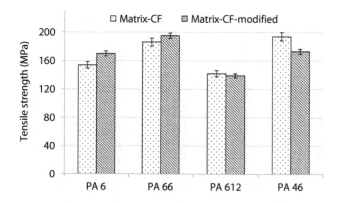

FIGURE 6.26 Tensile strengths of different PA based RCF composites.

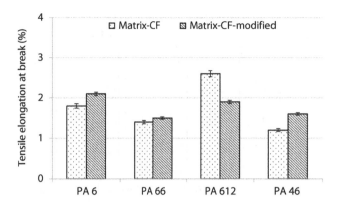

FIGURE 6.27 Tensile elongations at breaks of different PA based RCF composites.

best and polyamide 612 showed the lowest tensile strength modulus. On the other hand, the tensile strength of polyamide 6 and polyamide 66 showed almost a similar level. Polyamide 46 showed about 12% better flexural modulus than polyamide 66. Owing to the modification, the tensile strength reduced minimally but in the case of polyamide 46, the property reduced about 26%.

The tensile elongations at breaks of different PA based RCF composites is displayed in Figure 6.27. Polyamide 612 composites showed highest and polyamide 4.6 showed lowest flexural elongation at break. Polyamide 66 composites showed 29% lower tensile elongation at break than that of polyamide 6 composites. This property increased 16% for polyamide 6 composites, 5% for polyamide 66 composites, and 33% for polyamide 46 composites. This property reduced in the case of polyamide 612 composites.

6.6.4.4 Impact Properties

In impact different materials can behave in quite different ways when compared with static loading conditions. Ductile materials tend to become more brittle at high loading rates, and spalling may occur on the reverse side to the impact if penetration does not occur. The way in which the kinetic energy is distributed through the section is also important in determining its response. Projectiles apply a Hertzian contact stress at the point of impact to a solid body, with compression stresses under the point, but with bending loads a short distance away. Thus, most materials are weaker in tension than compression, this is the zone where cracks tend to form and propagate.

The impact deformations of PA 612 based RCF composites were found to be 24% higher (Figure 6.28) than that of PA6 based RCF composites. PA 6, PA 66, and PA 46 based RCF composites showed similar level of impact deformation. Due to the modification, this property improved 50% for PA 6 composites, 32% for PA 66 composites, 37% for PA 612 composites, and 38% for PA 46 composites.

The energy absorptions of PA based RCF composites are observed in Figure 6.29. Polyamide 6 showed 42% higher energy absorption than that of polyamide 66. PA 66 and PA 46 based RCF composites showed nearly similar level of impact energy absorption. Owing to the modification, this property increased 35% to 60%.

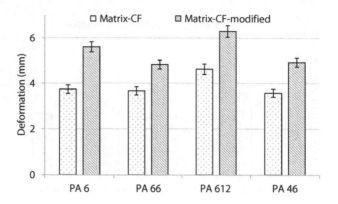

FIGURE 6.28   Impact deformations of different PA based RCF composites.

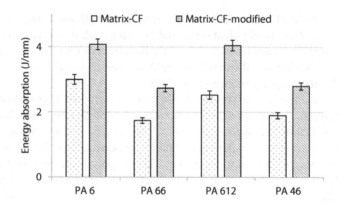

FIGURE 6.29   Impact deformations of different PA based RCF composites.

The increase of deformation and energy absorption are due to formation of sec-ondary phase which facilitates stress distribution ability and as well as boost up energy absorption ability [39].

The Charpy impact strengths of PA based RCF composites are seen in Figure 6.30. Polyamide 6 showed 25% higher and polyamide 46 showed 30% lower Charpy impact strength than that of polyamide 66. PA 66 and PA 612 based composites showed nearly similar level of Charpy impact strength. Due to the modification, this property increased 20% to 50%.

6.6.4.5   Heat Deflection Temperature

The heat deflection temperatures of PA based carbon fiber composites are presented in Figure 6.31. It was observed that the HDT of PA 6 composite was 200°C, PA 66 composite was 244°C, and PA 612 composite was 184°C. The heat deflection temperatures of PA 46 based carbon fiber composites was found to be higher than 250°C.

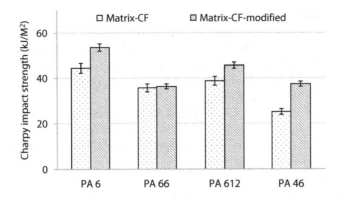

FIGURE 6.30 Charpy impact strengths of different PA based RCF composites.

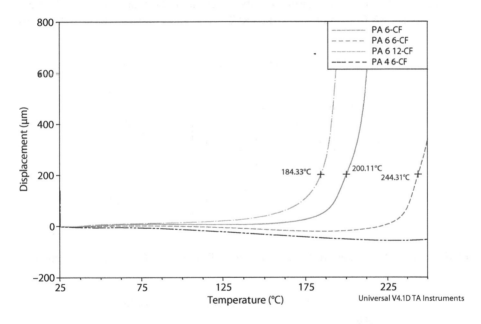

FIGURE 6.31 Heat deflection temperatures of different PA based RCF composites.

6.6.4.6 Heat Aging of Carbon Fiber Composites

Thermal aging testing of PA based composites was carried out using a weathering machine (Model UV-200 SB Simcon), collected from Weiss umwelttechnik GmbH, Germany. The weathering technique was done according to DIN EN ISO 4892-3. The aging consists of heating at 150°C. The samples were submitted to the heat weathering process for durations of 1000 h and after every 250 h, specimens were removed from the weathering cabinet and mechanical properties were measured.

6.6.4.7 Flexural Properties

The heat aging effects of different PA based RCF composites on flexural modulus are shown in Figure 6.32. The flexural moduli of PA 6 and PA 612 based RCF composites showed some extend increment of this property after 500 h of heat aging. This is because of recrystallization of polyamide. Afterward, this property reduced a little bit after 1000 h of heat aging.

In the case of PA 66 and PA 46 based RCF composites, this property reduced slowly with heat aging duration. The flexural modulus reduced 7% for PA 66 composites and 18% for polyamide 46 composites after 1000 h of heat aging. It could be degradation of polymeric material.

The heat aging effects of different PA based RCF composites on flexural strength are displayed in Figure 6.33. The flexural strength of PA 6 based RCF composites showed some extend increment of this property after 500 h of heat aging. This is because of the recrystallization of polyamide. Afterward, this property reduced a little bit after 1000 h of heat aging. On the other hand, the property reduced gradually for PA 66, PA 612, and PA 46 based RCF composites. It was reduced 14% by PA

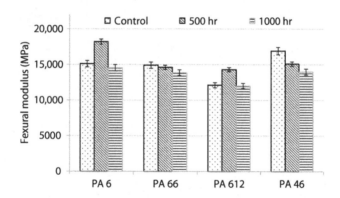

FIGURE 6.32   Heat aging effects on flexural moduli of different PA based RCF composites.

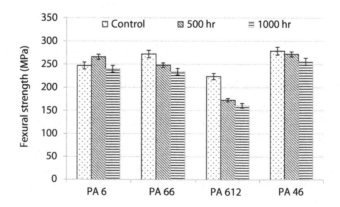

FIGURE 6.33   Heat aging effects on flexural strengths of different PA based RCF composites.

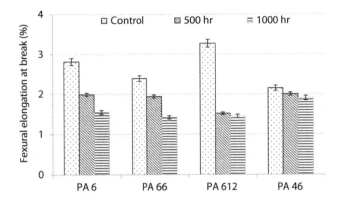

FIGURE 6.34 Heat aging effects on flexural elongations at breaks of different PA based RCF composites.

66 composites, 30% by PA 612 composites, and 9% by PA 46 composites after heat aging of 1000 h. The degradation of polymeric materials occurred and weakened the fiber-matrix adhesion due to heat aging of the PA composites.

The heat aging effects of different PAs based RCF composites on flexural elongations at breaks can be seen in Figure 6.34. The flexural elongations at breaks of all PAs composites showed continuous reduction during heat aging. The property reduced 45% by polyamide 6, 40% by PA 66 composites, 56% by PA 612 composites, and 12% by PA 46 composites after heat aging of 1000 h. This could be reduction of molecular weight of polyamide due to heat aging of the PA composites.

6.6.4.8 Tensile Properties

The heat aging effects of different PA based RCF composites on tensile strength are displayed in Figure 6.35. The similar tendency was observed as it is with flexural strength. This property remained unchanged for PA 6 and PA 46 based RCF composites after 1000 h of heat aging. On the other hand, the property was reduced 15% by PA 66 composites and 33% by PA 612 composites after heat aging of 1000 h.

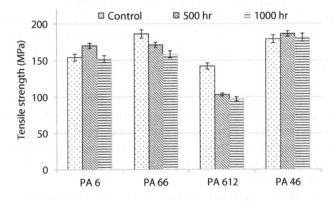

FIGURE 6.35 Heat aging effects on tensile strengths of different PA based RCF composites.

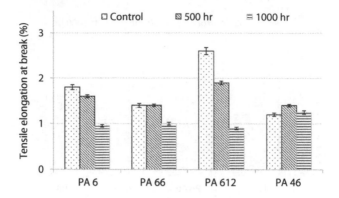

FIGURE 6.36   Heat aging effects on tensile elongations at breaks of different PA based RCF composites.

The heat aging effects of different PA based RCF composites on tensile elongation at break are observed in Figure 6.36. This property remained unchanged for PA 46 based RCF composites after 1000 h of heat aging. On the other hand, the property was reduced 46% by PA 6 composites, 29% by PA 66 composites, and 66% by PA 612 composites after heat aging of 1000 h.

6.7   RECENT APPLICATION AND FUTURE TRENDS

Carbon-fiber composites (CFRP) are materials with ever-higher strength-to-weight ratios and are able to replace metal in applications where light weight has outsized value (capable of supporting prices that can reach $140/lb), primarily for reducing fuel consumption.

As already explained in the introduction, CFRP products are well established in high-value sectors such as sporting goods, aerospace, military, and supercars, but priced out of most large-volume markets, particularly the mainstream automotive industries. The emerging methods and techniques from carbon fiber to CRFP production are being sped up and will bring down the high prices.

According to Lux Research, the estimated market of CFRPs application by 2020 will about $35 billion including $6 billion in automotive. The recent applications will not be limited to luxury and racing vehicles but excessively in truck and ordinary vehicles sometime after 2020. There are many applications in automobiles but some are explained below.

Ford Motor is producing front panel and rear seat structures and instrument panel beam with integrated HVAC ducts using injection molded CF (40 wt.%). The weight reduction is about 45 kg due to hte use of carbon fiber. Ford Motor also uses carbon fiber polyamide composites for front cover, oil pan, cam carrier, and wheels, which allows weight reduction by about 30 wt.%. Some parts produced from CFRP in Ford Motor can be seen in Figure 6.37.

Mitsui, Toray, and Hexagon Lincoln have entered the manufacture and supply of carbon fiber reinforced high-pressure hydrogen cylinders for vehicles in Japan.

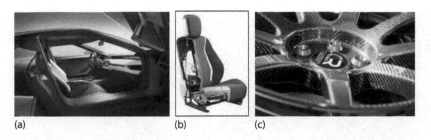

(a)   (b)   (c)

FIGURE 6.37 CFRP in Ford Motor: (a) front panel, (b) seat structure, and (c) wheel.

Among auto companies, BMW stands out as a pioneer of CFRP use. In 2009, a relationship with carbon-fiber maker SGL, BMW has invested $200 million more to push production capacity from 6000 tons to 9000 tons/yr. BMW will use the materials in its i-Series electric and plug-in hybrid cars, which are currently rolling off production lines at a rate of 100 cars per day. More recently, BMW and partner Boeing have aimed to improve CFRP production and recycling. Carbon fiber reinforced plastic is used in BMW i3, which provides a weight reduction of 250–350 kg.

BMW has made door cards, roof or hood panels from carbon fiber, mostly for limited production performance models. There are also million-dollar McLarens and Lamborghinis with CFRP bodies.

In order to correctly translate the lightweight ideology behind the 328, the majority of the exterior and interior of the 328 Hommage are made completely out of carbon fiber. The door cards, roof, center consulate cover trim, and exterior and interior parts of Hommage are seen in Figure 6.38.

There are also many automobile companies that are using CFRP for their limited version of car production, for instance, OSIR Design replacement double sided hood and spoiler for the VW Mk5 GTI, R32 and Jetta. Double sided hoods have finished carbon fiber on the underside of the hood, visible when the hood is up. Figure 6.39 represents CFRP in VW automation.

The OSIR Design full replacement vented fender kit for the Audi TT is a complete kit that includes that left and right side vented fenders, vent mesh, and custom windshield washer fluid tank.

Audi also offers a number of large and small CFRP components, primarily in the R8 model family. These range all the way to partially self-supporting structural components such as the side walls and the cover (Figure 6.40) for the top component in the R8 Spyder.

6.8 CONCLUSIONS

The demand of carbon fiber composites is increasing in automobile applications. However, there is still enormous work to be done to increase their cost performance to the desired level. The carbon content and surface properties of fiber strongly depend on the raw materials and production process. The performance of carbon fiber composites strongly depend on the surface modification of fiber. The development of a suitable modification process for carbon fibers is an ongoing process and depends on matrix material. Therefore, the proper optimization of each step used for long and

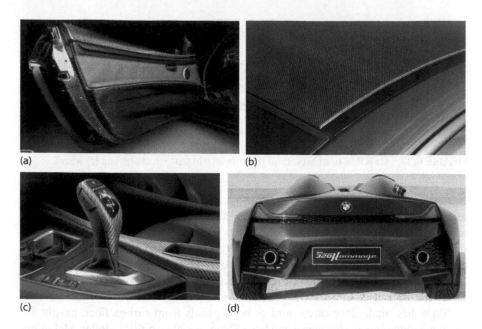

FIGURE 6.38   CFRP in BMW: (a) door cards, (b) roof, (c) center consulate cover trim, and (d) exterior and interior parts of Hommage.

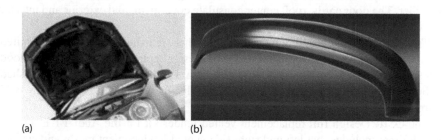

FIGURE 6.39   CFRP in VW: (a) double side hoods and (b) rear spoiler.

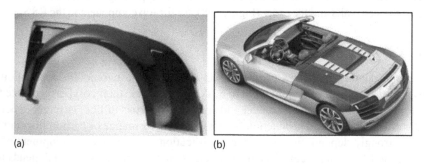

FIGURE 6.40   CFRP in Audi: (a) fender kit and (b) Audi R8 Spyder.

short carbon fiber thermoplastic composites processing could enable an enhanced performance and be used for advanced engineering applications. Recycling of carbon fiber is a change for low cost materials. The polypropylene and polyamide based recycling carbon fiber composites could be a cost-effective potential market.

REFERENCES

1. Fitzer, E., in *Carbon Fibers Filaments and Composites*, edited by J.L. Figueiredo, C.A. Bernardo, R.T.K. Baker, and K.J. Huttinger, Kluwer Academic Press, Dordrecht, 1990, pp. 3–41.
2. Askeland, D.R., *The Science & Engineering of Materials*, 2nd ed., PWS-kent, 1989, p. 591.
3. Chung, D.D.L., *Carbon Fibre Composites*, Butterworth-Heinemann, Boston, pp. 3–4.
4. Chand, S., Review: Carbon fibers for composites. *Journal of Materials Science*, 2000, 35(6), 1303–1313.
5. Callister, W. and Rethwisch, D., *Materials Science and Engineering: An Introduction*. 8th ed. John Wiley & Sons, Inc., Hoboken, NJ, 2010, pp. 646–649.
6. Gellerstedt, G., Sjöholm, E., and Brodin, I. The wood-based biorefinery: A source of carbon fiber. A short review discussing the wood-based biorefinary and different attempts to make a value-added product as carbon fibers from lignin, *Open Agric J*, 2010, 3, 119–124.
7. Lavin, J.G., Carbon fibre, in *High Performance Fibres*, edited by Harle, J.W.S., Woodhead Publishing Limited, Cambridge, England, 2001, pp. 156–188.
8. Baker, D.A., Gallego, N.C., and Baker, F.S. On the characterization and spinning of an organic purified lignin toward the manufacture of low-cost carbon fiber. *J Appl Polym Sci*, 2012, 124, 227–234.
9. Frank, E., Hermanutz, F., and Buchmeiser, M.R. Carbon fibers: Precursors, manufacturing, and properties. *Macromol Mater Eng*, 2012, 297, 493–501.
10. Frank, E., Steudle, L.M., Ingildeev, D., Spörl, J.M., and Buchmeiser, M.R. Carbon fibers: Precursor systems, processing, structure, and properties. *Angew Chem Int Ed*, 2014, 53, 2–39.
11. Sudo, K., Shimizu, K., Nakashima, A., and Yokoyama, J. A new modification method of exploded lignin for the preparation of a carbon-fiber precursor. *J Appl Polym Sci*, 1993, 48, 1485–1491.
12. Lallave, M., Bedia, J., Ruiz-Rosas, R., Rodriguez-Mirasol, J., Cordero, T., Otero J.C. et al. Filled and hollow carbon nanofibers by coaxial electrospinning of Alcell lignin without binder polymers. *Adv Mater*, 2007, 19, 4292–4296.
13. Biernat, U., First Carbon Fibre Reinforced Ultradur, October 14–18, Fakuma 2014, Friedrichshafen, Germany.
14. Pütz, A., Joint material research of innovative polyamide carbon fiber composites system: Carbon-fiber material system based on reactive polyamide developed, *JEC Europe Composits Show*, March 10–12, Paris, France.
15. Shury, J., Companies team up to develop Japanese hydrogen market, *Composites Today*, 2016.
16. Reithofer, N., Press release, "In Germany the BMW i3 Has Been the Best-Selling Electric Car Since It Was Launched. In the worldwide ranking it stands third," BMW Group, Munich, November 12, 2015.
17. Reithofer, N., Press release, "Annual Accounts Press Conference in Munich on 18 March 2015," BMW Group, Munich, March 18, 2015.
18. United States Environmental Protection Agency and U.S. Department of Energy, "Model Year 2014 Fuel Economy Guide—Electric Vehicles" Fueleconomy.gov, July, 25, 2014, pp. 33–35.

19. Crowe, P., "2014 World Green Car Finalists Revealed," HybridCars.com, February, 13, 2014.

20. King, D., "Tesla Model S, BMW i3 Among 2012 Green Car Vision Finalists," Autoblog Green, January 13, 2012.

21. Salama, M.M., Some challenges for deepwater development, *Proc. of Off-Shore Technology Conf., OTC*, 1997, p. 8455.

22. Sines, G.,Yang, Z., and Vickers, B.D., Creep of a carbon-carbon composite at high temperatures and high stresses, *Carbon*, 1989, 27, 403–415.

23. Kogure, K., Sines, G. and Lavin, J.G., Creep behaviour of a pitch-based carbon filament, *J Am Ceram Soc*, 1996, 79(1), 46–50.

24. Lavin, J.G., Boyington, D.R., Lahijani, J., Nysten, B., and Issi, J.-P., The correlation of thermal conductivity with electrical resistivity in pitch-based carbon fibre, *Carbon*, 1993, 31(6), 1001–1002.

25. Monthioux, M., Soutric, F., and Serin, V., Recurrent correlation between the electron energy loss spectra and mechanical properties for carbon fibres, *Carbon*, 1997, 35(10/11), 1660–1664.

26. Guigon, M., Oberlin, A., and Desarmot, G., Microtexture and structure of some high tensile strength, PAN-base carbon fibres, *Fibre Science and Technology*, 1984, 20, 55–72.

27. Hearle, J.W.S., Lomas, B., and Cooke, W.D. (Editors), *Atlas of Fibre Fracture and Damage to Textile*, 2nd ed., Woodhead Publishing, Cambridge, England, 1998, p. 65.

28. Oberlin, A., Carbonization and graphitization, *Carbon*, 1984, 22, 521–540.

29. Jones, C., The chemistry of carbon fibre surfaces and its effect in interfacial phenolmena in fibre/epoxy composites. *Comp Sci Technol*, 1991, 42(1–3), 275–298.

30. Dilsiz, N., Plasma surface modification of carbon fibers: A review, *J Adhesion Sci Technol*, 2000, 14(7), 975–987.

31. Mortensen, H., Kusano, Y., Leipold, F., Rozlosnik, N., Kingshott, P., Sørensen, B.F., Stenum, B., and Bindslev, H., Modification of glassy carbon surfaces by an atmospheric pressure cold plasma, *Jpn J Appl Phys*, 2006, 45(10B), 8506–8511.

32. Hughes, J.D.H., The carbon fibre/epoxy interface: A review. *Comp Sci Technol*, 1991, 41(1), 13–45.

33. Kusano, Y., Mortensen, H., Stenum, B., Goutianos, S., Mitra, B., Ghanbari-Siahkali, A., Kingshott, P., Sørensen, B.F., and Bindslev, H, Atmospheric pressure plasma treatment of glassy carbon for adhesion improvement. *Int J Adhesion and Adhesives*, 2007, 27(5), 402–408.

34. Penel-Pierron, L., Depecker, C., Séguéla, R., and Lefebvre, J.-M., Structural and mechanical behavior of nylon 6 films part 1. Identification and stability of the crystalline phases. *Journal of Polymer Science Part B: Polymer Physics*, 2001, 39, 484–491.

35. Klata, E., Borysiak, S., Van de Velde, K., Garbarczyk, J., and Krucinska, L. Crystallinity of polyamide 6 matrix in glasfibre-polyamide 6 composites manufactured from hybrideyarns, *Fibre & Textiles in Eastern Europe*, 2001, 12(3), 64–69.

36. Ferry, J.D. Some reflections on the early development of polymer dynamics: Viscoelasticity, dielectric dispersion and self-diffusion. *Macromolecules*, 1991, 24(19), 5237–5245.

37. Mohanty, S., Verma, S.K., and Nayak, S.K. Dynamic mechanical and thermal properties of MAPE treated jute/HDPE composites. *Composites Science and Technology*, 2006, 66, 538–547.

38. Mamun, A.A. and Heim, H.-P. Modification of semi-crystalline PLA: Impact, tensile and thermal properties. *Journal of Biobased Materials and Bioenergy*, 2014, 8, 292–298.

39. Mamun, A.A., Nikousaleh, M.A., and Heim, H.-P. Rheological, mechanical, ESCR and gamma resistance properties of semi-crystalline PLA: Effect of catalysts on the reactivity of a chain extender, *Journal of Biobased Materials and Bioenergy*, 2015, 9, 1–10.

7 Glass Fiber Composite Materials

T. Palanisamy Sathishkumar

CONTENTS

7.1 INTRODUCTION

This chapter discusses the detailed view of a synthetic fiber, namely glass fiber, and its composite materials. It is a manufactured fiber and the glass fiber ore materials are taken from natural sources. The glass fiber-reinforced polymer (GFRP) composites production, characterization, and their utilization are briefly explained. The glass fiber is reinforced with a suitable polymer to prepare the GFRP composites by various manufacturing methodologies. GFRP composites have been commonly used for various applications. The mechanical behavior of a GFRP composite basically depends on the glass fiber strength and modulus, glass fiber orientation, matrix strength, and the interface bonding between the fiber/matrix to achieve the maximum stress transfer. The glass fiber is in the form of long fiber, short fiber, woven fabric, and random fiber mat. The suitable weight or volume fraction (VF) and orientation of glass fiber give the desired properties of the composites and also enhance the functional characteristics of GFRP composites. The GFRP composites have higher strength equal to steel, have a higher stiffness than aluminum, and have one-quarter of the specific gravity of steel. GFRP composite materials have a wide range of industrial applications and

239

largely laminated composite materials are used in the marine industry, construction, and piping industries because of good environmental resistance, and better damage tolerance for impact loading, high specific strength, and stiffness.

7.2 HISTORY OF GLASS FIBER AND GFRP

Glass fiber is an amorphous structure with the addition of various impurities Na^+, Mg^{2+}, Ca^{2+}, and Al^{3+}. By the addition of these impurities, the glass fiber is classified into many types. The main component of the glass fiber is silica (SiO_2) and the molecular structure of glass fiber is shown in Figure 7.1. Continuous glass fibers were first manufactured in the 1930s for high-temperature electrical application. For the generation and development, the glass wool is used as glass fiber today; this was invented by Russell Games Slayter, a researcher at Owens-Corning in 1932–1933. It has no true melting point, but it softens at 1200°C on heating.

The glass fiber (like fiberglass) reinforced with a plastic to produce composite materials was developed by DuPont in 1936. The first glass fiber composite boat was developed by Ray Greene in Owens-Corning in 1937. Also, the fiberglass passenger car was developed by Russia in 1937 and later a car body with fiber glass (Figure 7.2) was prototyped by Stout Scarab in 1946. For continuous development and usage of glass fiber, the United States developed the fuselage and wings of an aircraft for commercial use.

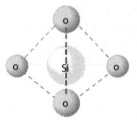

FIGURE 7.1   Molecular structure of glass fiber.

FIGURE 7.2   Stout Scarab car body made up of glass fiber polymer (in 1930s to 1940s).

Nowadays, glass fibers are being used in electronics, aviation, aerospace, civil, automobile sectors, and so on. They have excellent properties such as high strength, flexibility, stiffness, non-water absorption, stability, and resistance to chemical harm. Different types of glass fibers have unique properties and are used for various applications in the form of reinforced polymer composites.

7.3 CLASSIFICATION AND PROPERTIES OF GLASS FIBERS

Classification of glass fibers is based on the physical properties and chemical compositions. The physical properties of glass fibers are shown in Figure 7.3.

A-glass fiber is an ordinary glass made up of soda-lime silicate with little or no boron oxide and it is also known as alkali-lime glass. The letter A is derived from the word "alkali-lime." This is the first glass used in glass fibers. E-glass fiber is formed by an alumina-calcium-borosilicate with maximum alkali oxides content of

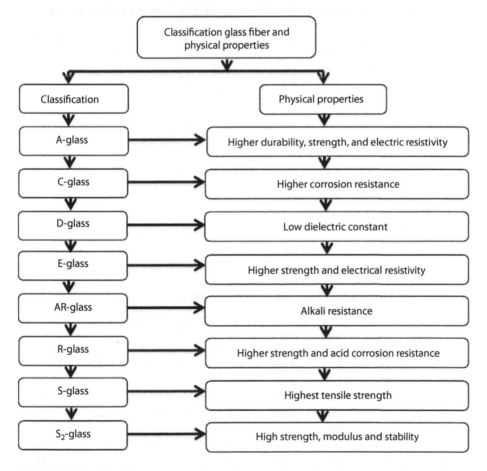

FIGURE 7.3 Classification and physical properties of various glass fibers. (From Sathishkumar, T.P., Satheeshkumar, S., and Naveen, J. *Journal of Reinforced Plastics and Composites* 2014, 33(13), 1258–1275.)

2% used as general purpose fibers where high electrical resistivity is needed. The letter E is originally derived from the word "electrical" because this is used in all electrical applications. This has a higher thermal expansion compared to other glass fibers. C-glass fiber is mainly used in the corrosive acid environment and it has higher chemical stability. It contains alkali-lime glass with high boron oxide. D-glass fiber contains silica and boron trioxide of borosilicate glass and it has a low coefficient of thermal expansion, making it resistant to thermal shock. S-glass contains alumina-silicate without CaO, but with high MgO content. It is mainly used in higher tensile strength and modulus applications. R-glass contains calcium aluminosilicate used for reinforcement, which has higher strength and acid corrosion resistance. The weight contents of chemical composition in various glass fibers are listed in Table 7.1.

TABLE 7.1
Chemical Compositions of Glass Fibers in Weight Content in Percentage

Type	SiO_2	Al_2O_3	TiO_2	B_2O_3	CaO	MgO	Na_2O	K_2O	Fe_2O_3	Softening Point
E-glass	55.0	14.0	0.2	7.0	22.0	1.0	0.5	0.3	–	840°C
C-glass	64.6	4.1	–	5.0	13.4	3.3	9.6	0.5	–	750°C
S-glass	65.0	25.0	–	–	–	10.0	–	–	–	950°C
A-glass	67.5	3.5	–	1.5	6.5	4.5	13.5	3.0	–	700°C
D-glass	74.0	–	–	22.5	–	–	1.5	2.0	–	720°C
R-glass	60.0	24.0	–	–	9.0	6.0	0.5	0.1	–	950°C
EGR-glass	61.0	13.0	–	–	22.0	3.0	–	0.5	–	840°C
Basalt	52.0	17.2	1.0	–	8.6	5.2	5.0	1.0	5.0	–

Source: Sathishkumar, T.P., Satheeshkumar, S., and Naveen, J. *Journal of Reinforced Plastics and Composites* 2014, 33(13), 1258–1275.

TABLE 7.2
Physical and Mechanical Properties of Glass Fiber

Fiber	Density (g/cm³)	Tensile Strength (GPa)	Young's Modulus (GPa)	Elongation (%)	Coefficient of Thermal Expansion (10^{-7}/°C)	Poisson's Ratio	Refractive Index
E-glass	2.58	3.445	72.3	4.8	54	0.2	1.558
C-glass	2.52	3.310	68.9	4.8	63	–	1.533
S_2-glass	2.46	4.890	86.9	5.7	16	0.22	1.521
A-glass	2.44	3.310	68.9	4.8	73	–	1.538
D-glass	2.11–2.14	2.415	51.7	4.6	25	–	1.465
R-glass	2.54	4.135	85.5	4.8	33	–	1.546
EGR-glass	2.72	3.445	80.3	4.8	59	–	1.579
AR glass	2.70	3.241	73.1	4.4	65	–	1.562

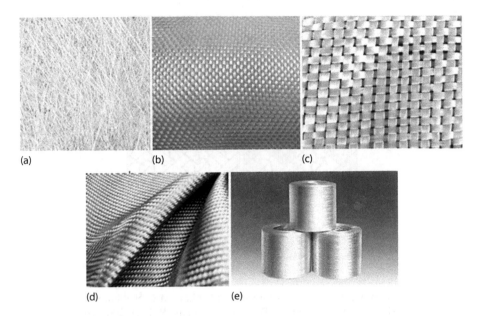

FIGURE 7.4 Forms of fiber glass. (a) Mat (chopped random mat), (b) cloth, (c) woven roving, (d) knits, and (e) roving.

The physical and mechanical properties of glass fibers are shown in Table 7.2. Among all fibers, the S_2-glass fiber is mainly used in automobile vehicles due to higher tensile strength and modulus with lower density. Glass fibers are in the form of mat, cloth, woven mat, knits, and roving (Figure 7.4). These are used to reinforce with the suitable polymer matrix to fabricate the glass fiber reinforced polymer composites.

7.4 PREPARATION OF GFRP COMPOSITES

The GFRP composites are prepared by adopting various manufacturing techniques which are discussed below. The various thermoset and thermoplastic polymer matrices, namely polyester, epoxy, vinyl ester, polypropylene, high density polyethylene, and low density polyethylene, are frequently used to prepare the various types of GFRP composites.

7.4.1 SILICONE RUBBER MOLD

The silica rubber is used as a mold for preparing the woven mat GFRP polyester composite. Initially the mold surfaces are coated with hard wax to act as a release agent. Unsaturated polyester resin containing curing additives is applied to the mold surface by using a brush. The glass fiber mat lays layer by layer and each layer is called the lamina. The steel roller is used to roll on each laid lamina to achieve the complete wetting of the glass fiber with the resin in the mold and finally the mold is closed. Over a period of time the laminated composite is fully solidified within the mold and the composites plated are removed from the silica mold.

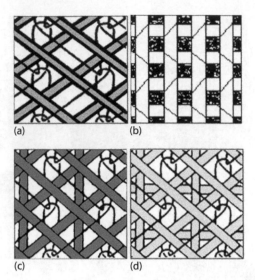

FIGURE 7.5   Multiaxial warp knit structure. (From Mathew, M.T., Naveen, V.P., Rocha, L.A., Gomesa, J.R., Alagirusamy, R., Deopura, B.L., and Fangueiro, R. *Wear* 2007, 263, 930–938.)

7.4.2 HAND LAY-UP TECHNIQUES

The directional oriented warp knit preforms are prepared, namely biaxial, biaxial non-oven, triaxial, and quadraxial, which are shown in Figure 7.5. The directionally oriented warp knit structures are having structural modifications of warp knitted fabrics with inlay yarns in horizontal (weft-90°), vertical (wale-0°), and diagonal (±45°) directions. The modification of warp knit structures is preparing the multiaxial warp knit structure such as ±45 directions (Figure 7.5a), 0°/90° directions (Figure 7.5b), ±45/90° directions (Figure 7.5c), and ±45/0°/90° (Figure 7.5d). The perform thicknesses are around 0.81, 1.22, 1.12, and 1.62 mm, respectively. Three thermoset resins, namely vinyl ester, epoxy, and polyester, are used to prepare the laminated composites. The additives are mixed with resin for improving the fast curing of the composites and better distribution of the resin on the fibers. Each lamina is wetted in the resin and it is manually placed on small steel plates and finally a steel plate is kept on the top of the impregnated preform under a load of 5 kg dead weight. These composite plates are cured for 24 h at room temperature.

7.4.3 HOT PRESS TECHNIQUE

A hot lamination press is used to fabricate the unidirectional E-glass fiber epoxy composite plates with two different stacking like [0°/90°/0°/90°] and [0°/90°/+45°/ −45°]. The weight density of 509 g/m² is used as reinforcing materials. Before curing, the laminated composite plates are kept in a heat assisted hot press at a constant pressure of 15 MPa and 120°C for 2 h. Curing of the composite plates is in controlled atmospheric conditions and hardner is not required for curing the composites. After curing, the composites are taken from the hot mold for studying the various properties. The nominal thickness of the plates is maintained as approximately 3 mm.

7.4.4 Filler Mixed Molding

The flyash filler is added during the mixing of epoxy resin and hardener. The ratio of epoxy resin and hardener mixed is around 100:10. The diameter of glass fiber is 10 μm and it is cut into 2.54 cm length for composite preparation. A steel mold is prepared with dimension of $154 \times 78 \times 12$ mm^3. Small size and spherical shape of the flyash particles facilitated the good mixing and wetting of fiber and matrix. The mixture is stirred for extended time at slow speed to ensure the uniform mixing quality. The mixture is poured into the mold and allowed to cure at room temperature for 24 h.

Two different fillers are mixed into vinyl ester resin such as 50 μm of graphite and 25 μm size of SiC. A dry hand lay-up technique is used to fabricate the composites. Initially a teflon sheet is placed on the mold and stacking the E-glass woven roving fabric layer by layer with well spread of resin. A porous teflon film again is used to complete the stacking sequence and ensures the uniform thickness of the sample. The whole assembly is kept in a hydraulic press at a constant pressure of 0.5 MPa and allowed to cure 24 h at room temperature.

7.4.5 Compression Molding Technique

The E-glass fiber mats are cut into small sizes and a hot air oven is used to heat the mats up to 150°C to make it moisture-free before composite preparation. The melt mixing technique adopted to blend the resin with 10% of poly(styrene-co-acrylonitrile) contains 25% of acrylonitrile content at 180°C with constant stirring. After getting the homogeneous mixture, the hardner of 4'-diaminodiphenyl sulfone is mixed with resin at 180°C before pouring it into the mold. Eight layers of E-glass laminas are added successively in the mold in order to get the 3 mm thickness for the composites and the blended resin is added successively in each layer. After placing all the layers, the mold is closed and the laminates are compressed. The laminate composite is cured at 180°C for 3 h and post-cured at 200°C for 2 h. Finally, the laminates are allowed to cool slowly at room temperature for solidification of the composite and it is removed from the mold.

The bidirectional E-glass woven fabric (360 g/m^2) is cut into the required size and placed successively layer by layer with resin. Both sides of the laminate are covered by cellophane membranes and sealed around the periphery with sealant. These members are used to absorb the excess resin and make the good surface finish of the composites. The weight percentage of the glass fabrics in composites is maintained around 60%. The H-type press is used to apply the constant pressure of 0.0965 MPa on the laminated composite in the mold for 24 h at room temperature of 100°C.

The developed GFRP composites are needed to study the mechanical, vibration, thermal, tribological, and environmental behaviors. These studies are very important for using the GFRP composites in various applications.

7.5 MECHANICAL PROPERTIES OF THE GFRP

The developed GFRP composites are subjected to study the various mechanical properties, namely tensile, flexural, impact, fracture toughness, and drop weight properties according to ASTM and IS standards. The dynamic mechanical properties are studied by varying the temperature and applied frequency. The GFRP composites are developed with various weight or volume fraction of the glass fiber.

Tensile properties of the woven mat glass fiber reinforced polymer composites with oligomeric siloxane (Slx) resin modifier are shown in Figure 7.6. It is shown that the resin modified by Slx in the concentration range of 1%–3% (w/w) enhanced the tensile strength of the composites. Polyester with 3% Slx gives higher tensile strength. This may be due to better adhesive between the glass fiber and polyester matrix. Because the alkoxy and alkyl functional groups are presented in the modifier, this may be well dispersed polyester into the glass fabric, which caused better wetting of the fabric by the matrix. The enhanced modulus of elasticity is found in the polyester with 3% Slx composites; it is around 18.0 ± 3.3 GPa.

The tensile properties of unidirectionally oriented continuous glass fiber reinforced epoxy composites with various fiber volume fraction (f) and average fiber diameters (D) are shown in Figure 7.7. These curves show a linear response up to a maximum stress level. The average glass fiber diameter plays an important role on the epoxy composites which significantly affects the maxim stress level. The fiber diameter of 18 μm and fiber volume fraction of 0.45 for composite resulted in higher tensile stress and strain. In case the average glass fiber diameter is 50 μm with 0.45 of fiber volume fraction in the composites, it showed lower tensile strength and strain. The quantity of glass fiber in the composites is higher at lower fiber diameter, which takes the higher stress to improve tensile stress of the short fiber composites.

Soypolyol 204 and polyol Jeffol G30-650 are mixed with polyurethane resin to prepare two types of E-glass fabric composites with 70% of the fiber weight content. When comparing the two polyurethane composites, the Jeffol composite has a little higher mechanical properties. Both composites have good adhesion between fiber and matrix. The chemical structure of both blends is shown in Figure 7.8.

Jeffol G30-650 is a polyether with glycerine–propylene oxide and it is a very short chain of average molecular weight. During the reaction with aromatic diisocyanates

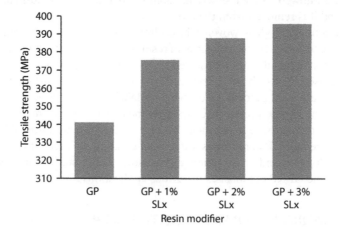

FIGURE 7.6 Tensile strength of glass fiber composites with resin modifier. (From Erden, S., Sever, K., Seki, Y., and Sarikanat, M. *Fibers and Polymers* 2010, 11(5), 732–737.)

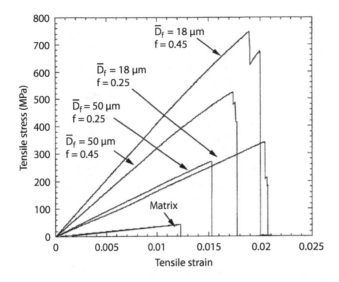

FIGURE 7.7 Typical stress–strain curves of the epoxy matrix and the composites. (From Iba, H., Chang, T., and Kagawa, Y. *Composites Science and Technology* 2002, 62, 2043–2052.)

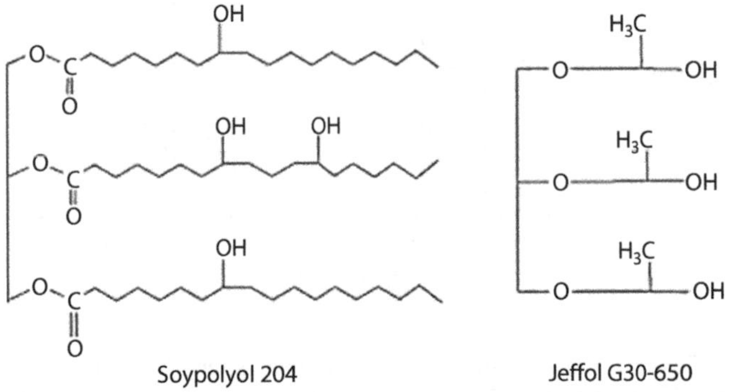

FIGURE 7.8 Structures of the Soypolyol 204 and Jeffol G30-650. (From Husic, S., Javnib, I., and Petrovic, Z.S. *Composites Science and Technology* 2005, 65, 19–25.)

it gives rigid, exceedingly cross-linked polyurethane net and the complete polyether chain net is generated. Soypolyol 204 has internal hydroxyl groups placed in the middle of an 18 carbon fatty acid chain. If the cross-linking is completed, a portion of the chain is not included in the network and is left dangling. It is increasing the free volume in the polymer net and it is acting as a plasticizer. For increasing the rigidity of the matrix, low molecular alcohol is added as a cross-linker. It increases the mechanical properties of the composites. The tensile, flexural properties and interlaminar shear strength (ILSS) of both composites are shown in Figure 7.9. The maximum strength and modulus are found in Jeffol composites.

Tensile and fracture behavior of chopped strand mat GFRP composites were analyzed with different fiber V_f such as 12%, 24%, 36%, 48%, and 60% according to

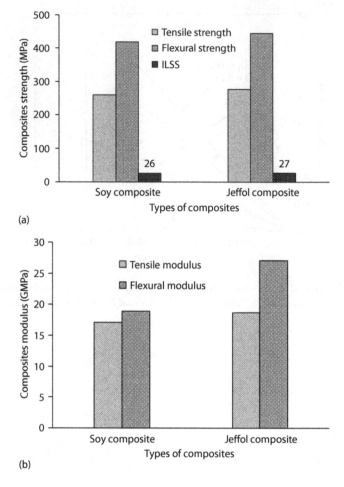

(a)

(b)

FIGURE 7.9 Flexural, tensile, and interlaminar shear strength of the composites. (From Husic, S., Javnib, I., and Petrovic, Z.S. *Composites Science and Technology* 2005, 65, 19–25.)

ASTM D638. The stress is calculated based on the nominal area. Stress versus strain is showing the brittle failure of the neat resin and GFRP composites. The curve reported that at 60% V_f of GFRP composite has a maximum tensile strength of 325 MPa, Young's modulus of 13.9 GPa, fracture toughness of 20-fold, and critical energy release rate of 1200-fold. The Young's modulus of the composites was continuously increased with fiber V_f [23]. The flexural properties of woven E-glass fiber polyester matrix composites with addition of various weight percentages of carbon nano filler (CNF) are studied according to ASTM D790-02. The weight content (wt) of the CNF is 0.1, 0.2, 0.3, and 0.4, respectively. The stress versus strain results show that 0.2 wt% of CNF-filled composite is obtained maximum flexural strength. This is due to excellent dispersion and better interfacial interaction between fiber and matrix [26]. An Na-MMT (sodium montmorillonite) is mixed with the glass fiber reinforced polyester composites in order to improve the mechanical properties.

At 3% weight of Na-MMT is found to have maximum tensile strength of 130.03 MPa, impact strength of 153.50 kJ/m², and flexural strength of 205.152 MPa [27]. The particulates namely Al_2O_3, SiC, and pine bark dust are reinforced with randomly oriented E-glass fiber reinforced epoxy composites and prepared the various particulate filler composites. The different composition of the specimen is prepared with GF (50 wt%)/ epoxy (50 wt%), GF (50 wt%)/epoxy (40 wt%)/alumina (10 wt%), GF (50 wt%)/epoxy (40 wt%)/pine bark dust (10 wt%), and GF (50 wt%)/epoxy (40 wt%)/SiC (10 wt%). The GF/epoxy composite has a maximum tensile strength of 249.6 MPa and flexural strength of 368 MPa than the other compositions. GF/epoxy/pine bark dust combination has the maximum interlaminar shear strength of 23.46 MPa, GF/epoxy/SiC combination has the higher impact strength of 1.840 J and higher hardness of 42 Hv [28].

The compression, bending, and shear behavior of E-glass woven mat reinforced epoxy composites are prepared with different fabrics like unstitched plain weave, biaxial non-crimp, and uniaxial stitched plain. The interlaminar shear strength is found maximum in biaxial stitching 5 mm composite than the plain weave and biaxial stitching 10 mm. After testing, the damage in the upper surfaces form compressive failure and lower surfaces form tensile failure. In the laminated composites, the shear failure or crack initiation is no indication. Concerning the bending results, the maximum bending and impact strength is obtained for plain weave composites. It is around 370 ± 13 and 267 ± 9. The Z-directional stitching fibers increase the delamination resistance as well as reduce the impact damage. Moreover, the compressive strength of the non-crimp laminate was 15% higher than woven fabric composite [29].

Typical tensile stress–strain curve of laminate glass fiber epoxy composites under various environmental conditions is presented in Figure 7.10. It exhibits linear elastic behaviors of the composites until complete breakage. The slop of the curve is increased as the temperature decreases and on the other hand the strain to failure decreases with temperature. The average value of the tensile strength is increased from 700.11 MPa (at RT) to 784.98 MPa (at −60°C). Also, the similar trend is obtained in the Young's modulus (23.05 GPa to 28.65 GPa) [30].

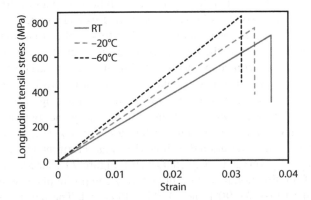

FIGURE 7.10 Longitudinal tensile stresses versus strain behavior of unidirectional laminate glass fiber epoxy composites at room temperature (RT), −20°C, −60°C. (From Torabizadeh, M.A. *Indian Journal of Engineering & Materials Sciences* 2013, 20, 299–309.)

The tensile behavior of plane woven E-glass reinforced polyester composites is prepared with various curing pressure of 35.8 kg/m^2, 70.1 kg/m^2, 104 kg/m^2, and 138.2 kg/m^2. The symmetrical and non-symmetrical layup is used to prepare the composites and the tensile modulus decreases with increasing curing pressure for both symmetrical and non-symmetrical lay-up. The symmetrical lay-up has less stiffness to the composites which has a lower strength than the non-symmetrical lay-up. The ductility increases with increasing curing pressure for non-symmetrical arrangement and for decrease in symmetrical arrangement [9].

The unidirectional continuous glass fiber reinforced epoxy composites are prepared with three fiber diameters, 18, 37, and 50 mm, respectively, and fiber V_f from 0.25 to 0.45. The longitudinal Young's modulus and tensile strength of the composite increased with increasing the fiber V_f. The mean tensile strength is increased with decreasing the fiber diameter. The maximum tensile strength and modulus are found at fiber diameter of 18 μm in 0.45 V_f of glass fiber [21].

The three-point bending method is used to investigate the Mode-I fracture behavior of sand particles associated in chopped strand GFRP composites with varying notch-to-depth ratios (a/b ratios of 0.38, 0.50, 0.55, 0.60, and 0.76). The crack depth and specimen width are represented by a and b. The composites are prepared with 0%, 1%, and 1.5% V_f of GF and various V_f of polyester resins like 13.00%, 14.75%, 16.50%, 18.00%, and 19.50%. According to resin content, the maximum flexural strength is obtained for the composites containing 16.5% of resin ratio at 0% fiber. Overall the maximum flexural strength is obtained for composite containing 19.5% of resin with 1.5% of glass fiber. A similar trend is obtained for the stress intensity factor under compliance method. The maximum stress intensity factor is found in the compliance method compared to initial notch depth method and J-integral method for all a/b ratios. The lower stress intensity factor is found in J-integral method compared to the initial notch depth method and compliance method for all of the above [24]. However, the mechanical properties of various GFRP composites are presented in Table 7.3.

The drop weight impact testing instrument is used to analyze the impact energy (E_i) and absorbed energy (E_a). These are two important parameters to measure impact response and resistance of composite structures under dynamic loading. The impact energy is defined as the total amount of energy applied to a composite specimen. The absorbed energy is the total energy absorbed by the composite structure through the impact occurrence by formation of damage in the structure. The typical energy profile diagram of a composite plate is shown in Figure 7.11. The region AB represented the non-penetrated tip of the impact hammer during the test. The region BC is the penetration range, whereas the maximum impact energy is absorbed by the composite plate and CD stands for plate perforation. E_e is an excessive energy stored in the composites. The penetration threshold (Pn) was determined by variation of the excessive energy (E_e) versus impact energy (E_i). Figure 7.12 shows the energy profile diagram of unidirectional E-glass fiber epoxy composite plates with two different stacking like [0°/90°/0°/90°] and [0°/90°/+45°/−45°]. The penetration threshold [0/90/+45/−45] laminate composite is found to be smaller than that of [0/90/0/90] laminate composite, the difference is about 3.4 J or 3.9 J. The excessive energies for both stacking are found to be very close due to the damages of the composites

TABLE 7.3

Mechanical Properties of GFRP Composites

Type of Glass Fiber	Resin	Curing Agent	V$_f$ (%)	Testing Standard	Tensile Strength (MPa)	Tensile Modulus (MPa)	Elongation at Break (%)	Flexure Strength (MPa)	Flexure Modulus (MPa)	Impact Strength	Interlaminar Shear Strength (MPa)	Ref.
Woven mat	Polyester	*MEKP/* Cobalt naphthalene	0.25	ASTM D412	1.601 (N/mm^2)	80.5 (N/mm^2)	20.0×10^{-3}	–	–	41.850 (J)	–	[2]
Woven mat	Polyester with (3% oligomeric siloxane)	*MEKP/* Cobalt naphthalene	37	ASTM D-3039, ASTM D 790, ASTM D 2344	395.8	18000	3.9	399.4	18800	–	44.7	[3]
Woven mat	Polyester	*MEKP/* Cobalt naphthalene	33	ASTM D 638-97, ASTM D 3479/D 3479M–96,	249	6240	–	–	–	–	–	[33]
Woven mat	Polyester	*MEKP/* Cobalt naphthalene	–	2810 E6	189.0	–	–	–	–	–	–	[32]

(Continued)

TABLE 7.3 (CONTINUED)
Mechanical Properties of GFRP Composites

Type of Glass Fiber	Resin	Curing Agent	V_f (%)	Testing Standard	Tensile Strength (MPa)	Tensile Modulus (MPa)	Elongation at Break (%)	Flexure Strength (MPa)	Flexure Modulus (MPa)	Impact Strength	Interlaminar Shear Strength (MPa)	Ref.
Chopped strand	Polyamide66 (PA66)/polyphenylene sulfide (PPS) blend	MEKP/Cobalt naphthalene	30	GB/T 16,421–1996, GB/T 16,419–1996, GB/T 16,420–1996	–	124	–	159	–	98.2 (kJ/m²)	–	[25]
Woven mat [0/90°]	Isophthalic/neopentyl glycol polyester	MEKP/Cobalt naphthalene	42	PS25C-0118	200	–	–	–	–	10 (J)	–	[34]
Woven mat Non symmetrically	Polyester	MEKP/Cobalt naphthalene	–	362F (BS, 1997)	220	7000	0.055	–	–	–	–	[9]
Woven	Polyurethanes	MEKP/Cobalt naphthalene	49	ASTM D2734, ASTM D3039, D2344 and D790M	278	18654	–	444	27075	–	27	[10]

(Continued)

TABLE 7.3 (CONTINUED)
Mechanical Properties of GFRP Composites

Type of Glass Fiber	Resin	Curing Agent	V_f (%)	Testing Standard	Tensile Strength (MPa)	Tensile Modulus (MPa)	Elongation at Break (%)	Flexure Strength (MPa)	Flexure Modulus (MPa)	Impact Strength	Interlaminar Shear Strength (MPa)	Ref.
Chopped strand mat	Polyester	MEKP/Cobalt naphthalene	60	ASTM D638	250	325	0.022	–	–	–	–	[23]
Woven mat	Polyester	MEKP/Cobalt naphthalene	–	ASTM D 2344	–	–	–	–	–	–	30	[35]
Chopped strand mat	Polyester resin	MEKP/Cobalt naphthalene	1% and 1.5% by weight of the composite	ASTM E 399	–	3000	–	16.5	–	–	–	[24]
Chopped strand + vertical roving	Polyester	MEKP/Cobalt naphthalene	–	ASTM D 3039, ASTM D 5379	103.4719	–	–	–	–	37.926 (J)	–	[36]
Virgin fiber	Polyester	MEKP/Cobalt naphthalene	–	ASTM D256, ASTM D2240	64.4	7200	1.8	–	–	645.1 J/m	–	[16]

(Continued)

TABLE 7.3 (CONTINUED)
Mechanical Properties of GFRP Composites

Type of Glass Fiber	Resin	Curing Agent	V_f (%)	Testing Standard	Tensile Strength (MPa)	Tensile Modulus (MPa)	Elongation at Break (%)	Flexure Strength (MPa)	Flexure Modulus (MPa)	Impact Strength	Interlaminar Shear Strength (MPa)	Ref.
Chopped strand	Epoxy	–	3.98	ASTM standard ($10 \times 10 \times 75$ mm^3)	–	–	–	–	–	17.6×10^{-3} J/mm^2	–	[6]
Woven (biaxial stritch)	Epoxy	–	–	ASTM D2677, ASTM D2355	450	–	3.5	–	–	–	18.2	[29]
Randomly oriented	Epoxy (10 wt% SiC)	Hardner	50 wt%	ASTM D 3039-76, ASTM D 256	179.4	6700	–	297.82	–	1.840 (J)	18.99	[28]
Woven	Epoxy	Hardner	73 wt%	ASTM D 2344	–	–	–	–	–	–	41.46	[37]
Woven	Epoxy	Hardner	60 wt%	ASTM D 3039	311	18610	3.8	–	–	–	–	[38]
Woven+ (35 wt% short borosili)	Epoxy	Hardner	–	–	355	43700	1.65	–	–	–	–	[39]

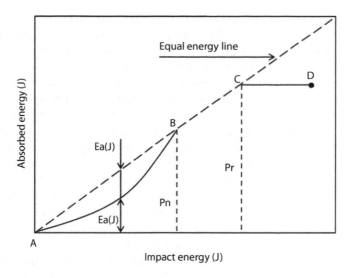

FIGURE 7.11 Typical energy profile diagram of a composite plate. (From Mehmet, A., Cesim, A., Bu lent Murat, I., and Ramazan, K. *Composite Structures* 2009, 87, 307–313.)

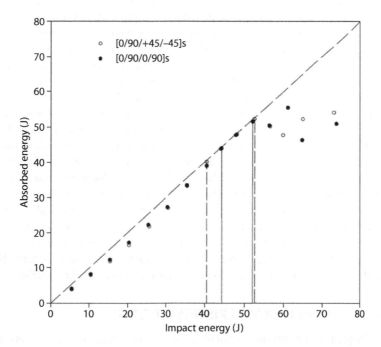

FIGURE 7.12 Energy profile diagram of stacking sequences chosen. (From Mehmet, A., Cesim, A., Bu lent Murat, I., and Ramazan, K. *Composite Structures* 2009, 87, 307–313.)

specimens being slightly different. In this diagram, the vertical solid and dashed lines represent the penetration and perforation thresholds. The penetration threshold of both composites is found approximately 40.5 J for [0/90/+45/−45] and 44.4 J for [0°/90°/0°/90°]. The difference between these thresholds is 3.9 J.

Impact response of woven mat glass fiber reinforced epoxy composites were analyzed by Atas and Liu [22]. This analysis was carried out with orthogonal fabric and non-orthogonal glass fabric of weaving angles of 20°, 30°, 45°, 60°, 75°, and 90°, respectively, from the vertical direction (warp direction). The energy absorption is based on the weaving angles of the fabrics, its absorption increased with decrease of the weaving angle between interlacing yarns. The composites with smaller weaving angles of 20° and 30° have lower peak force, larger contact area, larger deflection, and higher energy absorption than larger weaving angle of 60°, 75°, and 90°, respectively. The glass fiber woven of [0/20°] in composite absorbed higher impact energy compared to [0/90°] woven composite. The impact resistance and damage characteristics of an E-glass fiber reinforced polyester composite was analyzed according to various thicknesses of the laminates and three glass fabrics, namely multiaxial warpknit blanket (MWK), woven fabric (W), and non-woven mat (N). A guided drop-weight test rig is used to carry out the impact test. The maximum Hertzian failure force is found in MWK-13 code laminate composite structures. The maximum Hertzian failure energy is found in R800-7 code laminate composite structures [25].

The viscoelastic properties of the GFRP composites are analyzed with dynamic mechanical analyzer (DMA) with varying temperature and applied frequency in terms of load. Normally a three-point bending mode is used as a frequency of 1 Hz to analyze the storage modulus, loss modulus, and tanδ. The E-glass fiber reinforced with SAN (i.e., poly (styrene-co-acrylonitrile)) modified epoxy composites with the various V_f of fibers ranging from 10% to 60%. Under the three-point bending mode the viscoelastic properties of the composites are measured at the frequency of 1 Hz. The samples are heated up to 250°C at a heating rate of 1°C/min. In Figure 7.13, the E, S, and G represent epoxy resin, SAN, and glass fiber, respectively, and numbers 1 to 6 represent the V_f from 10% to 60%. It seems that by increasing the temperature, the storage modulus of all composites is decreased. The neat resin the storage modulus is below 3.5 MPa at starting and it is always low in all temperatures compared to composites. There is a prominent increasing in the storage modulus while incorporation of glass fiber in the composites as well as increasing the fiber V_f. This may be due to increase of the stiffness of the composites with more reinforcement effect by the glass fiber. The storage modulus does not drop too much up to a particular temperature; that state is the glass state. Therefore, the storage modulus of the composites has a sudden drop from the elevated temperature; that state is the rubbery state. This crossing state temperature is known as glass transition temperature of the composites. This curve reports 50% V_f (ESG5) of composites and has the maximum storage modulus (15.606 GPa).

7.6 VIBRATION CHARACTERISTICS OF GFRP

Natural frequency, excitation frequency, and mode shape measurements of the composite specimen have been carried out to determine the amplitude, modulus, and damping factors. A long specimen with cantilever method is used to study the

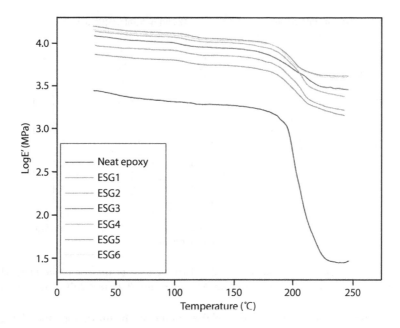

FIGURE 7.13 Storage modulus versus temperature plot of different epoxy/SAN/glass fiber composites. (From Nishar, H., Sreekumar, P.A., Bejoy, F., Weimin, Y., and Sabu, T. *Composites: Part A* 2007, 38, 2422–2432.)

vibration characteristic of the composite beam as shown in Figure 7.14. One end of the beam is fixed and other end of the beam is free to apply the impact hammer force. The impact hammer is directly interfaced with A/D (analog to digital convertor) interface card. A small size and weightless unidirectional piezoelectric accelerometer is fixed on the beam and it is interfaced with the A/D interface card. Fast Fourier transform (FFT) and modal analysis software are used to determine the resonance frequency of the tested beams after the signals are received from the A/D interface.

The effects of SLx concentration on the first, second, and third fundamental natural frequency of the composite plate are shown in Figure 7.15. The natural frequency increased with concentration of SLx. The modulus of elasticity increased the beam stiffness with the effect of SLx; it increased the natural frequency of the beam. So

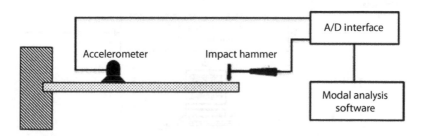

FIGURE 7.14 Cantilever method single mode vibration analysis setup. (From Erden, S., Sever, K., Seki, Y., and Sarikanat, M. *Fibers and Polymers* 2010, 11(5), 732–737.)

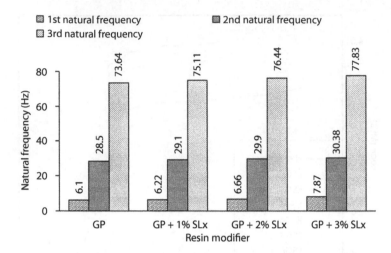

FIGURE 7.15   First three fundamental natural frequencies of the composite plate. (From Erden, S., Sever, K., Seki, Y., and Sarikanat, M. *Fibers and Polymers* 2010, 11(5), 732–737.)

that the stiffness and elastic moduli of the composite beam defines the change in fundamental natural frequency.

The up and down response of the cantilever beam is detected by using a laser vibrometer which measures the transverse vibration (displacement) of a beam at the free end (Figure 7.16). An impulse hammer is used to induce the excitation on the beam. A force transducer is fixed on the hammer and allows getting the excitation

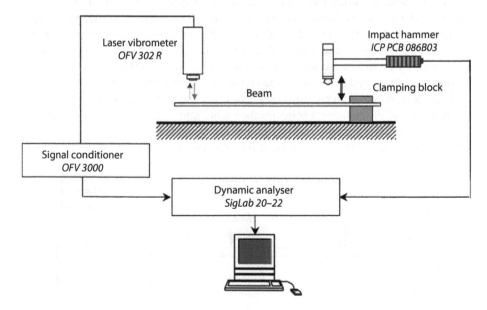

FIGURE 7.16   Transfer vibration measuring experimental setup. (From Berthelot, J.-M. and Sefrani, Y. *Composites Science and Technology* 2004, 64, 1261–1278.)

signal with a function of time. After applying the hammer force on the beam, the beam vibration response is detected by using a laser vibrometer, and the excitation signals are digitized and processed by a dynamic analyzer. These data are stored in a personal computer.

The flexural vibration of the unidirectional glass fiber epoxy composite beam is measured at the free end that is measuring point x, when the impulse excitation is applied at impact point x_1 near to the fixed point (Figure 7.17). After measuring the beam response by laser vibrometer, the amplitude versus frequency curve of the glass-epoxy composite beam is plotted (Figure 7.18). This response curve shows the various peaks of the natural frequencies of the bending beam. The different peaks are obtained by changing the measuring point on the beam.

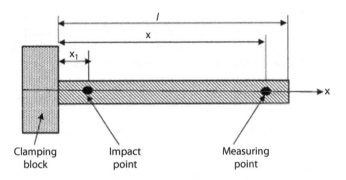

FIGURE 7.17 Vibration measuring and impact point on the cantilever beams. (From Berthelot, J.-M. and Sefrani, Y. *Composites Science and Technology* 2004, 64, 1261–1278.)

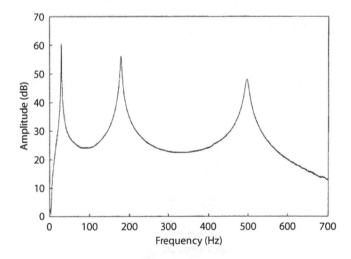

FIGURE 7.18 Typical frequency response to an impulse input of a unidirectional glass fiber/ epoxy beam. (From Berthelot, J.-M. and Sefrani, Y. *Composites Science and Technology* 2004, 64, 1261–1278.)

The E-glass fiber fabric laminated epoxy composite beam is fixed as a cantilever mode using a clamping block in an oven (Figure 7.19). The two thermocouples are fixed nearer to the composite beam which is used to measure the temperature inside the oven. The temperature inside the oven is set by the temperature regulation after fixing the beam. The light carbon rod is fixed in the vibration electromagnetic exciter, which induces the excitation force of the composite beam. The vibration exciter is controlled by impulse generator with the power amplifier. The laser vibrometer is used to measure the velocity of the transverse vibration of the beam and measured response data is transferred to dynamic analyses to convert the response signals into digital signals. These data are stored and analysis is completed with a personal computer to plot the input data to response data.

Figure 7.20 compares the impulses obtained by the impulse hammer and by the electromagnetic exciter. By electromagnetic exciter the width of the impulse is nearly two times lower compared to results obtained by the impulse hammer.

The frequency response of the unidirectional E-glass fiber composite beam is measured for three temperature levels (Figure 7.21). These response curves show different peaks which correspond to the fundamental natural frequency of the

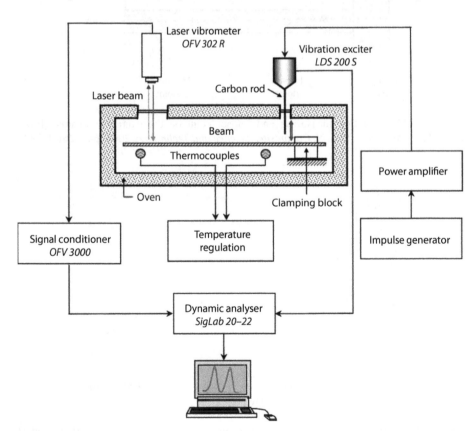

FIGURE 7.19 Vibration measurement of E-glass fiber fabric laminated epoxy composite. (From Sefrani, Y. and Berthelot, J.-M. *Composites: Part B* 2006, 37, 346–355.)

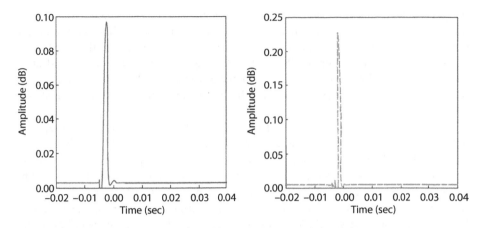

FIGURE 7.20 Comparison between impulses obtained (a) by impact hammer and (b) by electromagnetic exciter. (From Sefrani, Y. and Berthelot, J.-M. *Composites: Part B* 2006, 37, 346–355.)

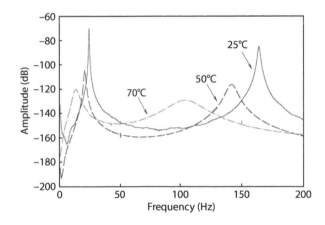

FIGURE 7.21 Temperature influence on the frequency responses of unidirectional E-glass fiber composites, for 30° fiber orientation. (From Sefrani, Y. and Berthelot, J.-M. *Composites: Part B* 2006, 37, 346–355.)

transverse vibration of the beams. At 25°C, the damping of the glass fiber beam is low and then the fundamental frequency peaks are separated from the other temperatures peaks. Increase in the temperature of the analysis increases the frequency bandwidth. In these pecks, the damping is derived from the half-power bandwidth.

7.7   THERMAL CHARACTERISTICS OF GFRP

The thermal properties of the composites are analyzed by thermogravimetric analysis (TGA). By using this method the changes in physical and chemical properties of the glass fiber, resins, and composite materials are measured as a function of increasing temperature with constant heating rate or as a function of time with constant temperature and/or constant mass loss. It is continuously measured as the weight loss of the sample and heated up to 2000°C for coupling with Fourier transform infrared spectroscopy (FTIR) and mass spectrometry gas analysis. The measurements are potted by weight loss in percentage on the Y-axis and temperature increments on the X-axis.

The epoxy resin glass fiber reinforced polymer composite (ESG) is prepared with various E-glass fiber volume factions such as 10% (ESG1), 20% (ESG2), 30% (ESG3), 40% (ESG4), 50% (ESG5), and 60% (ESG6), respectively. Before composites processing, the resin is blended with poly(styrene-co-acrylonitrile) at 10 h. TGA measurements are performed by taking samples of 6 to 10 mg from room temperature to 900°C in nitrogen atmosphere. The heating rate is constant at 5°C/min. Figure 7.22 shows the weight loss in percentage with the temperature effect of the ESG composites and resin. The pure resin starts to decompose at 345°C and is fully decomposed at 440°C. In the case of EGS5 composites, the decomposition starts at 354°C. Initial decomposition temperature (IDT) and maximum decomposition temperature from differential thermogravimetric (DTG) curve (Figure 7.23) with various composites are listed in Table 7.4. The maximum decomposition temperature of each sample is measured from the DTG and it is shown in Table 7.4. The thermal stability of the composite is more associated with pure resin. The composites degradation is shifted to a higher temperature due to the glass fibers slowing down the degradation process. Increasing the glass fiber content, the degradation is shifted to the high temperature region, at 10% fiber containing composites (ESG1) the degradation

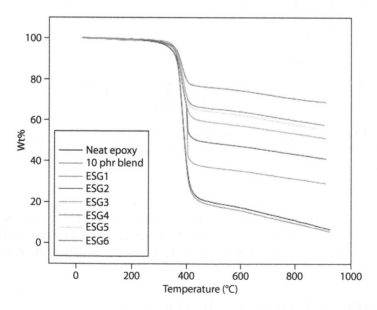

FIGURE 7.22 Thermograms of different epoxy resin/SAN/glass fiber composites. (From Nishar, H., Sreekumar, P.A., Bejoy, F., Weimin, Y., and Sabu, T. *Composites: Part A* 2007, 38, 2422–2432.)

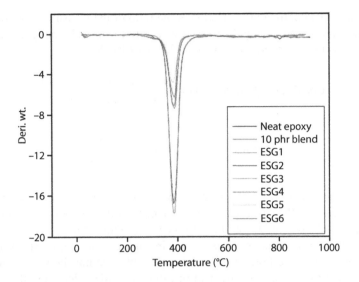

FIGURE 7.23 DTG curves of different epoxy resin/SAN/glass fiber composites. (From Nishar, H., Sreekumar, P.A., Bejoy, F., Weimin, Y., and Sabu, T. *Composites: Part A* 2007, 38, 2422–2432.)

TABLE 7.4

Thermogravimetric Analysis of Epoxy Resin and Glass Fiber Reinforced Epoxy Composites

Samples	Initial Decomposition Temperature (IDT) (°C)	T_{max} (°C)
Neat epoxy	345	386
ESG1	349	386
ESG2	350	387
ESG3	352	388
ESG4	355	389
ESG5	354	387
ESG6	357	390

starts at 349°C, at 30% fiber containing composites (ESG3) starts at 352°C. Also, the maximum degradation temperatures of resin to composites are shifted from 386°C to 387°C.

7.8 TRIBOLOGICAL BEHAVIOR OF GFRP

Tribological study is associated with wear and the friction behaviors of the glass fiber reinforced composites. The rate of wear, wear loss, and friction coefficients are analyzed according to various sliding distant and normal load applied during the experiments. The coefficient of friction is the ratio of the measured frictional force to normal applied load during the wear test. The following relations are used to calculate the specific wear rate (W_s) and coefficient of friction (μ):

$$Coefficient\ of\ friction\ (\mu) = \frac{Measured\ frictional\ force}{Normal\ applied\ load} \tag{7.1}$$

$$Specific\ wear\ rate\ (W_s) = \frac{W_I - W_F}{\rho \times F_N \times D} \tag{7.2}$$

where W_s = specific wear rate [mm³/N-m]; W_I = initial weight of the specimen [kg]; W_F = final weight of the specimen [kg]; ρ = density of the composite [kg/mm³]; F_N = normal applied load [N]; D = sliding distance of specimen [m].

The multipurpose wear tester is shown in Figure 7.24. It consists of two rollers, namely the top and bottom roller. The glass fiber composite specimen is fixed in a slot of the top roller with the help of mild steel (MS) block and the specimen (Figure 7.25) is kept on the bottom roller. The top roller is fixed on journal bearings and is controlled by the gear mechanism. This roller moves up and down. The normal load is applied on the top roller through the push pin. The dead weights are placed on the pan of the loading arm and the actual normal load is applied in the test specimen. By twisting the winding screw out, the sample comes in contact with the bottom roller. The bottom roller is rotated by the electric motor and it is controlled

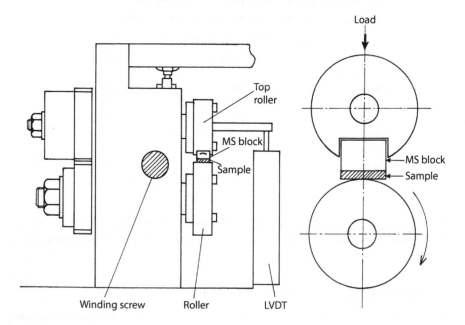

FIGURE 7.24 Multipurpose wear testing machine. (From Kishore, S.P, Seetharamu, S., Vynatheya, S., Murali, A. and Kumar, R.K. *Wear* 2000, 237, 20–27.)

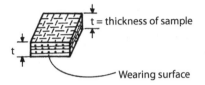

t = thickness of sample

t

Wearing surface

FIGURE 7.25 E-glass plain weave bi-directional fabric composites. (From Kishore, S.P., Seetharamu, S., Vynatheya, S., Murali, A., and Kumar, R.K. *Wear* 2000, 237, 20–27.)

by fixing the various testing parameters, namely the sliding velocity, applied normal load, and sliding distance. A controlling unit is used to control the above parameters.

The E-glass plain weave bi-directional fabric composites are prepared with rubber and oxide particles, and studied with the wear test. The weight loss for rubber and oxide blend composites shows (Figure 7.26) the effect of sliding distances and two different normal applied loads. The weight loss is increasing with sliding distance for both loads. The weight loss is found to be maximum for higher loads in both composites and more weight loss occurred in oxide bearing composites (Figure 7.27).

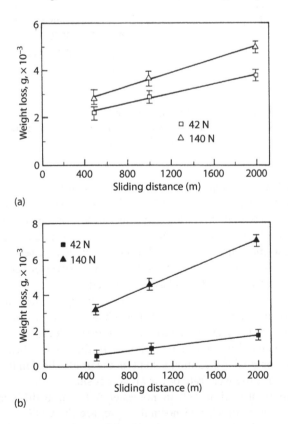

(a)

(b)

FIGURE 7.26 Weight loss versus sliding distance at two loads for a constant sliding velocity of 0.5 m/s in rubber and oxide bearing composite. (From Kishore, S.P., Seetharamu, S., Vynatheya, S., Murali, A., and Kumar, R.K. *Wear* 2000, 237, 20–27.)

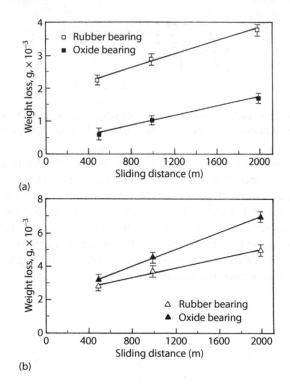

(a)

(b)

FIGURE 7.27 Wear loss versus sliding distance of two filler bearing composites at 0.5 m/s for loads 42 N and 140 N, respectively. (From Kishore, S.P., Seetharamu, S., Vynatheya, S., Murali, A., and Kumar, R.K. *Wear* 2000, 237, 20–27.)

Oxide blend E-glass composite resists the wear better at low loads compared to rubber blend composites. But in higher loads, it is in reverse. This may be due to the wear out of the matrix region and the releasing of the oxide debris with fiber breakage. The oxide is harder than epoxy, which has lower wear loss. For the reverse trend, the higher thermo-mechanical loadings loosen the oxide and getting to wear along with matrix material which makes a higher wear loss. Also wear debris formation is increased in the matrix by the fibers having inclined fracture.

The multi-pass two-body abrasive wear test on chopped strand mat R-glass fiber orthophalic unsaturated polyester composites (CGRP) is shown in Figure 7.28. The different abrasive grade sheets are fixed on the stainless steel counterface. The composite specimen is fixed in the specimen holder in the vertical beam which carries self-weight. The composite specimen is kept on the grade sheet in the rotating counterface and measures the wear rate of the all composites.

The glass fibers are reinforced in polyester with three different orientations, namely parallel, anti-parallel, and normal to prepare the CGRP. The three different SiC abrasive grade sheets are used with different grit sizes of 19 μm ≈ 1500, 29 μm ≈ 1000, and 53 μm ≈ 400 for wear test, and the testing speed is maintained as a constant of 50 rpm. In Figure 7.29, the composites have high initial wear rates

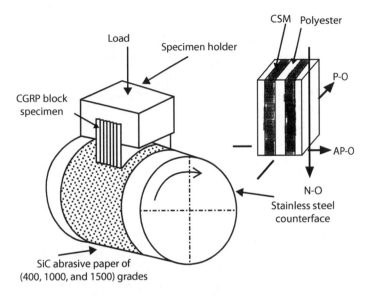

FIGURE 7.28 Schematic diagram of multi-body abrasive wear test. (From El-Tayeb, N.S.M. and Yousif, B.F. *Wear* 2007, 262, 1140–1151.)

and the gradual decreasing trend is observed by increasing the normal applied load. The glass fiber orientation and SiC grit size play an important role in the wear performance. Increasing the grade sheet grit size decreases the wear rate of the R-glass fiber composites. The normal orientation of fiber containing composites is found in higher wear rate except in wear at 400 grit size. The wear rate is high in parallel oriented fiber at the use of 400 grit size compared to other grit size. Low wear rate in antiparallel orientation is observed in all grit sizes. The continuous wear is formed in the crevices on the abrasive paper that clogged with wear debris. This reduces the abrasivity of the wear grits which lowers the wear in further sliding. The small grit size is formed with polymer film on the grade sheet by the combined effect of thermo-mechanical interaction.

The friction coefficients as a function of time for both parallel orientation (P-O) and anti-parallel orientation (AP-O) CGRP composites are shown in Figure 7.30. The friction coefficients are varied as a function of time and sliding velocities. The steady state friction coefficients are found after 16 min of wear testing. The friction coefficients are more or less at 0.25 friction. Using both wear techniques, there are no remarkable differences in both sliding velocities. Meanwhile, the friction coefficients determined by BOR techniques are found to be low when compared to the friction coefficients determined by POD techniques for AP-O composites.

Three body abrasive wear is analyzed with the effects of two fillers (graphite with 50 μm size and silicon carbide with 25 μm size) on the woven E-glass fabric vinyl ester composites. Three formulations are used to prepare the composites with 0% weight content of filler (G-V), 50% weight content of silicon carbide (SiC-G-V), and 25/25% weight content of both (Gr-SiC-G-V). The specimen is fixed in the specimen holder which is in a static position and the rubber wheel rotates against the specimen.

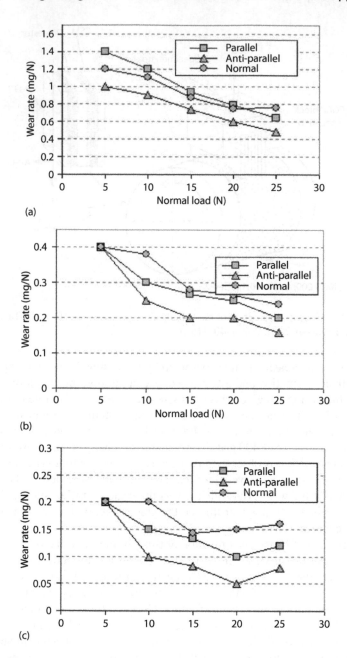

FIGURE 7.29 Wear performances of R-glass fiber polyester composites with different 400, 1000, and 1500 grit SiC abrasive paper. (From El-Tayeb, N.S.M. and Yousif, B.F. *Wear* 2007, 262, 1140–1151.)

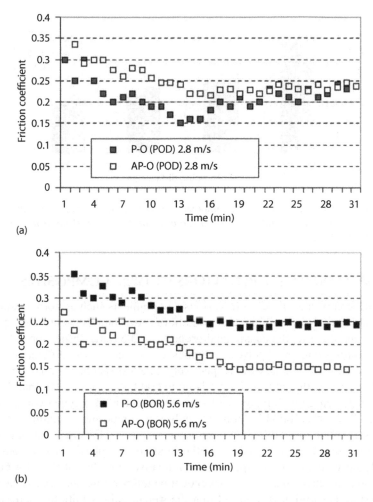

FIGURE 7.30 The friction coefficient as a function of time at 50N applied load with two different sliding velocities: (a) pin-on-disc (POD) and (b) block-on-ring (BOR). (From El-Tayeb, N.S.M. and Yousif, B.F. *Wear* 2007, 262, 1140–1151.)

The abrasive particles are fed in-between the specimen and rotating wheel to measure the wear performance. The specific wear rate is gradually reduced with increase of the abrasive distance (Figure 7.31) with normal applied load of 22 N. The highest specific wear rate is found more in G-V composites than the other two. The absence of filler in the glass fiber composite leads to more wear rate. The fillers are reinforced with GFRP composite that are reducing the wear rate of the materials. Moreover, the low wear rate is found for SiC-G-V GFRP composites. Here, the fillers have a higher resistance against wear.

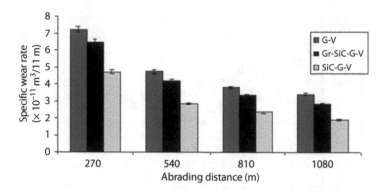

FIGURE 7.31 Specific wear rates of composites as a function of abrading distance at 22 N. (From Suresha, B. and Chandramohan, G. *Journal of Materials Processing Technology* 2008, 200, 306–311.)

7.9 ENVIRONMENTAL BEHAVIORS OF GFRP COMPOSITES

The GFRP composites are used in various environments such as normal water, sea water, and higher temperature water conditions. The composites are absorbing the water and are affecting the mechanical properties. The water absorption behavior of fiberglass wastes–reinforced polyester composites are studied with different glass fiber wastes like 20%, 30%, and 40%, respectively. The test specimens are immersed in distilled water at different time intervals up to 600 h. It was reported that the water sorption is decreased with increase of glass fiber content in polymer composite and the lowest water absorption is found for polyester composites containing the fiberglass wastes about 40% [16]. Under sea water conditions, the environmental behavior of woven mat GF-reinforced poly etherimide thermoplastic matrix composites are analyzed with varying temperature at a relative humidity of 90% for 60 days. The moisture absorption is mostly dependent on temperature and relative humidity. The moisture absorption is reported that the weight gain is initially increased linearly with respect to time. The maximum moisture absorption is found to be 0.18% after 25 days. It is lower water absorption compared to other testing conditions [17]. At elevated temperatures such as 45°C and 55°C, the environmental degradation of GFRP composites is examined in a normal water and sodium hydroxide (NaOH) bath after 1 and 2 months. The percentage weight gain increases with increase of the bath time and temperature. The NaOH bath is found to have a larger weight gain compared to a normal water bath [18].

The glass fiber epoxy composites are subjected to water aging in NaCL, 5M NaOH, distilled water, and 1M HCL solutions with different water aging time such as 1, 3, and 5, respectively. The maximum stress at failure is found in the control sample rather than water aging samples. The water aging is reducing the tensile strength and strain at failure of the composites; however, the elastics modulus is not affected by aging, except in the case of aging after 5 months in 1M HCL solutions. The reduction in ultimate tensile strength of the composites is around 73% after 5 months of immersion in hydrochloric acid by the result of degradation of glass fibers or fiber/matrix interface because the epoxy matrix is contributing very little to the total composite strength [19]. The glass fiber/isophthalic polyester, glass fiber/vinyl ester, and glass fiber/urethane-modified vinyl ester composites are tested at different temperatures from 20°C to 120°C for 30 days, 120 days, and 240 days under normal water and alkaline environments. The tensile strength and modulus were decreased in an alkali environment at maximum temperature. The rate of water absorption is found to be high [20].

The water absorption of the six layers of $[\pm60_3]_T$ glass fiber reinforced EPON 826 epoxy composites are measured using following formula:

$$M(t) = \frac{(m_t - m_{dry})}{m_{dry}} \times 100 \tag{7.3}$$

where m_{dry} is the initial mass of the glass fiber composites, m_t is the mass of the composites after immersion for time t in hours, and M(t) is the percentage of moisture of the composites.

The moisture absorption of the glass fiber epoxy tubular composite specimens as a function of the square root of time is plotted in Figure 7.32. The rate of water absorption is found to be maximum at higher temperatures of 50°C. The acceleration of diffusion of water molecules into the glass fiber composite is based on the time and temperature. The temperature increases the activation of diffusion, which takes place by the creation of microcracks or voids in the composite and leads to weakening of the fiber–matrix adhesion. This affects the mechanical properties of the composites (Figure 7.33). The hoop stress is found lower when the composites were immersed in water at 50°C that was compared to low temperature.

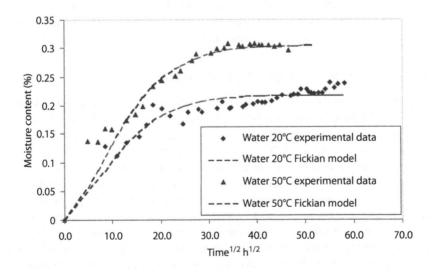

FIGURE 7.32 Fickian diffusion model fitted to water absorption of $[\pm603]_T$ glass fiber epoxy resin tubular specimens immersed in water at two temperatures. (From Ellyin, F. and Maser, R. *Composite Science and Technology* 2004, 64, 1863–1874.)

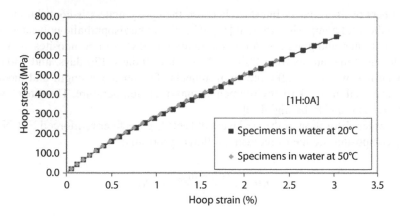

FIGURE 7.33 Hoop stress versus hoop strain of the $[\pm60_3]_T$ glass fiber epoxy tubular composite at two temperatures such as 20°C and 50°C. (From Ellyin, F. and Maser, R. *Composite Science and Technology* 2004, 64, 1863–1874.)

Normally, the water absorbed GFRP composites are reducing in tensile, flexural, and impact strength. These are directly affecting the effective utilization period of the composites. The maximum water absorption is found in boil water condition and minimum water absorption in composites is found in freezer condition of the water. Among all conditions, the NaOH solution treated glass fiber in GFRP composites is found in minimum water absorption compared to other absorbing conditions.

The moisture absorption behaviors of the E-glass fiber epoxy composite tubes are analyzed in distilled water at two temperature levels such as 20°C and 50°C for 4 months. The weight gain of the composites is around 0.23% at 20°C and 0.29% at 50°C. Temperature increase is gradually increasing the water absorption [31].

7.10 APPLICATION OF GFRP COMPOSITES

The GFRP composites are used in various applications according to the types of glass fibers. The electronics produce of E-glass reinforced GFRP composites are used in circuit board manufacture (PCBs), TVs, radios, computers, cell phones, electrical motor covers, and so on. The construction materials of GFRP composites are used in furniture, windows, sun shades, shoe racks, book racks, tea tables, spa tubs, and so on. The GFRP composites have been extensively used in aviation and aerospace parts such as airframe construction, engine cowlings, luggage racks, instrument enclosures, bulkheads, ducting, storage bins, and antenna enclosures. The boats and ships have been fabricated with GFRP composites to reduce water absorption and increase the life of the product. The GFRP composites have been widely used in automobile parts namely body panels, seat cover plates, door panels, bumpers, and engine covers.

7.11 CONCLUSION

The various GFRP composites are prepared by adopting manufacturing technologies. The mechanical, dynamics, tribological, thermal, and environmental behaviors are discussed according to various standards. Ultimate tensile, flexural, and impact strength of the fiber GFRP composites are increased with fiber content, the different glass fibers are reinforced either by volume fraction or weight fraction. The linear elastic strain of the composite increased up to 0.25 to 0.35 V_f of fiber and subsequently decreased with increase in fiber content. The elastic and flexural modulus of the composites increased with the fiber glass content.

Vibration analysis of the frequency and amplitude of the glass fiber composite beam is normally carried out by the cantilever mode. The laminating sequences and temperature are affected by the response of the amplitude and frequency. The water absorption is carried out in various environmental conditions and waters with different time periods. Increasing the time of water absorption, the composites absorb more water; this could affect the mechanical properties, and have faster propagation of cracks when the samples are treated. The wear and friction behaviors of the laminated and filler reinforced GFRP composites are analyzed with various sliding velocities, sliding distance, and normal applied load. The wear rate gradually increases with sliding distance and applied load. The lower wear rate and higher friction coefficient are obtained for more fiber incorporated in the polymers. Hence the GFRP composites have better alternate materials to replace the many parts in various applications especially in automobile parts.

REFERENCES

1. Sathishkumar, T.P., Satheeshkumar, S., and Naveen, J. Glass fiber-reinforced polymer composites—A review. *Journal of Reinforced Plastics and Composites* 2014, 33(13), 1258–1275.
2. Mathew, M.T., Naveen, V.P., Rocha, L.A., Gomesa, J.R., Alagirusamy, R., Deopura, B.L., and Fangueiro, R. Tribological properties of the directionally oriented warp knit GFRP composites. *Wear* 2007, 263, 930–938.
3. Erden, S., Sever, K., Seki, Y., and Sarikanat, M. Enhancement of the mechanical properties of glass/polyester composites via matrix modification glass/polyester composite siloxane matrix modification. *Fibers and Polymers* 2010, 11(5), 732–737.
4. Iba, H., Chang, T., and Kagawa, Y. Optically transparent continuous glass fibre-reinforced epoxy matrix composite: Fabrication, optical and mechanical properties. *Composites Science and Technology* 2002, 62, 2043–2052.
5. Mehmet, A., Cesim, A., Bu lent Murat, I., and Ramazan, K. An experimental investigation of the impact response of composite laminates. *Composite Structures* 2009, 87, 307–313.
6. Gupta, N., Brar, B.S., and Woldesenbet, E. Effect of filler addition on the compressive and impact properties of glass fibre reinforced epoxy. *Bulletin of Materials Science* 2001, 24(2), 219–223.
7. Suresha, B. and Chandramohan, G. Three-body abrasive wear behaviour of particulate-filled glass–vinyl ester composites. *Journal of Materials Processing Technology* 2008, 200, 306–311.

8. Nishar, H., Sreekumar, P.A., Bejoy, F., Weimin, Y., and Sabu, T. Morphology, dynamic mechanical and thermal studies on poly(styrene-co-acrylonitrile) modified epoxy resin/glass fibre composites. *Composites: Part A* 2007, 38, 2422–2432.
9. Faizal, M.A., Beng, Y.K., and Dalimin, M.N. Tensile property of hand lay-up plain-weave woven e glass/polyester composite: Curing pressure and ply arrangement effect. *Borneo Science* 2006, 19, 27–34.
10. Husic, S., Javnib, I., and Petrovic, Z.S. Thermal and mechanical properties of glass reinforced soy-based polyurethane composites. *Composites Science and Technology* 2005, 65, 19–25.
11. Kishore, S.P, Seetharamu, S., Vynatheya, S., Murali, A., and Kumar, R.K. SEM observations of the effects of velocity and load on the sliding wear characteristics of glass fabric–epoxy composites with different fillers. *Wear* 2000, 237, 20–27.
12. El-Tayeb, N.S.M. and Yousif, B.F. Evaluation of glass fiber reinforced polyester composite for multi-pass abrasive wear applications. *Wear* 2007, 262, 1140–1151.
13. El-Tayeb, N.S.M. and Yousif, B.F. Wear and friction characteristics of CGRP composite under wet contact condition using two different test techniques. *Wear* 2008, 265, 856–864.
14. Berthelot, J.-M. and Sefrani, Y. Damping analysis of unidirectional glass and Kevlar fibre composites. *Composites Science and Technology* 2004, 64, 1261–1278.
15. Sefrani, Y. and Berthelot, J.-M. Temperature effect on the damping properties of unidirectional glass fibre composites. *Composites: Part B* 2006, 37, 346–355.
16. Edcleide Araujo, M., Kasselyne Araujo, D., Osanildo Pereira, D. et al. Fiber glass wastes/polyester resin composites mechanical properties and water sorption. *Polimerso Ciencia E Tecnologia* 2006, 16, 332–335.
17. Botelho, E.C., Bravim, J.C., Costa, M.L. et al. Environmental effects on thermal properties of PEI/glass fiber composite materials. *Journal of Aerospace Technology and Management* 2013, 5, 24–254.
18. Chhibber, R., Sharma, A., Mukherjee, A. et al. Environmental degradation of glass fibre reinforced polymer composite. *Upwind* 2006, 6, 1–48.
19. Kajorncheappunngam, S., Gupta, R.K., and GangaRao, H.V.S. Effect of aging environment on degradation of glass-reinforced epoxy. *Journal of Composites for Construction* 2002, 6, 61–69.
20. Budai, Z., Sulyok, Z., and Vargha, V. Glass-fibre reinforced composite materials based on unsaturated polyester resins. *Journal of Thermal Analysis and Calorimetry* 2012, 109, 1533–1544.
21. Iba, H., Chang, T., and Kagawa, Y. Optically transparent continuous glass fibre-reinforced epoxy matrix composite: Fabrication, optical and mechanical properties. *Composite Science and Technology* 2002, 62, 2043–2052.
22. Atas, C. and Liu, D. Impact response of woven composites with small weaving angles. *Int J Impact Eng* 2008, 35, 80–97.
23. Leonard, L.W.H., Wong, K.J., Low, K.O., and Yousif, B.F. Fracture behavior of glass fiber reinforced polyester composite. *Journal of Materials Design and Applictions Part L* 2009, 223, 83–89.
24. Avci, A., Arikan, H., and Akdemir, A. Fracture behavior of glass fiber reinforced polymer composite. *Cement and Concrete Research* 2004, 34, 429–434.
25. Shyr, T.W. and Pan, Y.H. Impact resistance and damage characteristics of composite laminates. *Composite Structure* 2003, 62, 193–203.
26. Hossain, M.K., Hossain, M.E., Hosur, M.V. et al. Flexural and compression response of woven E-glass/polyester CNF nanophased composites. *Composite Part A* 2011, 42, 1774–1782.

27. Mohbe, M., Singh, P., and Jain, S.K. Mechanical characterization of Na-MMT glass fiber reinforced polyester resin composite. *International Journal of Emerging Technology and Advanced Engineering* 2012, 2, 702–707.

28. Patnaik, A., Satapathy, A., and Biswas, S. Investigations on three-body abrasive wear and mechanical properties of particulate filled glass epoxy composites. *Malaysian Polymer Journal* 2010, 5, 37–48.

29. Yang, B., Kozey, V., Adanur, S., and Kumar, S. Bending, compression, and shear behavior of woven glass fiber-epoxy composites. *Composite Part B* 2000, 31, 715–721.

30. Torabizadeh, M.A. Tensile, compressive and shear properties of unidirectional glass/epoxy composites subjected to mechanical loading and low temperature services. *Indian Journal of Engineering & Materials Sciences* 2013, 20, 299–309.

31. Ellyin, F. and Maser, R. Environmental effects on the mechanical properties of glass fiber epoxy composite tubular specimens. *Composite Science and Technology* 2004, 64, 1863–1874.

32. Awan, G.H., Ali, L., Ghauri, K.M. et al. Effect of various forms of glass fiber reinforcements on tensile properties of polyester matrix composite. *Journal of Faculty of Engineering & Technology* 2009, 16, 33–39.

33. Hussain Al-alkawi, J., Dhafir Al-Fattal, S., and Abdul-Jabar Ali, H. Fatigue behavior of woven glass fiber reinforced polyester under variable temperature. *Elixir Mechanical Engineering* 2012, 53, 12045–12050.

34. Yuanjian, T. and Isaac, D.H. Combined impact and fatigue of glass fiber reinforced composites. *Composite Part B* 2008, 39, 505–512.

35. Putic, S., Bajceta, B., Dragana, V. et al. The interlaminar strength of the glass fiber polyester Composite. *Chemical Industry & Chemical Engineering Quarterly* 2009, 15, 45–48.

36. Alam, S., Habib, F., Irfan, M. et al. Effect of orientation of glass fiber on mechanical properties of GRP composites. *Journal of the Chemical Society of Pakistan* 2010, 32, 265–269.

37. Liu, Y., Yang, J.P., Xiao, H.M. et al. Role of matrix modification on interlaminar shear strength of glass fibre/epoxy composites. *Composite Part B* 2012, 43, 95–98.

38. Mohan, N., Natarajan, S., Kumaresh Babu, S.P. et al. Investigation on sliding wear behaviour and mechanical properties of jatropha oil cake-filled glass-epoxy composites. *Journal of the American Oil Chemists' Society* 2011, 88, 111–117.

39. Godara, A. and Raabe, D. Influence of fiber orientation on global mechanical behavior and mesoscale strain localization in a short glass-fiber-reinforced epoxy polymer composite during tensile deformation investigated using digital image correlation. *Composite Science and Technology* 2007, 67, 2417–2427.

40. Aramide, F.O., Atanda, P.O., and Olorunniwo, O.O. Mechanical properties of a polyester fibre glass composite, *International Journal of Composite Materials* 2012, 2(6), 147–151.

8 Lightweight Nanocomposite Materials

Wojciech (Voytek) S. Gutowski, Weidong Yang,
Sheng Li, Katherine Dean, and Xiaoqing Zhang

CONTENTS

8.1   INTRODUCTION

Development and commercialization of strong lightweight materials and related new products design and novel concepts of rapid assembly of structures are the key elements of emerging advanced structures and vehicles for the aerospace and automotive industries.

The key materials used by the auto industry are high-strength steel (HSS), aluminum (Al), carbon fiber (CF) composites, and plastics which offer weight savings over traditional steel structures of up to 20% for HSS, 40% for Al, and up to 50% for

CF composites. Regardless of the above potential, the quest for lightweight vehicles is not entirely successful across a broad spectrum of global manufacturers who continually release consecutive models heavier than the proceeding ones. This trend is clearly demonstrated by data [1] presented in Figure 8.1.

A number of factors impede broader ingress of lightweight materials in automotive bodies, that is:

1. Their costs which, in comparison with steel, are higher by 15% for HSS, by 30% for Al, and by 500% for CF composites. Only nonstructural engineering plastics offer an approximately 20% weight and costs saving.
2. Long processing cycles in adhesive bonding of plastics, composites, and dissimilar materials, typically in excess of 5–8 min, compounded with tedious substrate surface preparation prior to bonding. This contrasts unfavorably with the rapid (seconds) and easy to robotize welding of metals.
3. Concerns regarding the lifetime structural integrity of bonded structures.

Engineering plastics, although offering 20% weight savings, exhibit mechanical properties such as the strength, fracture and impact properties, modulus of elasticity, and shear modulus which are significantly lower than those of metals and hence, are inadequate for structural applications such as frame, drive module, or life capsule of the vehicle. These deficiencies frequently limit the use of polymers as high-performance engineering materials.

An effective way of improving their properties is by reinforcing polymers with fillers in the form of micro-sized particulate, fibers, platelets, or by alternatively nano-sized materials.

Incorporating micro-sized fillers as a means for improving designated physical properties of polymers frequently introduces undesirable trade-offs [2]. Increasing, for example, the elasticity modulus frequently causes reduction of the composite's toughness due to lowering its fracture energy.

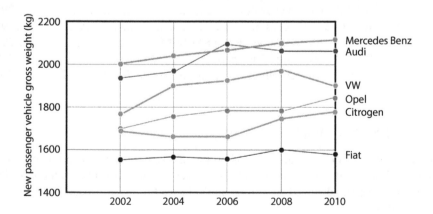

FIGURE 8.1 New passenger cars: vehicles' weight for selected global brands. (Based on data in M. Campestrini, P. Mock, *European Vehicle Market Statistics*, Pocket Book Edition, International Council on Clean Transportation (2011).)

8.2 QUEST FOR A LIGHTWEIGHT VEHICLE

Representing approximately 40% of total vehicle weight, the car body is the heaviest vehicle element. Consequently, the most common approach concerning design of lightweight vehicles is an integrated body structure utilizing combination of various materials based on their optimum engineering properties and functional characteristics. The most significant trend over recent years is the use of high-strength steel (HSS) contributing to approximately 50% to 65% of vehicle bodies in European compact cars. This is used in combination with increasing use of thermoplastics and thermoset composites in body cladding, hoods, bonded roof components, and some structural components of the body [3–8].

The most prominent achievement in lightweight vehicle design and manufacture is the BMW-i3 weighing just 1195 to 1390 kg. It is illustrated by a set of photos in Figure 8.2, which also provide comprehensive details of the BMW-i3 modular body assembly, including a complete vehicle.

The BMW i3's body comprises a carbon-fiber-reinforced plastic (CFRP) passenger cell (the Life Module, LM) integrated with an aluminum chassis (the Drive Module, DM).

The following interior and exterior components utilize CFRP with a polyurethane (PU) matrix:

1. Passenger cell ("Life Module") structure
2. Roof panel: recycled CFRP
3. Rear seat shells: CFRP using recycled carbon fiber fleece in a PU matrix

The light weight of the carbon composite structure is clearly demonstrable in Figure 8.3.

Preformed carbon fiber sub-components of the life module, fabricated with the use of high-pressure resin injection using resin transfer molding (RTM) process are assembled, by adhesive bonding, into a rigid and yet impact resistant three-dimensional supporting structure.

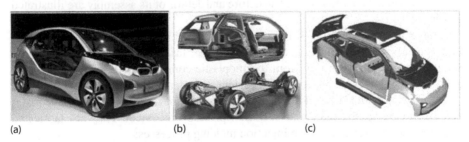

(a) (b) (c)

FIGURE 8.2 (a) BMW i3, the first mass manufactured lightweight car with a predominant part of its structure and body manufactured from carbon fiber reinforced composite; (b) life capsule and drive module; and (c) body panels fabricated from hemp-reinforced thermoplastic composites. (Composite illustration prepared using materials presented in (a) *"BMW i3"* in Wikipedia; (b) *"2016 BMW i3 Rescue Guide"*: in http://www.boronextrication .com/2016/01/31/2016-bmw-i3-rescue-guide/; (c) *"BMW i3 Service and Training Manual"* in http://www.partzine.com/bmw-i3-service-and-training-manual/.)

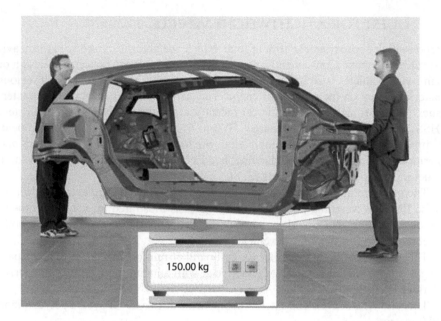

FIGURE 8.3   Demonstrable light weight of BMW i3 life module fabricated from carbon fiber composite. The total weight of the capsule is 150 kg. (Composite illustration prepared using material presented in G. Nica (2013), *"BMW i3 Might Be Cheaper to Live with Due to Carbon Fiber Construction"* in http://www.autoevolution.com/news/bmw-i3-might-be-cheaper-to-live-with-due-to-carbon-fiber-construction-73054.html.)

All structure subcomponents are robotically assembled to allow a uniform 1.5 mm adhesive gap to ensure no variation in adhesive bond quality. Due the rapid cure of a PU adhesive, all assembled components can be handled within 90 seconds while the total adhesive's cure is completed within 30 min. The total length adhesive bond is approximately 160 m.

The schematics of the BMW i3 structure and details of its assembly are illustrated in Figure 8.4.

The i3 model is the first BMW car with an exterior shell and other key components such as: door panels, tailgate panels, bumper fascias (bumper bars), and wing/fender manufactured entirely from thermoplastic composites (PP/EPDM elastomer modified copolymer: poly-propylene/ethylene propylene diene monomer).

The injection molded cladding panels utilize 25% of recycled or renewable material and weigh approximately 50% that of pressed sheet steel. They are fabricated with the use of three alternative injection molding processes:

- A standard injection molding process
- A "twin" two-component injection molding process, molding and bonding the outer skin and its substructure in separate, successive stages
- Parallel injection molding of the outer skin and substructure with bonding in one automated process

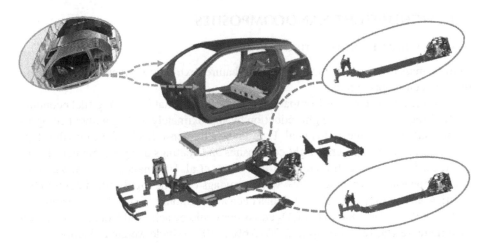

FIGURE 8.4 Schematics of BMW i3 structure assembly. (Composite illustration prepared using materials presented in L-H upper corner: From *"BMW i3 production rate increased to meet rising demand"* (15 Apr 2014); in http://bmwi.bimmerpost.com/forums/showthread.php?t=972579; middle section: From *"Inside the BMW i3 Body..."* http://www.autoindustryinsider.com/?p=6976. R-H elements: From: "Inside the BMW i3 Body..." http://www.autoindustryinsider.com/?p=6976.)

Due to the use of a nonpolar substrate material (PP/EPDM) for exterior body panels, products such as the bumpers, body panels, and the front and rear parts are painted individually and require flame pretreatment to ensure good paint adhesion.

Elimination of conventional cataphoretic dip priming essential for metallic body reduces the weight of the vehicle by an additional 10 kg.

The BMW i3 door composite panels are compression molded using 50% hemp or kenaf fiber-reinforced EPDM/PP (1800 g/m^2 weight) and are surface finished in a 200 μm thick (180 g/m^2) PP matt providing weight saving of each panel by ~10% in comparison with light gauge steel.

The dashboard cladding is fabricated from a kenaf fiber reinforced composite and utilizes open-pore eucalyptus veneer lamination. The air-conditioning areas on the dashboard (the underside area) are surface finished in either 100–400 μm PP film or a PES/PP nonwoven material.

Among nonvisible subcomponents of the structure significantly contributing to the vehicle lightweightness, the following needs highlighting: the trailing edge element of the body structure is fabricated from glass-fiber-reinforced plastic; 30% lighter than a conventional sheet steel component. The direct connection between the power electronics and electric motor in the rear of the BMW i3 reduces the length of cabling required; it cuts the overall weight of the drive train by 1.5 kg. Aluminum suspension links weigh 15% less than in conventional designs. The hollow composite drive shaft is 18% lighter than a conventional equivalent, and the forged aluminum wheels are 36% lighter than steel rims.

8.3 LIGHTWEIGHT NANOCOMPOSITES

8.3.1 VEHICLE LIGHTWEIGHTING GAINS

Vehicle mass is one of the most important features exhibiting direct correlation with the fuel consumption rate.

Historically Ford Motor Company (2010) estimated that improving fuel economy by 40% required vehicle weight reduction by approximately 340 kg without compromising safety. To achieve this goal, Ford launched intense R&D effort into developing and implementing lightweight composites and coatings using nanomaterials [9].

According to research completed by Garces et al. [10], nanocomposite automotive components offer up to a 25% weight savings over the highly filled commodity plastics, and up to 80% over steel. It has been also demonstrated [11] that an approximately 1.3%–1.8% reduction of CO_2 emissions could be achieved from 5% reduction in vehicle weight, reaching up to 2.7%–3.6% at 10% vehicle weight reduction.

The use of strong lightweight nanocomposites is additionally anticipated to facilitate optimized car body designs leading to lowering the drag coefficient due to an estimated gain of 10% in aerodynamic flow improvement which can, in turn, produce an associated 20% reduction of rolling resistance of tires and 7.5% increase in average power train efficiency [12].

8.3.2 POLYMERIC NANOCOMPOSITES

Polymer nanocomposites are a new class of hybrid materials in which the traditional micro-sized fillers are substituted by nano-sized inorganic equivalents. These additives produce a drastically greater level of improvement of designated structural and functional properties of polymers than their micro-sized analogues.

These are typically used in the form of nanoparticles, nanotubes, nanoplatelets, nanofibers, nanocubes, and other geometry materials dispersed in the host polymeric matrix. The resultant nano-composites exhibit superior mechanical, electrical, thermal, barrier properties, fire retardancy, impact resistance, and other physico-chemical properties not achievable through the addition of micro-sized fillers [13–16] making them suitable for replacing metals in automotive and other applications.

An additional advantage of nanocomposites is that while the targeted properties are improved, some other significant performance parameters such as the viscosity of polymer matrix during processing, or material shrinkage and warpage in most cases are not adversely affected.

Such significant improvements of specific physico-chemical properties of polymeric nanofilled composites in comparison with micro-sized fillers are due to a number of factors:

1. Nano-sized fillers, for example, individual platelets of exfoliated clays or individual single wall carbon nanotubes, have a large aspect ratio (1000:1), with the thickness (or diameter) of each individual element being approximately 1 nm.
2. Individual nano-platelet sheets, or nanotubes of exfoliated nanomaterial, exhibit high surface density of inherent surface chemical groups associated

with very high surface-to-volume ratios, which, in turn, are readily available for physico-chemical interactions between these surface groups and polymeric matrix.

3. As a consequence of (1) and (2), there is an exponential increase in the number of molecular bridges formed between the matrix and filler nanoparticles resulting in the drastic improvement of targeted properties of nanocomposites in comparison with a neat polymer or composite reinforced with micro-sized analogue reinforcements.

One of the key advantages of nanocomposites over conventional composites is the fact that the above improvements are achievable with significantly lower additions of nanofillers versus conventional composites reinforced with micro-sized reinforcement to achieve a similar level of improvement. Typically nanoclays replace talc or glass fillers at a 3:1 ratio, with 5%–8% of a nanoclay replacing 15% of glass filler [17]. As an example, a 3%–5% addition of nanofillers is needed in nanocomposites versus conventional talk-reinforced composites which require significantly higher filler levels, 10%–50% [p/w], to achieve a similar level of improvement [18].

Nanofillers can be designed (see Section 8.4) to achieve excellent dispersion and exfoliation providing not only improvement of mechanical properties, but also synergistic flame retardancy and the product weight reduction (up to approximately 20%–40% compared with commodity composites), consequently providing gains in fuel economy of up to 10%–20%. An even more spectacular example of future reinforcing materials are carbon nanofibers in the form of carbon nanofiber BuckyPaper composites, which promise a tensile strength of 150 GPa, about 50 times that of steel while exhibiting only 20% of its weight [16].

Table 8.1 provides an outline of gains in nanocomposites properties in comparison with the counterpart micro-filler reinforced composites.

The key categories of automotive applications of polymeric nanocomposites to date are [18] body frames, interior and exterior body parts, power train, suspension and breaking systems, exhaust systems, fuel and other fluid lines, paints and coatings, lubrication, tires, and electrical/electronic equipment.

TABLE 8.1
Gains in Nanocomposite Properties versus Standard Composites

	Filler Size/[% Addition]	
Property	Micro- [10%–40%]	Nano- [2%–5%]
Specific density	1	0.5–1.0
Tensile strength	1	1.5–2.0
Elasticity (Young's) Modulus	1	4–5
Impact strength (Izod)	1	0.5–1.0
Heat distortion temperature	1	1.5–2.0

Source: Adapted from Presting, H. and Koning, U. 2003. *J. Mater. Sci. Eng.* 23, 737.

Table 8.2 provides a brief outline of automotive applications of nanocomposites fabricated with a range of typically used matrix materials and currently available nanofillers. Their designated applications are also listed in this table.

Nanoclays such as montmorillonite (MMT) or synthetic layered double hydroxides (LDHs) are commercially most commonly used additives in the preparation of nanocomposites, constituting about 80% of the market consumption.

Polyolefins are the most important polymeric matrix in nanocomposites applications. Other thermoplastics used by automotive industry are: polyamide (nylon), polyphenylene sulfide (PPS), polyetheretherketone (PEEK), polyethylene terephthalate (PET), and polycarbonate (PC). Thermoplastic elastomers such as butadiene-styrene diblock copolymer are also used.

The use of thermoplastics as a matrix material in nanocomposites has been growing steadily, predominantly in automotive applications, due to material's low cost, high performance, low density, longer shelf life, easy processing with nanomaterials,

TABLE 8.2
Examples of Typical Areas of Automotive Applications of Nanocomposites

Material/Product	Application	Benefits
Polymeric Matrix		
Polyolefins	Facia (bumper bar), center console, instrument panel, trim panels, rocker panel, step bar, fender, seat backs, grill, sail panel, engine block cover	Lightness, toughness (impact resistance), easy recyclability, low price, no need for drying prior to molding, barrier properties
Nylon engine	Engine block cover, fuel houses and lines, timing belt, door handles, brackets,	High strength and modulus, high heat distortion temperature
Elastomers	Tires	Reduced rolling resistance, high barrier properties, weight reduction, improved durability
Nanofillers		
Nanoclays	Polymer reinforcement, improvement of key mechanical properties, improved electrical, thermal and barrier properties, fire retardancy, impact resistance, scratch resistance, tactile properties	Low cost, improved strength and moduli, improved toughness, barrier properties, flame retardancy, matrix compatibility of organic-modified clays
Carbon nanotubes	Electromagnetic shielding, electro-static painting through conductive coatings or bulk additive for product conductivity	Excellent electrical and thermal conductivity, low thermal expansion, polymer reinforcement
POSS (polyhedral oligomeric silsesquioxane)	N/A	Reinforcement, flame retardancy, impact properties, hardness, abrasion resistance

ability to regrind, and recyclability. Thermoplastics also offer enhanced mechanical, thermal, electrical, and barrier properties, and excellent fracture toughness over thermosets. The ease of their joining by mechanical means, adhesive bonding, and welding techniques is also important.

The most important category of thermoset materials used by the auto industry is epoxies and polyurethanes.

8.3.3 ELASTOMERIC NANOCOMPOSITES

Historically, nanofillers have been used as reinforcing materials in tire manufacturing for a long time; carbon black has been used in rubber compounding since 1904.

The current and emerging categories of elastomeric nanocomposites rapidly gaining importance in tire manufacturing are reinforced by clays and layered double hydroxides (LDHs) facilitating significant enhancement of tire properties and overall performance such as increased flexibility combined with excellent tensile strength, lower rolling resistance, increased traction, outstanding fatigue performance, and low cost-to-performance ratio. The most important types of host elastomeric matrix materials are natural rubber (NR), styrene butadiene rubber (SBR), butadiene rubber (BR), isoprene rubber (IR), and halogenated butyl rubber (HBR), with SBR being the most common material in tire applications of nanocomposites.

The key applications are tire inner tube, tire inner liner, off-the-road (OTR) tire tread; typically used in tires of construction vehicles such as haul tracks, wheel loaders, backhoes, graders, trenchers, and so on, and conveyer belts.

Elastomeric nanocomposites find increasing applications in the manufacture of automotive tires. The key drivers for rapidly growing demand for these materials and related products are

- Lower weight and subsequent improvements in fuel saving
- Lower rolling resistance and noise reduction
- Enhanced product durability
- The need for meeting increasingly demanding automotive standards for safety.

Elastomeric nanocomposite foams could also be seen as the next generation of energy dissipating and thermally insulating materials.

8.3.4 ELECTRO-CONDUCTIVE NANOCOMPOSITES

Carbon nanofibers, carbon nanotubes (mainly multi-wall carbon nanotubes, MWCNTs), exfoliated graphite, exfoliated graphene nanoplatelets (GnP), and carbon nanofiber Bucky Papers are the most important categories of electro-conductive fillers used in nanocomposites.

Carbon nanotubes and graphene impart outstanding electrical and thermal conductivity, which allows electrostatic painting (typically: PP, nylon, and PC/PPS blends) and provide electromagnetic shielding protecting against electrical discharge of static electricity in the fuel system [20]. Other applications are semiconductors in automotive electronics and electrical applications. However, their commercial development is hindered by their high price ($20/g), although they are available as master batch dispersions (15% to 20% contents of nanotubes) for about $100–120/kg.

Dispersion of commercially available MWCNTs master batches can be accomplished on commercial equipment without any additional surface treatment of CNTs.

It needs to be pointed out, however, that virgin CNTs, if not functionalized, are difficult to disperse, particularly in water-based systems since they exhibit hydrophobic properties. The generally accepted methods of compatibilizing CNTs with the water-based carrier involve, for example, the use of nonionic surfactants, such as Triton X-100 [21] and sodium lauryl sulfate [22] which improve their dispersibility in aqueous solutions.

Technological applications of CNT-filled materials rely on the effectiveness of interactions between CNTs and the adjacent matrix. These may include electron transfer in conducting materials, mechanical load transfer and control of fracture propagation mechanisms in composite materials, molecular absorbance in environmental applications, and others. The interactions between the CNTs and surrounding medium frequently counteract the mutual interactions (self-interactions) between adjacent CNTs. Those competing phenomena have significant importance in the process of effective dispersing of CNTs, either within a carrier liquid or within a polymeric matrix.

8.3.5 GREEN NANOCOMPOSITES

The European "End of Life Vehicle (ELV) initiative" recommended that by 2015 all vehicles were to be manufactured with 95% recyclable materials, 85% of which must be recoverable by either reuse/mechanical recycling or 10% through energy recovery (thermal recycling) [23]. Despite significant improvements regarding establishment of feasible systems regarding disposal and sorting recycled polymers and development of recycling technologies, the problems with recycled materials properties combined with environmental and societal concerns make petroleum-based nanocomposites relatively unattractive.

Considering the above drawbacks, an increased use of green nanocomposites fabricated from renewable and environmentally benign materials such as bio-fibers and bio-based resins and naturally occurring benign mineral reinforcements such as clays is anticipated to grow rapidly.

Green and biodegradable nanocomposites, which may potentially replace current materials based on synthetic petro-polymers, are the next generation of composite materials for automotive applications.

Biodegradable polymers which have been improved by the incorporation of nano-particles and micro-particles include starches [24], proteins [25], and polyesters [26,27] such as polycaprolactone, polylactic acid, and other aliphatic polyesters.

The major car manufacturers such as Daimler Chrysler, Mercedes, Volkswagen, Audi Group, BMW, Ford, and Opel steadily increase the use of biocomposites in various vehicle applications [28]. Typically, these are commodity composites reinforced with various types of nano-sized cellulose based plant fibers.

8.4 NANOFILLERS FOR NANOCOMPOSITE APPLICATIONS

8.4.1 STRUCTURE OF CLAY MINERALS: GENERAL OUTLINE

Natural and synthetic clay minerals attract growing interest from scientists and industry as a choice of nanofiller for nanocomposites. This is due to their unique properties, especially their capacity to adsorb inorganic and organic materials onto their surface. This, in turn, provides an attractive platform for engineering novel functional organic/inorganic hybrid materials for various end-applications, including advanced nanocomposites.

Particles of clays [29] exhibit a characteristically layered structure: see Figure 8.5 illustrating the structure of montmorillonite (MMT), one of the most technologically important nano-clays.

Montmorillonites are composed of structured lamellae units comprising two tetrahedral silica sheets and an alumina octahedral sheet. Their inter-sheet layers include exchangeable metal ions (e.g., sodium ions) neutralizing the net negative charges generated by partial substitution of Al^{3+} with Mg^{2+} at the octahedral sites. The number of exchangeable ions within the clay mineral determines the amount of organic guest ions which can be intercalated between the clay lamellae.

8.4.2 INTERCALATION BY SELF-ASSEMBLED ORGANIC COMPOUNDS

The efficiency of intercalation is dependent on the type of clay and that of guest molecules [30]. The adsorbed organic compounds form self-assembling molecular aggregates due to the two-dimensional lamellae surface. This feature has been confirmed by various spectroscopic analytical methods, for example, X-ray diffraction,

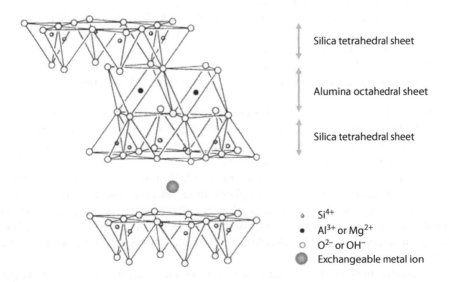

Silica tetrahedral sheet

Alumina octahedral sheet

Silica tetrahedral sheet

○ Si^{4+}
● Al^{3+} or Mg^{2+}
○ O^{2-} or OH^{-}
● Exchangeable metal ion

FIGURE 8.5 The lamellar structure of montmorillonite clay.

neutron scattering, high resolution electron microscopy, nuclear magnetic resonance (NMR) [31], electron spin resonance (ESR) [32], and polarized spectroscopy [33], which were able to clarify the orientation of guest molecules "sandwiched" between the layers.

The self-assembly interaction is variable depending on the following factors:

- The structure of clay minerals and guest organic compounds
- The type and amount of exchangeable metallic ions
- The amount of intercalated water in the interlayers.

Layered double hydroxides (LDHs), alternatively known as hydrotalcite compounds, are available as naturally occurring minerals and as synthetic materials. The structure of these materials is based on the stacking of brucite $[Mg(OH)_2]$-like layers with hydrated anions in the interlayer. These materials have been traditionally used in applications such as catalysis, adsorption, anion-exchange, medicine and, recently, as additives for nanocomposites [34,35].

8.4.2.1 Surface-Engineered Clays: Layered Double Hydroxides (LDH) Modified with Fatty Acids

The focus of this section is on the synthesis of organophilic LDHs as an emerging class of nano-additives for polymer reinforcement.

Pristine LDHs are not suitable for melt intercalation by large species such as polymer matrices because of their short intergallery space (about 7.6 Å) and their high layer charge density. It is therefore necessary to modify the LDHs prior to their inclusion in the polymer to increase the inter-gallery distance (basal distance) and to improve their compatibility. These can be achieved by incorporating suitable organic anion molecules into their inner structure.

Three main approaches for incorporating anionic molecules into LDHs [36] are

1. Anion-exchange of a precursor LDH
2. Direct synthesis by co-precipitation
3. Reconstruction through rehydration of a calcined LDH precursor.

The LDHs reconstruction with the use of fatty acids provides feasible means for effective surface modification of mineral fillers (e.g., calcium carbonates) for improved compatibility with polymeric matrix materials and related enhancing of mechanical properties of nanocomposites.

A conventional reconstruction method of LDHs involves calcination at 500–800°C to transform them into amorphous mixed magnesium aluminum oxides. Subsequent rehydration of the latter in the presence of water and designated anions results in intercalation of the anions into the LDH inter-layer space. This method has been demonstrated to be effective for intercalating small anions, for example, Cl^-, NO_3.

Below we present the outcome of our work on inserting significantly larger organic anions into the structure of LDHs, as also reported in other studies [37]. The focus of this work was intercalation of fatty acids exhibiting varying chain lengths and degree of unsaturation (e.g., sorbic acid, octanoic acid, lauric acid, oleic acid, and stearic acid).

A modified reconstruction method presented below is shown to provide effective means for intercalating carboxylic acids with LDHs resulting in fabrication of engineered, well-organized layered hydroxide structures. The method involves a separate rehydration step of the calcined LDH to enable formation of a controlled lamellar structure before incorporation of fatty acids.

Commercial Mg-Al LDH (Hycite 713, from Ciba Specialty Chemicals) with an Mg/Al ratio of 2.4 was used as a precursor material. LDH was first calcined at 500°C for 14 h, and then cooled down to room temperature under nitrogen. 100 ml of water to be used for rehydration was boiled for half an hour, and cooled to rehydration temperature.

5.5 g calcined LDH (Mg2.4AlO3.9, 0.0372 mol) was weighed and either directly added during agitation, or it was first rehydrated at 80°C for 3 h in water before a designated fatty acid (lauric acid, octanoic acid, oleic acid, stearic acid, and sorbic acid, all from Aldrich) was added and stirred for another 18 h. After filtering, washing, and drying, a fine white powder was obtained with nearly 100% yield in all cases.

The structure and properties of intercalated LDHs were confirmed by analyses carried out with the use of the following analytical instruments:

1. SEM: Philips XL 30 Field Emission Scanning Electron
2. Diffuse reflectance infrared Fourier transformation (DRIFT): Equinox 55 (Bruker) spectrometer (4000–600 cm^{-1}) with a diffuse reflectance attachment (64 co-added interferograms were scanned at 4 cm^{-1} resolution)
3. Thermogravimetric analysis (TGA): Perkin Elmer Pyris TGA thermogravimetric analyzer.

The characteristic lamellar peak of the Mg-Al LDH at $2\theta = 11.6°$, as determined by XRD, was used to optimize the rehydration step as a part of the modified reconstruction method.

At room temperature the rehydration process is very slow taking up to 24 h to restore the lamellar structure of the LDH. When the rehydration temperature is increased to 80°C, a large proportion of the lamellar structure of the Mg-Al LDH is restored after a time as short as 1 h. The reaction has been shown to be completed after 3 h. Consequently, the rehydration at 80°C for 3 h was selected for the subsequent experiments targeting fabrication of LDHs exhibiting structure suitable for intercalation with a broad range of carboxylic acids.

The photos in Figure 8.6 show SEMs illustrating the appearance of LDHs nano-platelets at consecutive stages of the above modification.

Figure 8.7 compares the XRD patterns of the "as received," calcined and rehydrated LDHs. The "as received" LDH exhibits an XRD pattern characteristic of a crystalline lamellar material. Subsequent to calcination, the crystalline diffraction peaks of the LDH completely disappear indicating that the layered structure has been destroyed by this process.

After rehydration in hot water (3 h at 80°C) the diffraction peaks of LDH appear at the same 2 theta angles as in the "as received" LDH suggesting that the lamellar

FIGURE 8.6 SEM photos illustrating the appearance of LDHs nanoplatelets at consecutive stages of modification.

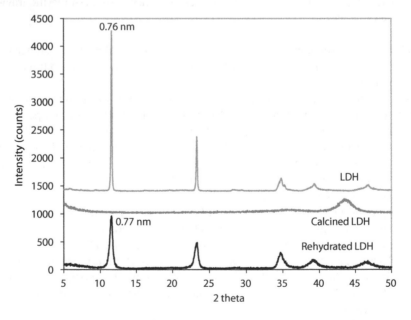

FIGURE 8.7 Comparison of XRD spectra of LDH at various stages of processing: an initial LDH, calcined LDH, and rehydrated LDH.

structure has been successfully restored. However, the rehydrated material appears to be less crystalline than before the heat treatment (reduced peak intensities and the broadened peak widths).

The LDH, intercalated with fatty acid, synthesized by our modified reconstruction method exhibits a well-organized layered structure in contrast to the disordered one in the material synthesized using the conventional method (Figure 8.8).

During the conventional reconstruction process the anions quickly adsorb onto the surface of the positively charged hydroxide lamellae once they are formed, and these subsequently become a platform for generating the next layer. While the presence of small anions (i.e., Cl⁻, NO_3, etc.) was shown to cause minimum disruption to the reconstruction of the layered structure of LDHs, the adsorption of relatively large and hydrophobic anions such as fatty acids onto the surface of hydrophilic hydroxide platelets is expected to lead to a loose and disordered layered structure due to incompatibility and steric hindrance.

The above problem is eradicated in our modified reconstruction process due to the fact that the rehydration step of the calcined LDH in water is known to first produce LDHs with a well-organized layered structure containing OH⁻ as interlayer anions [38]. Subsequent addition of fatty acid anions leads to their intercalations into the LDH structure by replacing the hydroxyl anions with minimum interruption of the layered structure.

Successful intercalation of all types of carboxylate anions investigated in this work is confirmed by their XRD patterns illustrated in Figure 8.9a. In all cases, the broadening and shifting of the basal reflection peaks toward a lower field is observed for intercalated carboxylate molecules in comparison with the "as received" LDH. The graph in Figure 8.9b, in turn, illustrates the influence of the number of carbon atoms in a fatty acid chain on the resultant basal distance of the reconstructed LDH.

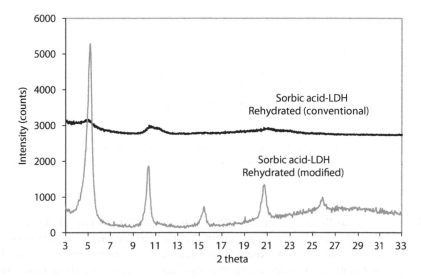

FIGURE 8.8   XRD spectra of LDHs intercalated with sorbic acid: rehydrated using a conventional method and our modified method.

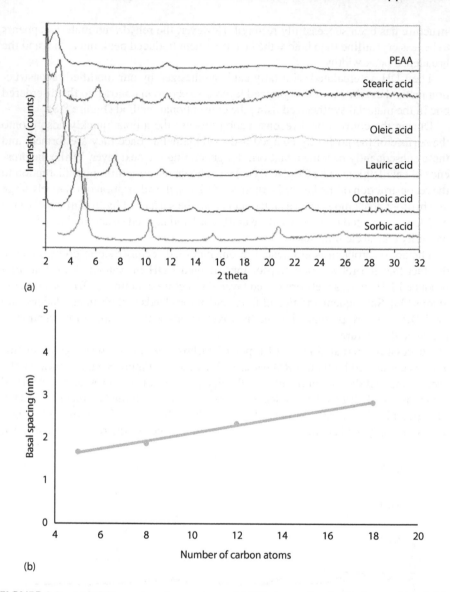

(a)

(b)

FIGURE 8.9 (a) XRD spectra of reconstructed LDHs intercalated with fatty acids of different chain lengths, and (b) the distance between LDH platelets in relation to the number of carbon atoms in the fatty acid chain.

It is seen that as the length of the carboxylate molecule is increased, the distance between the LDH platelets is increased in a linear fashion. This expansion of the space between the platelets is expected to facilitate their separation and dispersion into a polymer matrix during the compounding of nanocomposites. These results suggest also that the effectiveness of intercalation reaction is not significantly affected by the alkyl chain length or whether the alkyl chain is saturated or unsaturated.

This broadening of basal spacing indicates the presence of some turbostratic effect caused by a decrease of the ordering along the stacking axis due to the loss of van de Waals interaction between adjacent layers and the absence of a densely packed interlayer space formed by "high charge-density" anions such as carbonate or halides. Moreover, a broad scattering hump is observed between 20 and 30° in 2θ for the stearic acid modified LDHs suggesting that some stearic acid molecules are adsorbed on the surface of the LDH in similarity to the case of LDHs modified by polyethylene oxide derivatives and oleate ions.

FTIR analysis of modified LDHs confirms the presence of fatty acid anions in the LDH; see Figure 8.10. The "as received" LDH has an intense absorption peak at 1361 cm^{-1} due to the ν (asymmetric) stretching mode of the carbonate anion, which is not present in the spectra of any of the fatty acid modified LDHs. After incorporation of fatty acid anions into LDH structure one observes the characteristic peaks of the COO group at 1600 cm^{-1} (symmetric) and 1400 cm^{-1} (asymmetric) instead of the free acid (COOH) peak at 1705 cm^{-1}. Weaker bands arising from the alkyl C–H stretches of the fatty acid molecules are observed in the range 3000–2800 cm^{-1}.

The SEM analysis of layered hydroxides investigated in this work reveals (see Figure 8.11), that the LDHs intercalated with the smaller carboxylate anions such as sorbic acid, octanoic acid, and lauric acid maintain a well defined and regular-shaped platelet structure. On the other hand, as confirmed by XRD analysis, longer molecules such as stearic acid (not shown here) appear to be present on the surface of the LDH as well as being inserted into the platelet structure.

It can be concluded, based on the evidence presented in Section 8.4 that LDHs modified with fatty acids maintain well-ordered layered structure exhibiting significantly increased interlayer distances easily controlled by the length of intercalated fatty acid molecules.

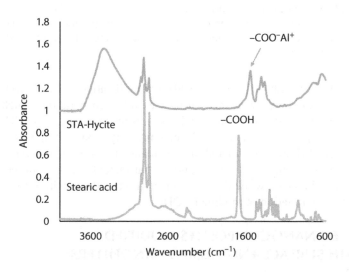

FIGURE 8.10   FTIR spectra of the "as received" LDH and that of LDH intercalated with sorbic acid.

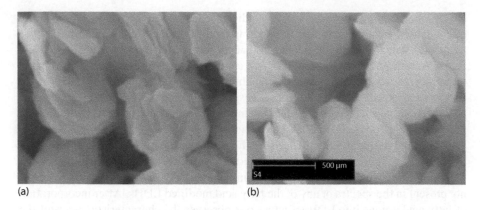

(a)   (b)

FIGURE 8.11   SEM images of layered double hydroxide: (a) rehydrated LDH at 80°C, 3 h and (b) sorbic acid modified LDH. The bar indicating dimensions in both images is 500 nm.

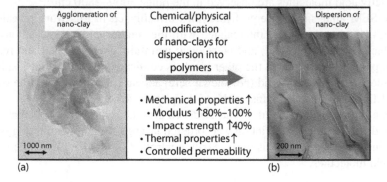

(a)   (b)

FIGURE 8.12   PP/LDH nanocomposite with the following forms of layered double hydroxide: (a) a tactoid (agglomerated cluster) of tightly bound virgin (unmodified) LDH platelets in a PP matrix, and (b) completely exfoliated nano-platelet sheets of LDH modified by intercalated fatty acid molecules.

The use of fatty acids (e.g., stearic acid) for modification of LDHs converts their surface into hydrophobic, hence facilitating a feasible incorporation of such modified LDHs as property modifying additives in polymeric nanocomposites. This feature is illustrated by photos in Figure 8.12 demonstrating a nanocomposite filled with unmodified LDH (Figure 8.12a) and that intercalated with fatty acid molecules (see Figure 8.12b). It is apparent in these photos that tactoids of unmodified LDH are present in the PP matrix, while the same material intercalated with fatty acid exhibits an excellent exfoliation of modified LDH.

8.5   GREEN NANOCOMPOSITES MODIFIED WITH SURFACE-ENGINEERED NANOFILLERS

Similar to petro-based composites and nanocomposite, nano-technology has been utilized to improve the properties (thermal, mechanical, barrier, and others) of biodegradable polymers including those that are based on starches, proteins, or bio-polyesters (polycaprolactone, polylactic acid, and other aliphatic polyesters) [24–27].

8.5.1 Examples of Green Nanocomposites and Their Properties

8.5.1.1 Input Materials

The following bio-polymers were investigated in this work with the aim of developing high-performance composite materials targeting automotive, packaging, and other industries:

1. Bio-polyesters: Polybutylene succinate adipate (PBSA, Bionolle™ 3001, supplied by Showa, Japan) and poly(lactic acid) (PLA, 7000, supplied by Natureworks USA)
2. Starch: High amylose corn starch (supplied by Penford, Australia) predominantly plasticized with water
3. Raw wheat gluten proteins: Supplied by Manildra Group Australia. Raw wheat gluten (WG) used in this work contained (on dry basis): ~80% of proteins, 15% of residual starch, 4% of lipids, and around 1% of fibers and other impurities.

As the candidate reinforcing materials, the following materials were used:

1. Commodity montmorillonite clays (MMT): Supplied by Southern Clay Products, including Cloisite 30B modified with methyl tallow bis-2-hydroxyethyl quaternary ammonium chloride.
2. Synthetic fluoro-hectorite clays:
   a. neat Na-FHT: supplied by Unicoop Japan
   b. modified MEE: di poly(oxyethylene) alkyl methyl ammonium
   c. MAE: dimethyl dialkyl ammonium (where R' and R" are hexadecyl and octadecyl, respectively)
   d. MTE: trioctyl methyl ammonium were used in this study of nanocomposite formation, structure and properties
3. Wood flour: Used for the micro-composite fabrication (supplied by American Wood Fibers, AWF 2010). Wood flour was used unmodified and modified using methylenediphenyl diisocyanate (MDI).

8.5.1.2 Fabrication of Green Nanocomposites

Composites fabricated in this study were extruded with the nanoclays using a Theysohn twin screw extruder (L/D 40) operating at 120 rpm using various barrel temperatures depending on the polymer matrix. The screw design included a number of specifically designed high shear sections to encourage exfoliation of platelets.

For the PBSA samples, each unmodified FHT and three modified FHT loadings of 1, 2, and 5 wt% were investigated. For the starch and protein samples, 1, 2.5, 3, and 5 wt% loading of MMT or FHT were investigated. Extruded PBSA pellets were subsequently dried at 70°C for 12 h and then either injection molded at 130°C into standard test pieces for tensile and impact testing or compression molded at 130°C to produce film (270–300 μm) for permeability studies and thermal analysis.

The starch nanocomposites were directly extruded into film using a film die attached to the extruder. The gluten/protein samples were blended using a high-speed and Banbury mixer, and finally compression molded at 130°C for 5 min.

8.5.1.3 Characterization

XRD was predominantly used to monitor the d_{001} spacing corresponding to the intergallery spacing of the Na-FHT. The XRD measurements were performed on the nanocomposite samples using a Bruker D8 Diffractometer operating at 40 kV, 40 mA, Cu Kα radiation monochromatized with a graphite sample monochromator. A diffractogram was recorded between 2θ angles of 1° and 10°.

The structure of nanocomposites was imaged using a Phillips CM30 TEM using an accelerating voltage of 100 keV at magnifications of up to 100,000 to study dispersions of Na-FHT and Na-MMT particles. The samples were sectioned using a 45° diamond knife at –100°C. The 70 nm thick ribbons were carefully sandwiched between two copper grids for imaging using transmission electron microscopy.

The composites and blends were also imaged using a FEI Quanta 200 ESEM. Imaging was performed in high vacuum mode at an accelerating voltage of 5 kV. The fracture surfaces: (1) formed during impact testing, and (2) freeze fracture were coated with iridium prior to imaging.

Mechanical testing was carried out in compliance with ASTM 638 standard using an Instron tensile testing apparatus utilizing a 1–30 kN load cell and a 10–50 mm/min strain rate depending on the sample. An external extensometer was used for independent modulus determination. The individual values provided are an average of 7 repeats.

Impact properties were determined according to the ASTM 256 standard on a Radmana ITR 2000 instrumented impact tester in Izod mode. The impact strain rate was 3.5 ± 0.2 m/sec. The individual values provided are the average of 10 repeats.

8.5.2 Results and Discussion

8.5.2.1 Morphology/XRD/TEM/SEM

8.5.2.1.1 PBSA Nanocomposites

TEM imaging (Figure 8.13) revealed that all samples containing modified clays exhibit good dispersion of FHT platelets (many single platelets and small tactoids). The highly dispersed and oriented Na-FHT-MEE and Na-FHT-MAE nanocomposites (as seen via TEM) had higher d_{001} spacings (3.15 nm and 3.71 nm, respectively) whereas the d_{001} spacing for the not so well dispersed Na-FHT-MTE remained relatively unchanged from the neat Na-FHT-MTE (2.61 nm as compared to 2.42 nm, respectively). Both chemical functionalities (presence of hydroxy groups in the Na-FHT-MEE), and high initial d_{001} spacing (presence of longer alkyl chain hexadecyl and octodecyl groups in Na-FHT-MAE) were proven to be advantageous in terms of clay dispersion.

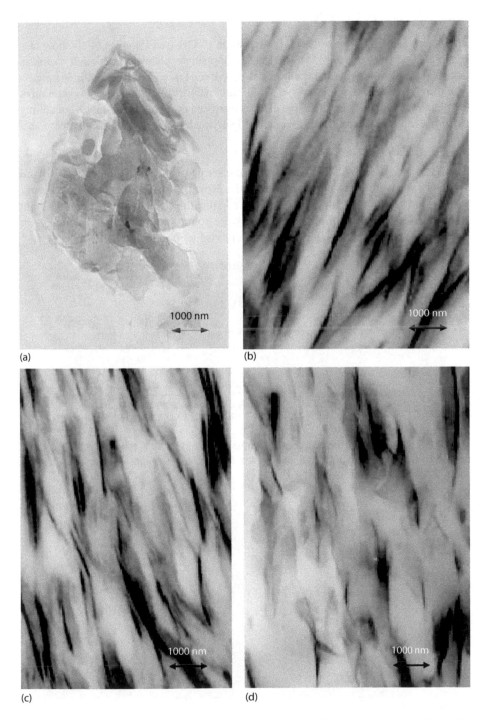

FIGURE 8.13 TEM of (a) PBSA with 5 wt% Na-FHT; (b) PBSA with 5 wt% Na-FHT-MEE; (c) PBSA with 5 wt% Na-FHT MAE; and (d) PBSA with 5 wt% Na-FHT MTE.

8.5.2.1.2 Starch Nanocomposites

The starch nanocomposites that did not contain PVOH (see Figure 8.14[a]) show reasonable dispersion of Na-MMT platelets (many single platelets and small tactoids). Conversely, the materials containing addition of PVOH (see Figure 8.15 sample initially containing 15 wt% water 2.5 wt% Na-MMT and 5 wt% PVOH) showed agglomeration of the platelets (numerous larger tactoids). It has also been observed that samples containing varying concentration of PVOH all exhibited varying levels of order or agglomeration.

8.5.2.1.3 Protein Nanocomposites

The TEM images of WG-1N (wheat gluten:water:glycerol [3 wt% Cloisite 30B]) and WG-3N (wheat gluten:water:glycerol:PVOH:glyoxal [3 wt% Cloisite 30B]) (Figure 8.15) clearly show that individual exfoliated silicate platelets are the predominant structure in the nanocomposites. There were some smaller tactoids comprising 3–5 particles, but no large agglomerates were visible in the systems.

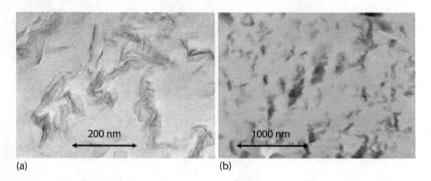

(a)　　　　　　　　　　　　(b)

FIGURE 8.14 (a) Starch nanocomposite sample containing 15 wt% water 2.5 wt% Na-MMT and no PVOH (high magnification) and (b) starch nanocomposite sample containing 15 wt% water 2.5 wt% Na-MMT and 5 wt% PVOH.

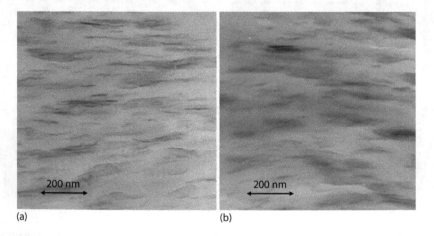

(a)　　　　　　　　　　　　(b)

FIGURE 8.15 The TEM images of (a) WG-1N and WG-3N, and (b) both containing 3 wt% Cloisite 30B.

8.5.2.2 Mechanical Properties

8.5.2.2.1 PBSA Nanocomposites

All three PBSA organically modified FHT nanocomposites showed significant increases in the elasticity modulus (Young's modulus): approximately 120% increase for all three systems at 5 wt% clay loading was observed. This improvement in Young's modulus is attributed to the high modulus of the individual clay layers, the excellent dispersion and orientation of the clays, and the shape and aspect ratios of these clays.

The yield elongation increased on addition of low levels of Na-FHT-MEE suggesting that strong interactions between hydroxyl groups from the Na-MMT and the PBSA backbone may have contributed to higher energy required for crack initiation and propagation. For the other systems studied here, yield elongation was similar to the neat PBSA.

The Na-FHT-MEE system exhibited the highest impact strength (30% higher than the neat PBSA for the system containing 5 wt% modified clay). Presumably this is due to the improved interfacial interactions between NA-FHT-MEE and PBSA.

8.5.2.2.2 Starch Nanocomposites

In the starch nanocomposites containing no PVOH, increasing the Na-MMT content was shown to significantly increase the modulus (approximately 50% increase). This result was not unexpected as significant increases in modulus have also been observed in conventional composites. This is due to the intrinsic modulus of the clay platelets themselves.

Increasing the Na-MMT content was also shown to increase the tensile strength (up to 67% increase) and a small drop in break elongation.

In the nanocomposites containing 2 or 5 wt% PVOH the Young's modulus and tensile strength were shown to increase with increasing Na-MMT content. It has been observed that composite samples containing PVOH exhibited an ordered Na-MMT structure (with agglomerated particles) which, as anticipated, results in a significant increase in nanocomposites tensile strength; up to 75% increase compared to the neat plasticized starch.

This is an interesting result in that although good dispersion (or complete exfoliation) of Na-MMT platelets is important in improving mechanical properties in nanocomposites, the interfacial interactions of filler and matrix appear to play just as important a role (the more agglomerated composites containing 2 and 5 wt% PVOH all showed significant increases in tensile strength as compared to the more well dispersed composites without PVOH).

8.5.2.2.3 Protein Nanocomposites

Exfoliation of nanoparticles in the protein matrices resulted in significant improvement in composites mechanical strength. Table 8.3 lists the mechanical properties of a series of plasticized wheat protein materials and their nanocomposites (denoted N) determined after conditioning the samples under two relative humidity conditions (RH50% and RH85%) for 7 days.

Under RH = 50% (moisture content approximately 12%), most of the nanocomposites showed significant improvement in tensile strength and modulus as

TABLE 8.3

Mechanical Properties of Wheat Protein Materials and Their Nanocomposites (Denoted N) after Conditioning the Samples Under Two Relative Humidity Conditions (RH = 50% and RH = 85%) for 7 days.

Sample Denomination	Plasticizer [%]	Tensile Strength [MPa]		Elongation at Break [%]		Young's Modulus [MPa]	
	RH = 50%	RH = 50%	RH = 85%	RH = 50%	RH = 85%	RH = 50%	RH = 85%
WG-1	29	5.8	3.0	106	239	65.4	43.4
WG1N	28	9.0	4.6	78	98	91.8	62.6
WG-2	26	10.8	4.9	74	128	100.8	62
WG-2N	25	12.6	6.4	68	89	105.2	68.7
WG-3	27	10.5	5.2	104	128	93.1	62.3
WG-3N	27	13.9	6.5	58	63	106.2	73.9

Note: WG-1: wheat gluten/water/glycerol; WG-2: wheat gluten/water glycerol/PVOH; WG-3: wheat gluten/water/glycerol/PVOH/glyoxal (crosslinker). All samples denoted "N" contain 3 wt% Cloisite 30B with respective formulations.

compared to their corresponding systems without nanoparticles. The elongation at break decreased, but still remained at a sufficient level for flexibility of the composites. The nanoparticles were very effective at improving the mechanical properties of all wheat protein systems studied in the work, even for the WG-2 and WG-3 systems where the strength had been already enhanced by blending with PVOH and further cross-linking.

8.5.3 CONCLUSIONS REGARDING PROPERTIES OF GREEN NANOCOMPOSITES

Nanocomposites and blends of bio-based and biodegradable polymers based on starches, proteins, and polyesters (including PLA and PBSA) have been prepared using a range of chemical and physical interfacial modification methods.

As a consequence of these modifications, superior improvements in a range of properties have been achieved. The most significant gains relate to the following:

1. Improved mechanical properties, for example:
   a. Up to 120% improvement in modulus of elasticity (E)
   b. 65%–70% increase in tensile strength
2. Improved barrier properties (up to 53% increase in oxygen barrier)
3. Improved thermal properties (not discussed in this chapter)

Improvements in properties of all investigated materials have been due to the intrinsic properties of the nano- or micro-additives themselves, for example, aspect ratios and internal structures. Improved compatibility of the matrix polymer with the nanofiller surface achieved through surface modification of these additives was shown to improve materials dispersion.

REFERENCES

1. M. Campestrini, P. Mock, *European Vehicle Market Statistics*, Pocket Book Edition, International Council on Clean Transportation (2011).
2. A.B. Morgan, C.A. Wilkie, *Flame Retardant Polymer Nanocomposites*. Wiley: Hoboken, NJ.
3. "BMW i3". From Wikipedia. In https://en.wikipedia.org/wiki/BMW_i3.
4. "2016 BMW i3 Rescue Guide". In http://www.boronextrication.com/2016/01/31/2016 -bmw-i3-rescue-guide/, (Anonymous document).
5. "BMW i3 Service and Training Manual". In http://www.partzine.com/bmw-i3-service -and-training-manual/, Ron (30 Dec. 2014).
6. G. Nica, "BMW i3 Might Be Cheaper to Live with Due to Carbon Fiber Construction". In http://www.autoevolution.com/news/bmw-i3-might-be-cheaper-to-live-with-due-to -carbon-fiber-construction-73054.html (2013).
7. "BMW i3 Production Rate Increased to Meet Rising Demand". In http://bmwi .bimmerpost.com/forums/showthread.php?t=972579 (15 April 2014), in Bimmerpost News (Anonymous).
8. A. Marsh, "Inside the BMW i3 Body…" In http://www.autoindustryinsider.com/?p=6976 (26 Oct. 2015).
9. R. Stewart, *Reinf. Plast.*, 53, 14, 18, 21 (2009).
10. J. Garcés, D. Moll, J. Bicerano, R. Fibiger, D. McLeod, *Adv. Mater.*, 12 (2000), 1835.

11. G. Fontaras, Z. Samaras, *Energy Policy*, 38 (2010) 1826.
12. M.C. Coelho, G. Torrão, N. Emami, J. Grácio, J. *Nanoscience and Nanotechnology*, 12 (2012), 1–10.
13. A. Lagashetty, A. Venkataraman, *Resonance*, 10 (2005), 49–57.
14. M.R. Bockstaller, R.A. Mickiewicz, L. Thomas, *Adv. Mater.*, 17 (2005), 1331–1349.
15. P. Iyer, G. Iyer, M.J. Coleman, *Membr. Sci.*, 358 (2010), 26–32.
16. P. Iyer, J.A. Mapkar, M.R. Coleman, *Nanotechnology*, 20 (2009) 325603–325603.
17. PolyOne Corporation. http://www.polyone.com/en-us/products/Pages/default.aspx.
18. V. Patel, Y. Mahajan, Polymer nanocomposites drive opportunities in the automotive sector. http://www.nanowerk.com/spotlight/spotid=23934.php (*Nanowerk Spotlight*, January 11, 2012)
19. H. Presting, U. Koning, *J. Mater. Sci. Eng.*, 23 (2003), 737.
20. F. Vautard, T. Honaker-Schroeder, L.T. Drzal, Sui, *Proc. of the SPE Automotive Composites Conference & Exhibition*, (2013), 11–13.
21. M.I.H. Panhuis, C. Salvador-Morales, E. Franklin, G. Chambers, A. Fonseca, J.B. Nagy, *Journal of Nanoscience and Nanotechnology*, 3(3), (2003), 209–213.
22. J. Sun, L. Gao, *Carbon*, 41(5), (2003).
23. T. Peijs, *MaterToday*, 6(4), (2003), 30–35.
24. L. Yu, K. Dean, L. Li, *Progress in Polymer Science*, 31(6), (2006), 576–602.
25. X. Zhang, M.D. Do, K. Dean, P. Hoobin, I. Burgar, *Biomacromolecules*, 8(2) (2007), 345–353.
26. K.M. Dean, S.J. Pas, L. Yu, A. Ammala, A.J. Hill, D.Y. Wu, *J. Appl. Polym. Sci.* Vol. 113 (2009), 3716–3724.
27. E. Petinakis, L. Yu, G. Edward, K. Dean, H. Liu, *Comp. Sci. and Techn.*, (2008).
28. M.J. John, S. Thomas, *Carbohyd. Polym.* 71(3), (2008), 343–364.
29. K.G. Theng, *The Chemistry of Clay-Organic Reactions*, Adam Hilger, London (1974).
30. R.M. Barrer, *Clays Clay Miner.*, 37 (1989), 385.
31. D. O'Hare, *Chem. Soc. Rev.*, 19 (1992).
32. R.M. Kim, J.R. Pillon, D.A. Burwell, J.T. Groves, M.E. Thompsom, *Inorg. Chem.*, 32 (1993), 4509.
33. S. Yamanaka, F. Kanamaru, M. Koizumi, *J. Phys. Chem.*, 79 (1975), 1285.
34. A. Vaccari, *Appl. Clay Sci.*, 14 (1999), 161.
35. S. Sinha, R. Okamoto, *M. Prog. Polym. Sci.*, 28 (2003), 1539.
36. T. Hibino, A. Tsunashima, *Chem. Mater.*, 10 (1998), 4055.
37. Y. Kameshima, H. Yoshizaki, A. Nakajima, K. Okada, *J. Coll. Inter. Sci.*, 298 (2006), 624.
38. F. Prinetto, G. Ghiotti, P. Graffin, D. Tichit, *Micro. Meso. Mater.*, 39 (2000), 229.

9 Tribology of Aluminum and Aluminum Matrix Composite Materials for Automotive Components

Sandeep Bhattacharya and Ahmet T. Alpas

CONTENTS

9.1 INTRODUCTION

Automobile companies are responding to progressively stringent fuel economy requirements by applying several lightweighting strategies. Consumers, on the other hand, demand improved interior comforts and advanced electronic systems for safety, navigation, and entertainment; most of which contribute to weight increase. New materials are being considered for incorporation into vehicle designs if they provide benefits at an affordable cost. To meet these challenges, automotive manufacturers are increasingly opting for lightweight metals with higher strength-to-weight ratios [1–3]. Light metals add considerable value by improving fuel economy, driveability, and performance. However, before a new material can be specified by a product engineer, several issues require resolution, including the effects on vehicle dynamics, durability, damageability, repair, and crash worthiness. These attributes relate to the effect of metallurgical characteristics and the impact of manufacturing practice on material and product performance. Engine blocks, suspension components, body panels, and frame members manufactured from Al are increasingly common. Meanwhile, replacing the traditional roles of current automotive materials with advanced metal-matrix micro- and nano-composites not only reduce mass, but also improve reliability and efficiency [4–6]. Engineering improvements that reduce emissions, such as using materials that reduce piston/cylinder bore clearance or by using composite inserts that enable reducing piston upper land thickness, are associated with an increased cost but, at the same time, help to improve air quality. A comprehensive review of the current and potential application for Al and Al matrix composites (AMCs) in the automotive industry has been provided here and primarily discusses the tribological behavior of AMCs. As a background, the sliding wear behavior of unreinforced Al alloys is also discussed. It is to be noted that wear resistance is not a material property; the mechanism of wear of a particular material, and the associated rate of wear, depend critically on the precise conditions to which it is subjected [7].

9.2　APPLICATIONS OF ALUMINUM ALLOYS IN VEHICLES

There are a broad range of opportunities for employing Al in automotive powertrain, chassis, and body structures. Table 9.1 [2] lists examples of weight reductions in the automotive industry.

9.2.1　Powertrain Applications

Reduction of parasitic friction losses is an effective way to improve engine efficiency, complementing other methods like applications of direct fuel injection, variable valve timing, turbocharging, and cylinder-shutdown systems [1]. In recent years, considerable progress has been made in reducing the friction losses resulting from piston movement. Lowering of friction in automobile engines could be attained by the use of friction-reducing coatings applied on AMC substrates, or by ultrafine polishing. Plasma transferred wire arc (PTWA) thermally sprayed coatings applied on the aluminum cylinder bores of Ford's GT500 Shelby Mustang 5.4-liter V8 helped to shed approximately 3.85 kg of steel cylinder liners. Mercedes-Benz has developed a friction-reducing cylinder-coating technology called Nanoslide, first used in 2006 on the 6.3-liter AMG V8. In the Nanoslide coating process, the cylinder walls are sprayed with an ultrathin layer—100–150 mm—of a molten Fe-C alloy. A special finishing process puts a smooth surface on this extremely hard coating, at the same time opening tiny pockets in the metal that retain oil for lubrication. In a diesel V6 engine, Nanoslide has reduced fuel consumption by 3%. A molybdenum treatment has been applied in a polka-dot pattern in pistons in the 1.8-liter engine of the 2012

TABLE 9.1

Potential Weight Reductions vs. Steel (%) in the Automotive Industry

		Body Structure	Body Closure	Chassis
High strength steel		25	15	25
Aluminum		40	45	50
Magnesium		55	55	60
Polymer	Carbon	>60	>60	60
composite	Glass	25	25	35
Titanium		Not applicable	Not applicable	50
Metal-matrix composite		Not applicable	Not applicable	60

Source: M.W. Verbrugge, P.E. Krajewski, A.K. Sachdev, *Proceedings of the Technical Sessions Presented by the TMS Aluminium Committee at the TMS 2010 Annual Meeting & Exhibition* (Ed. J. A. Johnson), The Minerals, Metals & Materials Society (TMS), Seattle, WA (2010)

Honda Civic. Such friction-reducing coating have also been used in engines like the 7-liter V8 of the Chevrolet Corvette Z06. Other automakers have taken different approaches to producing cylinder surfaces that are as smooth and wear-resistant as possible. For example, Honda has developed a technique called "plateau-honing" for the cylinder walls of its engines. Rather than a standard single-step machining process with a honing tool—an abrasive that smoothens the cylinder to the required finish—plateau-honing uses two stages of grinding to produce a surface that is ultra-smooth yet leaves a pattern of very fine grooves to hold oil. Advanced Powertrains at Chrysler have applied a technology that used a laser to burn a honing pattern into cylinders. Application of diamond-like carbon (DLC) coating and improved surface finish on piston ring by Ford offered friction reduction under boundary lubrication condition. Nissan uses a hydrogen-free DLC coating on piston rings, piston pins and cam followers. In presence of oil additives consisting of polyalpha-olefin (PAO) and glycerol mono-oleate (GMO), the coating produces a firm ultra-low friction film that binds well with the engine oil and, according to Nissan, reduces overall engine friction by 25%.

9.2.2 Chassis Applications

Over the past several years, Al castings have begun to be used in chassis applications, replacing heavier stamped steel wheels and stamped steel and cast ductile iron suspension components. A critical requirement is that the Al castings must not have defects that reduce fatigue and impact resistance. Therefore, the microstructure of the costing must be carefully controlled. This includes dendrite arm spacing, grain size, and eutectic Si morphology. In addition to alloy chemistry and melt temperature, dissolved gas and nonmetallic inclusions must be controlled to limit porosity and stress-raising oxide films. Foundry practice to eliminate turbulence during pouring that can cause such films can be enhanced by computer-based heat-flow fluid flow and solidification modeling to design the location and geometry of sprue, in-gates, and risers.

Road wheels are the most widespread application of Al in the chassis. The low-pressure die-casting process has been preferred because of its ability to produce a high-quality/high-performance product having a smooth/lustrous surface appearance. Cast Al wheels have largely replaced steel stampings in most luxury-class automobiles. Where the loading conditions are more extreme and where higher mechanical properties are required, as in the light-duty and medium-duty truck market, forged Al wheels are being increasingly considered.

9.2.3 BODY STRUCTURE

In an automobile body structure made from steel, the vehicle consists of stamped body panels spot welded together (body-in-white) to which stamped steel fenders, doors, hood, and deck lid are bolted. A lightweight Al body structure may be designed and fabricated in two ways: either a stamped alternative similar to current steel construction, or a "spaceframe" system consisting of castings, extrusions, and stampings welded together.

One requirement for Al sheets is sufficient formability such that complex stampings can be produced at economical rates. This requires understanding of the interactions between the sheet thickness, crystallographic texture, and stamping die/lubricant parameters. In addition, the Al alloys selected for exterior panels must satisfy the requirement that during the oven paint-baking operation, they age harden to provide sufficient strength for dent resistance.

Many OEMs worldwide have been examining the manufacturing and product durability issues associated with employing Al-body components. Some examples of vehicles (both production and prototype) in which large amounts of Al are used include:

1. Honda NSX, in production for several years, has a stamped-body structure and exterior panels weighing 210 kg, over 100 kg of Al chassis components, and 130 kg of other powertrain and drivetrain components. (The NSX also has titanium connecting rods—a first in a production automobile.)
2. Audi A8 is an Al-intensive spaceframe vehicle that is reduced in weight from 1348 to 1121 kg. The 385 kg Al components comprise 125 kg sheet products, 70 kg extrusions, 150 kg castings, and 40 kg other Al forms.
3. Ford Motor Company's Lincoln Mark VIII has 220 kg of Al components: 114 in the engine, 56 in the chassis, 32 in the transmission, 11 in climate control, and 7 in the body. Ford Motor Company also has built 20 Al-intensive Sable vehicles to study the manufacturing and performance of stamped Al-body structures. The body and exterior panels are 175 kg lighter than the conventional steel model.
4. General Motors, Volvo, Chrysler, and others also have built Al-intensive prototypes and/or concept cars.

9.3 ALUMINUM MATRIX COMPOSITES

The key benefits of application of AMCs in the transportation sector are lower fuel consumption, less noise, and lower airborne emissions. With increasing stringent environmental regulations and emphasis on improved fuel economy, use of AMCs in the transport sector will further increase in the future. Depending on the type of reinforcement, AMCs can be classified into four types [8]: (1) particle-reinforced AMCs, (2) whisker-or short fiber-reinforced AMCs, (3) continuous fiber-reinforced AMCs, and (4) monofilament-reinforced AMCs.

Particle reinforced AMCs generally contain equi-axed ceramic reinforcements with an aspect ratio less than 5. Ceramic reinforcements are generally Al_2O_3, SiC, or TiB_2, and present in volume fraction <30% when used for wear resistance applications. Mechanical properties of particle-reinforced composites are inferior compared to whisker/short fiber/continuous fiber reinforced AMCs but far superior compared to unreinforced Al alloys. Short fiber- and whisker-reinforced AMCs contain reinforcements with an aspect ratio of greater than 5, but are discontinuous. Short alumina fiber reinforced AMCs are one of the earliest AMCs to be developed and used in pistons. Mechanical properties of whisker reinforced composites are superior compared to particle or short fiber-reinforced composites. However, in recent years, usage of whiskers as reinforcements in AMCs is fading due to perceived health hazards and, hence, of late commercial exploitation of whisker-reinforced composites has been very limited. The reinforcements in continuous fiber-reinforced AMCs are in the form of continuous fibers (of alumina, SiC, or carbon) with a diameter < 20 um. The fibers can either be parallel or pre-woven, braided prior to the production of the composite. Monofilament reinforced AMCs contain monofilaments that are large diameter (100 to 150 um) fibers, usually produced by chemical vapor deposition (CVD) of SiC into a core of carbon fiber or tungsten wire. Bending flexibility of monofilaments is low compared to multifilaments.

In continuous fiber-reinforced and monofilament-reinforced AMCs, the reinforcement is the principal load-bearing constituent, and the Al matrix binds the reinforcement and contributes to load transfer/distribution. These composites exhibit directionality. Low strength in the direction perpendicular to the fiber orientation is characteristic of continuous fiber-reinforced and monofilament-reinforced AMCs. In particle- and whisker-reinforced AMCs, the matrix is the major load-bearing constituent. The role of the reinforcement is to strengthen and stiffen the composite by preventing matrix deformation by mechanical restraint. An additional variant of AMCs, known as hybrid AMCs, has been developed that contains more than one type of reinforcement. For example, hybrid AMCs may contain either a mixture of particle and whisker, fiber and particle, or hard and soft reinforcements. Al matrix composite containing a mixture of carbon fiber and alumina particles used in cylinder liner applications is a typical example of hybrid composite.

9.4 SLIDING WEAR APPLICATIONS OF ALUMINUM-MATRIX COMPOSITES

AMCs have experienced increasing use in the manufacturing of engine components, including those operating under sliding contact due to their ability to withstand high tensile and compressive stresses by the transfer and distribution of the applied load from the ductile matrix to the reinforcement phase. In some cases, the sliding motion is intentional: for example, in an internal combustion engine piston or cylinder liner, an automotive brake disk, or in the processing of material by machining, forging or extrusion. The enhanced wear resistance of particulate-reinforced AMCs over monolithic alloys at low contact loads, and their increased resistance to severe wear and seizure at high loads, outlines the conditions under which these composites can be considered for tribological applications in which they will be exposed to sliding wear. The design engineer must recognize that little or no advantage can be gained from using AMCs in intermediate load ranges. However, AMCs can show significantly lower wear rates than unreinforced alloys over significantly wider ranges of load and sliding speeds.

Typical applications that have received attention recently have mainly been associated with the automotive sector where lightweight alternatives to steel and cast iron components have been sought, such as replacement of cast iron brake disks and internal combustion engine cylinder liners. In the latter application, hybrid composites containing both SiC and graphite particulates in Al-Si alloy matrices have proved to be successful candidates, with advantageous low-load wear and overall seizure resistance. There have been studies [9,10] on the wear behavior of such hybrid composites under simulated engine conditions against piston ring materials. The presence of SiC in these composites could cause increased wear of traditional gray cast iron piston ring materials, leading to premature losses in engine compression and increased piston slap and scuffing effects. The use of piston rings with harder surfaces (e.g., plasma nitrided high Cr steels or Cr coatings) has reduced ring wear to within the range experienced by traditional cast iron rings in sliding contact against cast iron liners. Solid lubricants could be used to address scuffing and seizure problems during the oil starvation periods by incorporating graphite in Al-Si alloys reinforced with SiC or Al_2O_3 particles [11–13]. It was shown that the addition of graphite flakes or particles in Al alloys increased the loads and velocities at which seizure took place under the boundary lubricated [14,15] and dry sliding conditions [15–18].

Controlling friction and wear of the piston assembly is crucial to successful engine performance as shown in Figure 9.1. Several research opportunities in piston ring/cylinder bore contact can be explored by investigating low friction coatings, wear resistant coatings for Al bores, surface texture control, and better surface finish or treatments. Manufacturers have come to rely upon early life wear of the piston rings and cylinder wall to modify the profile and roughness of the interacting surfaces to achieve acceptable performance as part of the running-in process. However, a clear understanding of the complex interactions between lubrication and wear of these components is important for lubrication engineers [19].

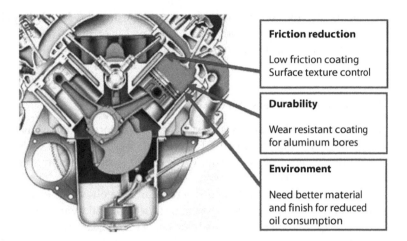

FIGURE 9.1 Research opportunities for improving friction and wear reduction in an engine design. (From S.C. Tung, M.L. McMillan, *Tribology International*, 37 (1991) 517–536.)

9.5 WEAR REGIMES AND TRANSITIONS IN ALUMINUM ALLOYS

Al alloys are known to exhibit two types of sliding wear behavior, which can be classified as "mild" and "severe" [20]. The general aspects of the mild and severe wear are well known. Severe wear involves massive surface damage and large scale material transfer to the counterface. Sliding contact at low loads and sliding speeds usually results in mild wear. In this type of wear surface damage is less extensive and the worn surfaces are generally covered by tribological layers. Sliding wear was studied as a function of the applied load and it was shown that for 99% pure Al [21] and Al-Si alloys [22–25], the transition from mild to severe wear occurred abruptly when a certain level of load was exceeded. The transition load depends on the properties of the material subjected to wear, such as the Si content of cast Al [22] and also on the mechanical and thermal properties of the counterface. The effect of the test speed on the wear behavior of Al was also studied. It was commonly observed that increasing the test speed initially causes a decrease in the rate of wear, but then wear rates start to increase again followed by a transition to severe wear and seizure [26–28]. Ling and Saibel [29] postulated that severe wear commences when the surface temperature of the contact area exceeds the recrystallization temperature of the material. A physical model for transition was developed on the basis of the heat dissipation processes occurring in the tribo-system. The "galling limit curves" constructed by Ling and Saibel [29] for steel and titanium provide early examples of model-based wear diagrams. Lim and Ashby [30] conducted an extensive literature survey on the dry sliding wear of steels and summarized the published data in the form of a wear mechanism map which shows the regimes of principal rate controlling wear mechanisms on normalized load and sliding speed axes. The proposed wear mechanism map [30] provided an effective approach to determine the dominant dry sliding wear mechanisms in steels as a function of the operating conditions.

Some issues involved in the construction of wear mechanism maps for Al alloys have been discussed by Antoniou and Subramanian [31] and Liu et al. [32]. However, the conditions under which wear transitions occur in these alloys remain to be further characterized.

9.5.1 Mild and Severe Wear in Aluminum Alloys

Zhang and Alpas [20] studied mild and severe wear behavior of 6061 Al as a function of applied load and sliding velocity. Experiments were performed under unlubricated conditions using a block-on-ring (SAE 52100 steel) configuration within a load range of 1-450 N and a sliding velocity range of 0.1–5.0 ms^{-1}. Three different forms of transition between mild and severe wear were observed: load-, sliding velocity–, and sliding distance–induced transition. An empirical wear transition map has been constructed to delineate the conditions under which severe wear initiated. The role of the contact surface temperature on wear transitions was also studied. It was observed that the transition to severe wear occurred when the bulk surface temperature exceeded a critical temperature. The dominant wear mechanisms in each wear regime were identified and classified in a wear mechanism map. A wear mechanism map, such as the one shown in Figure 9.2, can be a useful tool to predict

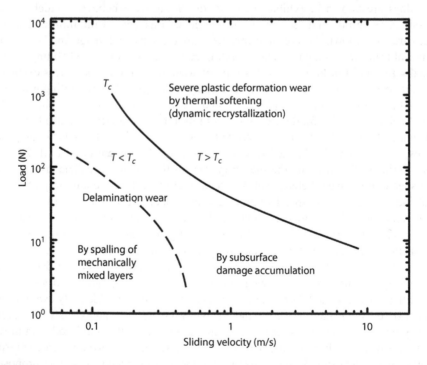

FIGURE 9.2   Wear mechanism map for Al alloys. The dotted line between the two mechanisms in the mild wear delineates load and velocity conditions where the wear debris consists of 50% plate-like and 50% mechanically mixed particles. (From J. Zhang, A.T. Alpas, *Acta Materialia*, 45 (1997) 513–528.)

the conditions under which a tribosystem can operate safely. A wear mechanism map can also serve as a guideline to select wear-resistant materials and suitable counterfaces for them. It was suggested that severe wear initiates when the average surface temperature or the bulk temperature, T_b, of Al becomes equal to a critical transition temperature, T_c, that is, $T_b = T_c = 395$ K. The range of conditions under which a tribological system operates safely can be improved either by making use of techniques that prevent the contact surface temperature of Al reaching the T_c, or improving the material's microstructure and surface conditions in order to increase T_c itself. For dry sliding contact conditions, these techniques include subjecting the tribosystem to a forced cooling process or using a counterface material with high thermal conductivity. Both of these methods will help to increase the rate of heat dissipation from the contact surfaces (Equation 9.1 [33]) and maintain the contact surface temperature below the T_c, so that the field of mild wear can be extended to higher loads and velocities:

$$T_b - T_o = \frac{\mu F v}{A_n} \left[\frac{k_s}{l_{sb}} + \frac{k_c}{l_{cb}} \right]^{-1} \tag{9.1}$$

Assuming that the heat is injected uniformly through a nominal contact area A_n, T_o is the temperature of the heat sink where the heat flows, k_s and k_c are the thermal conductivities of the sample and the counterface material, respectively, l_{sb} and l_{cb} are the equivalent heat diffusion distances, F is the applied load, v is the sliding velocity, and μ is the coefficient of friction.

There have been speculations about the wear mechanisms and exact composition of the transfer layers formed in the mild wear regime. Jahanmir and Suh [34] maintained that the formation of flakes by a delamination mechanism was dominant in their experiments in the mild regime, and that there was no evidence for oxidation. However, evidence for some Al oxidation was reported by Clarke and Sarkar [35], who found a-Al_2O_3 in X-ray analyses of mild wear debris from Al-Si alloys. Antiniou and Borland [36] confirmed the presence of Al_2O_3 in Al-Si mild wear debris by X-ray and selected area diffraction, but emphasized that oxidation played an insignificant role in mild wear. Instead, mechanical mixing between the Al alloy and the opposing steel counterface material, and compaction of the resulting debris into a dark transfer layer on the contact surfaces, was found to be the dominant mechanism. The presence of oxygen has a marked effect on mild wear rates and mechanisms in Al alloys, and has also been shown to play an important role in the wear of Al-based MMCs [37].

In general, an increase in sliding speed during dry sliding of Al alloys at low contact loads leads to a transition, from the production of black-colored powdery debris formed by the mechanical mixing and oxidation, to the formation of metallic debris. When the debris formation is metallic, wear mechanisms are commonly referred to by terms such as "galling" or "scuffing," although there is some uncertainty whether these mechanisms are associated with sever or mild wear rates and/ or seizure [38].

9.6 SLIDING WEAR OF ALUMINUM-SILICON ALLOYS

Al-Si alloys exhibit several features of tribological behavior which are similar to those of particulate MMCs. Several investigations on the wear of Al-Si alloys [25,39,40] have discussed two main wear mechanisms in Al-Si alloys: oxidational and metallic, and the effects of the sliding parameters and alloy composition on transitions between them. In the oxidational wear regime, wear rates were low, with the worn surfaces covered by a dark compacted transfer layer, presumed to consist of Al oxide and some transferred steel, although its exact nature was not investigated. The onset of metallic wear occurred above a characteristic load at which massive deformation of the Al alloy surface occurred, accompanied by the formation of metal fragments that tended to adhere to the steel counterface. Wear rates in this severe wear regime were at least an order of magnitude greater than those in the mild wear regime.

Zum Gahr [41] summarized the effect of Si content in Al-Si alloys in dry sliding wear. Wear resistance is improved in mild wear conditions if the Si content in Al-Si alloys is increased. In addition to the enhancement of mild and severe wear resistance with increased Si content, the presence of the hard Si phase contributes toward increased wear of the steel counter-material against which the Al-Si alloy is worn. This effect is also prominent in the sliding wear of Al-matrix composites and is discussed in the next section. The abrasive action of the Si phase increases the proportion of steel and ferrous oxides in the mechanically mixed debris and transfer layers produced on worn surfaces of Al-Si alloys under mild conditions [25].

For 300 series Al-Si alloys, laboratory-scale wear tests revealed a complex tribological behavior when tested under dry and lubricated sliding conditions [25,36,42–44]. Typically they show two distinct wear regimes, namely mild and severe wear. The wear rates in the mild wear regime vary between 10^{-4} and 10^{-3} mm^3/m, while in severe wear rates typically exceed 10^{-2} mm^3/m [20,45]. The micromechanisms that control wear rates in each regime are significantly different. In the mild wear regime, damage has been reported to occur by surface oxidation, plastic deformation, delamination, and material transfer to counterface [46–50]. A common attribute of the mild wear regime in Al-Si alloys and Al-based composites is the formation of tribolayers, or mechanical mixed layers (MML) on the contact surfaces [51–54]. The use of transmission electron microscopy (TEM), with complementary analyses techniques, revealed that MML consisted of fragmented Si particles, oxidized Al and Fe, and plastically deformed Al grains [55,56]. While wear rates are stable in the mild wear regime, accelerated wear rates (with sliding distance and load) are observed in the severe wear regime due to large-scale plastic deformation, and in some instances, local melting of the surface layers [24,57].

Under laboratory conditions, mild wear roughly simulates the wear of sliding components that operate under dry contact, while severe wear corresponds to metal-to-metal contact due to oil starvation as occurs in cold start, which causes scuffing in the Al-Si alloys [50,58]. Under normal lubricated sliding, wear of piston-cylinder bore assemblies should not exceed a few nanometers per hour to maintain the long-term durability requirements of cylinder bore surfaces, as revealed by radiotracer experiments run in conventional cast-iron engine blocks [59]. Accordingly, lightweight Al-Si alloys used in engine components must satisfy the same durability conditions. Minimal wear rates define a new regime in which wear rates are typically less than 10^{-6} mm^3/m or at least two orders of magnitude lower than the mild wear range. This regime is referred to as the ultra-mild wear (UMW) regime [60,61]. The UMW rates tend to be below the sensitivity limits of mass, volume, and dimensional change measurements used to determine the wear rates in the mild and severe wear regimes. These difficulties have made laboratory-scale research on UMW in Al-Si alloys almost nonexistent compared to the large body of studies conducted on mild and severe wear. Dienwiebel et al. [61] used a radio nuclide technique to measure the wear rate of an Al-17% Si cylinder bore made from an alloy operating in fully formulated engine oil and reported the presence of a surface layer consisting of a mixture of embedded Si particles and an Al matrix. For sliding tests performed under lubricated conditions that used engine oil, AES elemental composition analysis detected engine oil elements like Ca, O, C, and P on the worn cylinder bore surface [62]. Chen et al. [63] investigated UMW mechanisms in eutectic Al-Si alloys tested against a hard steel counterface using conventional SEM and white-light optical interferometery. They showed that the Si particles extending above an etched surface gradually lost their load-bearing ability as they fractured and sank in the Al matrix with increasing the sliding distance. While an infinitesimal amount of material appeared to be removed and the wear rates were extremely low, there was extensive damage localized within a narrow depth below the contact surface [60,61,63].

9.6.1 Micro-Mechanisms of Ultra-Mild Wear (UMW)

One characteristic feature of the Al-Si alloys tested under these conditions was that neither the mass of debris particles nor the material removed from the contact surface was large enough to be measured using a balance with high sensitivity (10^{-5} g). An alternative method for the estimation of very small quantities of volumetric wear loss was developed and used by Chen et al. [63]. Material loss was found to be associated with the formation of long grooves that extended along the contact surfaces in the sliding direction as shown in the 3-dimensional optical profilometry image in Figure 9.3a. Elevated portions of the Al matrix (due to Al pile-up formation) were particularly susceptible to wear. Accordingly, the volume of material loss after sliding for a given number of cycles is defined as the cross-sectional area (removed area) that falls below the reference position of the elevated plateau of the Al matrix, $A_{i,j}$, multiplied by the perimeter of the wear track. Thus, the volumetric material loss, W, from a wear track with radius R_w is determined using:

$$W = \frac{2}{n} \pi R_W \left[\sum_{i=1}^{n} \sum_{j=1}^{k} A_{ij} \right] \qquad (9.2)$$

where k is the number of grooves per section (Figure 9.3b). The surface damage inside the wear track was not always uniform, so the total removed area was determined by averaging measurements taken from 24 different sections along a wear track, that is, $n = 24$. The contact surface became covered with a tribolayer (an oil residue layer) after sliding for many cycles, and some of the scratches formed at low sliding cycles were filled as a result, causing the average cross-sectional area to decrease. This, in turn, caused the

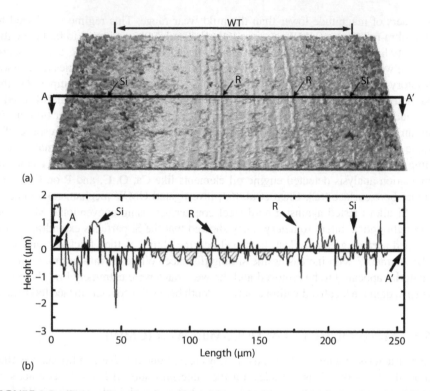

FIGURE 9.3 Three-dimensional surface profilometry images showing surface damage evolution in Al-25% Si at (a) 2.0 N after sliding for 5×10^4 cycles and (b) a cross-sectional profile scanned along a horizontal direction AA′ in (a) showing material loss. Matching locations of Si particles and Al ridges (R) are marked in (a) and (b). (SD indicates the sliding direction. WT is the width of wear track. The dimensions of each image shown are 250×190 μm). (From M. Chen, X. Meng-Burany, T.A. Perry, A.T. Alpas, *Acta Materialia*, 56 (2008) 5605–5616.)

volumetric wear to drop accordingly. Figure 9.4 illustrates the volumetric material loss from the wear tracks of Al-25% Si at different applied loads [63]. As expected, while no discernible material loss was detected at low loads of 0.5 and 1.0 N, after sliding for 5×10^4 cycles at 2.0 N, a volume loss of 2.24×10^{-4} mm^3 was measured. Increasing the sliding cycles to 6×10^5 cycles prompted the material loss to increase to 3.34×10^{-4} mm^3. Material loss in UMW was therefore not a linear function of sliding cycles, rather the volumetric wear rate loss between 10^4–10^5 cycles was the highest (at 2 N) and corresponded to a wear rate of 2.00×10^{-10} mm^3 cycle^{-1} (or 1.63×10^{-8} mm^3 m^{-1}). The slope of volume loss versus sliding cycles curve declined after sliding for 10^5 cycles.

In summary, UMW mechanisms consisted of the following sequence of damage events:

1. Initially wear to the top surfaces of Si particles occurred along with particle fracture and particle sinking-in. This is the period described as UMW I in Figure 9.4 where damage was limited to the Si particles. During UMW I, the Al matrix remained protected. UMW I was the only regime observed at low loads.

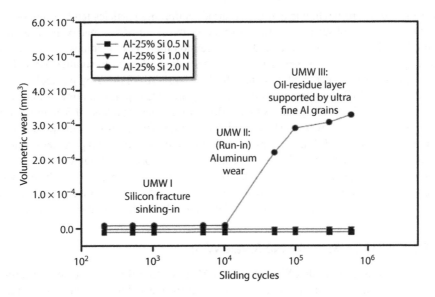

FIGURE 9.4 Variation of volumetric loss with sliding cycles showing the three stages of UMW in Al–25% Si (From M. Chen, X. Meng Burany, T.A. Perry, A.T. Alpas, *Acta Materialia*, 56 (2008) 5605–5616.)

2. The Al was no longer sheltered and measurable quantities of material loss occurred in UMW II, which is a period of rapid wear (or run-in).
3. UMW II was not persistent, and eventually the rate of damage decreased. This decrease was coincident with the formation of an oil residue layer on the contact surface as well as the formation of a subsurface structure consisting of ultra-fine Al grains. This is the regime depicted in Figure 9.4 as UMW III.

9.7 SLIDING WEAR OF ALUMINUM MATRIX COMPOSITES REINFORCED BY CARBIDE AND OXIDE PARTICLES

In this section, the effects of discontinuous ceramic reinforcement on the wear behavior of Al-based MMCs are considered. Additions of SiC, Al_2O_3, ZrO_2, and other ceramic particles, fibers, and whiskers to Al alloys generally improve their wear and seizure resistance. The effects of reinforcement material, volume fraction, matrix alloy strength and particulate size on wear mechanisms in Al MMCs will initially be discussed. The influence on wear transitions of frictional heating effects and the bulk temperature of the sliding surface will then be examined in relation to the construction of wear mechanism maps for these materials.

9.7.1 MILD AND ULTRA-MILD WEAR

Numerous investigations of the dry sliding wear of Al matrix composites against steels have reported significant increases in their wear resistance compared with unreinforced Al alloys. When contact loads and sliding speeds are kept very low, the frictional heating effects are negligible and the ceramic reinforcing particles tend to support the contact stresses. Subsurface plastic deformation and shear of the matrix alloy are prevented by the constraint introduced by the reinforcing phase. Composites containing larger particulates show superior wear resistance than alloys with smaller particulates and equivalent reinforcement content. The load support provided by particulate reinforcement in UMW was demonstrated by Zhang and Alpas [64] in sliding wear tests of 6061 Al–20 vol.% Al_2O_3 against SAE 52100 bearing steel. At a load of 3 N and sliding speed of 0.2 ms^{-1}, the worn surface exhibited Al_2O_3 particulates standing proud of the matrix alloy. During sliding the exposed portions of these reinforced particles created a localized abrasive action on the steel counterface. The worn steel fragments were transferred to the composite surface to form a protective, Fe-rich transfer layer which tended to oxidize, producing a reddish brown Fe_2O_3 wear scar. Fe_2O_3 shows a low COF during sliding against steel and thus the transfer layer provided an in situ lubricating effect.

9.7.2 Wear Transitions and Severe Wear

It is interesting to note that SiC-reinforced MMCs generally outperform those reinforced with Al_2O_3 in terms of resistance to severe wear. SiC particulates are more resistant to fracture due to their higher hardness, fracture toughness, and elastic modulus compared with Al_2O_3. This is readily observed in the empirical wear and temperature maps developed for an A356 Al-20 vol.% SiC_p composite in Figures 9.5a and b, respectively [49]. Severe wear arises in the composite at a critical bulk surface temperature of approximately 338°C (Figure 9.5b) that exceeds the critical temperatures measured for Al_2O_3 reinforced alloys by at least 100°C.

9.8 FRACTURE OF SECOND PHASE PARTICLES

Low wear resistance of Al-Si alloys, compared to cast iron, impose a limitation on their applications in powertrain components. Dienwiebel et al. [61] have observed that, while initially the engine piston ring slides over the protruding second phase Si particles on the Al-Si cylinder bore surface, the ring eventually comes into contact with a new bore surface consisting of a surface layer of fractured Si particles mixed with Al. Riahi and Alpas [65] have shown that in a chemically etched Al-Si alloy subjected to sliding contact with a Vickers indenter, that is, where tangential and normal forces were applied together, below a critical particle size-to-indenter contact width ratio (1:8) the protruding second phase particles fractured at their root, that is, at their intersection with the Al surface. For large ratios (>>1), the particles' contact surfaces exhibited plastic deformation, rather than fracture. At high loads such as 5.0 N, Si particles fractured and became embedded (particle sinking-in) in the Al matrix [66]. During the course of the sliding process, the fractured Si particles were comminuted to nano-size fragments and mixed with the organic components from the oil forming protective layer that helped to prevent further wear to the Al surface [63]. Under actual engine running conditions, it was reported that formation of micron size angular Si fragments intensified to the overall damage process by causing abrasive wear to the engine's cylinder surface [66]. Insight into the deformation and fracture characteristics of Si particles can be gained through the application of static indentation experiments. When Vickers or Berkovich-type sharp indenters were

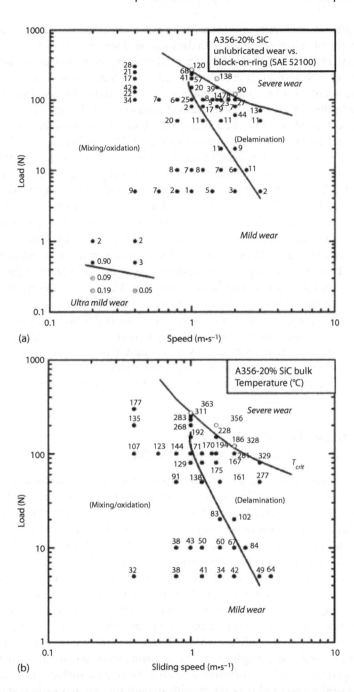

FIGURE 9.5 (a) Quantitative wear map for A356 Al-20 vol.% SiC_p MMC sliding against hard steel, showing experimentally measured wear rates (in units of 10^{-4} mm^3 m^{-1}) and regimes of wear mechanisms. (b) Temperature map corresponding to the conditions of (a); the points show experimental measurements of bulk temperature at the specimen surface. (From S. Wilson, A.T. Alpas, *Wear*, 212 (1997) 41–49.)

used, the greatest concentration of tensile stress was observed to occur directly below the indenter point, which coincided with the location of subsurface crack nucleation [67]. Bhattacharya et al. [68] performed Vickers indentations on individual Si particles in a cast Al-18.5wt.% Si alloy to observe damage microstructures below the residual indents. During the indenter loading stage, radial cracks were observed to form from the diagonals of the residual indent [69,70] while lateral cracks nucleated in the subsurface during unloading [67]. A plane-view SEM image of a Vickers micro-indentation after loading to 475 mN on an Si particle is shown in Figure 9.6a and indicates well-developed radial cracks emanating from each of the four corners of the indentation impression. The corresponding three-dimensional surface profilometry image of the same indentation, revealing the existence of pile-up adjacent to the indentation impression is shown in Figure 9.6b. The position of the FIB-milled trench is marked on Figure 9.6b. Cross-sectional investigations of the indentations revealed the damaged zone and subsurface crack patterns responsible for Si particle fracture. The SEM image of the resultant ion-milled cross-section, given in Figure 9.6c, depicts typical features of indentation induced subsurface damage. Accordingly, a semi-circular plastic core (of slightly lighter contrast) exists underneath the residual indentation impression. The radius of this plastic core was measured to be approximately 2.8 µm. A well-developed median crack, of about 9 µm long, propagated perpendicular to the contact plane starting from immediately beneath the plastic core. Lateral cracks, also originating from the base of the plastic core, were also prominent, and extend on both sides in a "saucer-like" manner, almost parallel to the surface. A localized amorphous Si zone at the median crack boundary was thought to cause volume mismatch with crystalline Si, thus contributing to subsurface crack formation.

9.9 ROLE OF TRANSFER LAYERS ON THE SLIDING WEAR BEHAVIOR

Scuffing and seizure problems that may occur during the oil starvation periods can be addressed by incorporating solid lubricants, namely, graphite in Al-Si alloys reinforced with SiC or Al_2O_3 particles [11–13]. It was shown that the addition of graphite flakes or particles in Al alloys increased the loads and velocities at which seizure took place under the boundary lubricated [14,15] and dry sliding conditions [15–18]. The high seizure resistance of graphitic Al matrix composites has been attributed to the formation of graphite layers on the contact surfaces that act as solid lubricants which reduce metal-to-metal contact between the sliding pairs [12–16]. Rohatgi et al. [17] observed that the lubricating film was very thin, about 10–20 nm thick, and consisted of a mixture of mainly graphitic carbon, carbonaceous oxides, and Al. According to Ames and Alpas [18], under severe loading and sliding speed conditions, graphite served as a temporary safeguard by keeping the initial temperature rise low, therefore allowing the contact surfaces enough time to generate a hard compacted tribo-layer with smooth surface and sufficient thickness (>10 µm) to protect the material underneath from excessive subsurface damage. The tribo-layers in graphitic SiC reinforced Al matrix composites consisted of graphite, fractured ceramic particles mixed with Al and iron oxides transferred from the steel counterface.

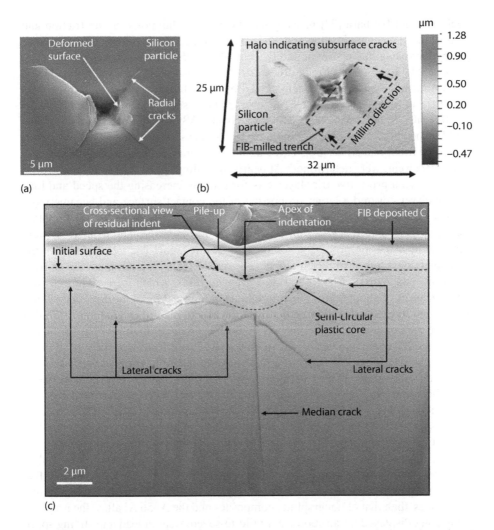

FIGURE 9.6 (a) An SEM image of Vickers indentation on an Si particle. (b) A three-dimensional surface profilometry image of the same indentation in (a), where the soft halo encircling the indentation was due to subsurface lateral cracks. The location of the FIB-milled trench is indicated by a dotted frame. (c) A cross-sectional SEM image of the FIB-milled region of the indentation (as in (a) and (b)) showing a semi-circular plastic core below the residual indent and the subsurface crack pattern of the Si particle. (From S. Bhattacharya, A.R. Riahi, A.T. Alpas, *Materials Science and Engineering A*, 527 (2009) 387–396.)

The formation of tribolayers has also been commonly observed in Al alloys without graphite. Razavizadeh and Eyre [71,72] indicated that a mechanically alloyed tribolayer, which was harder than the bulk material, contained compacted Al oxide particles and was rich in iron oxide emanating from the counterface steel. Antonionu and Borland [36] observed the presence of a very fine mechanical mixture of Al, Si, and Fe particles in an Al-Si alloy. Tandon and Li [55] reported that Fe-Al intermetallic phases were formed within the tribolayers in an Al-Si alloy worn against M2 steel. Yen and Ishihara [73] investigated the effect of humidity on the friction and wear of an Al-Si eutectic alloy and Al-Si based graphitic composites. They reported that by increasing the relative humidity from 70% to 95%, the wear rate of Al-Si alloy decreased by two orders of magnitude due to the in-situ formation of an Fe_2O_3-rich film on the alloy surface. The role played by the tribolayers that form on the contact surfaces during the sliding wear of graphitic cast Al matrix composites, A356 Al-10% SiC-4% Gr, A356 Al-10% SiC-4% Gr, and A356 Al-5% Al_2O_3-3% Gr, which could be potential candidates for cylinder liners in cast Al engine blocks, was investigated by Riahi and Alpas [50]. Three main wear regimes, namely, ultra-mild, mild, and severe wear, were determined. At nearly all sliding speeds and loads in the mild wear regime, a protective tribolayer was formed. By increasing the speed and load the tribolayer covered a larger proportion of the contact surface and became more compact and smoother. The hardness of the tribolayers increased with the applied load and speed and reached values as high as 800 kg/mm^2. The tribolayers were removed by extrusion process at the onset of severe wear. The topmost part of the tribolayer consisted of iron-rich layers. The rest of the tribolayer consisted of fractured SiC and Al_3Ni particles and thin graphite films, which were elongated over long distances of about 0.5–1.0 mm, as shown in Figure 9.7a, in the direction of sliding. A high magnification microstructure of the tribolayers is given in Figure 9.7b that depicted the lamellar structure formed by several graphite films running parallel to each other, at a depth of about 40–60 µm below the contact surface. The typical thickness of graphite films was 1–5 µm. It was suggested that the graphite layers serve to reduce the magnitude of shear stresses transferred to the matrix material underneath the tribolayers. Most severe damage within the tribolayers is concentrated between the iron oxide layers on the top and graphite layers. As shown in Figure 9.7b, the size of the Al_3Ni particles was reduced to about 2 µm. Similarly, ceramic particles (in this case SiC) were comminuted to a variety of sizes, the smaller ones around 5 µm.

A schematic description of the general features of the microstructure of tribolayers in a graphitic metal matrix composite is given in Figure 9.8. It was shown that because of the thicker and more stable tribolayers on the contact surfaces of graphitic composites, than that of nongraphitic composites and the A356 Al alloy, the graphitic composites displayed a transition from mild-to-severe wear at load and sliding speed combinations that were considerably higher than those of the A356 Al alloy and the non-graphitic A356 Al-20% SiC composite. A negative effect of the hard constituents in the tribolayers was the scuffing damage that they inflicted on the counterface.

9.10   LOAD-BEARING CAPACITY OF TRIBOLAYER

Dienwiebel et al. [61] analyzed a hypereutectic Al-Si engine cylinder bore surface with analytical tools including noncontact profilometer, atomic force microscope, and FIB techniques. The initial protrusion of Si primary particles is believed necessary to direct the energy in put into the Si grains in order to initiate a sintering process with Si wear particles and to separate the piston ring from the initial nascent Al surface. It was suggested that Al-Si surfaces obtain their wear resistance in the course of rubbing when the Si grains and the Al matrix are at the same height level after a short running time. In fact, the contact pressures and thus the local energy

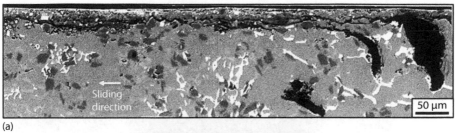

(a)

(b)

FIGURE 9.7 (a) Back scattered SEM micrograph of the cross-section of the A356 Al–10% SiC–4% Gr sample worn at 159 N and 2.0 m/s showing formation of graphite films from the graphite particles beneath the contact surface. (b) High magnification microstructure of the tribo-layer revealing the lamellar structure formed by several graphite films running parallel to each other. (From A.R. Riahi, A.T. Alpas, *Wear*, 251 (2001) 1396–1407.)

dissipation are highest on the raised Si crystals. This could lead to increased local wear on these spots. The released Si wear particles then, together with wear particles from other sources, were re-embedded into the relatively ductile Al matrix. This mechanism was described as friction-induced dispersion hardening (Figure 9.9). At this stage, the piston ring was not only supported by the Si grains but could interact with the Al matrix as well. Together with embedding of wear particles, the Al matrix was plastically deformed and foreign elements were introduced.

In an engine grade hypereutectic Al-18.5% Si alloy [74], UMW damage was found to be restricted to the top surfaces of large Si particles that acted as load bearing "asperities" on which a contact pressure as large as 1080 MPa was applied (at 2.0 N) [66]. For alloys with smaller Si particles like a spray deposited Al-25% Si alloy [63], the Si particles experienced higher contact pressure of 1680 MPa resulting in sinking-in of Si into the Al matrix. Consequently an earlier transition occurred, from the first stage of UMW where the protruded particles mitigated the matrix wear, to the UMW II stage where the Si particles were sunken in the matrix and hence the UMW II stage defined a running-in period where high volumetric wear occurred. For high sliding cycles a reduction in wear rate was observed (UMW III stage).

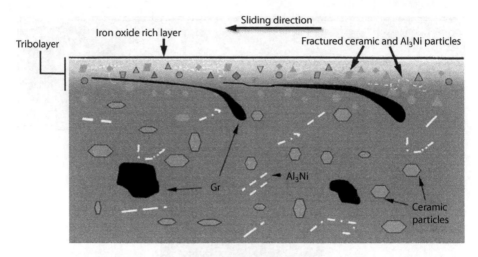

FIGURE 9.8   A schematic representation of the main constituents of the tribolayers in graphitic metal matrix composites: the topmost part is a layer rich in iron oxide. Fractured ceramic particulates and Al₃Ni intermetallics are mixed with the Al matrix. The graphite particles underneath the contact surface are elongated in the sliding direction and embedded in the tribolayer. The material under the tribolayer remains relatively damage free. (From A.R. Riahi, A.T. Alpas, *Wear*, 251 (2001) 1396–1407.)

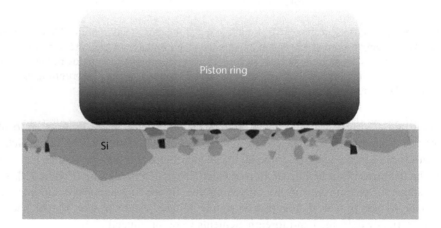

FIGURE 9.9   Schematic of the proposed wear model that describes the embedding and mixing of wear particles and foreign elements from the engine oil into the ductile Al matrix that results in an increase in the shear strength of the material. (From M. Dienwiebel, K. Pöhlmann, M. Scherge, *Tribology International*, 40 (2007) 1597–1602.)

This was attributed to the formation of a solid tribolayer called an oil residue layer (ORL) that covered the wear tracks and can be observed in the cross-sectional SEM image of the wear track in Figure 9.10a.

The ORL consisted of a mixture of C and other constituents, including Ca, S, Zn, and Al₂O₃, that originated from the synthetic oil used during sliding tests.

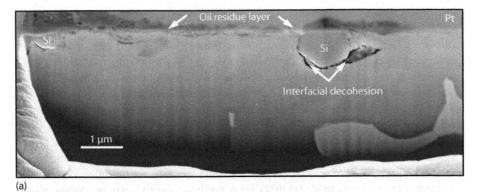

(a)

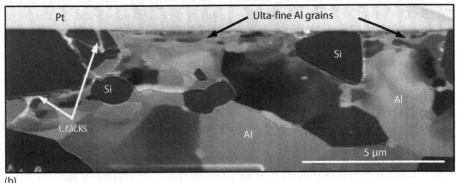

(b)

FIGURE 9.10 Cross-sectional SEM image of the wear track of Al-25% Si showing (a) the oil residue layer (ORL) on the contact surface, where damage to Si and at Al/Si interface are also indicated and (b) the microstructure of material under the wear track revealing ultra-fine Al grains and the ORL. (From M. Chen, X. Meng-Burany, T.A. Perry, A.T. Alpas, *Acta Materialia*, 56 (2008) 5605–5616.)

A closer inspection of the material immediately below the contact surface revealed a remarkable microstructural refinement process that was induced by sliding: specifically, the formation of a deformation microstructure consisting of ultra-fine Al grains. The cross-sectional FIB/SEM image in Figure 9.10b revealed the morphology of the Al grains in the deformed region within a depth of 2 μm below the contact surface and compares it to the original grain size of 3–5 μm [63]. Therefore, direct evidence was found for localized severe plastic deformation underneath the ORL that manifested itself by the formation of a nano-crystalline Al grain structure that was harder than the initial surface and thought to support the ORL. Thus, once the ORL was established, it acted as an anti-wear layer allowing Al-Si alloys to operate at near zero wear rates.

Wear tests performed on Al-Si alloys under boundary lubrication conditions have shown that a tribofilm [75] formed on top of Si particles at high temperature (100°C) improved the wear behavior of the Al-Si alloys. The tribofilm on top of Si particles having a 16.4% lower composite modulus ($E^* = 64.58$ GPa) at 100°C, compared to 77.24 GPa at 25°C, acted as an energy absorbing layer which reduced the amount of load transfer through Si particles to the Al matrix, the severity of Si particle fracture, and sinking-in. Bhattacharya and Alpas [76] studied the role of tribofilm on reducing the fracture in Si particles. Crack formation in Si was induced by performing Vickers micro-indentation experiments. Prior to the indentation experiments, the tribofilm was formed on the particles in an Al-18.5 wt.% Si alloy by conducting sliding wear tests under boundary lubricated conditions at 100°C as can be seen in Figure 9.11a.

A cross-sectional TEM image of the tribofilm is shown in Figure 9.11b. The tribofilm was well-adhered to Si at the interface. The thickness of the tribofilm in this section was 81.6 ± 2.9 nm. The variation of Vickers microhardness values of Si particles, with and without the tribofilm, is shown in Figure 9.12a. In the absence of tribofilm, indentation size effect was observed at low loads that resulted in increased elastic recovery and hardness values. On the other hand, lower hardness values were recorded at lower loads (e.g., 50 mN) for tribofilm-covered particles. When tested at higher indentation loads, the effect of tribofilm was diminished and the hardness values obtained at 300 mN were comparable (~2750 Kgf/mm^2) to those particles that were devoid of the tribofilm.

Incremental multi-cycle indentation experiments were performed on the tribolms to assess the variation in hardness with the relative indentation depth (RID). The RID has been defined as [77]: RID = h_c/t, where h_c is the indenter contact depth and t is the film thickness. According to Figure 9.12b, the hardness values can be classified into two regimes within a depth that was approximately equal to the tribofilm thickness (~82 nm). At $h_c/t \leq 0.24$, higher hardness values were indicative of an elastic-only response from the tribofilm. The decrease in the hardness values was attributed to the onset of an elasto-plastic [77] regime of the tribofilm. At $h_c/t > 0.24$, a gradual increase in the hardness values indicated an Si-dominated response. The interface of the tribofilm and Si (at $h_c/t = 1$) has been marked in Figure 9.12b as a 5 nm thick shaded region. Beyond this region, the rate of escalation of the hardness values decreased slightly, which could be attributed to the

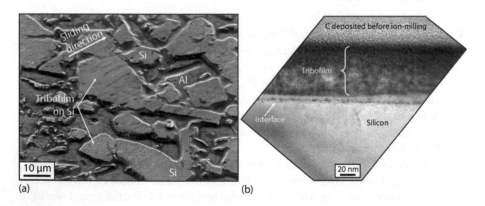

(a)   (b)

FIGURE 9.11 (a) Back-scattered (BS)-SEM image showing the distribution of tribofilm-covered Si particles in a wear tested Al-18.5 wt.% Si alloy after 5 × 104 sliding cycles at 100°C. The tribofilm appeared as bright areas in back-scattered electron images. (b) Cross-sectional TEM image of tribofilm formed on Si. (From S. Bhattacharya, A.T. Alpas, *Wear*, 301 (2013) 707–716.)

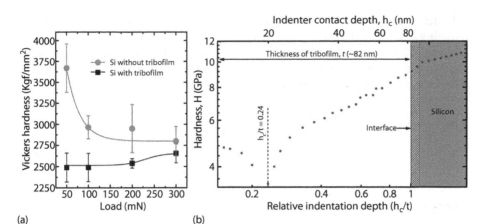

FIGURE 9.12 (a) Vickers hardness values for Si in wear tested Al-18.5 wt.% Si at 100°C, plotted as a function of the indentation load; hardness values being plotted both in the presence and the absence of tribofilm on Si. (b) Change in hardness of the tribofilm on Si measured with respect to the relative indentation depth and contact depth. Hardness measurements were performed using incremental multi-cycle indentation experiments using a Berkovich tip. (From S. Bhattacharya, A.T. Alpas, *Wear*, 301 (2013) 707–716.)

commencement of the substrate, that is, Si particle. Thus, the hardness and elastic modulus of tribofilm, determined at penetration depths less than 19.3 nm (for $h_c/t \leq$ 0.24), was devoid of any influence of the Si substrate. Accordingly, mechanical properties of the tribofilm were performed on the tribofilm at $h_c/t < 0.24$. The tribofilm had an elastic modulus of 116.30 ± 6.64 GPa and the hardness being 3.45 ± 0.17 GPa. A fracture mechanics argument was developed for justification of the crack growth and instability conditions in Si particles with and without tribofilm. The tribofilm acted as an energy-absorbing entity and reduced the driving force that is necessary for crack growth in Si. Hence, chipping fracture in Si occurred only at higher loads.

9.11 CONCLUSIONS

The growing demand to improve fuel economy and reduce environmental emissions is having a significant effect on the selection of the materials used in the automotive industry. In recent years, a number of Al matrix composites (AMCs) have been developed for potential tribological applications in automotive engines, particularly for lightweight cylinder liners. One of the most important aspects of the tribological performance of a cylinder liner is its ability to resist seizure and scuffing during adverse engine running conditions such as lubricant starvation. The wear of cylinder bore surfaces during normal engine operation should not exceed a few nanometers per hour to maintain the long-term durability requirements. Lightweight near-eutectic Al-Si alloy provides very low wear rates observed in the ultra-mild wear regime due to generation of tribolayers. The expected power increase of the next generation of combustion engines calls for a further improvement of the wear resistance of AMC cylinder bores and liners. Future research should focus on the application of friction-reducing coatings compatible with cast Al-Si alloys and AMCs to increase engine efficiency. The development of carbon-based coatings that mitigate adhesion of Al and Mg during contact between the sheet and die surfaces would enhance the quality of automobile lightweight automotive body panels.

ACKNOWLEDGMENT

The financial support provided by the Natural Sciences and Engineering Research Council (NSERC) of Canada is gratefully acknowledged.

REFERENCES

1. P. Stenquist, Rubbing out friction in the push for mileage, (2011); Available at http://www.nytimes.com/2011/10/16/automobiles/rubbing-out-friction-in-the-push-for-mileage.html?pagewanted=2&_r=0
2. M.W. Verbrugge, P.E. Krajewski, A.K. Sachdev, *Proceedings of the technical sessions presented by the TMS Aluminium Committee at the TMS 2010 Annual Meeting & Exhibition* (Ed. J. A. Johnson), The Minerals, Metals & Materials Society, Seattle, WA (2010).
3. J. Hirsch, T. Al-Samman, *Acta Materialia*, 61 (2013) 818–843.
4. Y. Yang, J. Lan, X. Li, *Materials Science and Engineering A*, 380 (2004) 378–383.
5. D. Carle, G. Blount, *Materials and Design*, 20 (1999) 267–272.
6. A. Macke, B.F. Schultz, P.K. Rohatgi, N. Gupta, *Advanced Composite Materials for Automotive Applications: Structural Integrity and Crashworthiness* (Ed. A. Elmarakbi), John Wiley & Sons Ltd., Chichester, UK (2013).
7. I.M. Hutchings, S. Wilson, A.T. Alpas, *Comprehensive Composite Materials 3*, Elsevier, Oxford, UK (2000) 501–519.
8. M.K. Surappa, *Sadhana*, 28 (2003) 319–334.
9. N.Ch. Kaushik, R.N. Rao, *Tribology International*, 96 (2016) 184–190.
10. O. Carvalho, M. Buciumeanu, S. Madeira, D. Soares, F.S. Silva, G. Miranda, *Tribology International*, 90 (2015) 148–156.
11. A.K. Senapati, R.I. Ganguly, R.R. Dash, P.C. Mishra, B.C. Routra, *Procedia Materials Science*, 5 (2014) 472–481.
12. S.V. Prasad, P.K. Rohatgi, *Journal of Metals*, 11 (1987) 22–26.
13. P.K. Rohatgi, S. Ray, Y. Liu, *International Materials Reviews*, 37 (1992) 129–149.
14. S. Das, S.V. Prasad, T.R. Ramachandran, *Materials Science and Engineering A*, 138 (1991) 123–132.
15. S. Das, S.V. Prasad, *Wear*, 133 (1989) 173–187.
16. S.V. Prasad, B.D. McConnell, *Wear*, 149 (1991) 241–253.
17. P.K. Rohatgi, Y. Liu, T.L. Barr, *Metallurgical and Materials Transactions A*, 22 (1991) 1435–1441.
18. W. Ames, A.T. Alpas, *Metallurgical and Materials Transactions A*, 26 (1995) 85–98.
19. S.C. Tung, M.L. McMillan, *Tribology International*, 37 (1991) 517–536.
20. J. Zhang, A.T. Alpas, *Acta Materialia*, 45 (1997) 513–528.
21. W. Hirst, J.F. Archard, *Journal of Applied Physics*, 27 (1956) 1057–1065.
22. A.D. Sarkar, *Wear*, 31 (1975) 331–343.
23. O.P. Modi, B.K. Prasad, A.H. Vegneswaran, M.L. Vaidya, *Materials Science and Engineering A*, 151 (1992) 235–245.
24. A.T. Alpas, J. Zhang, *Metallurgical and Materials Transactions A*, 25 (1994) 969–981.

25. A. Somi Reddy, B.N. Pramila Bai, K.S.S. Murthy, S.K. Biswas, *Wear*, 171 (1994) 115–127.
26. C. Beesley, T.S. Eyre, *Tribology International*, 9 (1976) 63–69.
27. C. Subramanian, *Wear*, 151 (1991) 97–110.
28. A. Wang, H.J. Rack, *Materials Science and Engineering A*, 147 (1991) 211–224.
29. F.F. Ling, E. Saibel, *Wear*, 1 (1957) 80–91.
30. S.C. Lim, M.F. Ashby, *Acta Metallurgica*, 35 (1987) 1–24.
31. R. Antoniou, C. Subramanian, *Scripta Metallurgica*, 22 (1988) 809–814.
32. Y. Liu, R. Asthana, P. Rohatgi, *Journal of Material Science*, 26 (1991) 99–102.
33. M.F. Ashby, J. Abulawi, H.S. Kong, *Tribology Transactions*, 34 (1991) 577–587.
34. S. Jahanmir, N.P. Suh, *Wear*, 44 (1977) 17–38.
35. J. Clarke, A.D. Sarkar, *Wear*, 54 (1979) 7–6.
36. R. Antoniou, D.W. Borland, *Materials Science and Engineering A*, 93 (1987) 57–72.
37. R.A. Saravanan, J.M. Lee, S.B. Kang, *Metallurgical and Materials Transactions A*, 30 (1999) 2523–2538.
38. P. Wycliffe, *Wear*, 162–164 (1993) 574–579.
39. K. Mohammed Jasim, E.S. Dwarakadasa, *Wear*, 119 (1987) 119–130.
40. H. Torabian, J.P. Pathak, S.N. Tiwari, *Wear*, 172 (1994) 49–58.
41. K.-H. Zum Gahr, *Microstructure and Wear of Materials*, Tribology Series, Elsevier, Amsterdam, vol. 10 (1987).
42. J. Clarke, A.D. Sarkar, *Wear*, 69 (1981) 1–23.
43. B.N. Pramila Bai, S.K. Biswas, *Wear*, 120 (1987) 61–74.
44. J. Zhang, A.T. Alpas, *Materials Science and Engineering A*, 160 (1993) 25–35.
45. J. Lasa, M. Rodriguez-Ibabe, *Materials Science and Engineering A*, 363 (2003) 193–202.
46. D.A. Rigney, *Wear*, 245 (2000) 1–9.
47. W.M. Rainforth, *Wear*, 245 (2000) 162–177.
48. S.G. Caldwell, J.J. Wert, *Wear*, 122 (1988) 225–249.
49. S. Wilson, A.T. Alpas, *Wear*, 212 (1997) 41–49.
50. A.R. Riahi, A.T. Alpas, *Wear*, 251 (2001) 1396–1407.
51. B. Venkataraman, G. Sundararajan, *Acta Materialia*, 44 (1996) 461–473.
52. X.Y. Li, K.N. Tandon, *Wear*, 225–229 (1999) 640–648.
53. S.K. Biswas, *Wear*, 245 (2000) 178–189.
54. X.Y. Li, K.N. Tandon, *Wear*, 245 (2000) 148–161.
55. K.N. Tandon, X.Y. Li, *Scripta Materialia*, 38 (1997) 7–13.
56. J. Li, M. Elmagdali, V.Y. Gertsman, J. Lo, A.T. Alpas, *Materials Science and Engineering A*, 421 (2006) 317–327.
57. J. Clarke, A.D. Sarkar, *Wear*, 82 (1982) 179–195.
58. H.H. Yoon, T. Sheiretov, C. Cusano, *Wear*, 237 (2000) 163–175.
59. M. Scherge, K. Pöhlmann, A. Gerve, *Wear*, 254 (2003) 801–817.
60. M. Chen, T. Perry, A.T. Alpas, *Wear*, 263 (2007) 552–561.
61. M. Dienwiebel, K. Pöhlmann, M. Scherge, *Tribology International*, 40 (2007) 1597–1602.
62. M.A. Nicholls, P.R. Norton, G.M. Bancroft, M. Kasrai, *Wear*, 257 (2004) 311–328.
63. M. Chen, X. Meng-Burany, T.A. Perry, A.T. Alpas, *Acta Materialia*, 56 (2008) 5605–5616.
64. J. Zhang, A.T. Alpas, *Materials Science and Engineering A*, 161 (1993) 273–284.
65. A.R. Riahi, A.T. Alpas, *Materials Science and Engineering A*, 441 (2006) 326–330.
66. S.K. Dey, T.A. Perry, A.T. Alpas, *Wear*, 267 (2009) 515–524.
67. B.R. Lawn, E.R. Fuller, *Journal of Materials Science*, 10 (1975) 2016–2024.
68. S. Bhattacharya, A.R. Riahi, A.T. Alpas, *Materials Science and Engineering A*, 527 (2009) 387–396.

69. B.R. Lawn, R.F. Cook, *Journal of Materials Science*, 47 (2012) 1–22.
70. B.R. Lawn, M.V. Swain, K. Phillips, *Journal of Materials Science*, 10 (1975) 1236–1239.
71. K. Razavizadeh, T.S. Eyre, *Wear*, 79 (1982) 325–333.
72. K. Razavizadeh, T.S. Eyre, *Wear*, 87 (1983) 261–271.
73. B.K. Yen, T. Ishihara, *Wear*, 198 (1996) 169–175.
74. M. Chen, A.T. Alpas, *Wear*, 265 (2008) 186–195.
75. S.K. Dey, M.J. Lukitsch, M.P. Balogh, X. Meng-Burany, A.T. Alpas, *Wear*, 271 (2011) 1842–1853.
76. S. Bhattacharya, A.T. Alpas, *Wear*, 301 (2013) 707–716.
77. A.M. Korsunsky, M.R. McGurk, S.J. Bull, T.F. Page, *Surface and Coatings Technology*, 99 (1998) 171–183.

10 Magnesium and Its Alloys

D. Sameer Kumar and C. Tara Sasanka

CONTENTS

10.1 INTRODUCTION TO MAGNESIUM

The name magnesium originated from the Greek word for a district in Thessaly called Magnesia. It was first discovered by Sir Humphrey Davy in 1808 and in metallic form by Antoine Bussy in 1831. Davy's first suggestion was magnium, but later it became magnesium [1]. Its chemical symbol is Mg.

Magnesium is found to be the eighth most-abundant element in the Earth's crust by mass and the ninth most-abundant element in the universe as a whole. It occupies the fourth position among the elements that contribute to earth's mass as a whole followed by iron, oxygen and silicon. It is ranked the third most-abundant element dissolved in seawater [1,2].

Magnesium is an alkaline earth metal having atomic number 12 with oxidation number +2. It has a hexagonal close packed (HCP) crystalline structure. The free element (metal) is not found naturally on earth, as it is highly reactive. Magnesium is a light, strong metal that gives a white brilliant light when exposed to the atmosphere.

Magnesium is a silvery white metal that is similar in appearance to aluminum but weighs one-third less. With a density of only 1.738 g per cubic centimeter, it is the lightest structural metal known. Because of its low density, many companies prefer magnesium as a potential substitute to conventional materials in weight-critical applications.

Magnesium is tougher than plastic and has better damping capacity as compared to cast iron and aluminum. It has good electro-magnetic interference (EMI) shielding and higher heat dissipation than that of plastics. Magnesium absorbs vibration energy effectively. Recyclability also makes magnesium a frontrunner. According to the combination of specific Young's modulus and high specific strength, magnesium alloys show similar or even better values than aluminum and many commercial steels [3]. The properties of magnesium are given in the following section.

10.1.1 PROPERTIES OF PURE MAGNESIUM

Magnesium can be used in both its pure form and as an alloy. Depending on the composition of the alloys, there can be remarkable differences in many properties. The following data is primarily regard pure magnesium [4,5]. However, several common alloys and their properties are also mentioned in the later parts of the study.

10.1.1.1 Atomic Properties and Crystal Structure

Symbol	Mg
Element classification	Group II A, Alkaline earth metal
Atomic number	12
Atomic weight	24.3050
Atomic volume	14.0 cm^3/mol
Atomic radius	0.160 nm
Ionic radius	0.072 nm
Orbital electron states	$1s_2$, $2s_2$, $2p_6$, $3s_2$
Electrons per shell	2, 8, 2
Most common valence	2+
Crystal structure	Hexagonal close-packed (HCP)
Vander Wall radius	173 pm
Electro negativity	Pauling scale 1.31
No. of isotopes	3 with mass numbers 24, 25, 26

10.1.1.2 Physical Properties

Property	Value
Density	1.738 g/cm^3
Melting point	923 K (650°C, 1202°F)
Boiling point	1363 K (1091°C, 1994°F)
Thermal Properties	
Thermal expansion	24.8 µm/(m · K) (at 25°C)
Thermal conductivity	156 W/(m · K) (at 27°C)
Specific heat capacity	1.025 kJ/kg K (at 20°C)
Latent heat of fusion	360–377 kJ/kg
Enthalpy	107 kJ/kg
Entropy	1.34 kJ/kg K (at 20°C)
Electrical Properties	
Electrical resistivity	43.9 nΩ · m (at 20°C)
Electrical conductivity	38.6% IACS
Magnetic Properties	
Magnetic susceptibility	0.00627 to 0.00632 mks (mass)
Magnetic permeability	1.000012
The Hall constant	-1.06×10^{-16} Ω · m/(A/m)
Optical Properties	
Reflectivity	0.72 (at $\lambda = 0.500$ µm)
Solar absorptivity	0.31
Emissivity	0.07 (at 22°C)

10.1.1.3 Mechanical Properties

Property	Value
Young's modulus	45 GPa
Shear modulus	17 GPa
Bulk modulus	45 GPa
Poisson ratio	0.290
Brinell hardness	30–50

Source: http://www.magnesium.com/w3 /data-bank/index.php?mgw=153.

10.1.2 Extraction

There is meager information shared by the industries for the production of magnesium metal unlike aluminum industries. The reason is that the extraction methods of magnesium are costlier when compared to other materials. This may be a reason restricting the usage of magnesium globally.

The primary sources for the production of magnesium are [2]

- Magnesite (28.8% weight of Mg) [$MgCO_3$]
- Dolomite (28.8% weight of Mg) [$MgCO_3 * CaCO_3$]
- Bischofite (11.96% weight of Mg) [$MgCl_2 * 6H_2O$]
- Carnallite (8.75% weight of Mg) [$MgCl_2 * KCl * 6H_2O$]
- Serpentine (26.36% weight of Mg) [$3MgO * 2SiO * 2H_2O$]
- Olivine (33% weight of Mg) [$(Mg, Fe)_2SiO_4$]
- Seawater (0.038%–20.8% weight of Mg) [$Mg_2+(aq)$]

This metal is now obtained mainly by pidgeon process, thermal reduction method, and electrolysis of magnesium salts obtained from brine.

The pidgeon process is the oldest and a relatively simple process than the others. It does not require skilled labor and is versatile (adjustable based on demand) with low capital cost. This process is high in energy consumption and has low productivity. In this process, magnesium is collected from a condenser on the outside of a furnace. High purity magnesium can be attained from the condenser, since vapor pressure of impurities that may be in the magnesium are low under the conditions of the tank.

In the thermal-reduction method, dolomite, magnesite, and other magnesium containing ores are broken down through the use of ferrosilicon to reduce magnesium oxide in a molten slag. The mixture is heated at a temperature of 1200°C–1600°C in a vacuum chamber forming magnesium vapors which later condense into crystals. The crystals are melted, refined, and poured into ingots for further processing.

On the other hand, the electrolytic method uses direct current electricity. Seawater and Lime are mixed in settling tanks so that the precipitates of magnesium hydroxide fall to the bottom of the tanks. They are filtered and mixed with hydrochloric acid. This resulting solution is then exposed to electrolysis which produces magnesium metal.

There are a few more methods like the dow process, AM process, silico thermic process, mintek process, magnetherm process that use different sources for the magnesium but use principles of electrolytic or thermal reduction methods [2,6]. An understanding of the extraction methods is still needed for higher production and the safe handling of magnesium.

The commercial production of electrolytic magnesium began in Germany in 1886 and Germany was the only country to produce magnesium this way until 1916. The production of magnesium for flares and tracer bullets was used in the United States, Britain, France, Canada, and Russia for their military applications. The worldwide production of magnesium dropped off between the World Wars. Germany's production increased to 20,000 tons by 1938, accounting for 60% of global production and the United States started 15 new magnesium production facilities by 1943; a capacity of over 265,000 tons has been produced [7]. In 2006, the production reached 726,000 metric tonnes in the world. Presently, China accounts for 75% of the world's magnesium production to become a leading supplier of it.

10.1.3 UTILIZATION

Historically, aluminum alloys with some magnesium content have been primarily used worldwide. The packaging industry is the largest market for magnesium in aluminum alloys, followed by transport, construction, and consumer durables.

The automotive industry is by far the largest user of magnesium die-cast components. Die-cast magnesium alloys are used for housings, assemblies, brackets, and other components for all sections of motor vehicles.

Magnesium die-cast housings for communication devices such as cell/mobile and smart phones, laptops, tablet and notebook computers, and other electronic equipment form the next largest uses after automobiles. Manufacturing of titanium sponge (i.e., crude titanium metal) is the third largest use of magnesium and desulfurizing of steel the fourth largest use. The use of magnesium in steel manufacture has slowed in recent years, due to the global economic downturn and resultant slowdowns (or declines) in steel output in many countries. On average, magnesium is used globally at a rate of about 50 g/ton of steel.

Magnesium is also used in other applications, such as the nodularization of cast iron and cathodic protection, a method of preventing corrosion by forcing all surfaces of a metal structure to be cathodes through the provision of external anodes of active metals.

Apart from the applications mentioned above, magnesium is considered to be a good choice material in the areas of defense and aerospace engineering for aircraft and missile components, aircraft engine mounts, control hinges, fuel tanks, and wings. Recent advances in electronics also resulted in an uptrend in the magnesium usage. Medical (orthopedic biomaterial used for implants) and sporting goods (like tennis rackets and the handles of archery bows) made of magnesium were prized over the past few years. The usage of magnesium in automotive applications can provide more than just a weight savings and the detailed applications of magnesium alloys in the automotive sector are discussed in the next sections.

10.1.4 Advantages and Disadvantages of Magnesium [8]

The advantages are

- Lowest density of all metallic constructional materials.
- High specific strength.
- Most alloys have high fluidity and so good castability.
- Suitable for high pressure die casting.
- Ability to be turned/milled at high speed.
- Good weldability under controlled atmosphere.
- Best strength-to-weight ratio of commonly used structural metals.
- Excellent damping capacity.
- Much improved corrosion resistance using high purity Mg.
- Availability of Mg is high.
- Compared with polymeric materials,
    - Better mechanical properties.
    - Better resistance to aging.
    - Better electrical and thermal conductivity.
    - Recyclable.

The disadvantages are

- Low elastic modulus.
- Limited cold workability and toughness.
- Limited high strength and creep resistance at elevated temperatures.
- High degree of shrinkage on solidification.
- Cost can be affected by volatile markets considerably.
- High chemical reactivity.
- Limited corrosion resistance.

Magnesium offers good strength with low density. The recyclability with reduced CO_2 emissions is also an added attraction for magnesium. It can also be used as a primary source, if the limitations are overcome.

10.2 ALLOYING OF MAGNESIUM

With the advantages discussed in an earlier section, magnesium has an edge over others to be considered a good metal in many applications. Pure magnesium is rarely used due to its poor mechanical properties and high reactivity. It can be mixed with other alloying elements to alter the properties. Especially for all structural applications, magnesium is alloyed with other metals to provide proper strength, corrosion resistance, formability, and so on. Magnesium alloy development started in the early days of 1945 [9]. Research is being done extensively on the manufacture of various products by different combinations of alloys and their suitability and the association of one alloying element over the other. This section briefly discusses the elements of alloying and types of alloys along with the properties.

10.2.1 ALLOYING ELEMENTS AND DESIGNATIONS

A list of the most common alloying elements that can be added to magnesium, along with abbreviation letters as per ASTM standard ASTM B275, is given in Table 10.1. Each alloy is labeled by letters followed by figures to indicate the elements and compositions present in the particular alloys.

The designation of a typical magnesium alloy consists of four parts.

- Part 1: Alphabet of two main alloying elements in two abbreviation letters. The first one is the major alloying ingredient; the second alphabet is a minor ingredient. (Refer to Table 10.1.)
- Part 2: Amounts (in weight percentage terms) of the two main alloying elements.
- Part 3: Distinguishes among the different alloys with the same percentages of the two main alloying elements. It is made up of a letter of the alphabet assigned in order as compositions become standard:
    - A First compositions, registered with ASTM
    - B Second compositions, registered with ASTM
    - C Third compositions, registered with ASTM

TABLE 10.1
Code Letters for Magnesium Alloying Designation

Alloying Element	Designation
Aluminum	A
Bismuth	B
Copper	C
Cadmium	D
Rare earths	E
Iron	F
Thorium	H
Strontium	J
Zirconium	K
Lithium	L
Manganese	M
Nickel	N
Lead	P
Silver	Q
Chromium	R
Silicon	S
Tin	T
Gadolinium	V
Yttrium	W
Calcium	X
Antimony	Y
Zinc	Z

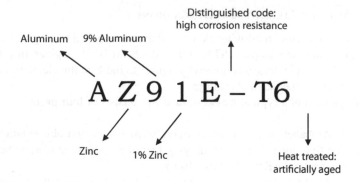

FIGURE 10.1 ASTM designations of magnesium alloys.

- D High purity, registered with ASTM
- E High corrosion resistance, registered with ASTM
- X Experimental alloy, not registered with ASTM
- Part 4: Temper designations in accordance with ASTM B296- 03. A dash is used to separate the alloy designation from the temper designation. H and T have subdivisions for the better understanding of processed conditions.
  - F As fabricated
  - O Annealed, recrystallized (wrought products only)
  - H Strain hardened
  - T Thermally treated to produce stable tempers other than F, O, or H
  - W Solution heat treated (unstable temper)

ASTM magnesium alloy designation is made clearer through Figure 10.1.

10.2.2 The Effects of Alloying Elements

Alloying of magnesium with a good amount of additives improves the strength, hardness, castability, workability, corrosion resistance, and weldability in a well-balanced way [9]. A summary of observations made by researchers on mechanical properties with the effect of various alloying elements in magnesium is as follows [4,5,10–13]:

- Aluminum: Aluminum is the most common alloying element. It improves strength, hardness and corrosion resistance, but reduces ductility.
- Beryllium: Beryllium is only added to the melt in small quantities to reduce the surface melt oxidation significantly during the casting, melting, and welding processes. It can also result in coarse grains.
- Calcium: The addition of calcium can assist in grain refinement and creep resistance. It can also enhance corrosion resistance, and thermal and mechanical properties of magnesium alloys.
- Copper: Studies have indicated that addition of copper to magnesium helps in increasing room temperature and high temperature strength. However, the ductility is compromised.

- Iron: Iron is a harmful addition in magnesium alloys, and even in small quantities, its presence is detrimental to the corrosion resistance.
- Lithium: Lithium reduces density and can improve ductility. Adding lithium significantly increases cost and reduces corrosion resistance.
- Manganese: Manganese slightly increases yield strength but has no effect on tensile strength. Its most important function is to improve the corrosion resistance of Mg–Al based alloys.
- Nickel: Inclusion of nickel in magnesium leads to an increase in room temperature strength. However, a decrease in ductility was observed.
- Rare earth metals (RE): Rare earths are added to increase the high temperature strength, creep resistance, and corrosion resistance.
  - Rare earth additives in the past were La, Ned, and Pr, but now 15 elements from La to Lu can be individually added as an alloy material. Baikov Institute of Metallurgy has applied almost all the materials in alloying and found the microstructural abilities with properties [14,15].
- Silicon: Silicon increases fluidity and slightly improves creep strength.
- Silver: Silver, used in conjunction with rare earths, improves the high temperature strength and creep resistance.
- Thorium: Addition of thorium leads to an improvement in creep strength up to 370°C. For alloys containing zinc, the addition of thorium also enhances the weldability.
- Tin: Tin together with aluminum in magnesium improves ductility. It also assists in reducing cracking tendency during forging.
- Yttrium: It is incorporated with other rare earth metals to enhance high temperature strength and increase creep resistance.
- Zinc: It is normally used in conjunction with aluminum to increase the strength without reducing ductility. The increase in strength to comparable levels is not achievable if only aluminum content is increased.
- Zirconium: Zirconium is an important grain refiner for sand and metal mold, gravity and low pressure casting of magnesium alloys.

Statistics show that the AZ series of alloys is the most commonly used and popular for good room temperature strength and ductility. AZ91B, AZ91C, AZ91D, and AZ91E are the more recent versions of the AZ91 series and so on. Table 10.2 provides the variations in percentages of alloying elements in the AZ series. These versions may vary in the range and amounts of secondary alloying elements, to satisfy cost targets or to provide some subset of properties or processing benefits. The main difference between AZ91C and AZ91D is not in the addition of alloying elements [10], but in the imposition of maximum values for the impurities—copper, iron, and nickel. AZ91E has shown good corrosion resistance among the AZ series of alloys [16].

The Mg-Al-Zn alloys were used in Germany from 1914 to the 1930s. Mg-Al-Zn-Mn was developed through the 1950s. Later on, various improvements were developed in the combination of alloy elements during the past decade. Mg–Al–RE alloys, Mg–Al–Ca alloys, Mg–Al–Ca–RE alloys, Mg–Zn–Al–Ca alloys, Mg–Al–Sr alloys, Mg–Al–Si alloys, Mg–RE–Zn alloys, and Mg-Zr, Mg-Y-RE-Zr have been extensively used for creep resistant applications [5,17]. Magnesium Elektron Ltd (MEL),

TABLE 10.2

Compositions of AZ91 Series of Magnesium Alloys

Element	AZ91A	AZ91B	AZ91C	AZ91D	AZ91E
Al	8.3%–9.7%	8.3%–9.7%	8.1%–9.3%	8.3%–9.7%	8.1%–9.3%
Mn	0.13%	0.13%	0.13%	0.15%	0.17%–0.35%
Zn	0.35%–1.0%	0.35%–1.0%	0.40%–1.0%	0.35%–1.0%	0.4%–1.0%
Si	0.50%	0.50%	0.30%	0.1%	0.20%
Fe	Not specified	Not specified	Not specified	0.005%	0.005%
Cu	0.10%	0.35%	0.1%	0.030%	0.015%
Ni	0.03%	0.03%	0.01%	0.002%	0.001%
Other metals	0.30%	0.30%	0.30%	0.02%	0.30%

Source: http://mg.tripod.com/asm_prop.htm.

a United Kingdom based company, extended its services to all over the world with the newly developed alloys.

10.2.3 TYPES OF MAGNESIUM ALLOYS

Magnesium alloys can be categorized mainly into two groups, namely, cast alloys and wrought alloys, based on the process of operations. Magnesium contains a hexagonal lattice structure (HCP) which resists plastic deformation. Hence, the majority of magnesium alloys are cast. Wrought alloys came into existence in 2003.

Tensile and other mechanical properties of the cast alloys are determined on separately poured test bars conforming to standard ASTM procedures, while in wrought alloys, the mechanical properties are determined on specimens cut from the actual manufacturers—extrusions, forgings, or rolled products. The selection of suitable alloy type depends on how the product will be made (cast or wrought), the strength required, and the conditions of the work environment.

10.2.3.1 Cast Magnesium

Casting is a complex process involving melting of the metal and pouring the melt into a die. The solidification of the melt gives a cast product. The melting point of pure magnesium is 650°C. Most cast magnesium alloys are subjected to oxidation in molten state. These oxides may produce a negative effect on the mechanical properties during the casting and should be avoided. So, in general, flux are added and/or passed through filters before entering the casting cavity in the mold. The detailed information on the preparation of cast magnesium products is explained in later sections.

As mentioned earlier, the AZ alloy series (Mg–Al–Zn) is the most well known of the magnesium alloys because of its wide use in both gravity and low pressure casting and in high pressure die casting. The AZ series increases fluidity for casting and also provides strength to the cast product. At lower aluminium levels, that is, 5%–6%, the ductility of the alloys (AM family) makes them especially attractive for applications requiring crash worthiness, for example, steering wheels, instrument panels, and seat frames. However, these alloys do not perform well at elevated temperatures (>125°C) due to their poor creep resistance [10]. Along with AZ, AM series is also being widely used because these can be cast using a wide range of processes, from gravity and low pressure sand casting to high pressure die casting and also form the basis of important wrought magnesium alloy families.

The AS series of cast alloys was developed with reduced aluminum content but with silicon replacing zinc. First AS41 was developed and later AS21 with better high-temperature properties than AS41. Both alloys rely on the reduced Al content and the formation of Mg_2Si to impart creep strength to the alloy [18].

Some magnesium and zirconium alloys do not contain aluminum. If aluminum is present, it will react with the zirconium and precipitate it from the melt. Zirconium is added during casting of these alloys for grain refinement. None of these alloys are used for high pressure die casting, since grain refinement is necessary only at the lower solidification rates of gravity and low pressure casting processes. Zirconium-refined alloy families have found extensive use in aerospace and military applications, but not in the automobile industry due to their high cost (as the alloying elements yttrium, silver, and rare earths are very expensive and the zirconium grain refiner is costly).

With due considerations of metallurgy and mechanical properties, both material as well as process innovations in casting have been developed. Table 10.3 provides much more information on cast magnesium alloys and their applications in the automobile sector.

10.2.3.2 Wrought Magnesium

Wrought magnesium alloys have better microstructural homogeneity than cast products and generally have enhanced mechanical properties. Wrought alloys have been made by ingot metallurgy and subsequent hot rolling. This can be a very expensive process, as magnesium has limited formability at room temperature. So the reduction of the ingot to the final sheet is done at elevated temperatures. As a result, the conversion cost (to turn raw magnesium material into sheet product) is higher than for converting steel or aluminum into sheets. However, recently some global research is trying to cut the cost of wrought alloys by casting them to near net shape and with minimum rolling.

The first developments in wrought magnesium include the modification of the Mg-Al alloys like AZ31 (for forming sheets) and AZ61 (for extrusion products). AZ31 is almost 50 years old. Modifications of AZ31 and AZ61 were effected via trace additions of RE. The other elements like Ca, Sr, and Sb were also added to optimize the performance of AZ31 alloy. The elimination of Zn from AZ31 for extrusion led to the development of the AM30 alloy. AM30 had slightly better formability than AZ31 at room and moderate temperatures, due to higher strain hardening rate and exponent. Other alloys, which are currently commercially available, are HM21, HK31, and ZM21.

AZ61 wrought alloy contains twice the amount of Al compared to AZ31. Wrought magnesium alloys containing various RE have recently been of keen interest, mainly for the improvement of strength in wrought Mg. RE as well as Sn additions to

TABLE 10.3

Cast Magnesium Alloy Families

Alloy Families	Casting Process	Properties	Applications

• AZ (Mg–Al–Zn–Mn)
  • AZ91D
  • AZ91E
	• Gravity or low pressure
	• Sand or metal mould
	• High pressure die casting
	• Squeeze casting
	• Thixo-molding
		High strength, low ductility	• Brake brackets, clutch brackets, transfer cases, gear box housing, intake manifolds, valve covers, manual transmission cases, wheels, CVT transmission cases, air bag housings

• AM (Mg–Al–Mn)
  • AM50
  • AM60
	• Gravity or low pressure
	• Sand or metal mould
	• High pressure die casting
	• Squeeze casting
	• Thixo-molding
		Low strength but higher ductility	• Steering wheels, seat frames, instrument panels, cross-car beams, trim, roof frame

• AS (Mg–Al–Si–Mn)
  • AS21
  • AS31
  • AS41
	• High pressure die casting	Improved creep strength	• Volkswagen air cooled engine in 1970s
			• Currently used in Mercedes automatic transmission case

• AE (Mg–Al–E–Mn)
  • AE42
  • AE44
	• High pressure die casting	High creep strength in power train operating environment to 150°C, but expensive due to E content	• Corvertte engine cradle

• Zr refined (Mg–Zn–Zr)	• Gravity and low pressure casting in sand or metal molds	High strength and creep resistance, but higher cost	• Used in aerospace and military applications

Source: P.K. Mallick, *Materials, Design and Manufacturing for Lightweight Vehicles*, CRC Press, 2010.

magnesium have also been developed. It has been reported that Sn addition into Mg-RE alloys improves the roll ability of the alloys.

Lithium has been observed to change the crystallographic texture by altering the balance of deformation mechanisms, which in turn influences the texture. Conventional Mg-Li alloys have usually depended on large Li additions that change

TABLE 10.4

Current Development and Progress in Wrought Magnesium Alloy

Alloy Families	Grain Structure	Second Phases	Properties
AZ31+Ca (hot extruded)	Refined grains	Ca Intermetallic	High strain rate
AZ31+Sr (hot extruded)	• Mixed mode (250°C), (Fine + elongated grains) • Fine uniform (350°C)	Al_4Sr Intermetallic	• High elongation at 0.8% Sr • High strength at 0.05% Sr
AM30+Ce (hot rolled)	Refined grains	$Mg_{12}Ce$ Dispersion	Improved extrudability
Mg-Gd-Y-T6 (hot extruded)	Refined grains	β Phase	High elongation
AZ31+Li+Al+Zn (hot rolled)	Deformed grain structure	α, β dual structure	• Improved rollability • Good combination of strength and ductility
Mg-RE-Sn (Al) (hot rolled)	Refined grains	Al_2MM	High elongation, formability, and low strength
Mg–Zn–Li	Refined grains	Solid solution alloys	Reduced edge cracking
Mg-Zn-Gd (rolled)	Refined grains	Tiny particles	• High ductility, good formability • High strain hardening

Source: C. Bettles, M. Barnett, *Advances in Wrought Magnesium Alloys-Fundamentals of Processing, Properties and Applications*, Wood Head Publishing, 2012.

the structure of magnesium from HCP to BCC and this is advantageous as these alloys can be made wrought.

The improved formability combined with the need for adequate strength are the driving forces for investigating new alloy development strategies. The use of two alloy systems, Mg-Mn (ME and MJ Series) alloys and novel Mg-Zn (ZK, ZE, ZW series) alloys have progressively increased over the past few years [19]. An overview of wrought alloys and their benefits are mentioned in Table 10.4.

10.2.4 PROPERTIES OF MAGNESIUM ALLOYS

Magnesium alloys with different combinations of alloying elements can be tried and evaluated in both cast and wrought forms. Some of the properties of the most commonly used magnesium alloys are mentioned in Table 10.5 [20]. Researchers doing

TABLE 10.5
Some of the Magnesium Alloys and Their Properties

Material	Density (g/cm³)	Thermal Conductivity (W/mK)	UTS (MPa)	YTS (MPa)	Fatigue Strength (MPa)	Impact (J)	Hardness (BHN)	% Elongation in 50 mm	Specific Heat (J/g-°C)	Coeff. of Thermal Expansion (μm/m-C)
AZ91	1.81	72.7	230	150	97	2.7	63	3	0.8	26
AM60	1.79	62	241	131	80	2.8	65	13	1	26
AM50	1.77	65	228	124	75	2.5	60	15	1.02	26
AZ31	1.771	96	260	200	90	4.3	49	15	1	26
ZE41	1.84	113	205	140	63	1.4	62	3.5	1	26
EZ33	1.8	99.5	200	140	40	0.68	50	3.1	1.04	26.4
ZE63	1.87	109	295	190	79	2.3	75	7	0.96	27
ZC63	1.87	122	240	125	93	1.25	60	4.5	1	26
ZK60	1.83	120	340	260	140	2.5	75	11	1	26
AS41	1.78	68	185	140	NA	4.1	75	4.5	1	26.1
AE42	1.78	68	225	140	NA	5.8	58	8–10	1	26.1
AM20	1.76	60	220	105	70	NA	47	8–12	1	26

Source: http://www.azom.com/; http://www.matweb.com/; D.S. Kumar, C.T. Sasanka, K. Ravindra, K.N.S. Suman, *American Journal of Materials Science and Technology*, 4(1), 12–30, 2015. http://dx.doi.org/10.7726/ajmst.2015.1002.

work on the above materials are advised to use standard material databases for more accurate and elaborate information.

10.3 PROCESSING OF MAGNESIUM ALLOYS

10.3.1 PROCESSING OF CAST MAGNESIUM ALLOYS

A wide variety of processing methods and technologies has been developed for magnesium alloys. Magnesium and its alloys can be processed either through solid or liquid phase. In spite of benefits in solid phase processing, it is used less commonly compared to liquid phase processing. This is due to the high processing cost of the solid phase, limitations on thickness, lower ductility and fracture toughness, and the required tedious handling of fine powders. Sand casting, high pressure and gravity die casting, squeeze casting, semi-solid metal (SSM) casting, spray forming, and melt infiltration methods are all examples of liquid phase processes. In spite of the availability of literature on various methods [4,5], this section briefly describes the various processes. Figure 10.2 shows various methods of obtaining cast magnesium alloys.

10.3.1.1 Sand Casting

Sand casting is the process in which molten metal is poured into a disposable mold created through compacting sand. The solidified metal is removed from the mold once it is cooled. Magnesium reacts with many molding materials and therefore precaution must be taken to reduce metal-mold reactions. This can be done through the

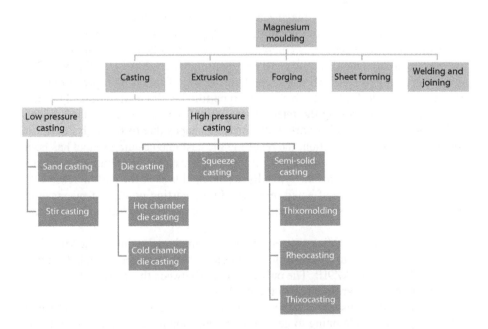

FIGURE 10.2   Various processes to make cast magnesium alloys.

use of minimal moisture content within the sand as well as using inhibitors in the sand mixture used to create the mold and cores. Depending on the temperature of the molten metal, the type of alloy cast and the thickness of the casting section, the amount of inhibitor added will vary. Inhibitors can be used alone or in combination with one another depending on the alloy that is being cast. Heat treatment can be done on sand and permanent mold castings [4].

Magnesium alloy sand castings are used in aerospace applications because they offer a clear weight advantage over aluminum and other materials. A considerable amount of research and development on these alloys has resulted in some spectacular improvements in general properties compared with the earlier AZ types.

Although there has been, and still is, a large volume of castings for aerospace applications being produced with the older, conventional AZ-type alloys, the trend is toward the production of a greater proportion of aerospace castings with the newer Zr types.

10.3.1.2 Stir Casting

Stir casting is a very popular liquid state process for magnesium alloys and composite fabrications. In the case of composites, where the matrix metal is heated to the liquid temperatures, reinforcement particles are introduced and distributed into molten matrix phase by mechanical/ultrasonic stirrer to overcome poor wettability of matrix and reinforcement phase [21]. The melt is cooled down to room temperature to get the final solid product. The arrangement of bottom pouring stir casting process with automatic stirrer is shown in Figure 10.3.

10.3.1.3 Die Casting

In a majority of applications, magnesium alloys are fabricated using high pressure die casting. This is due to the possibility for higher production volumes and low production costs. Die casting is completed by forcing molten metal through a narrow opening to fill a mold at a very fast rate (approximately 20m/s). A highly intense pressure of about 40–1000 MPa is applied during solidification. This pressure in the final stages of solidification enables areas to be filled with metal, ultimately reducing porosity and improving the internal integrity of the part. Die casting is widely used to produce thin-walled parts with intricate shapes due to the high fluidity of molten magnesium. In addition, the resulting die cast magnesium product has good part strength, requires minimum machining, and is capable of providing good surface and dimensional precision. Typically, as the thickness of the part increases, the strength and ductility are affected inversely. In die casting process of magnesium, about 40%–60% of the metal becomes process scrap via the runner system and overflow [22].

The alloys from which die castings are normally made are mainly of the Mg-Al-Zn type. Two versions of this alloy, from which die castings have been made for many years, are AZ91A and AZ9IB. The only difference between these two versions is the higher allowable copper impurity in the AZ91B.

A typical 400 ton magnesium hot chamber machine can make parts that weigh up to 2.5 kg. It has a clamping force of 400 ton, and applies a maximum pressure of about 35 MPa on the metal. Due to the short cycle time (up to six parts per minute),

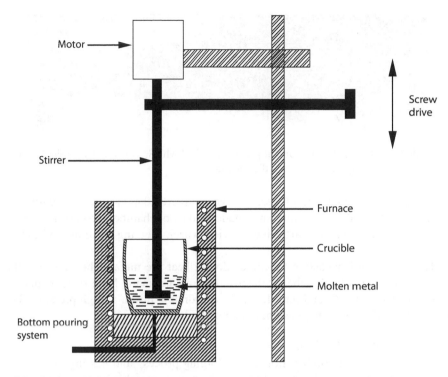

FIGURE 10.3 Processing of magnesium alloys using stir casting. (From D.S. Kumar, C.T. Sasanka, K. Ravindra, K.N.S. Suman, *American Journal of Materials Science and Technology*, 4(1), 12–30, 2015. http://dx.doi.org/10.7726/ajmst.2015.1002.)

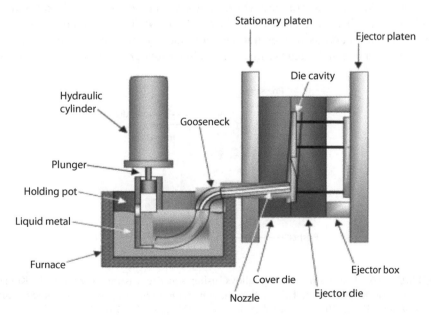

FIGURE 10.4 Schematic of hot chamber die casting.

FIGURE 10.5   One-piece instrument panel beam for GMC Savana and Chevrolet Express, 12 kg, the world's largest magnesium die casting.

the hot chamber die casting process, shown in Figure 10.4, is very competitive for small parts. Typical magnesium parts made by the hot chamber die casting method include small automotive parts like steering wheel, steering column, and airbag housing.

The cold chamber casting machines do not heat the metal. The molten metal is ladled into the cold chamber manually or by an automatic ladle system, and the molten metal is then forced into the die by a hydraulic piston at high pressure. The largest die casting with magnesium is shown in Figure 10.5.

10.3.1.4   Squeeze Casting

Squeeze casting is a combination of the forging process and casting process. In direct squeeze casting, shown in Figure 10.6, molten metal is poured slowly with a minimal amount of turbulence into the lower half of a die. An upper punch is pressed down on the metal, once the die cavity is filled. The metal solidifies under this high, unidirectional pressure which in turn reduces any internal defects during solidification. Direct squeeze casting, however, allows for the trapping of impurities within the metal as it does not have a runner system. Incidentally, this also results in high internal integrity material. Indirect squeeze casting involves molten

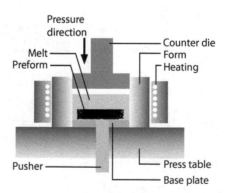

FIGURE 10.6   Processing of Mg alloys using squeeze casting. (From D.S. Kumar, C.T. Sasanka, K. Ravindra, K.N.S. Suman, *American Journal of Materials Science and Technology*, 4(1), 12–30, 2015. http://dx.doi.org/10.7726/ajmst.2015.1002.)

metal being poured into an encasement. A plunger is then used to control the speed at which the metal flows into the mold to eliminate the gas bubbles within the casting. There is a lower material yield with indirect squeeze casting due to greater material loss [23].

10.3.1.5   Semi-Solid Metal (SSM) Casting

Many of the disadvantages that result from high pressure die casting are overcome by SSM casting. Thixomolding, rheocasting, thixocasting and thixoforming are all variations of this processing method. SSM allows for the production of complex shapes with low porosity levels and longer die life. A semi-solid slurry having lower heat content and higher viscosity than the liquid form of the same alloy is used. This makes it possible to have faster cooling rates, greater thermal efficiency, and prolonged die life. The higher viscosity also allows for less turbulent flow during filling, resulting in lower levels of contained gas. The downside to this method is the high cost of the feedstock that is necessary.

Thixomolding, a type of SSM casting, is a relatively new method used specifically for processing magnesium alloys. This process is similar to that of injection molding for plastics. Magnesium chips at room temperature are put into a feeder that leads to a heated screw, which slowly heats them to just below their liquid temperature. The screw slowly pushes the chips while heating them to the semi-solid temperature range. This heat combined with the shear forces induced by the screw creates a semi-solid slurry, which is then injected into the die.

Rheocasting is another type, which uses molten metal as feedstock. Reinforcements are added to the alloy in the semi-solid stage. A mechanical stirrer is used to create homogeneous slurry, which is then poured into a mold. Rheocast components have several advantages including homogeneous distribution of any porosity that may occur, lower amounts of shrinkage, a lower tendency for micro- and macro-segregation, and a fine grain structure [4].

10.3.1.6   Melt Deposition Technique

This process can be used in two ways—either by producing droplet stream by molten bath (osprey process) or by continuous feeding of cold metal into a zone of rapid heat injection (thermal spray process). The disintegrated melt deposition (DMD) process is a combination of dispersion and spray process. Two jets of gas are passed for the preparation of the final product from the molten metal. The product thus formed can be sent to hot extrusion for further processing.

Table 10.6 provides a comparison of fabrication processes of magnesium alloys.

10.3.2   Processing of Wrought Alloys

Magnesium is also used in wrought product form for extrusions, forgings, sheets, and plates. Wrought alloys have high dimensional stability and ease of machining as well as typically better overall mechanical properties. Applications for these milled products range from bakery racks, loading ramps, and hand trucks to concrete finishing tools, computer printer plates, and nuclear fuel element containers [4].

TABLE 10.6
Comparison of Cast Magnesium Fabrication Processes

Route	Cost	Application	Comments
Die casting	Low/medium	Automotive industries, LCDs, laptops etc.	Used for thin walled products, high production volumes, intricate shapes
Squeeze casting	Medium	Widely used in automotive industry like connecting rods	Lower porosity, expensive, molds needed, large capacity presses needed
Stir casting	Low	Basic process for Mg MMCs, used in automotive and aerospace industry	Applicable for mass production, low volume fractions up to 30%
Melt deposition technique	Low/medium	Used to produce structural shapes such as rods, beams, etc.	Uniform distribution, high strengths

10.3.2.1 Extrusion Processes

10.3.2.1.1 Conventional Direct Extrusion Process

In conventional direct extrusion process, magnesium alloys can be warm or hot extruded in hydraulic presses to form bars, tubes, and a wide variety of profiles [12]. It is well known that a tubular section is significantly stiffer than a solid beam of the same mass. While hollow extrusions of magnesium can be made with a mandrel and a drilled or pierced billet, it is generally preferable to use a bridge die, where the metal stream is split into several branches, which recombine before the die exit.

10.3.2.1.2 Hydrostatic Extrusion Process

The hydrostatic extrusion process, typically used for copper tubing fabrication, is a much faster extrusion process compared with the conventional direct extrusion. It was reported that seamless magnesium tubes were extruded using the hydrostatic process at speeds up to 100 m/min, due to the absence of friction between the billet and container, as the billet is suspended in a hydraulic oil [19]. Although the process is capable of extrusion rates up to 700, the outer diameter of tubes produced by this process is limited to about 45 mm, even with a large 4000-ton press [19].

10.3.2.2 Forging Processes

Commercial extrusion alloys (AZ31, AZ61, AZ80, and ZK60) can also be forged into high-integrity components using hydraulic presses or slow-action mechanical presses. Forging is normally done within 55°C of the solidus temperature of the alloy. Corner radii of 1.6 mm, fillet radii of 4.8 mm, and panels or webs of 3.2 mm thickness can be achieved by forging. The draft angles required for extraction of the forgings from the dies can be held to 3° or less [12].

Forged magnesium has weight savings of up to 5%–10% than cast magnesium and up to 10%–15% than forged aluminum in the case of wheels. The cost of the process is also low when compared with the other two. Pressure sealed components have very good properties due to the forging process as it prevents a porous microstructure.

10.3.3 Sheet Production and Forming Processes

10.3.3.1 Sheet Production Processes

The direct-chill (DC) casting process is generally used to make magnesium slabs (about 50 mm thick), which are then hot-rolled at 315–370°C to produce magnesium sheets and plates. Unlike aluminum, for which a cold-roll is usually the final step in sheet production, a warm finish roll is applied to magnesium sheet products. Large grains of 200–300 μm in the slab can be reduced by warm rolling to fine recrystallized grains between 7 and 22 μm for a sheet gauge of 1.3–2.6 mm [24,25].

Recently, the twin-roll continuous casting (CC) process has been investigated for the production of low-cost magnesium sheets. Pilot plants have been established in Germany, Austria, and Australia, and production plants are being built in China and Korea. In this process, molten magnesium is poured into a gap between two rolls to produce a continuous strip about 2.5–10 mm thick, which is then warm-rolled to a final gauge. Currently, magnesium sheets using this process are available from Salzgitter (Germany) with a maximum width of 1850 mm and a minimum gauge of 1.0 mm [25].

10.3.3.2 Sheet Forming Processes

The majority of the processes like stamping, flanging, bending, hemming, and trimming used to convert sheet metal into automobile components occur at room temperature. The processes are very robust for high-formability materials such as mild steel but, with some concessions on draw depth and corner radii, have been successfully used with less formable materials such as aluminum and high-strength steel. Unfortunately, the limited formability of magnesium due to its HCP structure makes the use of these processes very difficult [10]. The magnesium sheets that are widely used as of now are "inner" panels. The only current production application of magnesium sheet is the center console in the low volume Porsche Carerra as shown in Figure 10.7.

10.3.4 Welding and Joining Techniques

10.3.4.1 Welding Processes

Various welding processes can be used for joining magnesium and magnesium alloys. Most magnesium alloys are readily fusion-welded with higher speeds than aluminum due to the lower thermal conductivity and latent heat of magnesium [12]. While gas metal arc welding (MIG or GMAW) is used for joining magnesium sections ranging from 0.6 to 25 mm, gas tungsten arc welding (TIG or GTAW) is more suited for thin sections up to 12.7 mm [12]. It should be recognized that hot-shortness may produce cracks in welded magnesium alloys containing more than 1% zinc, which can often be overcome by using proper filler wires. Magnesium sheets and extrusions ranging from 0.5 to 3.3 mm can be joined by all types of resistance

FIGURE 10.7 Center console cover in Porsche Carrera GT automobile made of sheet magnesium. (From P.K. Mallick, *Materials, Design and Manufacturing for Lightweight Vehicles*, CRC Press, 2010.)

welding, including seam, projection, and flash, but the most common type is spot welding [12]. A higher welding speed for thin magnesium sheets can be achieved with laser or electron beam welding, due to the narrow heat affected zones and lower weld distortion.

10.3.4.2 Joining Processes

Fusion welding of magnesium die castings can be challenging due to the presence of porosity and the formation of a brittle inter metallic phase ($Mg_{17}Al_{12}$) in the welds [26]. Solid-state welding techniques, such as friction-stir welding and magnetic pulse welding can be used to improve the weld quality of magnesium die castings. While these solid-state welding techniques can potentially be used to join magnesium to dissimilar materials such as aluminum, mechanical joining (self-piercing rivets, clinching, and hemming) and adhesive bonding are preferred for dissimilar material joining involving magnesium to aluminum or steel.

Adhesive bonding of magnesium parts (with or without dissimilar materials) requires proper pretreatment of the joint surfaces, which includes cleaning, etching, and wet chemical passivation. Many adhesives including epoxy and polyurethane can be used as long as they are chemically stable and have aging stability [19].

In addition to the methods discussed above, in situ techniques also show great significance for the processing of magnesium alloys now [5,27]. Even though there are different methods of fabrication, the selection of effective process will have a good impact on microstructure, properties, and cost of the product.

10.4 AUTOMOTIVE APPLICATIONS OF MAGNESIUM ALLOYS

Consumer's preference for vehicle performance is increasing day by day. Global trends of environmental protection have also forced the automotive industry to work on new developments by using alternate fuels, power train enhancements, aerodynamic modifications, and weight reduction methodologies. Among these, weight reduction of a vehicle by alternate materials is the simplest and a cost-effective solution.

10.4.1 HISTORICAL DEVELOPMENT

In the 1920s, magnesium parts made their way into racing cars. However, it was not until the 1930s that magnesium was used in commercial vehicles. Volkswagen (VW) was the first to apply magnesium in the automotive industry on its Beetle. The Beetle contained more than 20 kg of magnesium alloy in the transmission housing and the crankcase. Porsche first worked with a magnesium engine in 1928. The developments in the usage of magnesium are clearly demonstrated in Figure 10.8.

Magnesium Historical Milestones in Automotive applications [3]

- **1920–1922:** Studies commenced for engine applications (patent for magnesium piston—Opel: race motorcycles with magnesium piston and block).
- **1926–1927:** German company, Adler, manufactured many car parts with magnesium.
- **1947:** Ted Halibrand (United States) made magnesium wheels for race cars.
- **1949:** Magnesium was used for the hubs of the wheels that used drum brakes.
- **1954:** Jaguar D used a central body made of magnesium with front and back structures in aluminum tubes.
- **1955:** Mercedes Benz 300 SLR sports car uses a magnesium sheet metal body.
- **1960s:** Widely used by Porsche (911 race version, base of magnesium alloy RZ 5 = ZE 41 A with a weight of 22.3 kg made by sand casting). The series production was made of AZ 81.
  - Volkswagen used the magnesium alloys for the base of its Boxer engine and for the gear box. Since 1960, these components were made with die-casting. The finished base weight is 9.4 kg. The AZ 81 alloy is used.
- **Between 1960 and 1961:** Volkswagen uses more than 70,000 ton of magnesium.
- **1967:** Major portion of the F1 Eagle was made of magnesium metal plates.
- **At the end of the 1960s:** Ferrari carried out various experiments in the engine field with a magnesium base for racing engines. The alloys taken into consideration were ZRE 1, RZ 5, and Z5Z.
- **1970s:** Wheels of racing cars were made of magnesium alloy. They were initially melted and then forged. The AZ 80A alloy was mostly used. The weight of a wheel is lower by 25% than that of an aluminum wheel.
  - Famous companies such as Morini, Jawa, Bianchi, Parilla, Suzuki, and Kawasaki also used bases made from magnesium alloys.

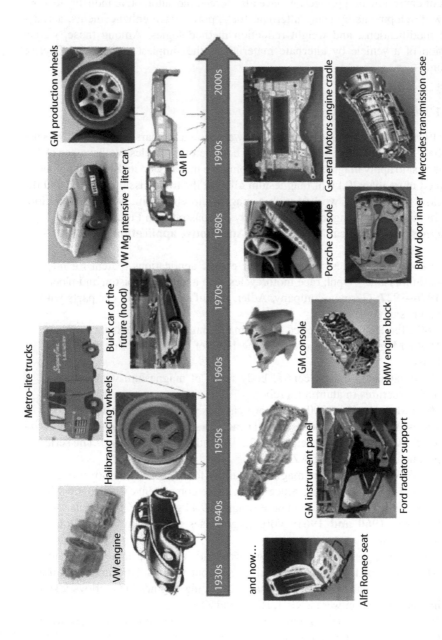

FIGURE 10.8 Summary of past and current automotive applications of magnesium. (From P.K. Mallick, *Materials, Design and Manufacturing for Lightweight Vehicles*, CRC Press, 2010.)

- **1973:** Porsche 917 K/30 used a magnesium tubular frame that weighed 45 kg.
- **1990s:** BMW M12 engine used magnesium for the two mountings with brackets for the cam shafts and the housing of the tappets with related covers as well as the casing of the distribution control.
- **2000s:** The magnesium alloys have seen considerable growth with many internationally reputed car manufacturers. Many components were fabricated with the application of Mg alloys. Magnesium's average usage and projected usage growth per car are given as 3 kg, 20 kg, and 50 kg for 2005, 2010, and 2015, respectively [28]. The growth of magnesium consumption over the last decade is at an annual rate of 15%. This growth is predicted to continue at a rate of at least 12% per year for the next 10 years [29].
- **By 2020:** It is expected that the magnesium usage will increase exponentially, leading to fuel savings of 9%–12%, according to the U.S. Automotive Partnership.
- **By 2050:** Magnesium will be able to occupy the majority weight of the total weight of a vehicle. The awareness on protection of Earth plays a crucial role so that all new cars are expected to have zero emissions.

10.4.2 MAGNESIUM AS A MATERIAL IN AUTOMOBILES

As mentioned earlier, magnesium has an advantage of density with strength to weight ratio. As the design of components and selection of materials needs so many factors to control, this section overviews the advantages of magnesium as a material in the automotive sector considering density, cost, mechanical properties, and recyclability.

10.4.2.1 Lightweight

It has been estimated that for every 10% of weight eliminated from a vehicle's total weight, fuel economy improves by 7% [30]. This also means that for every kilogram of weight reduced in a vehicle, there is about 20 kg of carbon dioxide reduction. Thus, weight reduction of a vehicle by alternate materials is the simplest and most cost-effective solution to reduce fuel consumption and environmental pollution.

Most of the castings in the automotive industry are made with steel/cast iron. It is a well-known fact that magnesium is a powerful weight saving option as its density is 36% of that of aluminum and it is 74% lighter than zinc and 79% lighter than steel. So replacement of components with magnesium alloys results in enormous saving in weight.

For example, a study by Lotus Engineering concludes that a vehicle mass reduction of 38% over a conventional mainstream vehicle can be achieved at only 3% cost. Comparison of weight reduction and associated costs for two future models are shown in Table 10.7 (Lotus Eng. Co., 2010).

Based on several studies, the distribution of weight in a vehicle is shown in Figure 10.9. The weight reduction in various areas of an automobile by magnesium compared with Al/Fe alloys was shown in Figure 10.10.

As can be seen from Figure 10.10, almost all areas of a vehicle contributed a weight reduction of 20%–70% when replaced by magnesium and its alloys. Material

TABLE 10.7

Assessment of Mass Reduction Opportunities, 2017–2020 Model Year Vehicle Program by Lotus

| | Base Toyota Venza excluding Power Train | Lotus Engineering Design | | | |
| | | 2020 Venza | | 2017 Venza | |
System	Weight (kg)	% Mass Reduction	% Cost Factor	% Mass Reduction	% Cost Factor
Body	383	42%	135%	15%	98%
Closures/fenders	143	41%	76%	25%	102%
Bumpers	18	11%	103%	11%	103%
Thermal	9.25	0%	100%	0%	100%
Electrical	23.6	36%	96%	29%	95%
Interior	252	39%	96%	27%	97%
Lighting	9.90	0%	100%	0%	100%
Suspension/chassis	379	43%	95%	26%	100%
Glazing	43.7	0%	100%	0%	100%
Misc.	30.1	24%	99%	24%	99%
Totals	1290	38%	103%	21%	98%

Source: Lotus Engineering, 2010.

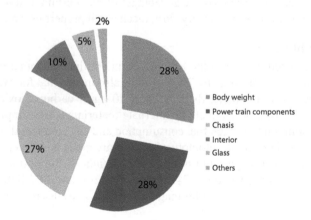

FIGURE 10.9 Mass reduction in various parts of an automobile using magnesium alloys.

alteration can be done in a vehicle in three major areas like body, power train, and chassis components. Body is the major contributor of total weight and also the first choice for a structural material. There have been many opportunities for researchers to use interior, exterior, and seat frames made of magnesium alloy materials. The power train is also another important element in a vehicle, where the transmission and engine systems are linked to work together by various mechanical couplings.

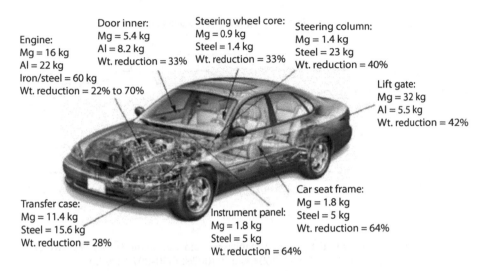

Engine:
Mg = 16 kg
Al = 22 kg
Iron/steel = 60 kg
Wt. reduction = 22% to 70%

Door inner:
Mg = 5.4 kg
Al = 8.2 kg
Wt. reduction = 33%

Steering wheel core:
Mg = 0.9 kg
Steel = 1.4 kg
Wt. reduction = 33%

Steering column:
Mg = 1.4 kg
Steel = 23 kg
Wt. reduction = 40%

Lift gate:
Mg = 32 kg
Al = 5.5 kg
Wt. reduction = 42%

Transfer case:
Mg = 11.4 kg
Steel = 15.6 kg
Wt. reduction = 28%

Instrument panel:
Mg = 1.8 kg
Steel = 5 kg
Wt. reduction = 64%

Car seat frame:
Mg = 1.8 kg
Steel = 5 kg
Wt. reduction = 64%

FIGURE 10.10 Mass distribution in an automobile. (From M.K. Kulekci, *Int J Adv Manuf Technol*, 39, 851–865, 2008. doi 10.1007/s00170-007-1279-2.)

10.4.2.2 Cost

One of the most important consumer driven factors in automotive industry is the cost. Since the cost of a new material is always compared to that presently employed in a product, it is one of the most important variables that determine whether any new material has an opportunity to be selected for a vehicle component [31].

Aluminium and magnesium alloys are certainly more costly than the currently used steel and cast iron that they might replace. The ability to approach the total cost depends on component manufacturing costs. Compared to cast iron and steel, cast aluminium and magnesium components are potentially less costly on the whole. This is based on their reduced manufacturing cycle times, better machinability, ability to have thinner and more variable wall dimensions, closer dimensional tolerances, and reduced number of assemblies. However, wrought aluminium and magnesium components are almost always more costly to produce than their ferrous counterparts. Even though the cost may be higher, the decision to select light metals must be justified on the basis of improved functionality [32]. The use of magnesium and its alloys in automotive components was limited in the early 1960s and 1970s because of high cost of Mg than Al. But the cost of magnesium has been decreasing below the cost of aluminium since 2004 as shown in Figure 10.11.

10.4.2.3 Mechanical Properties

Table 10.8 summarizes the mechanical and physical properties of typical cast and wrought magnesium alloys in comparison with other competing materials. The die-cast magnesium alloy AZ91 has the same yield strength and ductility as A380. Even though AZ91 has lower fatigue strength, it replaced A380 aluminium in nonstructural and low-temperature components such as brackets, covers, cases, and housings, providing essentially the same functionality with significant weight savings.

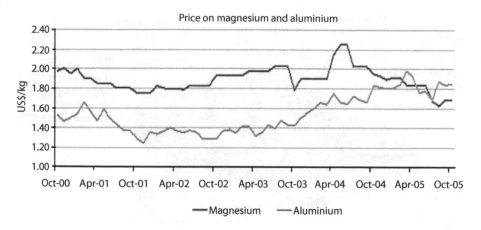

FIGURE 10.11   Changes in the prices of magnesium and aluminum. (From M.K. Kulekci, *Int J Adv Manuf Technol*, 39, 851–865, 2008. doi 10.1007/s00170-007-1279-2.)

For structural applications where crash worthiness is important, such as instrument panels, steering systems, and seating structures, magnesium die cast alloys AM50 or AM60 offer unique advantages of high ductility (10% to 15% elongation) and impact strength over aluminum die cast A380 alloy [29].

For extrusion parts, magnesium alloy AZ80 provides comparable tensile strength as aluminum alloy 6061, but less ductile. Magnesium sheet metal alloy AZ31 offers slightly lower strength, but a higher ductility, than commonly used 5xxx series aluminum sheet alloys. Although slightly heavier than plastics, magnesium components are much stiffer due to the fact that magnesium's elastic modulus is almost 20 times as high as a plastic material such as PC/ABS (a blend of polycarbonate and acrylonitrile-butadiene-styrene). Plastic materials are generally not suitable for elevated temperature applications due to their low softening temperatures (e.g., 143°C for the PC/ABS material).

Although the general corrosion resistance of high-purity magnesium alloys is even better than A380 aluminum alloy, galvanic corrosion remains a concern for design and applications engineers in developing magnesium components. The degree of galvanic corrosion attack on magnesium is strongly influenced by the compatibility of the second metal in contact with the magnesium parts, with aluminum alloy 5052 being the most compatible metal and steel the least compatible.

Another disadvantage of current magnesium alloys is that they have poor creep resistance at temperatures above 125°C which is inadequate for most power train applications; transmission cases can operate at up to 175°C, and engine blocks up to 200°C. Creep resistance is a major requirement for the use of magnesium in automotive power trains, which are currently made of aluminum or cast iron. The poor creep strength of such components can result in clamping load reduction in bolted joints and poor bearing/housing contact leading to oil leakage and/or increased noise, vibration, and harshness in power trains. New magnesium alloys with improved creep resistance are being developed for elevated temperature power train applications. Creep resistant alloys like rare earth added Mg alloys have a good potential to work in this area.

TABLE 10.8
Comparison of Mechanical and Physical Properties of Various Materials for Automotive Use

Material	Cast Mg		Wrought Mg		Cast Iron	Steel	Cast Al		Wrought Al	Polymers (PC/ABS)	GFRP	CFRP
Alloy/grade	AZ91	AM50	AZ-80 T5	AZ-31	Class 40	Mild steel grade 4	A355-T6	A 380	6061-T6	Dow pulse 2007	Structural (50% uni-axial)	Structural (58% uni-axial)
Process/product	Die cast	Die cast	Extrusion	Sheet	Sand cast	Sheet	Permanent mold cast	Die cast	Extrusion	Injection molding	Liquid molding	Liquid molding
Thermal conductivity (W/m K)	51	65	78	77	41	46	159	96	167		0.6	0.5
Melting temperature (°C)	598	620	610	630	1175	1515	615	595	652	143	130–160	175
Density (g/cm^3)	1.81	1.77	1.80	1.77	7.15	7.80	2.76	2.68	2.70	1.13	2.0	1.5
Elastic modulus (GPa)	45	45	45	45	100	210	72	71	69	2.3	48	189
Yield strength (MPa)	160	125	275	220	NA	180	186	159	275	53	1240	1050
Ultimate tensile strength (MPa)	240	210	380	290	293	320	262	324	310	55	–	–
Elongation (%)	3	10	7	15	0	45	5	3	12	5 at yield 125 at break	<1	<1
Fatigue strength (MPa)	85	85	180	120	128	125	90	138	95	NA	NA	NA

Source: C. Bettles, M. Barnett, *Advances in Wrought Magnesium Alloys-Fundamentals of Processing, Properties and Applications*, Wood Head Publishing, 2012.

The creep-resistant AE42 alloy was developed in the 1970s. Rare earth additions have been shown to impart creep resistance in non-Al containing magnesium alloys because the rare earths precipitate at the grain boundaries. However, the rare earth additions increased the cost of the alloy and made it susceptible to hot-cracking.

New alloys were subsequently developed that substituted the lower cost alkaline earth elements Ca, Sr, and Ba for rare earth additions. This development was subsequently extended to the development of the AXJ series of alloys for creep resistance, specifically AXJ530 (5% Al, 3% Ca, and 0.2% Sr), which demonstrated creep resistance approaching that of aluminum 380 [10,29].

10.4.2.4 Recyclability

Recycling of materials is always a priority, in view of renewable energies. As raw materials are scarce, the largest possible recovery from waste gives cost-effective utilization. Generally, metals are more recyclable than plastics because they can be remelted and reused. In particular, magnesium has lower specific heat and low melting point than other metals. This gives the advantage of using less energy in recycling, with recycled magnesium requiring as little as about 4% of energy required for manufacturing new material [33].

10.4.3 Current and Potential Applications of Magnesium Alloys

The global automotive applications of magnesium have been reviewed by a number of researchers and Table 10.9 shows that magnesium has made significant gains in worldwide interior applications, replacing mostly steel stampings in instrumental panels, steering wheels, and steering column components. In the power train area, General Motors and Ford Motors lead in the application of magnesium four-wheel-drive transfer cases in high-volume truck production; while Volkswagen is aggressively expanding the use of magnesium in manual transmission cases produced in both Europe and China. Only a limited number of body and chassis components are currently made of magnesium. Some examples of components are shown in Figure 10.12.

TABLE 10.9

Magnesium Alloys Applications in Car Components

Engine and parts transmission	Engine block, gear box, crank case, oil pump housing, cylinder head cover, transfer case, covers, cams, bed plate, engine cradle, clutch, and engine parts
Interior parts	Steering wheel covers, seat components, instrument panel, brake and clutch pedal, air bag retainer, and door inner
Chassis components	Wheels, suspension arms, engine cradle, rear support, tailgate, bumper, brake system, and fuel storage system
Body components	Cast components, radiator support, sheet components, extruded components, exterior and interior components, seats, instruments, and controls

Source: M.K. Kulekci, *Int J Adv Manuf Technol*, 39, 851–865, 2008. doi 10.1007/s00170-007-1279-2.
L. Gaines, R. Cuenca, F. Stodolsky, S. Wu, Automotive Technology Development Conference, October 28–November 1, 1996, http://www.transportation.anl.gov/pdfs/TA/101.pdf.

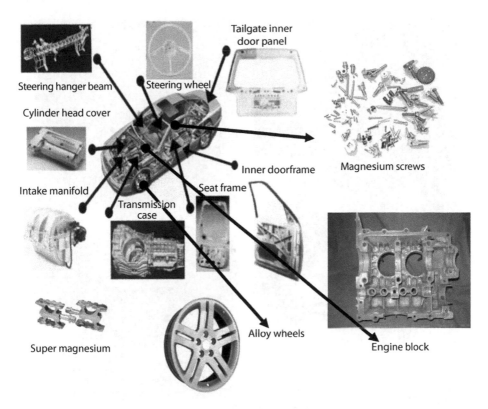

Steering hanger beam

Steering wheel

Tailgate inner door panel

Cylinder head cover

Magnesium screws

Inner doorframe

Intake manifold

Seat frame

Transmission case

Super magnesium

Alloy wheels

Engine block

FIGURE 10.12 Examples of components made with magnesium alloys. (From D.S. Kumar, C.T. Sasanka, K. Ravindra, K.N.S. Suman, *American Journal of Materials Science and Technology*, 4(1), 12–30, 2015. http://dx.doi.org/10.7726/ajmst.2015.1002.)

10.4.3.1 Interior

Since corrosion is of less concern in interior applications, this area has seen the most magnesium applications (Table 10.9), with the greatest growth in the instrument panels (IP) and steering structures. In addition to the IP beams, other interior magnesium applications at General Motors are shown in Figure 10.13. The design and die casting of magnesium IPs has advanced dramatically in the last decade.

The use of magnesium seat structures began in Germany, where Mercedes used magnesium die castings in the integrated seat structure with a three-point safety belt in the SL Roadster. This seat structure consisted of five parts (two parts for the seat back frame and three parts for the cushion frame) with a total weight of 8.5 kg and varying wall thickness of 2 to 20 mm. In the seat material selection, magnesium was chosen over plastic, steel sheet, and aluminum gravity casting designs. Magnesium die castings made of the high-ductility alloys of AM50 and AM20 offered the best combination of high strength, extreme rigidity, low weight, and cost.

10.4.3.2 Body

The use of magnesium in automotive body applications is very limited. GM has been using a one-piece die casting roof frame in its C-5 Corvette (Figure 10.14).

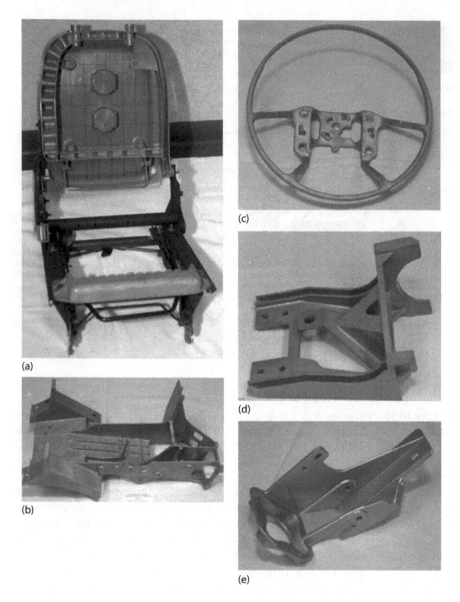

FIGURE 10.13 Interior parts made of magnesium at General Motors. (a) Seat structure, (b) steering column bracket, (c) steering wheel, (d) ABS bracket, and (e) pedal bracket. (From A.A. Luo, *JOM*, 42–48, 2002.)

European carmakers, especially Volkswagen, are pioneering the use of thin-wall magnesium die castings in body panel applications. The use of magnesium door inners and tailgates in VW's three-liter Lupo contributes to its potential 3 l/100 km fuel economy. The key to making those thin-walled (around 1.5 mm) castings lies in designing proper radii and ribs to smooth die filling and to stiffen the parts. These thin-walled die castings (door inners) can offer cost savings over steel sheet-metal

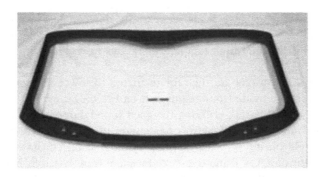

FIGURE 10.14   GM's magnesium root frame for the Chevrolet Corvette. (From A.A. Luo, *JOM*, 42–48, 2002.)

constructions due to part consolidation. In body panel applications where bending stiffness is frequently the design limit, magnesium sheet metal can offer as much as 62% weight saving [29].

10.4.3.3   Chassis

GM has been offering cast magnesium wheels for Corvette since 1998, which is the first original equipment option of magnesium wheels in North America. However, significantly high cost and the potential corrosion problems of magnesium wheels prevent their use in high-volume vehicle production. The production of low-cost and high integrity magnesium wheels and other chassis components, such as control arms, depends on the improvement of magnesium casting processes. Various casting processes have been developed for the production of aluminum wheels and chassis parts. These processes include permanent mold casting, low pressure casting, squeeze casting, and SSM casting. The successful adaptation of these processes to magnesium alloys will make magnesium castings more competitive with aluminum in the chassis area. The development of low cost, corrosion-resistant coatings and new magnesium alloys with improved fatigue and impact strength will also accelerate the further penetration of magnesium in chassis applications [29].

10.4.3.4   Power Train

Magnesium alloys that will withstand higher temperatures are being developed for engine blocks and transmission housings. At present, nearly 1 million of GMs GMT800-based full size trucks and sport utility vehicles produced annually have two magnesium transfer cases (total weight of 7 kg) per truck. At Volkswagen, around 600 manual transmission cases are produced daily for use in VW Passat and Audi A4/A6, and magnesium transmission cases are also used in the Santana model built in China. The operating temperatures for these applications are below 120°C, and AZ91 is the alloy of choice due to its excellent combination of mechanical properties, corrosion resistance, and castability.

In the power train part, earlier all of the transmission case was produced using aluminium die-casting, but recently magnesium manual transmission cases and engine head covers have been successfully developed using AZ91D alloy. Automotive transmission cases, engine oil pans and cylinder blocks for V6 engines are being developed using high temperature alloys to reach 12 kg weight reductions [30].

Power train applications are thus far restricted to lower temperature applications. Cam covers are made from magnesium on vehicles ranging from the Dodge Viper to the new Ford F-150, utilizing the good sound dampening characteristics. Audi has some more applications for magnesium in the power train, such as the air intake module on its W12 engine, and cylinder head covers on its V8. The company's Multitronic CVT and five-speed manual transmissions both have magnesium housings. BMW uses magnesium housings for the fully variable intake manifold featured on BMW's 8-cylinder power units and also VW's W12 engine for the Phaeton has a magnesium inlet manifold. Engine blocks (after the end of VW's air-cooled engine Beetle) are used in racing cars only, as with more conventional engine applications corrosion is a problem, because the coolant tends to react with the magnesium.

GM R&D developed the Mg-Al-Ca-Sr based alloys as a research project and is in testing condition for power train applications.

The chassis components are highly individualistic and show diverse characteristics. Wheel material is the first to be replaced with magnesium [34]. A variety of cast products are available to serve the purpose. The detailed areas of application of magnesium and its alloys in motor components by various leading automotive companies are given in Table 10.10.

Magnesium was used only in racing cars in the early 1920s but the usage of magnesium in automotive applications is expanding day by day as structural lightweight material. AM50 and AM60 alloys are extensively used in making interior parts of a car. General Motors (GM) used 26 kg of Mg alloy in Savanna and Express vans [35].

Ford Company uses AZ91 B alloys in its Ranger, Aerostar, and Bronco models in clutch housing, steering column and transfer case applications, respectively. GM used AZ91 series of alloys in valve covers, air cleaners, clutch pedals, brake pedals, steering column brackets, and many more. Porsche AG used AZ91D magnesium alloy for wheels in 944 Turbo Model. Alfa–Romeo is using 45 kg of magnesium in various components. Chrysler is using magnesium alloys in drive as well as in steering column brackets and in oil pans [36].

AZ91D magnesium alloys are popular and offer 20%–25% weight savings over aluminum in transmission casings. Kumar and Suman [37] made an interesting attempt with mathematical MADM methods to select a good material for alloy wheels. The result is in accordance with References 36 and 37. AM 50A and AM60B are more ductile and are used in seats, wheels, instrument panels, cylinder head covers, and so on [11,35]. AS41 is used for crank case and in transmission housings because of better fluidity [11]. ZE41 and AC63 are low pressure die casting alloys used in engine blocks [38]. Using extrusion process AZ31B is widely used in the preparations of bumper support beams, valve covers, electric motor frames and oil pans [39]. Volkswagen Group is leading the other companies like Mercedes Benz, BMW, Ford, Jaguar, and Audi.

TABLE 10.10
Magnesium Applications Globally in Automobiles

Product	Company
	Interior
Instrument panel	GM, Chrysler, Ford, Audi (A8), Toyota (Toyota Century)
Knee bolster retainer	GM
Seat frame	Chrysler, GM (Impact), Mercedes-Benz (Mercedes Roadster 300/400/500 SL), Lexus (Lexus LS430)
Seat riser	GM, Ford, Chrysler
Seat pan	Ford, Chrysler
Console bracket	Ford
Airbag housing	Chrysler
Steering wheel	Ford (Ford Thunderbird, Cougar, Taurus, Sable), Chrysler (Chrysler Plymouth), Toyota, BMW (MINI), Lexus (Lexus LS430)
Key lock housing	GM, Ford, Chrysler
Steering column parts	GM, Ford, Chrysler
Radio housing	Ford
Glove box door	GM
Window motor housing	Ford
	Power Train
Valve cover/cam cover	GM, Ford, Chrysler
4WD transfer case	GM, Ford
Clutch housing and piston	GM
Intake manifold	GM (V8 North Star motor), Chrysler
Alternator/AC bracket	Chrysler
Transmission stator	GM
Oil filter adapter	GM
Electric motor housing	GM
Crank shaft box	Volkswagen
	Body
Door frame	Chrysler
Roof frame	GM
Sunroof panel	GM, Ford
Mirror bracket	GM, Ford, Chrysler
Door handle	Volvo

(Continued)

TABLE 10.10 (CONTINUED)
Magnesium Applications Globally in Automobiles

Product	Company
	Chassis
Wheel	GM, Toyota (Toyota 2000GT, Toyota Supra), Alfa Romeo (GTV), Porsche AG (911 Series)
ABS mounting bracket	GM, Chrysler
Brake pedal bracket	GM, Ford, Chrysler
Brake/accelerator bracket	GM, Ford
Brake/clutch bracket	Ford
Brake pedal arm	Ford

Source: M.K. Kulekci, *Int J Adv Manuf Technol*, 39, 851–865, 2008. doi 10.1007/s00170-007-1279-2; A.A. Luo, *JOM*, 42–48, 2002.

10.4.4 TECHNICAL PROBLEMS AND CHALLENGES FOR USE OF MAGNESIUM ALLOYS IN THE AUTOMOTIVE INDUSTRY

The disadvantages of magnesium alloys are the high reactivity in the molten state, limited corrosion resistance, and inferior fatigue and creep compared to aluminum. The problems in using magnesium alloys extensively stem from their low melting points (650°C) and their reactivity (inadequate corrosion resistance). The main problem magnesium alloys encounter during fabrication and usage is the fire hazard/risk, especially in machining and grinding processes due to their relatively low melting point. The machinability aspects of magnesium are also to be enhanced.

Magnesium is a reactive metal, so it is not found in the metallic state in nature. It is usually found in nature in the form of oxide, carbonate or silicate often in combination with calcium. Because of this reactivity, the production of magnesium metal requires large amounts of energy. This situation makes magnesium an expensive metal [40].

Welding of magnesium alloys can also present a fire risk, if the hot/molten metal comes in contact with air. To overcome this problem, the welding region must be shielded by inert gas or flux. A larger amount of distortion relative to other metals may arise due to high thermal conductivity and coefficient of thermal expansion in welding of magnesium alloys, if required precautions are not taken. AZ91C and AZ91E can be readily welded by the gas shielding arc process [41].

The low creep properties of magnesium alloys limit the application of these alloys used for critical parts such as valve covers. The following are the main issues that need special attention to increase creep properties of magnesium alloys: stress relaxation in bolted joints, the potential for creep at only moderately elevated temperatures, corrosion resistance, and the effects of recycled metal on properties.

Different coating methods are used to increase the corrosion resistance of magnesium alloys. Problems with contact corrosion can be minimized, on the one hand, by constructive measures and, on the other hand, by an appropriate choice of material couple or the use of protective coatings [42]. Chromate coating of magnesium alloys

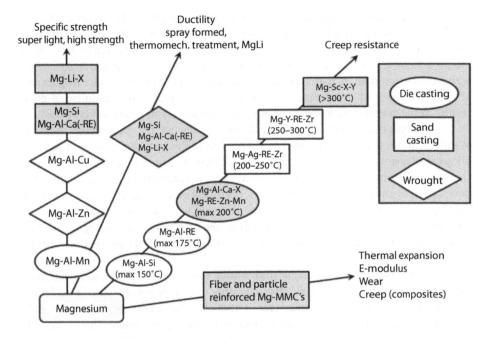

FIGURE 10.15 Future directions of magnesium alloy development for automotive applications. (From B.L. Mordike, T. Ebert, *Materials Science and Engineering*, A302, 37–45, 2001; C. Blawert, N. Hort, K.U. Kainer, *Trans. Indian Inst Met*, 57(4), 397–408, 2004.)

is hazardous and not environmentally friendly. A newly developed Teflon resin coating has been developed for magnesium alloys [28], which is a low cost chromium-free corrosion resistant coating. The coating not only has corrosion resistant properties but also has good lubricity, high frictional-resistance, and non-wetting properties.

Research is being carried out to increase the fatigue resistance of wheels and to improve the corrosion behavior of various magnesium alloys. Kim et al. discussed recent developments in magnesium alloys, the research activities and their successful applications in Hyundai and Kia Motor Corporation [43]. Research in developing the magnesium alloys for high temperature applications has just been started. But significant research is still needed on magnesium processing, alloy development, joining, surface treatment, corrosion resistance, and mechanical properties improvement. Integration of multimaterials in the automotive sector is also a good thought for commencing research. Figure 10.15 may be helpful to researchers working in the area of development on new alloys.

10.5 CONCLUSIONS

The economic growth of any country heavily depends on the automobile industry. The usage of optimum materials, reduced CO_2 emissions, fuel efficiency, cost, and many factors are directly or indirectly associated with the automotive sector. So it is necessary to keep in mind to use the resources effectively.

Magnesium has an ample number of qualities compared to other lightweight metals. But many people still believe that aluminum is the greatest replacement of steel. Magnesium alloys provide a lot of scope and opportunity for researchers to work in the area to provide the necessary processes and methods of fabrication. A 20%–70% weight reduction with increased power performance can be obtained with magnesium alloys. A number of developments are coming every year with a promise of good results in the magnesium alloys in view of the considerations discussed previously.

Cast magnesium alloys find general applications in the automotive industry. Wrought alloys are currently used to a very limited extent, due to lack of suitable alloying elements and some technological restrictions imposed by the HCP crystal structure. Significant research is still needed on magnesium processing, new alloy development, joining, surface treatments, corrosion resistance, and improvement in mechanical properties to achieve the future goals to reduce the vehicle mass and the amount of greenhouse gases. Production and application technologies must be cost-effective to make magnesium alloys economically viable alternatives for the automotive industry.

It can be concluded that magnesium alloys show a great potential as alternate materials in the automotive sector. But a number of challenges are to be faced for the successful use of automotive structures. These challenges require the collaboration and involvement of experts among various industries, governments, and academia all over the world. The automakers are thinking of using 40–100 kg of magnesium alloys in a car. The amount of usage of magnesium alloys is going to increase by 300% in the near future. It is expected that future developments will enhance the benefits of magnesium, the lightest structural metal.

ACKNOWLEDGMENTS

The authors acknowledge all the persons who rendered their services in the beautiful field of magnesium. They also express their sincere thanks to all who contributed information on the web for the benefit of aspiring researchers.

REFERENCES

1. G.J. Simandl, H. Schultes, J. Simandl, S. Paradis, *Magnesium—Raw Materials, Metal Extraction and Economics—Global Picture*, Proceedings of the Ninth Biennial SGA Meeting, Dublin 2007, pp. 827–830. http://www.empr.gov.bc.ca/Mining/Geoscience/IndustrialMinerals/Documents/Magnesium.pdf.
2. D. Guillen Abasolo, *Magnesium—The Weight Saving Option*, University of Burgos, Spain, http://www.fisita.com/students/congress/sc08papers/f2008sc031.pdf.
3. B.B. Buldum, A. Sik, I. Ozkul, Investigation of magnesium alloy's machinability, *International Journal of Electronics, Mechanical and Mechatronics Engineering*, 2(3), 261–268, 2012.
4. Information from http://www.intlmag.org/.
5. M. Gupta, N.M.L. Sharon, *Magnesium, Magnesium Alloys, and Magnesium Composites*, John Wiley & Sons, 2011.
6. W. Wulandari, G.A. Brooks, M.A. Rhamdhani, B.J. Monaghan, *Magnesium: Current and Alternative Production Routes*, http://ro.uow.edu.au/cgi/viewcontent.cgi?article=2295&context=engpapers.

7. Information from http://metals.about.com/od/properties/a/Metal-Profile-Magnesium.htm.

8. B.L. Mordike, T. Ebert, Magnesium, properties, applications, potential, *Materials Science and Engineering*, A302, 37–45, 2001.

9. M. Josefa, F. Gándara, Recent growing demand for magnesium in the automotive industry, *Materials and Technology*, 45(6), 633–637, 2011.

10. P.K. Mallick, *Materials, Design and Manufacturing for Lightweight Vehicles*, CRC Press, 2010.

11. S. Fleming, *An Overview of Magnesium-based Alloys for Aerospace and Automotive Applications*, Project Report of M.E. in Mechanical Engineering, Rensselaer Polytechnic Institute, Hartford, CT, August 2012. http://www.ewp.rpi.edu/hartford/~ernesto/SPR/Fleming-FinalReport.pdf.

12. M.M. Avedesian, H. Baker, *Magnesium & Magnesium Alloys*, ASM International, 1999.

13. H.E. Friedrich and B.L. Mordike, *Magnesium Technology—Metallurgy, Design Data and Application*, Springer-Verlag, Berlin-Heidelberg, 2006.

14. L.L. Rokhlin, *Advanced Magnesium Alloys with Rare-Earth Metal Additions, Advanced Light Alloys and Composites*, 59, 1998, 443–448, doi: 10.1007/978-94-015-9068-6_58.

15. T.A. Leil, Development of New Magnesium Alloys for High Temperature Applications, Doctoral Thesis, Clausthal University of Technology, 2009. file:///C:/Users/SREE%20E/Desktop/db109602.pdf.

16. B. Gwynne, P. Lyon, *Magnesium Alloys in Aerospace Applications, Past Concerns, Current Solutions*, Triennial International Aircraft Fire & Cabin Safety Research Conference, October 29–November 1, 2007. http://www.fire.tc.faa.gov/2007conference/files/Materials_Fire_Safety/WedAM/GwynneMagnesium/GwynneMagnesiumPres.pdf.

17. A. Luo, M.O. Pekguleryuz, Cast magnesium alloys for elevated temperature applications, *Journal of Materials Science*, 29(20), 5259–5271, 1994.

18. F. Hollrigl-Rosta, Magnesium in Volkswagen, *Light Metal Age*, 22–29, 1980.

19. C. Bettles, M. Barnett, *Advances in Wrought Magnesium Alloys—Fundamentals of Processing, Properties and Applications*, Wood Head Publishing, 2012.

20. D.S. Kumar, C.T. Sasanka, K. Ravindra, K.N.S. Suman, Magnesium and its alloys in automotive applications—A review, *American Journal of Materials Science and Technology*, 4(1), 12–30, 2015. http://dx.doi.org/10.7726/ajmst.2015.1002.

21. R. Anish, G.R. Singh, M. Sivapragash, *Techniques for Metal Processing—A Survey*, International Conference on Modeling, Optimization and Computing (ICMOC2012), Proceedia Engineering, 38, 3846–3854, 2012.

22. A.A. Luo, Magnesium casting technology for structural applications, *Journal of Magnesium and Alloys*, 1(1), 2–22, 2013. doi:10.1016/j.jma.2013.02.002.

23. H. Hu, Squeeze casting of magnesium alloys and their composites, *Journal of Materials Science*, 33(6), 1579–1589, 1998, 10.1023/A:1017567821209.

24. A. Moll, M. Mekkaoui, S. Schumann, H. Friedrich, Application of Mg sheets in car body structures, in *Magnesium—Proceedings of the 6th International Conference on Magnesium Alloys and Their Applications*, ed. K.U. Kainer, Wiley-VCH, Weinheim, Germany, 2003, pp. 935–942.

25. S. Braunig, M. During, H. Hartmann, B. Viehweger, Mg sheets for industrial applications, in *Magnesium—Proceedings of the 6th International Conference on Magnesium Alloys and Their Applications*, ed. K.U. Kainer, Wiley-VCH, Weinheim, Germany, 2003, pp. 955–961.

26. A. Stern, A. Munitz, G. Kohn, Application of welding technologies for joining of Mg alloys: Microstructure and mechanical properties, in *Magnesium Technology*, ed. H.I. Kaplan, TMS, Warrendale, PA, 2003, pp. 163–168.

27. T. Kupec, I. Hlavačová, M. Turňa, Metallurgical joining of magnesium alloys by the FSW process, *Acta Polytechnica*, 52(4), 2012.
28. M.K. Kulekci, Magnesium and its alloys applications in automotive industry, *Int J Adv Manuf Technol*, 39, 851–865, 2008. doi 10.1007/s00170-007-1279-2.
29. A.A. Luo, Magnesium: Current and potential automotive applications, *JOM*, 42–48, 2002.
30. L.A. Dobrzański, Structure and properties of magnesium cast alloys, *Journal of Materials Processing Technology*, 192–193, 567–574, 2007.
31. E. Ghassemieh, *Materials in Automotive Application, State of the Art and Prospects, New Trends and Developments in Automotive Industry,* ed. M. Chiaberge, 2011. In Tech, doi: 10.5772/13286. Available from: http://www.intechopen.com/books /new-trends-and-developments-in-automotive-industry/materials-in-automotive-appli cation-state-of-the-art-and-prospects.
32. A.H. Musfirah, A.G. Jaharah, Magnesium and aluminum alloys in automotive industry, *Journal of Applied Sciences Research*, 8(9), 4865–4875, 2012.
33. H. Watarai, Trend of research and development for magnesium alloys, *Science and Technology Trends*, 84–97, 2006.
34. L. Gaines, R. Cuenca, F. Stodolsky, S. Wu, Potential automotive uses of wrought magnesium alloys, Automotive Technology Development Conference, October 28–November 1, 1996, http://www.transportation.anl.gov/pdfs/TA/101.pdf.
35. Information from http://www.magnesium-elektron.com/markets-applications.asp?ID=7.
36. L. Cizek et al., Mechanical Properties of the model casting Magnesium alloy AZ91, 11th International Scientific Conference—Achievements in Mechanical and Materials Engineering, AMME 2002, pp. 47–50, http://www.journalamme.org/papers_amme 02/1112.pdf.
37. D.S. Kumar, K.N.S. Suman, Selection of magnesium alloy by MADM methods for automobile wheels, *International Journal of Engineering and Manufacturing*, 2, 31–41, 2014. doi: 10.5815/ijem.2014.02.03.
38. Information from http://iweb.tms.org/Communities/FTAttachments/Mg%20Alloys % 20for%20Automotive.pdf.
39. M. James, J.M. Kihiu, G.O. Rading, J.K. Kimotho, Use of magnesium alloys in optimizing the weight of automobile: Current trends and opportunities, *Sustainable Research and Innovation, Proceedings*, 3(49), 2011. http://elearning.jkuat.ac.ke/journals/ojs/index .php/sri/article/view/49.
40. Z.M. Shi, G.L. Song, A. Atrens, Influence of anodising current on the corrosion resistance of anodised AZ91D magnesium alloy, *Corros Sci*, 48(8), 1939–1959, 2006.
41. Information from http://mg.tripod.com/asm_prop.htm.
42. C. Blawert, N. Hort, K.U. Kainer, Automotive applications of magnesium and its alloys, *Trans. Indian Inst Met*, 57(4), 397–408, 2004.
43. J.J. Kim, D.S. Han, Recent development and applications of magnesium alloys in the Hyundai and Kia Motors Corporation, *Materials Transactions*, 49(5), 894–897, 2008.

11 Thermoplastics Foams
An Automotive Perspective

Sai Aditya Pradeep, Srishti Shukla,
Nathaniel Brown, and Srikanth Pilla

CONTENTS

11.1 INTRODUCTION

The automotive industry has witnessed a massive shift in terms of materials used, ranging from being a metallic heavyweight in the 1950s to employing a hybrid sandwich of multiple material systems. This apparent shift can be attributed to achieving improvements in performance, safety, and fuel efficiency, along with responding to the various environmental regulations imposed by different governments. The recent advocacy of Corporate Average Fuel Economy (CAFE) standard of 54.5 MPG by 2025 by the US Environmental Protection Agency (EPA) to reduce greenhouse gas (GHG) emissions [1] has spurred the sector at large toward the use of lightweight materials.

Thermoplastic foams possess enhanced strength-to-weight ratio, impact resistance and acoustic properties, reduced permeability to water vapor/air compared to their solid counterparts, and have increasingly become an indispensable commodity in today's industrialized, mechanized world. Foamed thermoplastics from polyolefins to polycarbonates can be best utilized in the automotive sector to fulfill the demand for lightweight materials, given that 75% of fuel consumption is directly related to vehicular weight [2], and that a 6%–8% increase in fuel economy can be realized for every 10% reduction in vehicular weight [3].

A thermoplastic foam typically consists of two phases (solid and gas), wherein the solid phase is the polymer matrix and the gaseous phase is air trapped in inter-connected or isolated cell-like structures within the matrix. Furthermore, foams can also be classified on the basis of cell size, structure, rigidity, structure of strut, and blowing agents employed as exemplified in Figure 11.1. Typically, during foam processing, gas is either blown into the molten polymer (physical foaming) or chemical compounds that evolve gases under different processing conditions either due to chemical reactions or thermal decomposition (chemical foaming). However, obtaining thermoplastic foams is challenging as it encompasses the effective utilization of knowledge base of various scientific fields, including polymer chemistry, physics, engineering—chemical, and mechanical—and equipment, design and operations.

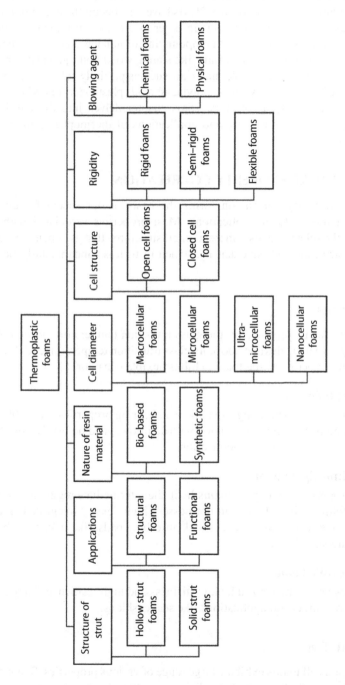

FIGURE 11.1 Classification of thermoplastic foams on the basis of various criteria.

The automotive industry is currently experimenting with and has successfully commercialized some biodegradable thermoplastic foams in an effort to combat large-scale pollution. Few automotive OEMs have also begun the use of bio-sourced polymeric foams in premium segments in order to promote sustainable bioplastics. Examples of such foams include the compounding of nondegradable polymers such as polyolefins with degradable materials like starch, wood flour, jute and hemp, and/ or the use of inherently degradable materials such as partially substituted cellulose, starch, aliphatic/aromatic polyesters, polylactic acid, plasticized polyvinyl alcohol, polyesteramide, and polycaprolactone. This chapter delves into the physical and mechanical properties, processing and applications of thermoplastic foams in the automotive industry.

11.2   STRUCTURE-PROPERTY CORRELATION

The structure property correlation forms the basis of our successful utilization of foams in a multitude of applications. Most properties associated with foams are endemically related to its inherent microstructure, that is, open/closed cells, solid/hollow strut, and cell size/density, which in turn is fundamentally tied to its processing.

11.2.1   STIFFNESS

Stiffness is primarily governed by the glass transition temperature of the polymers [4]. Generally, a foam is flexible above its glass transition temperature, so if the latter is below room temperature, the foam exhibits flexible behavior.

11.2.1.1   Rigid Foams

Rigid foams typically exhibit high specific load bearing properties, over 90% closed cell content, and permanent deformation with definitive yield points. Generally, such foams are used for load bearing applications.

11.2.1.2   Semi-Rigid Foams

In terms of properties, semi-rigid foams lie in the intermediate region between rigid and flexible foams. These foams do not possess high density integrated skins, and in the automotive sector, are used as shock absorbing pads, like in console box lids, door trims, and sun visors.

11.2.1.3   Flexible Foams

Flexible foams are characterized by low stiffness and high resilience. They are typically used in nonstructural applications like seat cushions.

11.2.2   CELL TYPE

Open and closed cell foams exhibit a large range of various properties. Some of these properties, along with their structural differences, have been summarized in subsequent sub-sections. Figure 11.2 shows both open and closed cell foams.

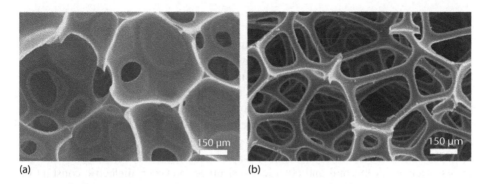

(a) (b)

FIGURE 11.2 Open cell foam (a); closed cell foam (b). (From Y.J. Lee, K.B. Yoon, *Microporous Mesoporous Mater.* 88 (2006) 176–186. doi:10.1016/j.micromeso.2005.08.039.)

11.2.2.1 Closed Cell Foams

In closed cell foams, gas is trapped in the polymer matrix in individual isolated cells. Compared to open cell foams, closed cell foams are denser, less permeable to heat and vapor, exhibit higher R-values and rigidity, and require more material for unit volume, making them more expensive.

11.2.2.2 Open Cell Foams

Open cell structures contain interconnected cells, are permeable to vapor, and insulate sound by capture of sound waves which get repeatedly reflected and absorbed between interconnected cell networks. They are softer, mechanically weaker, less dense, and exhibit lower R-values compared to closed cell foams.

11.2.2.3 Cell Diameter

Generally, cell size varies throughout the foam due to two factors: random nucleation of cells in the polymer matrix, and growth of cells by diffusion of gas. Based on cell diameter (size), foams are classified as macrocellular, microcellular, ultra-microcellular, and nanocellular foams. Cell diameter has a significant influence on multiple foam properties—for example, cell size is inversely proportional to foam resilience. The effect of cell diameter on important foam properties has been described in Table 11.1.

TABLE 11.1
Variation of Foam Properties with Cell Diameter

	Average Cell Diameter	Mechanical Strength	Foam Density
Macrocellular foams	$d > 100 \ \mu m$		
Microcellular foams	$1 \ \mu m < d < 100 \ \mu m$	↓	↓
Ultra-microcellular foams	$0.1 \ \mu m < d < 1 \ \mu m$		
Nanocellular foams	$2.1 \ nm < d < 100 \ nm$		

Source: M.B. Gadi, *Mater. Energy Effic. Therm. Comf. Build.* (2010) 681–708. doi:10.1533/978184569 9277.3.681.

11.2.3 STRUCTURAL FOAMS

Currently, structural foams are of special interest to the automotive sector, especially to the manufacture of trucks. These foams are constituted of a foamed inner core, and a high density skin whose porosity is negligible compared to that of the foamed core [7]. This variation in porosity is due to the fact that the foam skin cools faster than the inner core as it is in immediate contact with the atmosphere. Consequently, more cells get accommodated in the inner core, with fewer cells making it to the already stiffened surface.

Generally, during formation of structural foams, the resin retains most of its properties, such as its thermal and chemical resistance; however, dielectric constant of the foam increases radically when compared to that of the solid resin [8]. Structural foams are suitable for structural applications due to a number of properties:

1. High strength-to-weight ratios: generally 2–5 times that of any metal [4]
2. High rigidity due to higher wall thickness
3. High stiffness imparted by high density skin, along with superior impact absorption properties imparted by the foamed core
4. Significant sound dampening characteristics

Example of structural foams that are of interest to the automotive sector include modified phenylene oxide, polyoxymethylene, polystyrene, polycarbonate, and in some cases polyethylene and polypropylene foams.

11.2.4 STRUCTURE OF STRUT

The strut, or typically the cell wall structure, influences multiple foam properties. Depending on the strut, foams are broadly classified into two categories described in the following sub-sections.

11.2.4.1 Solid Strut Foams

Solid strut foams constitute the bulk of manufactured foams, and their properties have been extensively studied. They possess a cell wall made of homogenously distributed material across the strut thickness.

11.2.4.2 Hollow Strut Foams

In the case of hollow strut foams, as Figure 11.3 shows, the cell wall itself constitutes an outer wall while the inner core of the wall is empty. This foam structure does not naturally result from any conventional foaming technique; however, specific manufacturing methods have been developed to fabricate them. Their macroscopic deformation is largely governed by strut geometry and structure, with many recent studies showing improved mechanical properties (Young's modulus, creep performance, and plastic collapse strength) compared to solid strut foams with the same relative density [9]. This can be explained using the fact that the bending stiffness of any hollow cross-section is higher than that of a solid cross-section prism, thus augmenting the bending stiffness of the strut, and resulting in enhancement of macro-mechanical properties.

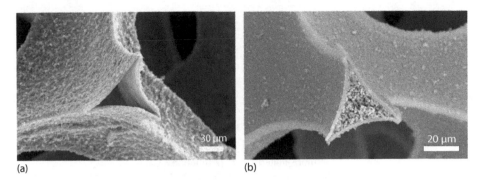

FIGURE 11.3   Hollow strut foam (a); solid strut foam (b). (From Y.J. Lee, K.B. Yoon, *Microporous Mesoporous Mater.* 88 (2006) 176–186. doi:10.1016/j.micromeso.2005.08.039.)

Young's modulus between the two strut structures—hollow and solid—was compared by Gibson and Ashby [10], showing an insignificant difference in favor of hollow strut foams. Steady state creep resistance and plastic collapse strength of both types of foams was compared by Andrews [9], indicating a significantly higher creep resistance and less-significant but higher plastic collapse strength of hollow strut foams compared to solid strut foams; the former difference easily exceeding an order of magnitude [9].

11.2.5   Blowing Agent

The means by which foaming is induced must be examined in order to predict structural changes brought about by such means. Physical and chemical foaming remain the two primary methods for foaming polymers. Due to ease and simplicity in processing, chemical foaming has been in existence for quite some time, while physical foaming has been undergoing development and continuous modification since the 1980s to date [11]. Both foaming processes are typically performed independently, though cases exist of concurrent performance to achieve greater foam expansion. Key aspects of both processes are discussed in Figure 11.4 and subsequent subsections [11].

11.2.5.1   Chemical Foaming

In chemical foaming processes, solid chemical foaming agents, often in the form of a master batch, are mixed into the resin prior to additional processing. Master batches are made of a combination of a chemical foaming agent (up to 70% of total mass) and a carrier polymer [12]. At processing temperatures, the foaming agent decomposes to generate fluids such as N_2, CO_2, and water [11]. These fluids dissolve and get homogenized throughout the polymer melt due to the mixing action of equipment used in the process. Following the decomposition of the solid foaming agents, the residues act as nucleation points for foaming, enabling the formation of a fine cell structure [12]. This decomposition reaction may be endothermic or exothermic—a critical factor to consider in evaluating the

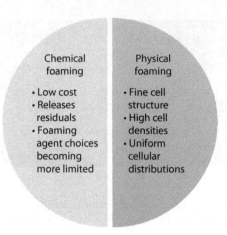

FIGURE 11.4 Key points on chemical and physical foaming. (From J. Xu, *Microcellular Injection Molding*, Wiley, 2010.)

effectiveness of a given foaming agent. Commonly used chemical foaming agents are listed in Table 11.4.

11.2.5.2 Physical Foaming

Unlike chemical foaming, physical foaming does not involve chemical decomposition of the foaming agent. Instead, a fluid (compressed gas, volatile liquid, or supercritical fluid) is injected directly into the polymer melt and homogenized throughout the melt using distributive and dispersive mixing [12]. Upon this dispersion, melt viscosity is lowered significantly to levels lower than those achievable in chemical foaming. Increased temperatures or drop in pressure causes the fluid state to change, evolving a gas that results in formation of cells. A critical aspect to physical foaming is the selection of a foaming agent, as the latter has a significant effect on cellular structure obtained due to differences in solubility of different foaming agents in different polymer matrices. Historically, chlorofluorocarbons (CFCs) have been used as physical foaming agents, but are being phased out in favor of N_2 and CO_2 to avoid their associated environmental impacts and satisfy the norms set out in the Montreal Protocol [11,14]. Lower density parts are generally created using physical foaming instead of chemical foaming primarily due to the significant quantity of residuals released on the use of large quantities of chemical foaming agent [11]. Commonly used physical foaming agents are listed in Table 11.2 [11].

11.2.6 PROPERTIES OF FOAMS

11.2.6.1 Thermal Conductivity

Thermal conductivity (λ_t) of a foam can be defined as

$$\lambda_t = \lambda_g + \lambda_s + \lambda_c + \lambda_r \tag{11.1}$$

TABLE 11.2
Frequently Used Physical and Chemical Foaming Agents

Common Chemical Foaming Agents	Common Physical Foaming Agents
Azodicarbonamide	Isobutane
Modified ADC	Cyclopentane
4,4'-Oxybis (benzene-sulfonylhydrazide)	Isopentane
5-Phenyltetrazole	CFC-11
p-Toluenesulfonyl-semicarbazide	HCFC-22
p-Toluenesulfonyl-hydrazide	HCFC-142b
Sodium carbonate	Nitrogen
Citric acid	Carbon dioxide

Here λ_g represents thermal conductivity through the gaseous phase, λ_s refers to conductivity through the solid phase, λ_c denotes the thermal convection term, and λ_r refers to thermal radiation heat transfer.

In general, if pore size falls below 4 mm in closed cell pores [15] and below 2 mm in open cell pores [16], the convection effect can be neglected due to the small cell size. However, for cell diameters below 0.1 μm, it is assumed that the radiation wavelength is smaller than pore size. Based on this concept, it has been established that the radiation effect can be neglected when relative density is above 0.2 [17]. However, this model cannot be extended to nanocellular foams where cell size and radiation wavelength are comparable. Recent studies on nanoporous materials have shown though that radiation effect can be neglected even at lower densities [18]. Thus, a shift in scale from microporous polymer foams to nanoporous polymer foams leads to observable significant variation only in gaseous and solid phase conductivities.

With regard to conduction, thermal conductivity of gaseous phase is expected to reduce, as can be explained using the Knudsen effect [19,20]—when cell size is smaller or comparable to mean free path of gas particles, gas particles collide more frequently with the surrounding solid walls compared to other gas molecules, leading to lower conductivity through the gaseous phase. Equation 11.2 is known as the Knudsen equation and is used to describe the effective thermal conductivity (λ') of a gaseous phase comprising air in porous media as:

$$\lambda' = \frac{\lambda'_{g,o}}{\left(1 + \beta\left(\frac{l_g}{\phi}\right)\right)} \tag{11.2}$$

Here, $\lambda'_{g,o}$ is thermal conductivity of free air (0.0260 W/m-K at room temperature and 1 atm pressure), β accounts for energy transfer between gas molecules and the surrounding solid (2 for air), l_g is the mean free path of air molecules (70 nm at room temperature), and ϕ is the average pore diameter. It is obvious from the nature of the Knudsen equation (Equation 11.2) that if cell diameter falls below 0.1 m, a significant reduction can be obtained in the thermal conductivity of a gaseous phase as well as of the foam.

11.2.6.2   Mechanical Properties

While polymeric foams have gained attention in the automotive industry as light-weight and less energy sensitive materials, the industry has struggled to compensate for their poor mechanical properties. Specific properties of common automotive foams are illustrated in Figure 11.5. Dependence of physical properties of foams on their relative density is well described by the Gibson and Ashby equation [21] as shown in Equation 11.3. In Equation 11.3, P_f refers to physical property of the foam, P_s is the physical property of resin/solid phase, ρ_f refers to foam density, ρ_s represents the density of parent resin, n describes another constant with a value between 1 and 2, and C describes a constant with a value of 1 for most polymer foams. It has been observed that mechanical properties of nanocellular and microcellular foams are far better than those of conventional foams, with the former having the best properties.

$$P_f = CP_s \left(\frac{\rho_f}{\rho_s} \right)^n \tag{11.3}$$

11.2.6.3   Response to Compression

Figure 11.6 shows the stress–strain curve to explain the behavior of foams on being subjected to uniaxial compression. The curve can be broadly divided into three sections; changes in cell and foam structure corresponding to these three sections have been described below.

- Region 1 illustrates the linear Hooke's behavior, with the reciprocal of slope of the curve used to calculate Young's modulus. This region is characterized by bending of struts in open cell foams, and by simultaneous stretching and bending of adjacent elements of cell walls along with slight compression of cells in case of closed cell foams. This region represents the elastic region, and any strain observed is reversible.
- Region 2 shows dramatic increase in strain for slight increment in stress. This is credited to permanent strut buckling in case of open cell foams, and to rupture of cell walls in closed cell foams due to which deformation is rendered plastic. This is true for all thermoplastic elasto-plastic foams. However, an exception to this rule is elastomeric foams, where all deformations tend to be elastic even in the plateau region.
- Region 3 is marked by the complete rupture of cell walls resulting in the opposite sides of cell walls coming into contact. At this stage, any further application of stress leads to compression of the solid. Thus, application of a large amount of stress does not lead to substantial increase in strain, but does contribute to growth in the density of solid. Hence, this region is termed the "densification region."

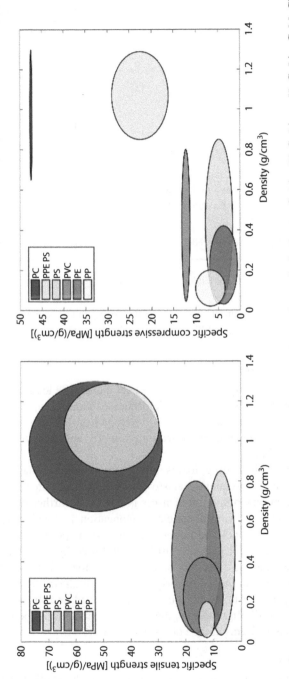

FIGURE 11.5 Specific mechanical properties of various thermoplastic foams. (From C.-C. Kuo, L.-C. Liu, W.-C. Liang, H.-C. Liu, C.-M. Chen, *Compos. Part B Eng.* 79 (2015) 1–5; D. Rosato, *Designing with Plastics and Composites: A Handbook*, Springer Science & Business Media, 2013; K.A. Arora, A.J. Lesser, T.J. McCarthy, Compressive behavior of microcellular polystyrene foams processed in supercritical carbon dioxide, *Polym. Eng. Sci.* 38 (1998) 2055–2062; S. Doroudiani, M.T. Kortschot, *J. Appl. Polym. Sci.* 90 (2003) 1421–1426; M.L. Berins, ed., *Plastics Engineering Handbook of the Society of the Plastics Industry, Inc.*, 5th ed., Kluwer Academic, Boston, 1991; S. Doroudiani, M.T. Kortschot, *J. Appl. Polym. Sci.* 90 (2003) 1427–1434; W. Gong, J. Gao, M. Jiang, L. He, J. Yu, J. Zhu, *J. Appl. Polym. Sci.* 122 (2011) 2907–2914.)

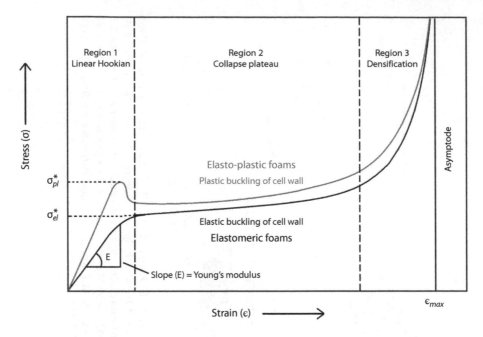

FIGURE 11.6 Stress–strain relationship curve, deformation of foams under compression. (From S. Wijnands, Volumetric behavior of polymer foams during compression and tension, (2010) 19.)

11.2.6.4 Response to Tension

Foams largely exhibit different responses to compressive and tensile loading (Figure 11.6); for example, a foam exhibiting plastic deformation under compression can undergo brittle fracture under tension as cracks developed during tensile loading lead to stress concentration at a single point, leading to more rapid fracture in comparison to compressive loading.

On being subjected to tensile loading, foams initially undergo linear elastic deformation obeying Hooke's law. Subsequently, elasto-plastic foams escalate to the plastic yield point, where temporary reduction in stress is observed; further increase in load leads to increase in stress. During linear elastic elongation, cell walls stretch in the opposite direction to that of buckling in compression. At the plastic yield point, cell walls rupture and plastic deformation is initiated, followed by a region of cell wall alignment along the loading direction, at which point foam density tends to increase until it becomes equal to that of the original resin at high strain values.

In elastomeric foams, the plastic yield point is not observed and the foam progresses directly to the cell wall alignment region. The difference in the behavior of elastomeric and elasto-plastic foams can be observed in Figure 11.7.

11.2.6.5 Melt Strength

Melt strength is defined as the maximum tension force that can be applied to a polymer resin in its molten state, and is essentially a measure of extensional or elongation viscosity of a polymer. Thermoplastic polymer resins constitute long polymer

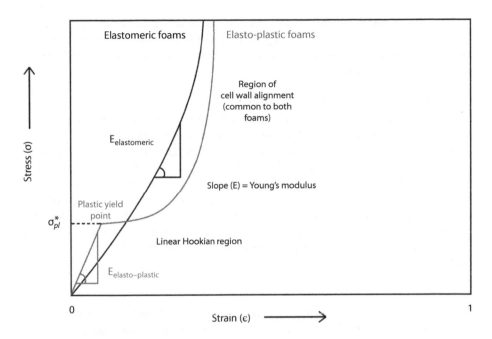

FIGURE 11.7 Stress–strain relationship curve, deformation of foams under tension. (From S. Wijnands, Volumetric behavior of polymer foams during compression and tension, (2010) 19.)

chains entangled with each other even in the molten state. On application of strain to the melt, these chains become disentangled and slide. At the molecular scale, melt strength can be understood as the resistance offered by polymer chains to being disentangled and sliding over each other. Important polymeric properties that affect such entanglement or disentanglement of polymeric chains, and thereby its melt strength, are molecular weight, molecular weight distribution, and molecular branching of the polymer. With regard to all three properties, increase in any of these properties leads to augmentation of melt strength. For this reason, linear polymers like polypropylene and high density polyethylene (HDPE), and polymers with short branches like linear low density polyethylene (LLDPE) have low melt strength, while heavily branched polymers like low density polyethylene (LDPE) have high melt strength. This intrinsically affects foaming as polymers that possess low melt strength have cell walls separating the cells do not have enough strength to bear extensional force resulting in rupture and decreased foaming.

11.2.6.6 Fire Retardancy

The prime function of fire retardants is to either impact fire ignition or to reduce the rate of spreading of a flame. Owing to the global problem of fire emergencies, the global market for fire retardants has escalated to a multi-billion-dollar market. Polymers do not burn, but disintegrate into low molecular weight, inflammable, volatile compounds, and combustible gases; these in turn ignite and result in a flame. The flame retardancy of any polymer is generally measured in terms of its material flammability, or the ability of a material to ignite and/or sustain combustion. Flammability is expressed using various measurable parameters and ratings; this text describes two most commonly used criteria to measure flammability in subsequent subsections.

11.2.6.6.1 Limiting Oxygen Index (LOI)

Limiting oxygen index (LOI) can be defined as the minimum concentration of oxygen required to support the combustion of a polymer. To evaluate LOI, a long prismatic sample, supported in a vertical glass column, is ignited at the top and allowed to burn downward. A mixture of air stream, containing O_2 and N_2, is then passed through the column with a gradual reduction in percentage of O_2 in the air stream. The minimum concentration required to just support combustion is recorded and quoted as a percentage. Although LOI is used as a measure of flammability of a material, it cannot explain the behavior of the same material when burnt in open atmosphere. LOI values for various thermoplastic resins have been mentioned in Table 11.1.

11.2.6.6.2 UL94

UL94 standard primarily classifies a material on the basis of its ability to resist initiation and propagation of flame in various widths and orientations. Originally released by Underwriters Laboratories, USA, it has now been assimilated into many international standards for flammability. UL classification system consists of many methods, of which the two most important methods have been listed below as follows:

a. Horizontal Burning Test (94HB)

In this method, a horizontal strip of specimen is ignited by a flame for either 30 seconds or until the flame reaches a certain specified starting mark on the specimen, following which the specimen is allowed to burn freely. If the flame continues to a second mark made on the specimen, the time taken by the flame to reach the second mark is observed. If the flame extinguishes before reaching the second mark, the damaged length and burn time are observed. A material is classified as 94HB either when the burning rate is less than 76 mm/min for a specimen of length less than 3 mm (if burning continues to the second mark), or if the damaged length is less than 100 mm (if flame extinguishes prior to reaching the second mark).

b. Vertical Burning Test (94V)

In this method, a sufficiently long specimen is clamped vertically and burnt at its lower end by a 20 mm long blue flame, such that half of the flame makes contact with the specimen. After 10 seconds of contact with the specimen, the flame is removed and the specimen response is recorded in terms of two aspects: time taken by the flame to extinguish (after-flame time 1), and any dripping of material which ignites cotton balls placed below it. The specimen is again subjected to a similar flame for another 10 seconds, and the time taken by the flame to extinguish is similarly recorded (after-flame time 2) along with any dripping that may ignite the cotton

TABLE 11.3

Criterion for Classification as V-0, V-1, and V-2

After-Flame Time 1	After-Flame Time 2	Ignition of Cotton Balls	Dripping of Material	
Less than 10 seconds	Less than 10 seconds	Not allowed	Dripping of non-flamed material allowed	V-0
Less than 30 seconds	Less than 30 seconds	Not allowed	Dripping of non-flamed material allowed	V-1
Less than 30 seconds	Less than 30 seconds	Allowed	Dripping of flamed material allowed	V-2

Flammability

indicator. It is important to note that in any case, the flame should not reach the top of the specimen. Based on this test, materials can be classified into three categories depending on the response recorded, with the classification criterion summarized in Table 11.3.

Broad differences are observed in properties with changes in various aspects of cell structure, processing techniques, and other parameters.

11.2.6.7 Dielectric Constant (k)

With efforts on curtailing size of circuits and chips in the microelectronics industry, current research focuses on materials with low dielectric constants. With conventional materials like aluminum and SiO_2 (k = 4.5) still waiting to be replaced in the microelectronics industry, and taking into consideration the dielectric constants of polymers (generally less than 3 [30]) and air (close to unity for a broad range of temperatures and pressures), it is expected that polymeric foams will provide the desired low-k materials. Reflecting on the requirement of developing extremely small components in microelectronics production systems, it is generally preferred for the size of the cellular pores to be an order of magnitude smaller than the element thickness used for producing the component [31]. Nanocellular foams provide a possible solution to the problem.

To evaluate the dielectric constant of a two-phase porous system, we make use of the Lich-Terecker mixing rule, which provides us the upper and lower limits of dielectric constant. The general equation used to describe this mathematically is given as Equation 11.4, where k_t^α is the dielectric constant of the system, k_g^α refers to dielectric constant of the gaseous phase, k_s^α represents the dielectric constant of the polymer matrix, α is a dimensionless parameter governing the type of mixing used, and V_g is the volume fraction of inclusions. For $\alpha = -1$, the lower limit of dielectric constant is obtained, whereas at $\alpha = 1$, the upper limit of dielectric constant is achieved. A clear fall is observed in the value of dielectric constants from Equation 11.4 with rise in volume fraction of voids when the former is plotted against the latter.

$$k_t^\alpha = k_s^\alpha(1 - V_g) + k_g^\alpha V_g \qquad (11.4)$$

11.3   PROCESSING ROUTES

As foamed automotive components are required in a variety of shapes and need to possess different properties, there exist different thermoplastic processing methods (shown in Table 11.4) specifically tailored for their production. These processes generally use one among three types of equipment: injection molding machines, extruders, and compressors. Though the specific processes used to produce thermoplastic foams can vary in significant measure, all such processes involve the creation of a structure of cells or voids, reducing both the total quantity of polymer used to produce the component and the overall cost of the component. However, this simple reduction is generally not the primary benefit sought; rather, the prime benefits being sought are improvements in thermal, acoustical, and mechanical properties along with lower densities.

11.3.1   FOAM EXTRUSION

11.3.1.1   Free-Foaming Extrusion

The process of foam extrusion begins with the addition of resin into the extruder, often after being premixed with a chemical foaming agent. As the resin passes through the screw and is plasticized, either the foaming agent begins to thermally degrade and release a fluid (or a combination of fluids), or gas is injected into the plastic in the second half of the plasticizing barrel [32]. To ensure the obtainment of a uniform material, the gas and plastic must be completely homogenized through a series of distributive and shear mixing elements on the screw [33]. After this stage, the material is passed through a die with an orifice smaller than the size of final product. Nucleation begins when pressure within the die drops below the partial pressure of blowing agent, but visible foaming is not observed until the material has moved past the die exit. As the extrudate cools and expands after exiting the die, it is often attuned to its final dimensions through use of a cold die [25,33], or by passing it through a water-cooled vacuum calibrator to ensure consistency among parts [31]. Once gas pressure has been reduced and polymeric viscosity enhanced up to a certain point, foam expansion ceases.

11.3.1.2   Controlled Foam Extrusion

This process is a modified form of free-foaming extrusion process with the same initial stages, but requires a modified die with a core plate [32]. This setup results in the creation of a tube-shaped extrudate, which is then passed into a vacuum calibrator attached directly to the die face. The calibrator shapes the extrudate and creates a solid outer skin on the material surface due to rapid cooling, even as the material core expands normally to fill the void created by the core plate due to its cooling at a lower rate. Post the completion of this step, the material is said to have achieved its final shape [25].

11.3.1.3   Tandem Extrusion

The above-mentioned processes can be optimized for low-density foams (as low as 0.024 g/cm^3) through use of a tandem extrusion line with physical foaming agents.

TABLE 11.4
Thermoplastic Foaming Methods

Pressure Range	Process	Advantages	Disadvantages	Component Types	Ref.
High	High Pressure Injection Foam Molding	High surface definition and structural integrity, uniform cells, solid skin	Significant tooling costs, expandable mold required	Large parts, small parts, solid and foamed regions	[25,34]
High and Low	MuCell (Injection)	No warpage or sink marks, low tonnage, uniform cell distribution	Requires machine tailored for operation	Small, thin-walled parts (3 mm and below)	[33]
	Ergocell	Low tonnage, significant reduction in material usage, process separation	Designed for Ergotech machines	Common parts	[35]
	Co-Injection Mold Foaming	No swirls on surface, high surface definition, can use different materials for skin and core	Polymer compatibility must be considered	Solid and foamed regions	[25,33]
Low	Gas-Assisted Injection Molding	No swirls on surface, warpage reduced, low tonnage, voids up to 50% part volume	Difficult to control gas direction	Thick and thin parts (down to 0.007 in.), parts with internal voids	[25,33]
	MuCell (Extrusion)	Microcellular extrudate, improvement in surface finish	Limited geometry	Slabs, sheets, rods, profiles, roll stock	[14,33]
	Low-Pressure Injection Foam Molding	Low tonnage, minor machine modifications with chemical blowing agents, low cost molds, solid skin formation	Requires machine tailored for operation with physical blowing agents, swirl pattern on surface	Common parts	[25]

(Continued)

TABLE 11.4 (CONTINUED)
Thermoplastic Foaming Methods

Pressure Range	Process	Advantages	Disadvantages	Component Types	Ref.
	Low-Pressure Nitrogen Process	Low tooling costs, solid skin formation	Poor surface quality	Bendable parts	[11,25]
	Free-foaming Extrusion	Density reduction	Limited geometry	Slabs, sheets, rods, profiles, roll stock	[25,33]
	Controlled Foam Extrusion	Density reduction, solid skin formation	Limited geometry	Slabs, sheets, rods, profiles, roll stock	[25]
	Tandem Extrusion	Very low density foams, proper cooling of foam achieved	Additional capital requirements	Slabs, sheets, rods, profiles, roll stock	[32]
	Steam Chest Molding	Fine-cell structure, very low density possible, easy to control part density	Low surface definition, long treatment times	Blocks, common parts	[25,33,36]

In this process, two extruders are used: one to homogenize the foaming agent within the melt, and the other to cool the melt. The first extruder pumps the gas/polymer material directly into the feeding section of the second extruder, where the material then cools to a temperature range where optimal foam properties can be achieved. Without the second extruder, greater probability exists of overblowing and cell rupture due to high partial pressure of the foaming agent and low melt viscosity [32]. Although setup costs are significantly higher, this process is required in cases where high material throughputs (up to 250–500 kg/h) must be achieved [14].

11.3.1.4 Low-Pressure Nitrogen Process

Among the most commonly used extrusion foaming processes is the nitrogen process, which specifically uses N_2 as a physical foaming agent. In this process—a modification of the general foam extrusion process—resin is added to an extruder and mixed with N_2 gas into the extruder barrel. The resultant two-phase material, composed of plastic and gas, is then transferred to an accumulator and held at low pressure (14–21 MPa). Once the material content filled in the accumulator is enough to partially fill the mold, the nozzle opens to release the material into the mold where it is held at a lower pressure (1–2 MPa) than the accumulator, enabling the material to expand and fill the interior of the mold. Use of such lower pressure allows the use of low strength materials like aluminum instead of tool steel, leading to significant lowering of tooling costs [25].

11.3.1.5 MuCell™

Although commonly known as the microcellular injection-molding process, MuCell™ was originally designed as a foam extrusion process. In this process, the plastic is injected with a supercritical fluid as it passes through the mixing region of the screw. The fluid forms diffused bubbles within the melt, which are broken apart by shearing action inside the extruder. This leads to the formation of a saturated, single-phase gas-polymer solution that is more fully homogenized than a two-phase material as it passes through the remainder of the screw. Through careful control of melt temperature and pressure, optimal properties can be maintained for the single-phase solution until the material reaches the die. Significant pressure drop occurs at the die, inducing homogenous nucleation due to thermodynamic changes in the solution, and resulting in the evolution of gas from the supercritical fluid. In addition to homogenous nucleation, heterogeneous nucleation is also observed to occur, though at a comparatively lower rate. Accordingly, high cell density and small cell size can be achieved, resulting in classification of the foam as microcellular foam. Due to the importance of nucleation stage in foaming, die design is critical as pressure drop must occur in the proper fashion to avoid early foaming from taking place [14,33].

11.3.2 INJECTION MOLDING FOAM

A large number of processes exist for producing injection-molded foams when compared to extruded foams, with several such processes having undergone significant development over the past 20 years. In general, it is easier to control foams created using injection molding due to ease in maintaining mold temperature and pressure. Despite this, certain processes for injection foam molding have geometric limitations, requiring careful selection for the part at hand.

11.3.2.1 Low Pressure Injection Foam Molding

In this process, prior to injection molding, a chemical foaming agent is combined with resin in a processor. As the foaming agent–resin mixture is passed through the screw and is homogenized, the foaming agent begins to thermally degrade, releasing gas in the process. This polymer/gas mixture then accumulates in front of the screw to form the shot, which is then injected into a low pressure mold, only filling it partially. Gases released due to thermal degradation of the foaming agent cause polymeric expansion to completely fill the mold due to significant pressure drop following injection of the polymer, causing homogenous nucleation to occur [34]. This process can also be performed by injecting a physical foaming agent into the barrel instead of using a chemical foaming agent. Regardless, due to use of low pressures, often additional equipment (i.e., hydraulic booster circuits) is required to increase injection speed and prevent drooling during plasticization [25].

11.3.2.2 High Pressure Injection Foam Molding

In this process, polymer/gas mixture is injected into the mold at high pressures, leading to complete filling of the mold. As the mixture begins to foam and expand following pressure drop, the mold itself expands or mold inserts are removed, foam expansion continues, resulting in obtainment of a slightly larger-yet-lower density part. Through careful adjustment of melt and mold temperatures, this process can be used to create a solid skinned part with a cellular core, both of which fuse together as the cycle comes to completion [34]. Due to use of high pressure during filling and the packing phase that follows, the solid skin is capable of being relatively free of surface defects [11].

11.3.2.3   Gas-Assisted Injection Molding

Unlike previously discussed processes, the gas introduced in this process does not mix with the plastic at all. Instead, a polymer is first injected into a mold in a typical fashion, and a gas (i.e., N_2) is then injected into the polymeric melt while it is in mold, forming a continuous void within the polymer melt instead of numerous cells. This gas can be injected from gas pins or through a nozzle on the injector. Through addition of multiple gas channels to mold design, gas pressure can be spread evenly, enabling the creation of voids encompassing up to 50% of the volume of the part. These voids enable the creation of structural support within the part, thereby leading to significant increase in the stiffness-to-weight ratio of the part [25,33].

11.3.2.4   Co-Injection Mold Foaming

In this process, a part with two different foamed polymers can be formed through sequential injection of a skin and core using low- or high-pressure injection foam molding. This is done by keeping the mold hot and having a short cycle time. The first shot is only partially injected with partial filling of the mold, but the second shot forces the material from the first shot to edges of the mold. Through adjusting the quantity of foaming agent, the degree to which different shots fill the final part can be manipulated. Accordingly, unique parts containing cores and skins that are either foamed and/or solid can be created [33].

11.3.3   Microcellular Foaming

11.3.3.1   MuCell™

The most prominent microcellular foaming process for injection molding, MuCell™ enables the creation of thin-walled, low density parts without warpage or sink marks. As the resin travels through the screw and plasticizes, gas in its supercritical fluid state is injected into the polymeric melt in two locations as the screw moves backward during each cycle. Supercritical fluid and polymer, initially a two-phase solution, get homogenized into a single-phase solution in the mixing section by a specially designed screw. The accumulated shot at the front of the screw, containing the single-phase solution, is then injected into a low pressure mold at which point homogenous nucleation results in the creation of cells throughout the core of the material. These cells are generally distributed evenly and are similar in size, a property typically difficult to achieve. Due to fluid movement in the mold, polymer bubbles get destroyed by shear action at the mold surface, resulting in obtainment of a structural foam and, in turn, makes for a poor surface finish. However, when counter gas pressure is combined with MuCell™, significant reduction is observed in surface roughness [12]. Another unique aspect of this process is the elimination of pack and hold phases of injection molding, enabling reduction in cycle time and tonnage [33].

11.3.3.2 Ergocell™

Ergocell™ process is quite different from other injection molding foaming processes. In this process, the polymer leaves the injector gate to enter the injection module. Unlike other processes, polymer has already been plasticized prior to the addition of foaming agent. Inside the injection module, gas (generally CO_2) is injected via a nozzle and a mixing element is added to homogenize the gas/polymer mixture. A plunger then injects the mixed material into a mold where it expands into a microcellular form on drop in pressure [35]. This process occurs concurrently with normal injection molding process as polymer is delivered into the injection module forward of the screw at the same rate as the rate at which material is moved forward by the screw [12].

11.3.4 BEAD FOAMING

11.3.4.1 Bead Production

In order to create beads needed for steam chest molding, polymer is pelletized and impregnated with a hydrocarbon gas (such as pentane) at elevated temperature and pressure conditions within a slurry. These pellets are then expanded into foamed spheres with diameters ranging from 4–5 mm [14] to make them suitable for manufacture of bead foams.

11.3.4.2 Steam Chest Molding

Steam chest molding is used in producing foams from aforementioned virgin beads impregnated with a physical foaming agent. These beads are blown into a metal mold sealed and evacuated of air. Compressed steam is then injected into the mold with venting to heat and soften the beads, resulting in their further expansion. This is followed by injection of pressurized steam into the mold, resulting in fusion of beads. On cooling of the mold, the part takes its final rigid form. However, it often takes more than 24 hours to fully treat most parts, making this a poor method for producing certain components [14]. A unique advantage associated with this process is the occurrence of controlled expansion during steam injection, resulting in fine cell structure of the final product [25,33,36].

11.4 AUTOMOTIVE APPLICATION OF THERMOPLASTIC FOAMS

The future of the automotive sector is irrevocably tied to the design and manufacture of thermoplastic foams in multi-material or sandwich structures due to the latter's vital link with light weighting and fuel efficiency. Current automobiles house several components designed and manufactured using these foams, as shown in Figure 11.8.

Table 11.5 consists of foamed automotive components along with design and mechanical requirements of these foamed components.

Table 11.6 contains a summary of various foams used in the automotive sector and their properties with respect to flammability, melt flow, heat deflection, and mechanical properties. Table 11.7 is a summary of various thermoplastic foams used for current automotive components.

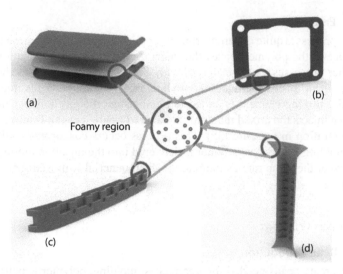

FIGURE 11.8 Foamed components from left to right: (a) sun visor, (b) gasket, (c) bumper beam, (d) reinforced pillar trim.

TABLE 11.5
Functional Need of Thermoplastic Foams for Automotive Applications

Component	Function and Need of Thermoplastic Foams
Bumper transverse beam Bumper core	Bumper beams are meant to transfer impact energy to crash boxes (placed to the left and right of the bumper beam) and to longitudinal chassis elements uniformly during frontal and frontal offset impacts, thereby increasing crashworthiness. During such impacts, a very small region undergoes bending, while the rest of the beam is subjected to rigid body rotation. Due to localized bending, resistance force and energy absorption abilities undergo a fall. Foam-filled bumper beams are structurally rigid and more capable of uniform load distribution [37].
Gaskets	Gaskets are wedged in between flanges and similar joints in order to avoid any leakage. A major application of gaskets in automobiles is to perform as mechanical seals between the piston and cylinder block, where it is known as head gasket. The material chosen for such gaskets should readily deform on application of a small amount of force in order to fill the defects of adjoining surfaces. It should also be impermeable to fluids [38].

(Continued)

TABLE 11.5 (CONTINUED)
Functional Need of Thermoplastic Foams for Automotive Applications

Component	Function and Need of Thermoplastic Foams
Instrument panel padding Instrument panel support	Materials chosen for instrument panels should not exhibit brittle nature even at low temperatures, while also possessing high energy absorption capacity given their direct interface with occupants and the possibility to come in contact with the latter in severe cases of airbag failure. The dashboard also contains several intricate structures, and so the material chosen for the same should be free flowing and fill the mold intricacies, while being easy-to-machine and possessing high heat resistance. It is also expected that the material does not show large variation in properties and has dimensional accuracy with differences in temperatures [39].
Head-rest cores	Head rest should be a relatively stiff structure with an ideal thickness of about 3 cm. The design should also be shaped intricately to accommodate the skull base and cervical spine, and should offer sufficient shock absorption to protect the occupant in the event of a rear collision. Head rest should also be closely interfaced with the head to avoid large skull movement in case of a jerk [39].
Door lock mechanisms	During front and side impacts, the door is expected to remain locked, but is also expected to open after the impact without use of any tools. Noise produced due to locking of doors is also a critical parameter in determining door quality, and in order to reduce such noise, thermoplastic foams are used as a suitable alternative to metal doors [39].
Seat cushion Seat cushion foam	The main aim of seat cushion foam and spring is to isolate the occupant's body from the vibrations transmitted to the cockpit through the vehicle's transmission, suspension, and propulsion systems. While a simple spring mass system is suitable only for absorption of high frequency resonant vibrations, foam behavior can be compared to the combination of series and parallel spring mass systems, providing for ideal damping conditions at both high and low frequencies [39].
Interior panels	Interior panel is required for pillar plates, relative joints, drilled and puckered components, and to address aesthetic deficiencies in plate pressing. Materials employed in interior panels must possess high energy absorption capacity to keep the occupants safe in the event of a collision, along with high thermal and acoustical insulation properties for a comfortable drive experience [39].
Exterior mirror housing core	Exterior mirror housing core should possess sufficient impact absorption properties to prevent any damage to the mirror housing from door slamming, road and bench vibrations. It should be heat resistant in order to account for any large dimensional change that could occur due to variation in temperature. It should also be impermeable to water and qualify the accelerated reliability test in the climatic chamber [39].

(Continued)

TABLE 11.5 (CONTINUED)

Functional Need of Thermoplastic Foams for Automotive Applications

Component	Function and Need of Thermoplastic Foams
HVAC flapper doors HVAC housing	Evaporator and radiator mass housing allows easy discharge of water through appropriate drain systems. The housing must be protected by a grid of appropriate material to avoid the clogging of drains by leaves and dirt. Stagnation must also be avoided in the design in order to avoid any putrefaction and accumulation of foul smell in the cockpit [39].
Car boot liner	Materials for car boot liners should be impermeable to water, resistant to oils and chemicals, durable and flexible so that they can be removed for washing as and when it is needed [39].
Knee bolsters	For knee bolsters, materials forming structural components should elastically deform for large limits and retain extremely high impact absorption properties in order to protect the occupant's knees from injury during crash [39].
Pillar trim and head panels	Material used for pillar trim should be chemically compatible with support materials in order to warrant adhesion between the two material sets. Further, there should not be any defects on the material surface for aesthetic reasons [39].
	Head panel or roof panel are exposed to snow load which exerts a pressure of about 1000–1500 N/m^2. The material is required to be mechanically stiff so that it does not undergo plastic deformation under aforementioned load. Mechanical stiffness should also be high in order to protect the occupant's neck and head in the event of a rollover [40,41].
Automotive body panel	Automotive body panel is inserted into housings, and so materials used for this component are required to be flexible and easy to elastically deform. The material must be resilient even at low temperature, resistant to scratches, UV rays, chemicals and corrosion for long-staying aesthetic finish. It should also not have metal sharp edges [39].
Sun visors	It is pertinent that materials used for sun visors be mechanically rigid to ensure they don't bend, crack, or crease along the midline due to repetitive opening and closing. Since a sun visor is not a structural component, thermoplastic foams can be used for weight reduction.
Door energy absorbing foam Door watershields Door interior trim covering	Energy absorbing foam, watershields, and interior trim covering, which make up the door panel, have a range of uses with respect to safety of occupants. It not only houses several features like armrest, side airbags, drawer, rear reflectors, and courtesy lights, but also is critical in avoiding injury to occupants in the event of side impacts. Armrest and material covering the interiors of the door panel should provide for aesthetic appearance and should possess high impact absorption properties, making foams ideal contenders for this application. Armrest should not be intrusive to thorax and abdomen of the occupant. Door panels also play a major role in absorbing noise and vibration due to external sources that may otherwise reach the cockpit.

TABLE 11.6
Properties of Thermoplastic Foams Used by Automotive OEMs

Thermoplastic Foam	Flammability (UL94)	Melt Flow Index (g/10 min)	Heat Deflect on Temperature (°C at 0.45 MPa)	Young's Modulus (MPa)	Compressive Strength (MPa)	Tensile Strength at Break (MPa)
Polycarbonate	V-0/5V (UL94) [4]	20 (300°C/1–2 kg load)	138 [4]	2068 [4]	37.9–62 [23,42]	22.7–50 [23,42]
Polystyrene	17.6%—18.3% (LOI, 23°C) [43]	9 [44]	85 [45]	1800 [46]	33.6 [46]	12.8 [46]
Expanded polystyrene					0.09–0.965 (10% E) [23,47]	0.2–1.3 [23,47]
Expanded poly-vinyl chloride	V-2 (UL94) 26% (LOI)	25 (Sibur PL-1GS polyvinyl chloride)	92 (at 1.88 MPa) [48]	993 (Thickness 10–25 mm)	0.5–5.8 [42]	0.9–20 [42]
Acrylonitrile butadiene styrene	V-0 (UL94) [4]	1–36 (220°C/10 kg load)	86 [4]	17,237 [4]	30.33 (10% D) [4]	27 [4]
Polyphenylene ether	32% (LOI) [49]	16 (280°C, 5 kg load) (NORYL PVX516 PPE)	191 (at 1.86 MPa) [50]	–	–	–
Polyphenylene ether-polystyrene blend	24% (LOI) [51]	9.2 (280°C, 5 kg load)	130 (at 1.8 MPa) [52]		20.7–27.6 [42]	28–29 [42]
High density polyethylene	HB (UL94) [53]	0.01–500 (Melt flow ratio 25–150) [51]	79–91 [54]		0.296–0.398 (Strain 10%–50%)	1.72
Polypropylene	18.6% (LOI)	0.25–0.70 [51]	96–120 [55]	545 [4]	19.3 [4]	13 [4]
Expanded polypropylene	HB (UL94)			10–28 (PS Bead foam)	0.1–2 [42]	2.1–9.1 (2%–8% E)
Polypropylene + glass fibers		6–10 (300°C, 1.2 kg load) EPICHEM Epilen PP 20 GF	149–166 (40% g ass) 143–146 (20%–30% glass) [45]	3000–5500 [56]	20 [57]	–
Polypropylene + talc	HB (UL94, 40% Talc) [4]	4–10 (10%–50% talc) [58]	102–121 (40% talc) [45]	–	–	–
Acetal or polyoxymethylene	HB (UL94)	8–13 [59]	170	692.23–1012.44 [59]	–	–
Nylon 6 foam	V2 (<1.6 mm, UL94, cast sheet) [60]	130 (BASF Ultramid® B27 E01 PA6 (Dry))	>160 [60]		0.280 (50% S) [61]	1.30 (70% E) [61]

TABLE 11.7

Use of Different Foams in Various Automotive Applications

	PP	PE	PS	PPE	EPP	EPS	PPE/PS	PP+talc	EPVC	PP GF	PA 6	PA 66	POM	ESI	PC	Ref.
Bumper beam	•															[36,62]
Sun visors				•	•											[36]
Door energy-absorbing foam			•	•	•		•									[36,63]
Head-rest cores					•											[36]
Instrument panel padding	•	•														[1]
Car trunk liner					•											[36]
Knee bolsters		•			•											[36]
Door watershields	•															[36]
Exterior mirror housing core	•				•			•	•							[36]
Interior door trim coverings					•			•	•							[36,64]
Spare wheel holder					•											[36]
Interior paneling																[65]
Fan shrouds										•	•					[64]
HVAC housing	•							•	•		•					[64]
Door lock mechanisms													•			[66]
Seat cushions	•															[66]
Instrument panel support	•															[67]
Pillar trim and head panels						•										[8]
Acoustical insulation	•	•														[68]
Gaskets														•		[69]
Automotive body panels									•						•	[70]
Roof liners	•					•										[36]

11.5 OUTLOOK

Current environmental concerns and governmental regulations are expected to transform the nature of plastics used in the automotive industry, particularly with respect to plastics sourced from practically nonrenewable resources like crude oil. An imminent transformation is on the horizon with regard to the use of sustainably sourced or biosourced plastics that are likely to be seen in three phases. The first phase is likely to focus on attempts toward developing bio-based precursors for synthesis of conventional polymers such as bio-nylon, bio-polyethylene, bio-epoxy, and bio-polyurethane. While these foams are structurally and chemically similar to petroleum-based plastics, they would also help in conserving crude oil for other significant applications while promoting sustainability.

The second phase is synthesis of new bio-based plastics such as polyhydroxy alkanoates and polylactic acid. These plastics are chemically and structurally different from conventional plastics, but are similar to the latter in terms of properties, making them potential alternatives. However, both these phases are likely to generate concerns about the use of food-based precursors for synthesis of plastics, as this will divert crucial food resources meant for human and animal consumption.

This would in due course of time lead to the third phase of transformation, wherein all precursors are derived from nonfood-based resources. A significant step in this direction will be the development of bio-based thermoplastics from CO_2 (e.g., polyalkyl carbonates). Overall, these transformations will not only help us conserve our environment via lowered carbon footprint, but also provide an expedited closed-loop system for producing plastics. Table 11.8 details the melt flow index, heat deflection temperature, compressive strength, and tensile strength at break of bio-based thermoplastic foams that are being investigated.

TABLE 11.8
Properties of Commonly Used Thermoplastic Foams

Thermoplastic Foam	Melt Flow Index (g/10 min)	Heat Deflection Temperature (°C at 0.45 MPa)	Compressive Strength (MPa)	Tensile Strength at Break (MPa)
PLA	1–32 [71,72]	55 [73]	0.06 [74]	33.3 [75]
PHBV	2.4–13.9 [71,76]	107 [77]	–	29.8 [75]
PBS	1.5–30.1 [71,78]	70–90 [75]	–	26 [79]
PPC	5.6–8 [73]	73–110 [42]	0.050–0.4 [80,81]	–
PLA/PHBV	–	60 [77]	–	30.4 [75]

11.6 SUMMARY

Thermoplastic foams have gained a major foothold in automobile manufacturing. Their ability to make lightweight cars, combined with a range of advantageous properties ranging from better strength-to-weight ratios, have made them a necessity in today's automotive sector. This chapter has presented an overview of thermoplastic foams, their properties, processing, and applications in various components housed in today's automobiles. The reasons and advantages associated with the use of foamed materials used in different parts of a car have also been discussed. Finally, this chapter presents a brief outlook on the future of thermoplastic foams vis-à-vis the automobile industry, opening new directions with the onset of responsibly sourced bio-based foams being viewed as a viable alternative. Hopefully, such enhanced usage can help improve fuel economy, reduce carbon emissions, and help contribute to humankind's fight against global warming and climate change.

REFERENCES

1. US EPA, Regulations and Standards: Light-Duty | Transportation and Climate, (n.d.). http://www.epa.gov/oms/climate/regs-light-duty.htm (accessed June 19, 2015).
2. B.B. (B/T Books), *Advanced Automotive Technology: Visions of a Super-Efficient Family Car*, Business/Technology Books (B/T Books), 1997. https://books.google.com /books?id=uYu-AAAACAAJ.
3. C.-K. Park, C.-D.S. Kan, W.T. Hollowell, Investigation of Opportunities for Light-Weighting a Body-on-frame Vehicle Using Advanced Plastics and Composites, in: n.d.
4. A.H. Landrock, *Handbook of Plastic Foams*, Landrock.pdf, Publication, Noyes, 1995.
5. Y.J. Lee, K.B. Yoon, Effect of composition of polyurethane foam template on the morphology of silicalite foam, *Microporous Mesoporous Mater.* 88 (2006) 176–186. doi:10.1016/j.micromeso.2005.08.039.
6. M.B. Gadi, Materials for energy efficiency and thermal comfort in buildings, *Mater. Energy Effic. Therm. Comf. Build.* (2010) 681–708. doi:10.1533/9781845699277.3.681.
7. M.J. Howard, *Foams*, edition 2, Desk-Top Data Bank, 1980.
8. S.S. Schwartz, S.H. Goodman, *Plastic Materials and Processes*, Van Nostrand Reinhold, New York, 1982.
9. E.W. Andrews, Open-cell foams with hollow struts: Mechanical property enhancements, *Mater. Lett.* 60 (2006) 618–620. doi:10.1016/j.matlet.2005.09.047.
10. L.J. Gibson, M.F. Ashby, *Cellular Solids: Structure and Properties*, Cambridge University Press, Cambridge, 1997.
11. W. Michaeli, A. Cramer, L. Flórez, Processes and process analysis of foam injection molding with physical blowing agents, in: *Polym. Foam.*, CRC Press, 2008: pp. 101–142.
12. A.K. Bledzki, O. Faruk, H. Kirschling, J. Kühn, A. Jaszkiewicz, Microcellular polymers and composites., *Polimery.* 51 (2006).
13. J. Xu, *Microcellular Injection Molding*, Wiley, 2010.
14. D. Eaves, T.L. Rapra, *Polymer Foams: Trends in Use and Technology*, Rapra Technology Ltd, Shawbury, Shrewsbury, Shropshire, UK, 2001.
15. J. Holman, *Heat Transfer*, McGraw-Hill, New York, 1981.
16. M. Alvarez-Lainez, M.A. Rodriguez-Perez, J.A. De Saja, Thermal conductivity of open-cell polyolefin foams, *J. Polym. Sci. Part B Polym. Phys.* 46 (2008) 212–221. doi:10.1002/polb.21358.

17. E. Solorzano, M.A. Rodriguez-Perez, J. Lázaro, J. de Saja, Influence of solid phase conductivity and cellular structure on the heat transfer mechanisms of cellular materials: Diverse case studies, *Adv Eng Mater.* (2009).
18. P. Ferkl, R. Pokorny´, M. Bobák, J. Kosek, Heat transfer in one-dimensional micro- and nano-cellular foams., *Chem Eng Sci.* (2013).
19. D. Schmidt, V. Raman, C. Egger, C. Du Fresne, V. Schädler, Templated cross-linking reactions for designing nanoporous materials., *Mater Sci Eng C.* (2007).
20. X. Lu, R. Caps, J. Fricke, C. Alviso, R. Pekala, Correlation between structure and thermal conductivity of organic aerogels., *J Non-Cryst Solids.* (1995).
21. L.J. Gibson, M.F. Ashby, *Cellular Solids: Structure and Properties,* Cambridge University Press, Cambridge, England, 1990. doi:10.2277/0521499119.
22. C.-C. Kuo, L.-C. Liu, W.-C. Liang, H.-C. Liu, C.-M. Chen, Preparation of polypropylene (PP) composite foams with high impact strengths by supercritical carbon dioxide and their feasible evaluation for electronic packages, *Compos. Part B Eng.* 79 (2015) 1–5.
23. D. Rosato, *Designing with Plastics and Composites: A Handbook,* Springer Science & Business Media, 2013.
24. S. Doroudiani, M.T. Kortschot, Polystyrene foams. II. Structure—Impact properties relationships, *J. Appl. Polym. Sci.* 90 (2003) 1421–1426.
25. M.L. Berins, ed., *Plastics Engineering Handbook of the Society of the Plastics Industry, Inc.,* 5th ed., Kluwer Academic, Boston, 1991.
26. S. Doroudiani, M.T. Kortschot, Polystyrene foams. III. Structure—Tensile properties relationships, *J. Appl. Polym. Sci.* 90 (2003) 1427–1434.
27. W. Gong, J. Gao, M. Jiang, L. He, J. Yu, J. Zhu, Influence of cell structure parameters on the mechanical properties of microcellular polypropylene materials, *J. Appl. Polym. Sci.* 122 (2011) 2907–2914.
28. K.A. Arora, A.J. Lesser, T.J. McCarthy, Compressive behavior of microcellular polystyrene foams processed in supercritical carbon dioxide, *Polym. Eng. Sci.* 38 (1998) 2055–2062.
29. S. Wijnands, Volumetric behavior of polymer foams during compression and tension, Thesis, University of Eindhoven, 2010.
30. R. Miller, In serach of low-K dielectrics, *Science* (80) (1999).
31. H.W. Ro, K.J. Kim, P. Theato, D.W. Gidley, D.Y. Yoon, Novel inorganic—Organic hybrid block copolymers as pore generators for nanoporous ultralow-dielectric-constant films, *Macromolecules.* 38 (2005) 1031–1034. doi:10.1021/ma048353w.
32. L.F. Sansone, Process design for thermoplastic foam extrusion, *Foam Extrus. Princ. Pract.* (2014) 267.
33. S.-T. Lee, C.B. Park, N.S. Ramesh, General foam processing technologies, *Polym. Foam. Sci. Technol.,* CRC Press, 2006: pp. 73–92.
34. A.H. Landrock, *Handbook of Plastic Foams: Types, Properties, Manufacture and Applications,* Elsevier, 1995.
35. B. Bregar, Demag Ergotech introduces variety of technology, *Plastics News,* (2001).
36. D. Mann, J.C. Van den Bos, A. Way, *Automotive Plastics and Composites,* Elsevier, 1999.
37. Z. Xiao, J. Fang, G. Sun, Q. Li, Crashworthiness design for functionally graded foam-filled bumper beam, *Adv. Eng. Softw.* 85 (2015) 81–95. doi:10.1016/j.advengsoft.2015.03.005.
38. J. Bickford, Gaskets and Gasketed Joints, CRC Press, Boca Raton, FL, 1997.
39. L. Morello, L. Rossini Rosti, G. Pia, A. Tonoli, *The Automotive Body,* Springer Publications, n.d.
40. D. Friedman, C. E. Nash, Measuring rollover roof strength for occupant protection. *Int. J. Crashworthiness* 8 (2003) 97–105.

41. K.F. Orlowski, R.T. Bundorf, E.A. Moffatt, Rollover crash tests—The influence of roof strength on injury mechanics, *SAE Tech. Pap. Ser.* (1985).
42. MatWeb: Material Property Database, MatWeb, 2016. http://www.matweb.com/search /datasheet.aspx?MatGUID=ff6d4e6d529e4b3d97c77d6538b29693.
43. C.J. Hilado, ed., *Flammability Handbook for Plastics*, 3rd ed., Technomic, 1982.
44. N.M. Bikales, H.F. Mark, G. Menges, C.G. Overberger, eds., *Encyclopedia Polymer Science Engineering*, 2nd ed., John Wiley & Sons, 1989.
45. B. Ellis, R. Smith, *Polymers: A Property Database*, 2nd ed., CRC Press, Taylor & Francis Group, 2013. doi:10.1017/CBO9781107415324.004.
46. S.B.H. Gmbh, *Integral/Structural Polymer Foams*, n.d.
47. Alliance of Foam Packaging Recyclers, properties, performance and design fundamentals of expanded polystyrene packaging, *Tech. Bull.* (2000) 0–3.
48. G. Pezzin, *Plastics and Polymers*, Plastics Institute, Michigan, 1969.
49. D.A. Kourtides, J.A. Parker, *Polymer Engineering & Science*, John Wiley & Sons, Hoboken, NJ, 1978.
50. M. Bauccio, ed., *ASM Engineered Materials Reference Book*, ASM International, 1994.
51. J.I. Kroschwitz, M. Howe-Grant, eds., *Kirk-Othmer Encyclopedia of Chemical Technology*, 4th ed., Wiley Interscience, 1993.
52. V.B. Elvers, *Ullmann's Encyclopedia of Industrial Chemistry*, VCH, 1992.
53. R. Kaps, R. Lecht, U. Schulte, *Kunststoffe*, Kunststoffe, Germany, 1996.
54. *Modern Plastics Encyclopedia: Guide to Plastics, Property and Specification Charts*, McGraw-Hill, New York, 1987.
55. F. Bai, F. Li, B.H. Calhoun, R.P.C. Quirk, S.Z.D. Cheng, *Polymer Handbook*, 4th ed., John Wiley & Sons, 1999.
56. M.R. Thompson, X. Qin, G. Zhang, A.N. Hrymak, Aspects of foaming a glass-reinforced polypropylene with chemical blowing agents, *J. Appl. Polym. Sci.* 102 (2006) 4696–4706. doi:10.1002/app.24770.
57. J. Yang, P. Li, Characterization of short glass fiber reinforced polypropylene foam composites with the effect of compatibilizers: A comparison, *J. Reinf. Plast. Compos.* (2015) 0731684415574142.
58. S. Kant, Urmila, J. Kumar, G. Pundir, Study of talc filled polypropylene- a concept for improving mechanical properties of polypropylene, *Int. J. Res. Eng. Technol.* 2 (2013) 411–415.
59. N. Mantaranon, S. Chirachanchai, Polyoxymethylene foam: From an investigation of key factors related to porous morphologies and microstructure to the optimization of foam properties, *Polymer (Guildf).* 96 (2016) 54–62. doi:10.1016/j.polymer.2016.05.001.
60. BASF, Ultramid S Polyamides, 1996.
61. ZOTEK® N B50 High Performance Nylon Foam, n.d.
62. E. Koricho, G. Belingardi, A. Tekalign, D. Roncato, B. Martorana, Crashworthiness analysis of composite and thermoplastic foam structure for automotive bumper subsystem, in: *Adv. Compos. Mater. Automot. Appl.*, John Wiley & Sons Ltd, 2013: pp. 129–148.
63. C.-K. Park, C.-D. Kan, W.T. Hollowell, S.I. Hill, Investigation of Opportunities for Lightweight Vehicles Using Advanced Plastics and Composites, (n.d.). http://trid.trb .org/view.aspx?id=1226276#.Vvs59yELgYk.mendeley (accessed March 30, 2016).
64. M. Berry, *Applied Plastics Engineering Handbook*, Elsevier, 2011. http://www.science direct.com/science/article/pii/B9781437735147100145 (accessed March 29, 2016).
65. J. Maxwell, *Plastics in the Automotive Industry*, Elsevier, 1994. http://www.science direct.com/science/article/pii/B9781855730397500015 (accessed March 28, 2016).
66. Frost & Sullivan, Automotive Lightweight Materials 360 Degree View, (n.d.).

67. J. Melzig, M. Lehner, Lightweight Construction Due to Thermoplastic Foams as Exemplified in an Instrument-Panel Support, in: 2006. http://papers.sae.org/2006-01 -1404/ (accessed April 5, 2016).

68. C. Park, M. Brucker, L. Remy, S. Gilg, S. Subramonian, New sound-absorbing foams made from polyolefin resins, International Congress and Exposition on Noise Control Engineering, (2000) 2–7.

69. M.S. Sylvester, Self-adhesive reinforced foam gasket, US Patent, no. US6190751 B1, 2001.

70. C.R. Billiu, Automotive body panel molded from polycarbonate foam, US Patent, no. US3903224 A, 1975.

71. K. Zhang, A.K. Mohanty, M. Misra, Fully biodegradable and biorenewable ternary blends from polylactide, poly (3-hydroxybutyrate-co-hydroxyvalerate) and poly (butylene succinate) with balanced properties, *ACS Appl. Mater. Interfaces.* 4 (2012) 3091–3101.

72. Prospector, UL, Northbrook, 2016.

73. Universal Selector, SpecialChem, Paris, 2016.

74. J.-F. Zhang, X. Sun, Biodegradable foams of poly (lactic acid)/starch. II. Cellular structure and water resistance, *J. Appl. Polym. Sci.* 106 (2007) 3058–3062.

75. H. Zhao, Z. Cui, X. Sun, L.-S. Turng, X. Peng, Morphology and properties of injection molded solid and microcellular polylactic acid/polyhydroxybutyrate-valerate (PLA/PHBV) blends, *Ind. Eng. Chem. Res.* 52 (2013) 2569–2581.

76. M. Boufarguine, A. Guinault, G. Miquelard-Garnier, C. Sollogoub, PLA/PHBV films with improved mechanical and gas barrier properties, *Macromol. Mater. Eng.* 298 (2013) 1065–1073.

77. M.R. Nanda, M. Misra, A.K. Mohanty, The effects of process engineering on the performance of PLA and PHBV blends, *Macromol. Mater. Eng.* 296 (2011) 719–728.

78. S.-K. Lim, S.-G. Jang, S.-I. Lee, K.-H. Lee, I.-J. Chin, Preparation and characterization of biodegradable poly (butylene succinate)(PBS) foams, *Macromol. Res.* 16 (2008) 218–223.

79. S.K. Lim, S.I. Lee, S.G. Jang, K.H. Lee, H.J. Choi, I.J. Chin, Synthetic aliphatic biodegradable poly (butylene succinate)/MWNT nanocomposite foams and their physical characteristics, *J. Macromol. Sci. Part B.* 50 (2011) 1171–1184.

80. J. Jiao, M. Xiao, D. Shu, L. Li, Y.Z. Meng, Preparation and characterization of biodegradable foams from calcium carbonate reinforced poly (propylene carbonate) composites, *J. Appl. Polym. Sci.* 102 (2006) 5240–5247.

81. L.T. Guan, F.G. Du, G.Z. Wang, Y.K. Chen, M. Xiao, S.J. Wang et al., Foaming and chain extension of completely biodegradable poly (propylene carbonate) using DPT as blowing agent, *J. Polym. Res.* 14 (2007) 245–251.

12 Lightweight Thermoset Foams in Automotive Applications

Numaira Obaid, Mark T. Kortschot, and Mohini Sain

CONTENTS

12.1 INTRODUCTION

The automotive industry is always in search of innovative materials that can reduce the weight of a vehicle without sacrificing performance. Automobile weight reduction is beneficial for consumers since lighter vehicles have a higher fuel economy, which decreases fuel costs over the lifetime of a vehicle. Weight reductions can also lower manufacturing costs for the automotive industry since lighter vehicles can use a smaller engine and have lighter wheels. "Lightweighting" can also be beneficial in auto-racing since a lightweight car can attain a higher maximum speed and requires less time to accelerate and decelerate to the desired speed.

The first Model T car made by Henry Ford in 1908 was mainly made from vanadium steel, and the automotive industry has come a long way in reducing the weight of cars since then [1]. With extensive research in metal alloys and polymers, the manufacturers now have several options to consider during the material selection for automotive components. In particular, an increase in the use of lightweight polymers and polymeric foams has led to reduced vehicle weight and increased fuel economy over the years [2].

The low density of polymer foams makes them an ideal choice when selecting weight-reducing materials for automotive components. The porosity of foams not only decreases their density but also affects mechanical properties, vibration attenuation, and flammability. By manipulating the properties of the base polymer and the foam microstructure, its mechanical and thermal properties can be altered to suit various applications.

Foams that are derived from thermoset polymers, such as polyurethane and epoxy, are extensively used in the automotive industry due to their desirable mechanical properties, ease of fabrication, and thermal stability. This chapter will discuss the applications of thermoset foams in automotive components. The chapter will also discuss the factors that influence the mechanical properties, vibration attenuation, and flammability of these foams. Currents trends in improving the sustainability of thermoset foams will also be presented.

12.2   THERMOSET FOAMS

Polymeric foams are cellular materials derived from either thermoset or thermoplastic polymers. Thermoset plastics and their foams are typically formed from the irreversible reaction of two components, making the final product very stable. The covalently bonded network structure makes the foam resistant to high temperatures and to chemicals in comparison to typical thermoplastic polymeric foams. Some examples of thermoset polymers suitable for foaming include polyurethanes (formed from the cross-linking of alcohols and isocyanates), epoxy (formed from the cross-linking of an epoxide and a curing agent), and phenolic formaldehyde (formed from phenols and formaldehyde). Although thermoset plastics and their foams are very stable, their biggest limitation is that their cross-linking is irreversible, making the final product difficult to recycle and increasing its environmental footprint.

Thermoset foams are formed by incorporating of a gas into a liquid polymer mixture during polymer cross-linking. As the polymer solidifies through cross-linking, the gas forms bubbles and either exits the polymer through an open network of pores (open-cell foams) or is retained in the final structure (closed-cell foams). The gas in a thermoset polymer can be introduced using either a physical or chemical blowing agent. A physical blowing agent is dissolved in the liquid polymer precursor and then vaporizes during foaming. This is a common method of generating thermoplastic foams. A chemical blowing agent generates a gaseous reaction product such as water vapor or CO_2 as part of the overall chemical reaction.

The behavior of gas within a polymer is controlled by thermodynamics. The generation of gas occurs at the same time as the polymerization reaction. Gas is generated by the blowing agent and coalesces into small bubbles. Additional gas diffuses through the polymer into these pockets causing the bubbles to grow until a thermodynamic equilibrium is established. The expansion of gas bubbles results in tensile stresses in the adjacent polymer, and if the viscosity of the polymer is high enough the foam expands without collapsing and the cellular structure is locked in place.

The cellular structure of thermoset foams is dependent on the relative rates of two parallel reactions: the gelling reaction, which is the formation of cross-links, and the blowing reaction, which is the production of gas within the polymer. If the blowing reaction is much faster than the gelling reaction, then the gas will leave the system before sufficient cross-links are formed, resulting in cell rupture and poor cellular structure. If the gelling reaction is much faster than the blowing reaction, the polymer solidifies too quickly and limits bubble expansion. Many aspects of the cellular structure and properties of foam can be adjusted by careful control of the chemical formulation and foaming conditions.

Two types of thermoset foams are generally used in the automotive industry: polyurethanes and epoxies.

12.2.1 POLYURETHANE FOAMS

Polyurethane foam is a highly cross-linked thermoset polymer formed by the exothermic addition reaction of isocyanate-rich polymers with hydroxyl-rich polymers known as polyols [3]. Polyurethanes can be foamed by adding water as a chemical blowing agent in the system. Water reacts with isocyanate to produce carbon dioxide gas during cross-linking. The gelling and blowing reactions are shown in Figure 12.1 and Figure 12.2, respectively.

In a polyurethane foam, isocyanate acts as a hard segment making the polymer rigid, while polyol acts as a soft segment and increases ductility. The formulation of polyurethane foam has a large effect on its molecular structure and resulting properties. Foams that have a higher ratio of isocyanate to polyol (the isocyanate index) have a higher degree of cross-linking and are more rigid. Comparatively, foams with a lower isocyanate index are more flexible.

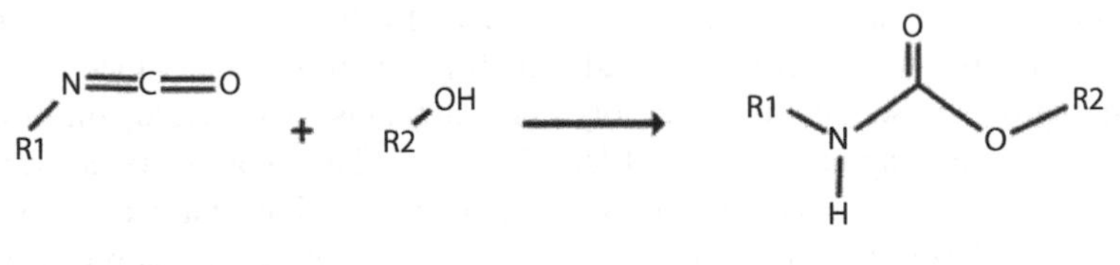

FIGURE 12.1 Reaction scheme for polyurethane gelling reaction.

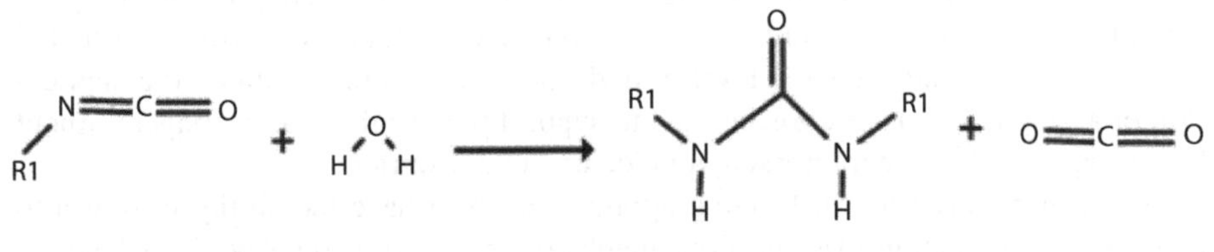

FIGURE 12.2 Reaction scheme for polyurethane blowing reaction.

12.2.2 EPOXY FOAMS

Epoxy refers to a subset of polymers that contain epoxide rings and can be formed using various monomers. The most common formulation to produce epoxy polymers is the step polymerization reaction between bisphenol A and epichlorohydrin. The large number of aromatic rings and oxirane rings in its molecular structure causes epoxy to have high strength, stiffness, hardness, and heat and chemical resistance. In addition, epoxy foams have very little shrinkage while curing, increasing their dimensional stability [4,5].

Epoxy foams are most commonly formed by adding chemical blowing agents that produce gases by reacting with the curing agent. For example, polysiloxanes can generate hydrogen gas by reacting with an amine-based curing agent. Since these reactions occur very quickly, the curing agent is often the last component added to the system. The decomposition of azodicarbonamide, which is a commercially available chemical blowing agent, can also be used to produce carbon dioxide and nitrogen for epoxy foaming. Last, epoxy can also be foamed by incorporating physical blowing agents such as carbon dioxide, hydrogen, and nitrogen directly in the matrix prior to curing [6].

12.3 APPLICATIONS OF THERMOSET FOAMS IN AUTOMOTIVES

The low density of thermoset foams makes them desirable for several automotive components including as cushioning in seats and steering wheels, in energy absorptive applications such as bumpers, and in specialty materials such as spoilers. The formulation of thermoset foams must be modified to produce materials that meet the individual requirements of each application.

12.3.1 CUSHIONING APPLICATIONS

One of the main applications of thermoset foams in automotive applications is as the cushioning material in seats and steering wheels. In the early years of automobile production, leather seats were attached to the frame with springs. Since then, in an effort to improve passenger comfort, polyurethane foams have become the dominant material for cushioning applications. Flexible polyurethane foams are desirable due to their low density, low cost, ease of fabrication, and good mechanical properties.

When considering polyurethane foams for automotive seating applications, the firmness, feel, and support must be kept in mind. Compressive strength and compressive modulus are two metrics used to evaluate the support provided by a foam. The compressive strength must be high enough for the seat to support the complete weight of a passenger. In addition, the compressive modulus (sometimes estimated by the sag factor) indicates the level of deflection induced in a foam on the application of a weight. Foams are formulated to support the weight of a passenger without becoming completely compressed, and leading to discomfort.

An ideal material for cushioning applications must have the ability to return to its original shape on unloading. Fortunately, polymer foams display viscoelasticity and tend to fully rebound over time, as opposed to permanent plastic deformation.

This hysteresis of polyurethane foams, which is the amount of energy dissipated by a foam during a compression/unloading cycle, is also critical. If a lower amount of energy is lost per cycle, the material tends to be more durable. However, materials with high energy damping can effectively absorb vibrations felt by passengers while traveling over uneven roads.

12.3.2 ENERGY ABSORPTIVE APPLICATIONS

Several automotive components, such as bumpers and headliners, require energy absorptive materials to improve passenger safety during a collision. Polyurethane foams are commonly used for these applications because of their ability to compress to large strains during collisions, leading to high energy absorption.

During an accident, a bumper is exposed to high loads, and the goal is to dissipate the kinetic energy of the vehicle in the collision without permanent damage, or in more severe collisions, without harm to the passengers. The bumper must be able to slow and eventually stop the impacting object. When a force is applied to a polyurethane foam, its cellular structure allows for large deformations, increasing the energy that it can absorb. Rigid materials may be stronger and stiffer, but foams can have a much larger area under the stress strain curve, leading to energy dissipation.

The most important function of an automotive bumper is to protect passengers against impacts. During impact with another vehicle, the bumper is subjected to compression over a large area. In this scenario, the bumper must have high compression modulus and strength. Bumpers must also be able to withstand impacts with smaller objects such as a pole, where the deformation is contained to one local area and may penetrate through the bumper. In these situations, it is important to evaluate the flexural and impact strength of the selected polyurethane foam. Since bumpers may be impacted at various strain rates, it is also important to understand how properties such as modulus and strength vary at various strain rates.

A few other properties that should be considered in the design of automotive bumpers and headliners are the resistance to environmental factors such as excessive heat, moisture, and ozone. In high-temperature climates, if the thermal expansion coefficient of the foams is too high, the foams can deform enough that dimensional tolerances are exceeded. In addition, the long-term durability such as fatigue life and creep resistance should also be considered.

12.3.3 SPOILERS

Epoxy-based thermoset foams are commonly used in automotive spoilers to improve the aerodynamics of a vehicle by redirecting the flow of air and decreasing drag forces. Epoxy foams are ideal for these applications since they can be very rigid, and have very little shrinkage during curing, resulting in high-dimensional stability. This ensures that spoilers are manufactured with dimensions, angles, and curvatures that are consistent with their initial design. Spoilers are exterior components that are exposed to various environmental conditions, making epoxy foams the ideal material for this application due to their high thermal stability and moisture resistance.

Automotive spoilers must be able to withstand various driving speeds and wind speeds and directions, making it critical to understand their strain-rate dependence. In particular, some polymers and their foams can become brittle at high deformation speeds, having adverse effects in auto-racing. Spoilers also undergo cyclic compressive and tensile stresses based on the direction and intensity of the wind, making it critical to understand and improve the fatigue properties of epoxy foams.

12.4 MATERIAL SELECTION FOR AUTOMOTIVE APPLICATIONS

12.4.1 MECHANICAL PROPERTIES OF THERMOSET FOAMS

The chemical formulations and processing methods must be optimized to produce foams that both meet the performance requirements of the application, and adhere to industrial requirements and regulatory standards. Very often, foam structure is adjusted to meet targets for mechanical properties such as modulus and strength. The properties of thermoset foams can be controlled by either modifying the cellular structure, such as density and cell size, or by changing the properties of the cell wall material using different chemical formulations or by fiber reinforcement.

12.4.1.1 Compressive Modulus

The cellular structure plays an important role in determining the properties of the foam. The effect of foam density on compressive modulus and strength has been extensively studied by Gibson et al. and it is well established that these properties increase with foam density [7]. The sag factor, which is a metric to estimate modulus, is calculated as a ratio of the compressive load of the material at 65% deflection to the load at 25% deflection of the foam [8]. Studies have shown that the sag factor increases with foam density.

In polyurethanes, the polymer modulus is a function of the cross-linking density, which is in turn a function of the formulation. As mentioned earlier, the cross-link density can be increased by increasing the isocyanate content in the foams. The isocyanate index is the most important factor in controlling the cross-link density and thus properties of the cell wall material in polyurethane foams. Several studies have examined the effect of isocyanate index on the properties of polyurethane foams and have found that increasing the isocyanate content results in an increase in modulus and strength [9,10]. Foams prepared with a low isocyanate index are flexible and often used for cushioning applications, while high-isocyanate index foams tend to be rigid and used in high-load applications such as bumpers.

Another method to change the polymeric cell wall material is to add load-bearing reinforcing fillers and fibers to the foam [11–13]. Gu et al. showed that the compressive strength of polyurethane foams was proportional to the fiber content for wood fibers, and roughly proportional to the fiber size; see Figure 12.3 [14].

Stefani et al. showed that the modulus of epoxy-based foams could be improved by the addition of small amounts of micro-cellulose fibers [15]. Other studies have also shown that reinforcement with glass fibers and aramid fibers can improve the rigidity of epoxy foams under both compressive and shear loading [16]. Another study found that glass fibers could increase the flexural modulus of polyurethane foams, as shown in Figure 12.4 [17]. The addition of particles such as graphite [18]

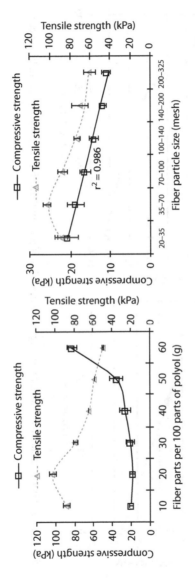

FIGURE 12.3 Effect of wood fiber content and aspect ratio on the compressive strength of polyurethane foams. (Obtained with kind permission from Bioresources.)

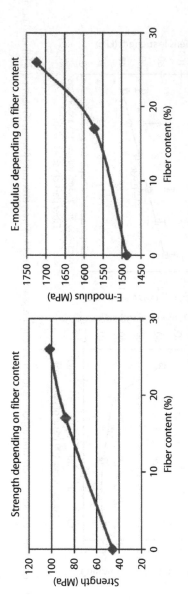

FIGURE 12.4 Effect of glass fiber content on the flexural strength and modulus of polyurethane foams. (Obtained with kind permission from Maney Publishing.)

and micro-silica [19–22] have also been shown to increase the foam modulus by measuring the sag factor and hardness, respectively.

The effect of fibers, however, on the properties of foams is quite complex. Although fibers can act as reinforcing agents, if the fiber content is too high or the fibers are too large, they can also have adverse effects on the cellular structure of the foam. In addition, high fiber content can increase the viscosity of the polymer, impeding bubble expansion and causing poor cell structure, which can be detrimental to the foam properties.

12.4.1.2 Energy Absorption

When a load is applied to a foam, the individual cells compress until the cellular structure collapses into a dense material. Energy can be absorbed by foams through various phenomena including deformation of the porous scaffold and the movement of entrapped air. Thus, the cellular structure plays an important role in the energy absorption of the foam.

Many groups have investigated the effect of foam density on the energy absorption capabilities of polyurethane foams. Naik et al. showed that increasing the density of polyurethane foams resulted in a decrease in the energy absorbed by the foams, as shown in Figure 12.5 and Table 12.1 [23]. In another study, Onsalung et al. showed the opposite: that polyurethane foams with higher densities had higher energy absorption [24]. This was also confirmed in a study by Subramaniyan et al. [25].

In contrast, Deb et al. showed that no specific trends could be drawn between the density of polyurethane foams and their energy absorption [26]. This conclusion been supported by other studies as well [27–29]. Polymer foams are viscoelastic, however, so

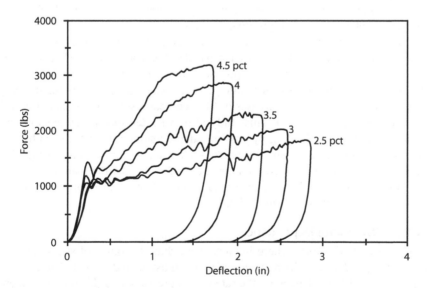

FIGURE 12.5 Effect of polyurethane foam density on the load–deflection curve under dynamic loading. (Image obtained by the courtesy of Sage Publications.)

TABLE 12.1

Effect of Foam Density on Energy Absorption of Polyurethane Foams during Dynamic Impact

Density (pcf)	Peak Load (lbs)	Maximum Penetration (in.)	Absorbed Energy (ft-lb)	Specific Energy Absorbed (ft-lb/lb)
2.5	1847	2.86	307	123
3.0	2030	2.59	303	101
3.5	2341	2.30	305	87
4.0	2874	1.95	290	72
4.5	3198	1.73	281	62

Source: Obtained with kind permission of Sage Publications.

it is reasonable to conclude that the energy absorption is also dependent on strain rate, and if studies are conducted at higher strain rates, they might have different conclusions.

Avalle et al. suggest the possibility of an optimal foam density to maximize energy absorption [28]. If the foam density is too low, the foam quickly collapses resulting in little energy absorption. Alternatively, foams that have a very high density also have a higher modulus. In this case, the total deformation at failure may be low, reducing the energy absorption capacity.

A number of studies have examined the effect of fiber reinforcement on the energy absorption of polyurethane foams; however, this effect is not well understood and requires further investigation. Obaid et al. [30] examined the effect of various amounts of glass fibers on the energy absorption of polyurethane foams at various strain rates. This study showed that increasing the fiber content has negligible effect on the energy absorption of polyurethane foams.

Sachse et al. investigated the effect of nano-clay filler on the energy absorption of polyurethane foams used in sandwich panels [31]. These sandwich panels were prepared with both polyamide and polypropylene face sheets. It was found that under low speed compression, low filler content resulted in a decrease in energy absorption; however, further increases in filler content resulted in an increase in the absorbed energy (Figure 12.6). This study also evaluated the energy absorption of the foams under impact loading by a point indenter; simulating a collision between an automotive bumper and a smaller object such as a pole. In this case, the addition of nano-clay filler generally increased the energy absorbed by the foam (Table 12.2). Other studies have also shown that the addition of fibers can improve the energy absorption of polyurethane foams [25,32].

12.4.1.3 Strain Rate Dependence

In automotive applications, where impacts can occur at a range of strain rates, the strain-rate dependence of the behavior of foams must be understood. The compressive properties of foam depend on the rate of compression, and this rate dependence stems from two main factors: the deformation of the cells and the inherent viscoelasticity of the base polymer.

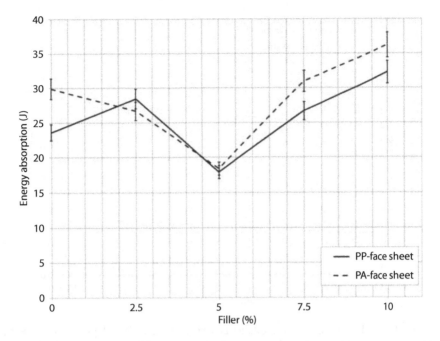

FIGURE 12.6 Effect of fiber content on the energy absorption of polyurethane foams under low-speed compression. (Image obtained by the courtesy of Sage Publications.)

TABLE 12.2

Effect of Nano-Clay Loading on the Energy Absorption of Polyurethane Foams Under Impact Loading with a Point Indenter

Polypropylene Face Sheets		Polyamide Face Sheets	
Filler Content	Energy Absorbed (J)	Filler Content	Energy Absorbed (J)
0%	3.73	0%	5.2
2.5%	5.76	2.5%	6.54
5.0%	6.25	5.0%	6.43
7.5%	6.69	7.5%	6.61
10.0%	6.36	10.0%	6.97

Source: Table obtained by the courtesy of Sage Publications.

The compression of open-cell polymeric foams induces intercellular transport of the entrapped air, and for rapid compression, the time is not sufficient to allow for pressure equalization, increasing the resulting load for a given strain. This makes it important to evaluate the effect of the open cellular morphology of the foam on its strain rate dependence.

Low-density foams have larger interconnected cells, which allow for air flow. In comparison, higher density foams have fewer and often smaller cells and this

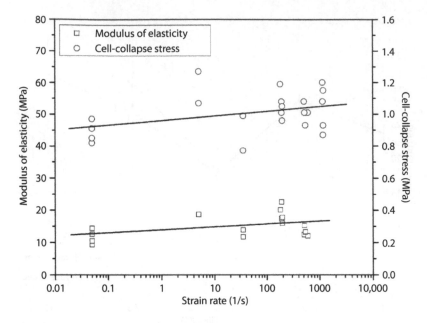

FIGURE 12.7 The compressive modulus and yield stress (cell-collapse stress) of epoxy foams was found to have a slight increase with strain rate. (Obtained with the kind permission of Springer.)

impedes the intercellular movement of air. The resistance to air transport leads to an increase of pressure within the cells, increasing the strength and modulus, especially at high strain-rates. Several studies have found that the strain–rate dependence of polyurethane foams increases at higher foam densities [33–36].

The viscoelasticity of the polymeric cell wall material also influences the strain-rate dependence of the foam. Several studies have shown that polymeric foams exhibit strain–rate dependence. A study conducted by Song et al. evaluated the compressive properties of epoxy foams at various strain rates and found that the modulus of elasticity ranged between 10 and 15 MPa, and increased with strain rate (see Figure 12.7) [37]. The strain–rate dependence of epoxy foams has also been investigated in other studies as well [33,38].

The strain–rate dependence of foams can also be altered by changing the viscoelastic nature of the base material by the addition of fillers [39]. However, research in this area is quite limited and the effect of fillers on the strain–rate dependence of foams requires further investigation.

12.4.2 Vibration Attenuation in Thermoset Foams

One of the most important properties to consider in the design of automotive components is noise, vibration, and harshness (NVH), which is mitigated by the use of materials that attenuate noise and vibrations. Materials selected for interior components must be able to attenuate vibrations caused by the rotation of the engine and by the travel over uneven surfaces [40]. These mechanical vibrations can also produce sounds that must be dissipated to improve the experience of passengers. Measures must be taken to either dampen these sounds or attenuate the vibrations themselves.

Polymeric foam contains entrapped air that absorbs sound energy and converts it into heat, allowing vibrations to be dissipated effectively. One of the most important parameters controlling the sound absorption capability of a foam is its resistance to air flow. If air flow is restricted, the foam has better sound absorption [41–43]. Air movement depends on the volume fraction, size, and connectivity of the pores. Low-density foams have a larger fraction of air and a lower resistance to air movement, and thus, have a lower sound absorption. The cell size and arrangement also have an important effect on sound absorption. When a sound wave hits a rigid surface, it is reflected; however, in a porous structure, the sound waves are continually reflected and refracted by the cell walls, and the frictional losses eventually dampen the wave [44]. A smaller cell size encourages more frequent propagation of the sound waves from the cell wall to entrapped air, and is known to increase sound absorption. In addition, smaller cells are less connected to each other and increase resistance to air flow. One study showed that decreasing cell size of polyurethane foams using nano-silica as a nucleating agent resulted in improved sound absorption [45]. Asadi and Ohadi confirmed this result by showing that better dispersion of the nano-silica, which resulted in smaller cell sizes, increased the sound absorption coefficient of the foams, even when the volume fraction of filler was fixed [46].

Polyurethane foams also dampen vibrations due to the intrinsic viscoelastic character of the cell wall material. Sound absorption can be improved by modifying these properties. Studies have shown that increasing the cross-link density of a thermoset polymer hinders the movement of the polymer chains, resulting in reduced damping. [47] The cross-link density in polyurethane foams can be decreased by decreasing the isocyanate content, which has been found to be effective in improving the sound absorption of the resultant foams [48]. A recent study, however, showed that although the addition of hyper-branched polymer (HBP) increased cross-linking in polyurethanes, it also resulted in increased sound absorption; see Figure 12.8 [49]. The study reported that the addition of HBP did not result in any morphological changes, and concluded that the increased sound absorption must have resulted from unknown changes in the viscoelastic character of the base polyurethane.

The sound-absorptive properties of the cell wall material can also be altered by the addition of fillers. Studies have shown that the sound absorption of polyurethane foams can be improved using both natural viscoelastic fibers [46,48,50] and rubber particles [51]. Since both natural fibers and rubber particles are viscoelastic, they can increase the intrinsic damping of the solid polyurethane contained in the cell walls, and improve the overall sound absorption of the foam.

12.4.3 Flame Resistance in Thermoset Foams

Serious collisions can result in vehicular fires, and as a result, flame resistance is an important property that must be considered in material selection for automotive components. Passenger safety can be increased by ensuring that a material does not ignite quickly in the presence of a flame and that once ignited, the flame spreads at a slow rate. The flammability of foam is dependent on both its cellular structure and chemical composition. The porous structure of polyurethanes helps the transport of air through the material, increasing flammability. Low-density foams have less resistance to air flow, typically resulting in faster flame spread rate [52] and lower thermal energy is required to begin the burning process; see Figure 12.8 [53,54].

Polyurethane is an organic material, making it important to investigate the effect of chemical composition of the polymer on its flammability. In the presence of thermal energy, the decomposition of a polymer occurs via the vibration and rotation of covalent bonds in the long molecular chains. The flame resistance of polymers can be improved by increasing chain rigidity, which increases the thermal energy required to induce bond breakage [55]. Studies have found that increasing the isocyanate index of foams can reduce their flammability by increasing cross-link density and molecular rigidity [56]. Studies have also shown the fire resistance of polyurethane foams to be related to its thermal stability. For example, increased concentration of aromatic groups resulted in improved fire resistance in one study [57].

Flame retardants can be used to increase the flame resistance of polyurethane foams by various mechanisms which absorb energy and dissipate heat. Retardants can expand and increase the surface area over which the heat is distributed (such as expandable graphite), can decompose in the presence of heat to form char preventing further degradation, and can act as heat sinks to absorb thermal energy by latent heating and thermal degradation (such as melamine). Although halogenated flame retardants have been found to effectively reduce flammability, health concerns have

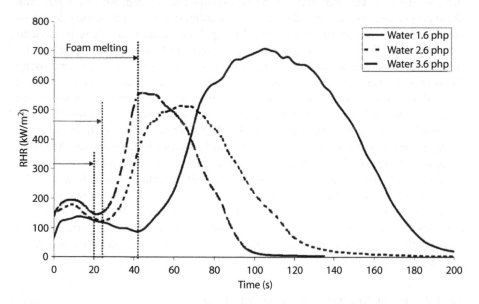

FIGURE 12.8   A higher blowing agent content (water) results in a lower density of the foam. It was observed that increasing the water content results in a reduction in the time required for the foam to burn. As well, a lower rate of heat release was required for each stage of the burning process. (Image obtained with kind permission from Elsevier.)

TABLE 12.3
Effect of Various Fillers on the Flame Resistance of Polyurethane Foams

	Burning Time (s)	Burning Length (mm)
Melamine	4	54
Expandable graphite (180 microns)	300	305
Expandable graphite (250 microns)	277	305
Acceptable threshold	<15	<203

Source: Obtained with kind permission Sage Publications.

prompted the evaluation of halogen-free fire retardants. One study compared the use of halogen-free fire retardants, melamine, and expandable graphite particles in polyurethane foams [18]. It was found that incorporating melamine additives in foams decreased the total burn time after the flame source had been removed (burning time) compared to foams with expandable graphite, as shown in Table 12.3.

12.4.4 OTHER PROPERTIES

A thermoset foam that is used in automotive components must be designed to meet performance requirements with passenger comfort in mind. This requires consideration of other properties that were not covered previously in this chapter. For example, cushioning applications may require consideration of hysteresis. This is a measure of the amount of energy that is lost by the foam during a compression cycle, and often indicates the durability of the foam [57]. Since foams are viscoelastic, the hysteresis loss is governed by the viscous behavior of the foam [58]. Increasing the elastic character of the foam by using particulates such as nano-silica can reduce the hysteresis loss in polyurethane foam by decreasing this viscous character [59].

Another important property that should be considered in automotive materials is moisture absorption. Excessive moisture in seats can result in mold growth and odor. On the other hand, poor water absorption by the foam can lead to retention of moisture in the fabric covering, compromising passenger comfort. Thermoset foams must be able to effectively transport moisture away from the contact surface, and allow vapor transfer back out so that they can dry quickly. It is well-established that moisture absorption in polyurethane foams is dominated by diffusion of water molecules through the base polymer [60]. High moisture uptake by the cell wall has also been shown to have detrimental effects on mechanical and thermal properties of thermoset foams [61]. Water molecules that have diffused into a polymer, such as polyurethane, can form strong hydrogen bonds due to their high polar affinity, often breaking the pre-existing bonds within the polyurethane matrix. Excessive moisture absorption must be prevented to ensure that the foam performance is not compromised.

The design of automotive components should also consider long-term durability of thermoset foams such as resistance to fatigue and creep. Creep is defined as the increase in strain over time under a fixed external load. In automotive applications such as seat cushioning, the foam should maintain its properties over the lifetime of the car, which can be approximately 10–15 years. It is important that foams be designed to minimize permanent deformation and distortion during this lifetime.

The response of a foam to fatigue loading is also an important consideration for its long-term durability. An automotive spoiler undergoes cyclic compressive and tensile stresses based on the direction and intensity of the wind. Although the addition of fibers can have positive effects on modulus and other mechanical properties, their effect on fatigue should also be considered. In particular, several studies have found that the addition of fibers actually has a negative effect on the fatigue life of epoxy foams due to the delamination and rupture of fibers [15,62,63].

12.5 TRENDS

Research in the area of thermoset foams has shifted in recent years toward improving their sustainability, primarily through the use of bio-sourced monomers. Several studies have attempted to produce polyols from vegetable oils such as soybean oil, linseed oil, and rapeseed oil. Foams prepared from bio-based polyols have been shown to have comparable mechanical properties to their petroleum-based counterparts. Tan et al. found that polyurethane foams, where petroleum-based polyol was partially replaced with soybean oil-based polyol, did not have a decrease in modulus [64]. Another study by Gu et al. also showed that the compressive strength of soy-based polyurethane foams was similar to those made with conventional polyols [65]. A few studies have also attempted to produce epoxy foams from epoxidized vegetable oils. These monomers can then be reacted with various curing agents to produce epoxy resins that are more sustainable than their petroleum-derived counterparts [66,67].

The cross-linking in thermoset foam leads to good mechanical properties; however, it makes these foams difficult to recycle, increasing their accumulation in landfills. One study has found that preparing polyurethane foams with vegetable oils can improve their biodegradability compared to their petroleum-based counterparts [68]. Although this was experimentally confirmed, it is important to understand the underlying mechanisms that accelerated biodegradation.

Although bio-based components can improve the sustainability of thermoset foams by shifting manufacturing away from petroleum-based components or by increasing their biodegradability, it is important to understand their properties. The processing of a bio-based monomer has an effect on its chemical properties including molecular weight and hydroxyl content, and its physical properties, such as viscosity. It is also important to optimize a scalable production method such that the properties of these novel monomers are reproducible. Inconsistencies in properties can have a significant effect on cellular structure and mechanical properties of foams, which is undesirable in any commercial application. Before bio-based foams can be used more extensively in the automotive industry, it is important that in addition to experimental studies, the underlying mechanisms are well-understood. Current research efforts are addressing this need, and bio-based polyurethanes are expected to be more widely used in the future.

12.6 CONCLUSION

Due to their low density, thermoset foams are prevalent as a "lightweighting" material in the automotive industry. The properties of polymeric foams can vary over a wide range based on the polymer from which the foam is derived and the morphology of the foam. Thermoset foams made from polyurethane and epoxy are used in several applications ranging from deformable cushioning materials to rigid energy absorptive materials such as bumpers to specialty materials such as spoilers.

Research in thermoset foams, for automotive applications, aims to improve the mechanical properties without increases in weight. Foam properties can be tailored to suit various applications by either modifying the molecular structure of the base polymer (by modifying cross-link density or by the addition of fibers) or modifying the morphological characteristics such as cell size and foam density. This chapter detailed the effect of these parameters on the mechanical properties of thermoset foams.

Material selection for automotive applications cannot be based solely on mechanical properties, and thermal and vibrational properties must be considered. Serious collisions can result in vehicular fires, and as a result, thermal properties and flame resistance of foams must also be examined. This is a particular issue in polymeric foams, which are derived from organic materials and often contain entrapped oxygen in the form of air due to their cellular structure. This chapter showed that the addition of flame retardants such as melamine and expandable graphite have been shown to decrease the flammability of polymeric foams. Another important property to consider in the design of automotive components is noise, vibration, and harshness (NVH). Materials selected for interior components must be able to attenuate vibrations caused by the rotation of the engine and by the travel over uneven surfaces. Measures must be taken to either dampen these sounds or attenuate the vibrations themselves. Foam contains entrapped air that absorbs sound energy, allowing vibrations to be dissipated effectively. As discussed in this chapter, the cellular structure of foams can be controlled to resist air flow and improve sound absorption.

The automotive industry's interest in "green" cars has shifted research toward improving foam sustainability, primarily through the use of bio-sourced monomers. The use of soy-based polyurethane foams for seat cushioning in the 2011 Ford Explorer has shown that foams derived from bio-based monomers are not limited to a lab-scale. Although bio-based components can improve the sustainability of thermoset, it is important to understand their properties. It is also important to optimize a scalable production method such that the properties of these novel monomers are reproducible. Additional experimental and theoretical knowledge must be developed before bio-based foams can be used more extensively in the automotive industry.

REFERENCES

1. Fountain, H. Many faces, and phases of steel in cars. *The New York Times.* 2009.
2. U.S. Department of Transportation, Research, and Technology Administration, Bureau of Transportation Statistics. National Transportation Statistics: Average Fuel Efficiency of U.S. Light Duty Vehicles, 2016.

3. Lee, S. T. and Ramesh, N. S. *Polymeric Foams: Mechanisms and Materials*. CRC Press, Boca Raton, FL, 2004.

4. Solorzano, E. and Rodriguez-Perez, M. A. Polymeric foams in Lehmhus, D., Busse, M., Herrmann, A., and Kayvantash, K. *Structural Materials and Processes in Transportation*. Wiley, Weinheim, Germany, 2013.

5. Pascault, J. P. and Williams, R. J. J. General concepts about epoxy polymers in Pascault, J. P. and Williams, R. J. J. *Epoxy Polymers: New Materials and Innovation*. Wiley, Weinheim, Germany, 2010.

6. Ren, Q. and Zhu, S. One-pack epoxy foaming with CO_2 as latent blowing agent. *ACS Macro Letters*. 2015; 4:693–697.

7. Gibson, L. J. and Ashby, M. F. *Cellular Solids: Structures and Properties*. Cambridge Solid State Science Series, Cambridge. 1999.

8. Casati, F. M., Herrington, R. M., Broos, R., and Miyazaki, Y. Tailoring the performance of molded flexible polyurethane foams for car seats. *Journal of Cellular Plastics*. 1998; 34:430–466.

9. Zhu, G., Wang, G., and Hu, C. Effect of crosslink density on the structures and properties of aliphatic polyurethane elastomer. *Acta Polymerica Sinica*. 2011; (3):274–280.

10. Edwards, B. H. Polyurethane structural adhesives in Hartshorn, S. R. *Structural Adhesives: Chemistry and Technology*. Plenum Press, New York, 1986.

11. Hussain, S. and Kortschot, M. T. Polyurethane foam mechanical reinforcement by low-aspect ratio micro-crystalline cellulose and glass fibres. *Journal of Cellular Plastics*. 2015; 51(1):59–73.

12. Li, Y., Ren, H., and Ragauskas, A. J. Rigid polyurethane foam reinforced with cellulose whiskers: Synthesis and characterization. *Nano-micro Letters*. 2010; 2(2):89–94.

13. Uddin, M. F., Mahfuz, H., Zainuddin, S., and Jeelani, S. Infusion of spherical and acicular nanoparticles into polyurethane foam and their influences on dynamic performances. *Proceedings of the International Symposium on MEMS and Nanotechnology*. 2005; 6:147–153.

14. Gu, R., Khazabi, M., and Sain, M. Fiber reinforced soy-based polyurethane spray foam insulation. Part 2: Thermal and mechanical properties. *Bioresources*. 2011; 6(4):3775–3790.

15. Stefani, P. M., Perez, C. J., Alvarez, V. A., and Vazquez, A. Microcellulose fibers-filled epoxy foams. *Journal of Applied Polymer Science*. 2008; 109:1009–1013.

16. Alonso, M. V., Auad, M. L., and Nutt, S. Short-fiber-reinforced epoxy foams. *Composites: Part A*. 2006; 37:1952–1960.

17. Mkrtchyan, L. and Maier, M. Fiber reinforced polyurethane foam to increase passive safety of vehicle occupants. *Plastics, Rubber and Composites*. 2007; 36(1):445–454.

18. Gharehbaghi, A., Bashirzadeh, R., and Ahmadi, Z. Polyurethane flexible foam fire resisting by melamine and expandable graphite: Industrial approach. *Journal of Cellular Plastics*. 2011; 47(6):549–565.

19. Javni, I., Zhang, W., Karajkov, V., Petrovic, Z. S., and Divjakovic, V. Effect of nano- and micro-silica fillers on polyurethane foam properties. *Journal of Cellular Plastics*. 2002; 38:229–239.

20. Johnson, M. and Shivkumar, S. Filamentous green algae additions to isocyanate based foams, *Journal of Applied Polymer Science*. 2004; 93:2469–2477.

21. Jang, S. Y., Kim, D. J., and Seo, K. H. Physical properties of garnet-filled polyurethane foam composites. *Journal of Applied Polymer Science*. 2001; 79:1336–1343.

22. Wolska, A., Gozdzikiewicz, M., and Ryszkowska, J. Thermal and mechanical behavior of flexible polyurethane foams modified with granite and phosphorous fillers. *Journal of Materials Science*. 2012; 47(15):5627–5634.

23. Naik, B. G. New polyurethane energy management forms for improved auto interior safety. *Journal of Cellular Plastics*. 1995; 31:479–498.

24. Onsalung N., Thinvongpituk C., and Painthong K. The influence of foam density on specific energy absorption of rectangular steel tubes. *Energy Research Journal*. 2010; 1(2):135–140.

25. Subramaniyan, S. K., Mahzan, S., Ghazali, M. I., Zaidi, A. M. A., and Prabagaran, P. K. Energy absorption characteristics of polyurethane composite foam-filled tubes subjected to quasi-static axial loading. *Applied Mechanics and Materials*. 2013; 315:872–878.

26. Deb, A. and Shivakumar, N. D. An experimental study on energy absorption behavior of polyurethane foams. *Journal of Reinforced Plastics and Composites*. 2009; 28(24):3021–3026.

27. Saha, M. C., Mahfuz, H., Charavarthy, U. K., Uddin, M., Kabir, M. E., and Jeelani, S. Effect of density, microstructure, and strain rate on compression behavior of polymeric foams. *Materials Science and Engineering*. 2005; A406:328–336.

28. Avalle, M., Belingardi, G., and Montanini, R. Characterization of polymeric structural foams under compressive impact loading by means of energy-absorption diagram. *International Journal of Impact Engineering*. 2001; 25:455–472.

29. Maiti, S. K., Gibson, L. J., and Ashby, M. F. Deformation and energy absorption diagrams for cellular solids. *Acta Metal*. 1984; 32(11):1963–1975.

30. Obaid, N., Kortschot, M., and Sain, M. Investigating the mechanical response of soy-based polyurethane foams with glass fibers under compression at various rates. *Cellular Polymers*. 2015;34(6):281–298.

31. Sachse, S., Poruri, M., Silva, F., Michalowski, S., Pielichowski, K., and Njuguna, J. Effect of nanofillers on low energy impact performance of sandwich structures with nanoreinforced polyurethane foam cores. *Journal of Sandwich Structures and Materials*. 2014; 16(2):173–194.

32. Khanna, S. K. Application of reinforced polymer foams for energy absorption in lightweight structures. *Proceedings of the SEM Annual Conference on Experimental and Applied Mechanics*. 2001.

33. Subhash, G., Liu, Q., and Gao, X. Quasistatic and high strain rate uniaxial compressive response of polymeric structural foams. *International Journal of Impact Engineering*. 2006; 32(7):1113–1126.

34. Chen, W., Lu, F., and Winfree, N. High-Strain-rate compressive behavior of a rigid polyurethane foam with various densities. *Experimental Mechanics*. 2002; 42(1):65–73.

35. Luong, D. D. and Gupta, N. Compressive properties of closed-cell polyvinyl chloride foams at low and high strain rates: Experimental investigation and critical review of state of the art. *Composites Part B*. 2013; 44(1):403–416.

36. Linuel, E., Marsavina, L., Voiconi, T., and Sadowski, T. Study of factors influencing the mechanical properties of polyurethane foams under dynamic compression. *Journal of Physics: Conference Series*. 2013. 451:1–4.

37. Song, B., Chen, W., and Lu, W. Y. Compressive mechanical response of a low-density epoxy foam at various strain rates. *Journal of Materials Science*. 2007; 42:7502–7507.

38. Li, P., Petrinic, N., Siviour, C. R., Froud, R., and Reed, J. M. Strain rate dependent compressive properties of glass microballoon epoxy syntactic foams. *Materials Science and Engineering A*. 2009; 515:19–25.

39. Woldesenbet, E. and Peter, S. Volume fraction effect on high strain rate properties of syntactic foam composites. *Journal of Materials Science*. 2009; 44:1528–1539.

40. Lemer, N. D., Kotwal, B. M., Lyons, R. D., and Gardner-Bonneau, D. J. *Preliminary Human Factors Guidelines For Crash Avoidance Warning Devices*. U.S. Department of Transportation, National Highway Traffic Safety Administration. 1996.

41. Imai, Y. and Asano, T. Studies of acoustical absorption of flexible polyurethane foams. *Journal of Applied Polymer Science*. 1982; 27:183–195.

42. Patten, W. N., Sha, S., and Mo, C. A Vibration model of open celled polyurethane foam automotive seat cushions. *Journal of Sound and Vibration.* 1998; 217(1):145–161.

43. Kinkelaar, M. R. and Cavender, K. D. Vibrational characterization of various polyurethane foams employed in automotive seating applications. *Journal of Cellular Plastics.* 1998; 34:155–173.

44. Broos, R., Sonney, J. M., Thanh, H. P., and Casati F. M. Polyurethane foam molding technologies for improving total passenger compartment comfort. *Proceedings of the API Polyurethane Conference.* Technomic, Lancaster. 2000; 341–353.

45. Lee, J., Kim, G. H., and Ha, C. S. Sound absorption properties of polyurethane/nanosilica nanocomposite foams. *Journal of Applied Polymer Science.* 2012; 123:2384–2390.

46. Asadi M. and Ohadi A. Improving sound absorption of polyurethane foams by the incorporation of nano-particles. *Proceedings of the 22nd International Congress on Sound and Vibration.* Florence, Italy. 2015.

47. Fay, J. J., Murphy, C. J., Thomas, D. A., and Sperling L. H. Effect of morphology, crosslink density, and miscibility on interpenetrating polymer network damping effectiveness. *Polymer Engineering and Science.* 1991; 31(24):1731–1741.

48. Sankar, H. R., Krishna, P. V., Rao, V. B., and Babu, P. B. The effect of natural rubber particle inclusions on the mechanical and damping properties of epoxy-filled glass fibre composites. *Journal of Materials: Design and Applications.* 2010; 224(2):63–70.

49. Anderson, A., Lundmark, S., Magnusson, A., and Maurer, F. H. J. Vibration and acoustic damping of flexible polyurethane foams modified with a hyperbranched polymer. *Journal of Cellular Plastics.* 2010; 46:73–93.

50. Geethamma, V. G., Kalaprasad, G., Groeninckx, G., and Thomas, S. Dynamic mechanical behaviour of short coir fiber reinforced natural rubber composites. *Composites, Part A.* 2005; 36:1499–1506.

51. Shan, C. W., Ghazali, M. I., and Idris, M. I. Improved vibration characteristics of polyurethane foam via composite formation. *International Journal of Automotive Mechanical Engineering.* 2013; 7:1031–1042.

52. Lefebvre, J., Bastin, B., Le Bras, M., Duquesne, S., Ritter, C., Paleja, R., and Poutch, F. Flame spread of flexible polyurethane foam: comprehensive study. *Polymer Testing.* 2004. 23:381–290.

53. Weil, E. D., Ravey, M., and Gertner, D. Recent progress in flame retardancy of polyurethane foams. 7th Annual BCC Conference on Flame Retardancy, Recent Advances in Flame Retardancy of Polymeric Materials, Stamford, CT, 1996.

54. Lefebvre, J., Bastin, B., Le Bras, M., Duquesne, S., Paleja, R., and Delobel, R. Thermal stability and fire properties of conventional flexible polyurethane foam formulations. *Polymer Degradation and Stability.* 2005; 88:28–34.

55. Chattopadhyay, D. K. and Webster, D. C. Thermal stability and flame retardancy of polyurethanes. *Progress in Polymer Science.* 2009; 34:1068–1133.

56. Zabski, L., Walczyk, W., and Weleda, D. Flammability and thermal properties of rigid polyurethane foams with additionally introduced cyclic structures. *Journal of Applied Polymer Science.* 1980; 25(12):2659–2680.

57. Wolfe, H. W. Cushioning and fatigue, in Hilyard N. C. *Mechanics of Cellular Plastics.* Macmillian, New York, 1982.

58. Chartoff, R. P., Menczel, J. D., and Dillman, S. H. Dynamic mechanical analysis (DMA), in Menczel, J. D. and Prime, R. B. *Thermal Analysis of Polymers, Fundamentals and Applications.* Wiley & Sons Ltd, Hoboken, NJ, 2009.

59. Kang, S. M., Kim, M. J., Kwon, S. H., Park, H., Jeong, H. M., and Kim, B. K. Polyurethane foam/silica chemical hybrids for shape memory effects. *Journal of Materials Research.* 2012; 27(22):2837–2843.

60. Schwartz, N. V., Bomberg, M., and Mumaran, M. L. Water vapor transmission and accumulation in polyurethane and polyisocyanurate foams, in Trechsel, H. R. and Bomberg, M. *Water Vapor Transmission Through Building Materials and Systems: Mechanisms and Measurement.* American Society for Testing and Materials, Philadelphia, PA, 1989.

61. Yu, Y. J., Hearon, K., Wilson, T. S., and Maitland, D. J. The effect of moisture absorption on the physical properties of polyurethane shape memory polymer foams. *Smart Materials and Structures.* 2012; 20(8):1–16.

62. Ferreira, J. A. M., Reis, P. N. B., Costa, J. D. M., and Richardson, M. O. W. Fatigue behaviour of kevlar composites with nanoclay-filled epoxy resins. *Journal of Composite Materials.* 2012; 47(15):1885–1895.

63. Shafiq, B. and Quispitupa, A. Fatigue characteristics of foam core sandwich composites. *International Journal of Fatigue.* 2006; 28:96–102.

64. Tan, S., Abraham, T., Ference, D., and Macosko, C. W. Rigid polyurethane foams from a soybean oil-based polyol. *Polymer.* 2011; 52(13):2840–2846.

65. Gu, R., Konar, S., and Sain, M. Preparation and characterization of sustainable polyurethane foams from soybean oils. *Journal of the American Oil Chemists' Society.* 2012; 89:2103–2111.

66. Mazzon, E., Habas-Ulloa, A., and Habas, J. P. Lightweight rigid foams from highly reactive epoxy resins derived from vegetable oil for automotive applications. *European Polymer Journal.* 2015; 68:546–557.

67. Gerbase A. E., Petzhold C. L., and Costa, A. P. O. Dynamic mechanical and thermal behavior of epoxy resins based on soybean oil. *Journal of the American Oil Chemist's Society.* 2002; 79(8):797–802.

68. Chen Q., Li R., Sun K., Li J., and Liu C. Preparation of bio-degradable polyurethane foams from liquefied wheat straw. *Advanced Materials Research.* 2011; 217–218:1239–1244.

13 Life Cycle Assessment of Lightweight Materials for Automotive Applications

Masoud Akhshik, Jimi Tjong, and Mohini Sain

CONTENTS

13.1 INTRODUCTION

When we talk about our environments as we only have one planet we should be very cautious because there is an upper limit for the amount of pollutions and emissions that we can put in our environment. These boundaries could be seen in each category that we are polluting or emitting and define the limit for emission and after that we are beyond the planetary boundary for that item, which is not considered in the safe operating mode anymore. For some of the categories like biodiversity loss and nitrogen cycle, we have already passed the safe boundary over 10 times and over 3 times, respectively, and there is nothing we can do about that. In the climate change we have just passed the safe boundary and all the international efforts like the Paris agreement is to stop more damage. Phosphorous cycle and ocean acidifications are next to pass the safe boundary. Another important category is ozone depletion, which was once passed the boundary and now it is coming back to the limits (Rockström et al. 2009).

Unfortunately, as we have seen before, people don't feel the urgency in the climate change. They simply think that this is for the future generations while now we know that statements like this are not true and if we don't take serious actions, there may be no future generations. The fear over climate change may cause a renaissance for bio-based materials (Pawelzik et al. 2013) and that may help our environments a lot.

But until humans reach the conclusion that this is seriously dangerous, nobody is even trying to change behaviors. When we notice that we depleted our resources, recycling ratio of plastics may change dramatically. As the ontological principle says, there is a vicious circle principle around all the human life aspects. If the early humans did not run out of food, they would have never looked for new resources. We have always seen an accelerated movement from the situations of scarcity and danger, to a huge technological innovation that may end up to the new relaxed situation and will stay there until we face the scarcity or danger again and this cycle will continue on.

There are regulations that enforce the emissions and also test methods to unify the emission testing on automobiles. The most important ones are the New European Driving Cycle (NEDC), United States Corporate Average Fuel Economy (CAFÉ), and Japanese Japan 10-15. Figure 13.1 depict the CO2 emissions and its equivalent fuel consumptions.

In order to reduce our emissions, we should use fewer resources and become more efficient; to do so there is a simple and common solution in the automotive and aviation industries and that is lightweighting. For every 10% weight that our car sheds we save up to 8% on fuel (Stans and Bos 2007; Van den Brink and Van Wee 1999). In case of aviation industries, releasing the CO_2 in the atmosphere will cause more damage than a car because the altitude is higher and therefore it may be more important. The concept of lightweighting is really easy; it is literally replacing materials with higher density with the materials that have lower density but the same strength. For example, glass fiber has a density of about 2.5 g/cm^3; therefore, whatever we put in our composite that has reinforcing fiber with lower density like natural fibers (1.4 g/cm^3) we could see a lightweighting effect (Wambua et al. 2003). There is a well-known concept in science called the rebound effect, which can explain why with all the improvements in the internal combustion engines in terms of efficiency our car's consumption has not been changed much. Rebound effect on the

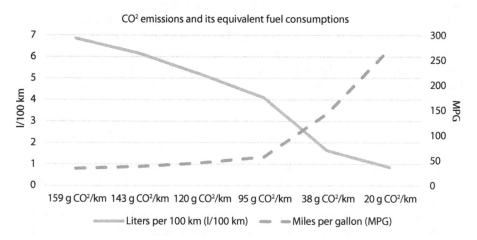

FIGURE 13.1 CO_2 emissions and their equivalent fuel consumptions. Some countries already suggest 20 g CO_2/km for 2050. (Data based on the US EPA and European commission climate action and Copenhagen accord 2009.)

auto industries offsets the advantages (Hertwich 2005) on engine performance by increasing the curb weight of the vehicle. As you can see in Figure 13.2, global car weight increased during the past decades.

Each auto manufacturer has its own strategy to reduce emissions, and some like Ford even try to adopt aluminum instead of heavier steel. Ford Motor Company has been very active in the utilization of new and innovative materials. In fact, Henry Ford was working on his bio-composite car (soybean) around 1941. Due to the cheap price of petroleum-based chemicals, his soy-based plastic never made it to the mass market (Songstad et al. 2011). Now with the emerging aluminum body trucks (Ford F-150), the first 3D printed car and so on, we are at the advent of the revolution in our automotive industries that will change many things around us.

One of the interesting milestones in lightweighting is the production of a multimaterial lightweight vehicle (MMLV) which was a joint project of Magna International, Ford Motor Company, and the US Department of Energy. This project used available technology and materials to make a base model Ford Fusion 2013 out of available lightweight materials without affecting safety or performance of the vehicle. As a result, a lightweight version of the vehicle (23.5% lighter) was made and was able to perform with a 1 L, three-cylinder engine instead of 1.6 L, four-cylinder engine. Both vehicles were compared through a life cycle assessment (LCA) and simulated to be driven for 250,000 km in North America. MMLV was able to save over 3600 L of fuel. What is not to love about this vehicle is the price tag. This concept vehicle is worth almost half a million dollars, which is not even comparable to the baseline model which is sold for around $25,000.

During recent years, every single automaker and parts manufacturer felt the pressure from the customer and government to reduce the weight, as the curb weights are 20%–30% increased during the past decades (IEA data). Lightweighting reduces "material service density," which is the mass of materials that is needed to do the same service (Xu et al. 2008), therefore it will end up using fewer resources,

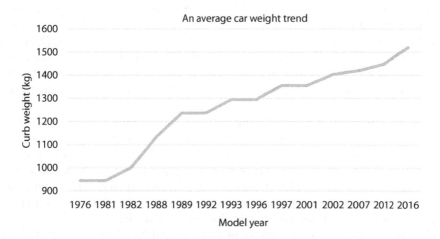

FIGURE 13.2   Vehicle weight trends from 1976–2016 data averaged and extracted from some of the manufacturers' released information to show the uptrends for the curb weight.

and also it will reduce fuel consumption, especially during the auto use phase and send less material to the landfill. Having a saved mass usually related to cost efficiency means that the manufacturer can afford a higher price for lightweight materials and still remain within the same price range. However, the lightweighting could be really expensive and as it is shown in Figure 13.3, the GHG emission is inversely correlated to the cost. As it is indicated by the image's iso-value lines, there is always a way to reduce greenhouse gases if the industry pays for it. Fortunately, in some cases these costs will be covered by the money saved due to mass saving. Production of lightweight materials in some cases may be very energy intensive, for example, carbon nanotube is very strong and light, but not really green to make, therefore, any claim that is not considering the holistic view of the lightweighted automobile may not even be as green as consumers think. The irony is that the materials that are natural are not necessarily environmentally friendly (Jolliet et al. 1994).

For the lightweighted parts, the saving of use phase (66%–97% of the energy demand) usually overweigh the material production, manufacturing, and end of life impacts even if this lightweighting does not lead to the secondary weight reduction (Koffler 2014). Secondary weight reduction or mass decompounding is the weight reduction that comes after we replace a heavy component with a lighter one, for example, if we reduce the weight of the engine we may not need the engine support as strong as before and we may be able to shed weight on the support as well. It could be assumed a 0.2–0.3 kg weight could be shed for every 1 kg primary weight reduction in the powertrain system (Lewis 2013).

The lightweighting in the engine component is more valuable as there should be a balance of 50%–50% weight for front and rear of the vehicle. Another aspect of fuel savings by lightweighting comes from the fact that we need less fuel to extract and bring all the way to the fuel station. This part is usually called well to tank (WTT) or well to pump (WTP) phase.

The fuel economy is being affected by a number of different factors like vehicle specifications and also fuel type. Fuel type is important as it may change the result of emissions, for example, using gasoline will emit 2.32 kg CO_{2eq}/L (EPA average) while using E85 (contains 85% ethanol) will only emit 0.65 kg CO_{2eq}/L (Kim and Dale 2006). As it is obvious by adding more ethanol to gasoline you can reduce the emission, but it may cause damage to the engine, therefore in many cases E10 (10% ethanol and 90% gasoline) considered to be safe for the vehicle is used as a primary source of fuel.

Although lightweight material is the way for manufacturers to meet CAFE, these materials are usually not cheap and the worst part is that there are some reports indicating that some of these materials are not really environmentally friendly. Production of the 1 kg carbon fiber will emit 14 times CO_2 than the production of 1 kg steel (Das 2011). Therefore, if we don't recycle our carbon fibers it is not even a good idea to use these materials as they only move the sources of emission from the use phase of a car to the manufacturing phase of the materials.

Making a vehicle is a huge source of emission and based on Argonne National Laboratory, for manufacturing a vehicle there would be an average emission of 2 ton CO_{2eq}. Lightweighting may help to alleviate these emissions as well by using

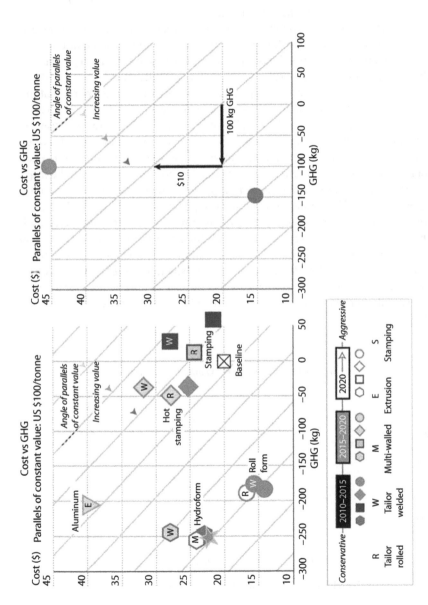

FIGURE 13.3 (Right) GHG emission and cost are co-related to the money. Iso-value lines help to estimate that if you want to reduce how much you should pay more. (Left) Rocker solution comparison graphs, cost versus GHG. (From Shaw, J., Kuriyama, Y. and Lambriks, M. 2011. Achieving a Lightweight and Steel-Intensive Body Structure for Alternative Powertrains (No. 2011-01-0425). SAE Technical Paper.)

less materials. The shift from metal to plastic and now to natural fiber reinforced bio-based plastics (Figure 13.4) will have some advantages aside from drawbacks. Although bio-based plastic has been around for 155 years now (Kaufman 1963), they weren't taken seriously by auto industries as global warming was not regarded as serious as it is now and regulations were not in place.

To solve the debate that lightweight materials are greener than our current materials the International Organization for Standardization provides a procedure to help us see the reality and make a wise decision. There is a series of ISO family standards 1404X that creates guidelines and rules to the true way of the LCA.

LCA of a vehicle could be very complicated, as many of the auto parts that are made by OEM may include outsourced steps or materials from Tier 1 and Tier 2 (Keoleiann et al. 1998; Kobayashi 1997; Sullivan et al. 1998). There are some aspects in lightweighting which are sometimes hard to cover in an LCA. For example, when we are using less materials the amount of materials that we save is no longer needed to be produced and therefore these materials do not need the energy and oil extraction in the first place. Another important thing is the fact that if we make an automobile lightweight even if it is minor, during the use phase, we could see fuel savings. This saved fuel not only reduces our emissions directly, but also indirectly as it doesn't need to be extracted, refined, and transported to the gas station and also although very minor the energy that is needed to pump the fuel from the gas station's tank to your car's tank will be less. These effects, which are often called "well to wheel," will be very huge when we multiply the amount of fuel savings for a single car to the total number of cars that are produced and could be lightweighted with the new materials.

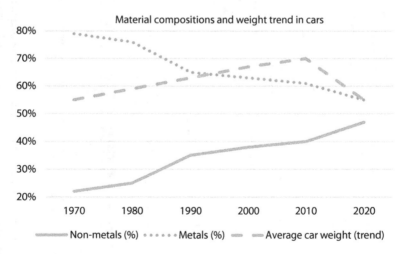

FIGURE 13.4 Material compositions in automotive industry show a shift from metal to non-metal materials like composites. The average vehicle weight was added to the picture to provide an idea of the curb weight trend from 1970–2020. (Data from A.T. Kearny analysis; Mckinsey & Company; and Economics & Statistics Department, American Chemistry Council.)

Researchers have been working on the life cycle of the automotive parts for years and there has been great publication on the topic. Currently a few reviews exist on the LCA topic. Table 13.1 provides a summary of the published articles on the LCA for the automotive parts.

There are also some simplified LCA methods in the literature like the matrix method (De Vegt and Haije 1997), which is favored because it is not as restricting as the ISO methods and it is faster and easier to perform.

Until we reach a better propulsion system or even a better infrastructure for hydrogen fuel cells or electric cars, the internal combustion engines will remain the main propulsion system for automotive industries. Therefore, our only option for now is replacing our heavier materials with lighter counterparts and the good news is that blending these materials with current technology is easy and cheap as most of the infrastructure does not even need to be changed. Simply, if we change an energy intensive fiber with less energy intensive ones we have reduced the life cycle energy.

The automotive industry is the major consumer of the fiber reinforced polymers and each year this industry alone consumes about 31% of total production (Anandjiwala and Blouw 2007). Natural fibers are considered carbon neutral by some of the researchers and have thermal and acoustic properties (Peijs et al. 2002) that are favored by the auto industry.

13.2   BIO-BASED CARBON STORAGE

Researchers have been treating natural fibers with three different approaches. Some see the natural fibers as a carbon negative process (Joshi et al. 2004), some believe that this process is carbon neutral in nature and although it may have short-term benefits for capturing carbon for several years, at the end it will emit the CO_2 to nature. The last group considers that the process is carbon negative and therefore you can reduce the number of carbons captured from the actual emission (European Union Directive 2009). There is even a clause in the greenhouse gas protocol mentioning that you can deduct the carbon in the ashes of wood product and landfill from emission (GHGP 2011).

ISO 14067 mentions that if these carbons will be emitted back within 10 years, they should be assumed as they are emitted right after and if they are going to stay for more than 10 years the effect of time should be calculated and reported (ISO 2013). As described in one of the literatures (Vogtländer et al. 2014), the delay in the release of CO_2 may end up to carbon credit. As we usually estimate the global warming potential in 100 years (GWP100), if a bio-composite keeps all the carbon intact for 50 years and then starts to release it back to nature slowly, whatever carbon that is released in 100 years minus the total amount of carbon that could be released from the object without 50 years delay is considered as credit (Lashof calculations) (Fearnside et al. 2000; Clift and Brandão 2008). If a biocomposite keeps carbon for 120 years, it won't even be considered in GWP100 and only if we look at the global warming potential in the longer window, like 500 years, we could see the effects. But as the imminent global warming problem will have its effect within 100 years, many researchers only look at the short term.

Natural fiber versus glass fiber studies show that aside from the lightweighting derived fuel saving, we have some savings in the production of natural fibers in

TABLE 13.1

Summary of Published LCA in the Automotive Industry

Year	Study/Part	System Boundary	LCA Details	Software	References
1998	EPDM/PP/GF vs. EPDM/PP/hemp fiber (30% wt.) Insulation component Ford	C2G	Simplified LCA	NA	(Schmidt and Beyer 1998)
1999	Underfloor panel, Mercedes-Benz A-class, FGRPP and flaxRPP	C2G	NA	NA	(Diener et al. 1999)
1999	Audi A3 panel, ABS (acrylonitrile-Butadiene-styrene) vs. hemp fiber	C2G	Eco-indicator 95	NA	(Wötzel et al. 1999)
1999	Propeller shaft Isuzu (aluminum alloy vs. glass and carbon fiber)	Non-ISO	NA	DFE	(Kasai 1999)
2001	Three aluminum casting processes: lost foam, semi-permanent mold, and precision sand		NA	NA	(Stephens et al. 2001)
2001	Body panels, NiMH and LiIon HEV (hybrid electric) batteries, and ICEVs (internal combustion engine vehicle) and FCVs (fuel cell vehicle)	C2G	DEAM lifecycle inventory database, APME data source, TRI	LCD Toolkit	(Schexnayder et al. 2001)
2004	ABS (acrylonitrile-butadiene-styrene) vs. hemp fiber	I/O-LCA	NA	NA	(Gärtner and Reinhardt 2004)
2004	Polymer nanocomposite	I/O-LCA	Hybrid LCA (process LCA)	NA	(Lloyd et al. 2003, 2005)
2005	Wood panel		NA	NA	(Cinar 2005)

(*Continued*)

TABLE 13.1 (CONTINUED)
Summary of Published LCA in the Automotive Industry

Year	Study/Part	System Boundary	LCA Details	Software	References
2006	Coconut, flax, cotton fiber, and wood (interior components)	C2G	NA		(Finkbeiner and Hoffmann 2006)
2007	Curaua/PP vs. GF/PP (interior component)	C2G	CML		(Zah et al. 2007)
2008	PTP vegetable resin vs. polyester resin + hemp (bus body component)	C2G	NA		(Mussig et al. 2008)
2008	Wood fiber reinforced PP	C2G	Eco-indicator 99	SimaPro	(Xu et al. 2008)
2008		C2C	NA	NA	(Kumar and Putnam 2008)
2008	LCI of ELV, dismantling and shredding	Gt2Gt	NA	NA	(Sawyer-Beaulieu and Tam 2008)
2008	CNF polymer composite, equivalent of a standard steel plate (122 × 224 × .05 cm)		NA	NA	(Khanna et al. 2008)
2008	Hemp/PTP vs. GF polyester (bus body component)	C2G	Eco-indicator 99	SimaPro	(Schmehl et al. 2008)
2009	Exterior door skin	C2G	GEMIS 4.3	NA	(Puri et al. 2009)
2009	CNF polymer composite, 150,000 vehicle miles of a 3300 lb body panel in a midsize car	C2G	NA	NA	(Khanna and Bakshi 2009)
2009	Rear body of truck (glass fiber/unsaturated polyester composites)	C2G	NA	NA	(Song et al. 2009)
2010	Sugarcane bagasse-reinforced polypropylene vs. talc-filled polypropylene (interior aesthetic covering component)	C2G	CML 2001	Gabi	(Luz et al. 2010)

(Continued)

TABLE 13.1 (CONTINUED)
Summary of Published LCA in the Automotive Industry

Year	Study/Part	System Boundary	LCA Details	Software	References
2010	ASR treatments	Gt2G	Eco-indicator 99	SimaPro 7.1	(Ciacci et al. 2010)
2010	Economic assessment of GHG reduction due to lightweighting		NA	NA	(Kim et al. 2010)
2010	Jute fibers vs. GFRP (buggy bonnet)	C2G	Eco-indicator 99	SimaPro 7.0	(Alves et al. 2010)
2011	Lightweighting costs and batteries	C2G	NA	NA	(Shaw et al. 2011)
2011	30.8 kg floor pan (conventional textile-type acrylic vs. lignin [through P4 and SMC])	C2C	GREET, MOBILE6,	SimaPro 2008	(Das 2011)
2011	Magnesium, steel vs. composite (glass mat thermoplastic, structural reaction injection mold (P4 and ACC))		Impact 2002+	SimaPro	(Witik et al. 2011)
2012	ELV 100 kg	Gt2C, Gt2G, Gt2Gt	EcoInvent ELCD and CML 2001	MicroSoft Excel®	(Fonseca et al. 2013)
2012	Fender and hood (aluminum alloy vs. cold roll carbon steel)	C2G	CML 2001	Gabi	(Marretta et al. 2012)
2013	Physical model (lightweighting) of mass induced fuel consumption		NA	NA	(Kim and Wallington 2013)
2014	Plastic engine cover, plastic engine cover with PU foam, plastic engine cover with shoddy materials, PU foam alone	Gt2G	NA	NA	(Miller et al. 2014)

(Continued)

TABLE 13.1 (CONTINUED)
Summary of Published LCA in the Automotive Industry

Year	Study/Part	System Boundary	LCA Details	Software	References
2014	Rim made of different materials and structure		Eco-OptiCAD	eVerdEE, an abridged	(Russo and Rizzi 2014)
2014	Remanufacturing truck gearbox		Eco-design ISO19439	NA	(Goepp et al. 2014)
2014	Steel, aluminum, and composite bodies-in-white	Dynamic LCI	NA	MicroSoft Excel®	(Stasinopoulos and Compston 2014)
2014	Grill shutter housing 30% GF vs. 30% cellulose fiber vs. 40% kenaf fiber composite	C2G	Energy saving and GHG emissions,	Gabi 6	(Boland et al. 2014)
2014	Bolster for hood latch (GFRP vs. alloy)	C2G	TRACI, NREL USLCI	Gabi 4	(Koffler 2014)
2014	Eco-sandwich panel bio-based epoxy resin and cork, hemp fiber vs. GF petroleum based resin	C2G	NA	SimaPro 7.2	(La Rosa et al. 2014)
2014	Drum brake system (US trucks)	C2G	Impact 2002+	SimaPro/MicroSoft Excel®	(Han and Choi 2014)
2015	Multimaterial lightweight vehicle (MMLV)	C2G	TRACI 2.1, GREET-1 & 2	SimaPro 8.03	(Bushi et al. 2015)
2015	GF (30%) reinforced PP vs. 40% kenaf fiber PP vs. 30% cellulose fiber reinforced PP console components	C2G	Energy demand and GHG	Gabi	(Boland et al. 2015)

Note: C2G: cradle to grave, Gt2Gt: gate-to-gate, Gt2G: gate-to-grave, Gt2C: gate-to-cradle.

comparison to glass fibers (Figure 13.5). When we use less materials (which means less emissions), we send less materials to the landfills, and this fact in some cases leads to a carbon credit (Pervaiz and Sain 2003).

It has been reported that using a natural fiber could be environmentally friendlier than other materials, but they may have worse water emissions like nitrate, phosphate, and nitrogen oxide due to fertilizer usage (Wötzel et al. 1999). Bos (2004) reported that while mechanical properties are caused by fibers, environmental pollution is caused by matrix (Bos 2004), but on the other hand, although they have a bad reputation, plastics are not very bad for our environment. Plastics use only 4% of current oil as an input (O'Neill 2003) which is only 0.3% if we focus on the automotive industries and also in comparison to metal and glass it has much less energy intensive production (Association of Plastics Manufacturers in Europe Ref No. 8041/GB/08/03). On average whenever we use plastic it reduces 2–3 times the weight of alternative materials (APME Ref No. 8041/GB/08/03). The replacement of metal with plastic was actually a good movement toward lightweighting and the truth is reinforcements are causing a bad reputation for plastic composites. Glass fiber production needs a temperature around 1550°C (Kellenberger et al. 2007) and for carbon fiber the temperature needs to reach up to 2000°C for production of high strength fibers (Subic et al. 2009) and to see the real picture after consuming this amount of energy the yield has never been 100%. Production of natural fiber, on the other hand, is less energy intensive than other fibers and this will help to reduce the total composite emissions. For example, kenaf reinforced bio-composites are 23%–24% less energy consumer and also have 6%–16% less GHG emissions (Kim et al. 2008) but if we need fertilizer it may cause some environmental issues. However, by using wood fiber or the fibers that do not require fertilizers, especially if it is a by-product or waste like wheat straw, we could claim that natural fibers are greener.

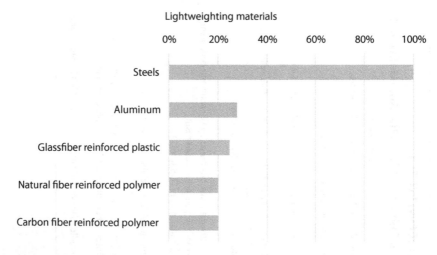

FIGURE 13.5 Some of the lightweigthing materials and their weight reductions, based on their average densities. Please note that in some cases like aluminum we may need to have higher volume of the materials to meet the designated duties. (Data from Klink et al. 2012 and Mainka et al. 2013.)

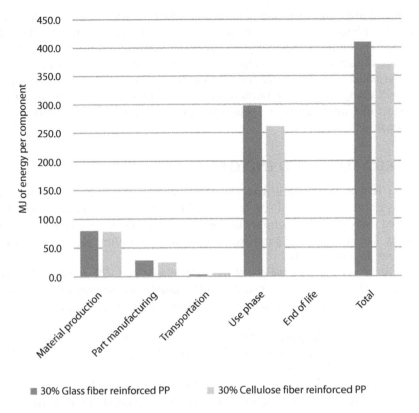

FIGURE 13.6 Comparable primary energy consumptions associated with production of glass fiber reinforced PP versus cellulose fiber reinforced PP. (From Boland, C., DeKleine, R., Moorthy, A., Keoleian, G., Kim, H.C., Lee, E. and Wallington, T.J., 2014. A Life Cycle Assessment of Natural Fiber Reinforced Composites in Automotive Applications (No. 2014-01-1959). SAE Technical Paper.)

Renewable biomaterials also have their own problems, for example, emission of N_2O is an important factor (Tufvesson et al. 2013) which is usually derived from using the fertilizer or decomposition of plant in soil (Tufvesson and Borjesson 2008).

Even if these biomaterials are produced by an energy intensive farming method and have a comparable production emission with the glass fiber (Figure 13.6), as they will have a lightweighting effect in the vehicle, during the use phase, they usually offset the production emissions and after the breakeven point they will be environmentally better choices.

13.3   LAND USE CHANGE PROBLEMS IN THE PRODUCTION OF THE NATURAL FIBERS

Most of the biomaterials simply need land to grow. Where these lands come from is important. If they are changed from food producing or jungle to biomaterials production it is called direct land use change (PAS 2050:2011) and there is also an indirect land use change because of material production. An example is that because an industry needs wheat straw we cultivate wheat in a land that normally was a corn field. Therefore, we need to grow our corn elsewhere to meet the demand and so on (Searchinger et al. 2008; Tipper et al. 2009). Land use change is not a case in every bio-based material production, for example, some of the bio-based materials are not changing land use because they use an agricultural and/or silvicultural by-product like wheat straw or sawdust from the mills, and some are produced by microorganisms.

13.4   RECYCLING AND END OF LIFE

Your car is recyclable, more than any other products (Steel Recycling Institute) around 95% of your car will be recycled after the use phase (Directive, E.L.V., 2000) and many car manufacturers use recycled materials, for example Ford Motor Company has made plastic parts from more than 50 million plastic soda bottles (Auto Recycling fact sheet).

Almost 10% of the curb weight of the car is plastic and ~7% is rubber (Sullivan et al. 2010). The current plastic needs strength; usually the first option is glass fiber reinforcement, which can be replaced with natural fibers to decrease environmental impacts.

What happens to the plastics at the end of life of a vehicle is not really the most environmental. Unfortunately, most of it ends up in landfills and in fact 41% of land-fills are plastics (Mirabile et al. 2002).

Not all bio-based materials are biodegradable, biodegradability will be affected by several factors, including chemical composition and constitution of the product (Nampoothiri et al. 2010; Hermann et al. 2011). The plastic will stay in the landfill for a long time. Currently a material is considered biodegradable if 60% (based on ASTM D6400) or 90% (based on EN 13432) of the materials degrade in 180 days. The good news is that bio-based thermoplastics are recyclable the same as their pet-rochemical equivalents (Shen et al. 2009).

When an automobile reaches its end of life (250,000 for passenger cars and 290,000 for trucks and SUVs) it is mostly recycled (up to 95% based on the new required regulations). Although cars are the most recycled product today (Steel Recycling Institute) and we have seen efforts from the car makers to recycle even more like reusing bumper covers (Ford, Toyota, and Mazda) and returning paints for recycling (General Motors), all of these steps are still not enough. What is happening to your car at the end of life is that after recovering most of the metals and recoverable items, the rest will be shredded and is called automotive shredder residue (ASR). Unfortunately, almost half of the ASR is plastic (Nourreddine 2007). Plastic at the end of its life is usually landfill in North America. Nowadays only around 150 kg plastic is being recovered from each vehicle in the United States, which is a sign of the difficulty in the recycling of plastic (Jenseit et al. 2011; Zhao and Chen 2011). And that 150 kg mainly comes from the bumper; it is one of the most important plastic parts (Tian et al. 2015). Aside from landfill, plastics could have four other scenarios as well: recycling, incineration in cement kiln, reuse, or standard incineration (18% efficiency). Landfill is usually the worst option as you are not recovering anything from the materials or the energy but it is incineration that is the worst in terms of greenhouse gas emissions (USEPA 2015).

United States Environmental Protection Agency has a waste reduction model (WARM) to reduce the wastes. WARM sees four different scenarios for plastic at the end of life:

1. Recycling (collect and transport > sorting > process > reuse)
2. Combustion (collect and transport > combustion > ashes to landfill)
3. Landfill (collect and transport > landfill)
4. Composting (collect and transport > composting > noncompostable parts to landfill) (USEPA WARM 2015)

Toward making recycling easier, there was a labelling system introduced a few years ago, which asked the manufacturer to check and label as follows if their design met the necessary criteria. One class of these materials could be labeled as "Design-for-environment" (DfE), which is essentially a class of materials that they will be labeled as one of the following:

* Design-for-dismantling (DfD)
* Design-for-recycling (DfR) (Bogue 2007)

This labelling system is not mandatory and even if it was, it doesn't mean that a part with DfD or DfR label will have a different fate other than landfill. This labeling system is no longer used frequently.

13.5 LIFE CYCLE ASSESSMENT

Life cycle assessment (LCA) is the powerful method to evaluate and assess the product in terms of environmental damage and harmfulness. It helps to compare different products or methods and calculate which one is greener. ISO defines a strict guideline for performing an LCA and if you don't follow them your LCA is not in compliance with ISO. The main outline of the true LCA will be discussed shortly.

An LCA can be done from cradle to gate, like extracting oil to bringing it to refinery, or gate to gate like evaluating two different means of transportation of goods from manufacturer to assembler, gate to grave, which is ending up in end of life scenarios or even cradle to cradle for a product from extraction of resources all the way to recycling and becoming a new material for another product!

Each LCA should contain the following.

1. Goal and scope definition: This includes functional unit, system boundary, reference flows, allocation, assumptions, selected impact categories and methodology, data requirement, limitations, critical review (if any) and simply answering the questions of (a) Who is the audience? (b) What is the purpose of the study? (c) Will these results be disclosed to the public? And even the format of the report.

2. Inventory analysis: It is obvious it is the list of inputs and outputs of the system. You may be required to go back and change your goal and scope definition based on the available data or the possibility of the analysis. There are several life cycle inventory analysis (LCIA) methods available globally. The most important ones are TRACI, IMPACT, CML, and Eco-indicator. Each country or region can have its own LCIA like LUCAS which is an LCIA method used for a Canadian-specific context. The impact assessment categories based on TRACI v2.1 are shown in Figure 13.7. There are six impact categories and each has a unit equivalent for ease of the measurements (Ryberg et al. 2013; Bare et al. 2003).

Each life cycle assessment should have an inventory indicator which literally means to show where does the energy come from (the source could be fossil fuel, nuclear, biomass, hydroelectric power, solar, geothermal, and wind) and also materials sources and water consumption with their units.

3. Impact assessment: It will assess the impact of the outputs. This is the part that tells us how our materials are doing on nature and what are the impacts.

4. Interpretation: The results are summarized here and to help a conclusion, recommendation, decision making, and maybe changing the goal and scope (Figure 13.8).

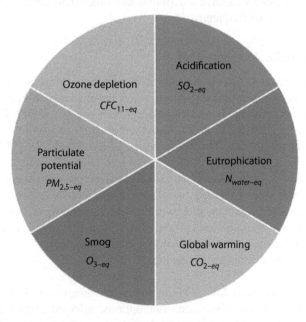

FIGURE 13.7 Life cycle impact assessment based on U.S. EPA TRACI v2.1. (From Ryberg, M., Vieira, M.D., Zgola, M., Bare, J. and Rosenbaum, R.K., 2014. *Clean Technologies and Environmental Policy*, *16*(2), pp. 329–339. Bare, J.C. et al., 2003. *Journal of Industrial Ecology*, *6*(3–4), pp. 49–78.)

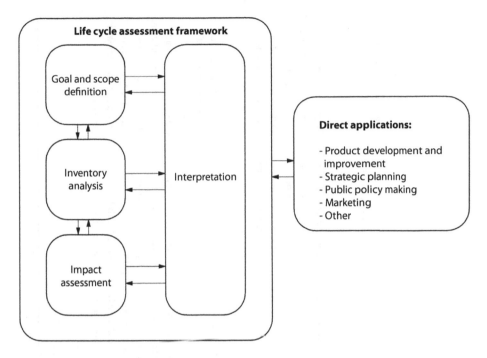

FIGURE 13.8 Stages of LCA. (From ISO 14040:2006.)

13.6 DATA QUALITY

There are several methods to evaluate the data quality. One of the simplest ways is the pedigree matrix method described by Weidema and Wesnaes (1996). Although it was introduced 20 years ago this method is still one of the methods that is used frequently around the world (Table 13.2).

This method includes 5 different quality indicators, among them, reliability is the most important factor and completeness is the least important after giving each process a score it will be matrix of scores, for final score you have to use the highest number of score for the total process. For example, if you are making a product and you have two process one with the data quality of (4,3,2,1,5) and the other one with score of (5,2,1,1,1) your final product score will be (5,3,2,1,5) and as it was mentioned earlier first score 5 is reliability and our data may have a coefficient of variance of up to 97%.

May and Brennan more recently (2003) recommended a newer approach that contains qualitative and quantitative assessments which is not essentially easier to evaluate the data quality (May and Brennan 2003).

13.7 LIFE CYCLE ASSESSMENT OF AUTO PARTS

When a new part is subject to a weight change due to material composition and manufacturing technology or design geometry, Canadian Standard Association (CSA)

TABLE 13.2

Data Quality Assessment Method

Data Quality Indicator Score	1	2	3	4	5
Reliability	Verified data based on measurement	Verified data partly based on assumptions or non-verified data based measurements	Non-verified data partly based on assumptions	Qualified estimate (e.g., by industrial expert)	Non-qualified estimate
Completeness	Representative data from a sufficient sample of sites over an adequate period to even out normal fluctuations	Representative data from a smaller number of sites but for adequate periods	Representative data from an adequate number of sites but from shorter periods	Representative data but from a smaller number of sites and shorter periods or incomplete data from an adequate number of sites and periods	Representativeness unknown or incomplete data from a smaller number of sites and/or from shorter periods
Temporal correlation	Less than three years of difference to year of study	Less than six years difference	Less than 10 years difference	Less than 15 years difference	Age of date unknown or more than 15 years of difference
Geographical correlation	Data from area under study	Average data from larger area in which the area under study is included	Data from area with similar production conditions	Data from area with slightly similar production condition	Data from unknown area or area with very different production conditions
Technological correlation	Data from enterprises, processes and materials under study	Data from processes and materials under study but from different enterprises	Data from processes and materials under study but from different technology	Data on related processes or materials but same technology	Data on related processes or materials but different technology

Source: Weidema, B.P. and Wesnæs, M.S., 1996. *Journal of Cleaner Production, 4*(3), pp. 167–174.

released a standard in 2014 to set a defined guideline to compare the products within the standard LCA (CSA 2014). There are usually three stages in this type of LCA: production phase, use phase, and end of life (EOL) phase (Figure 13.9).

As indicated by CSA, data shouldn't be older than 10 years and if they are between 5–10 years, it should be mentioned. Primary data should be reflecting one whole year without a gap in months.

13.8   EXCLUDED PROCESS AND CUT-OFF RULES

Excluded process could be human energy inputs, employee commute, capital infrastructure and production overhead like lighting, HVAC and so on.

In terms of cut-off, the data gap should be filled with generic data and insufficient inputs should be less than 1% of mass or energy for that process and for total cradle to grave should be within 5%.

13.9   SENSITIVITY ANALYSIS

Sensitivity analysis compares the result with the extreme situation and the influence on the result should stay within ±10%, it is usually done on the vehicle mileage, selection of different materials, design, process, electric power grid mix for production phase, fuel resources and emission profile, and finally end of life scenarios. For example, vehicle lifetime driving is usually 250,000 km for passenger cars and 290,000 for pickup trucks, SUVs, and vans. Sensitivity analysis needs to be done at 200,000 km for cars and 250,000 km for vans.

13.10   MASS INDUCE FUEL CHANGE

As we know, by changing the weight of a car, the fuel consumption will change. There are several reports indicating different methods and numbers as a fuel saved and among all CSA and PE International (2012a,b) provide the list of standard changes in fuel consumption.

There are formulas to calculate fuel change due to mass alteration:

Without powertrain adaptation:

$$C = (M2 - M1) * F * D$$

where C is fuel change in liters

$M2$ is new part weight

$M1$ is conventional part weight

F is constant range from 0.31 to 0.4 l/(100 km × 100 kg) depending on the type of engine and fuel

D is distance derived during the vehicle lifetime

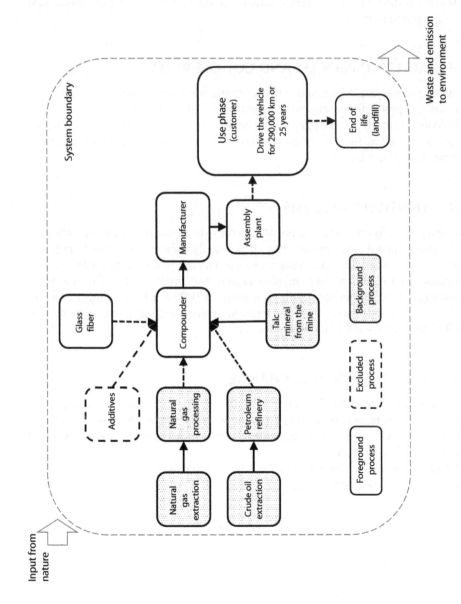

FIGURE 13.9 A simplified example of a system boundary in the lightweighted auto part.

With powertrain adaptation:

$$C = Cm + \sum_{i=1}^{n} (M1 - M2) * F * D$$

where C is fuel change in liters
  Cm is the fuel change in liters due to primary mass change
  $M2$ is new part weight
  $M1$ is conventional part weight
  n is the number of parts that exhibit mass changes
  F is constant range from 0.31 to 0.4 l/(100 km × 100 kg) depending on the type
    of engine and fuel
  D is distance derived during the vehicle life time

For more complex situations like secondary mass change with drive train adaptation, you can look at the CSA standard guideline.

All the environmental problems have been caused by humans and as a human is a complex being it is not a good idea to talk about an environmental issue without talking about the social and economic aspects of it. Socio-economic drives are deeply bound to human society and behavior; therefore, any complete impact assessment should consider social and economic aspects of the matter as well. Within the publications usually we can see graphs comparing these important aspects (Figure 13.10), but unfortunately as every researcher has his or her own means of weighting and scoring, you can't compare these graphs from paper to paper.

Recently, a more comprehensive and complete approach has been mentioned in the literature called "life cycle sustainability assessments (LCSA)." This method includes aspects of life cycle cost (LCC), cost benefit analysis (CBA), and also a social LCA into an environmental LCA, which makes it better and also more complicated to perform. This approach was introduced to contain economic, social, and environmental aspects of product or service (Hoogmartens et al. 2014), it may or may not comply with ISO 14040 family as one can do the LCC (financial, full environmental and social LCC) or CBA (financial, environmental, and social CBA) instead of LCA. If this technique improves, the results of all three analyses should be comparable.

LCA is becoming more and more important and everyday more publications could be seen in the literature (Table 13.1) on the topic what is even more interesting is that the nature of LCA has been changing in the past 6 decades from the company driven to the regulatory driven and in the past two decades to the policy driven, while in the early stage, a single product was important. Now, the policy development is important and also we see more and more socio-economic aspects coming out (McManus and Taylor 2015).

LCA, like any other tools, has its own cons, one of the most important drawbacks of the LCA is that its quality is highly tied to the available data, and we don't have

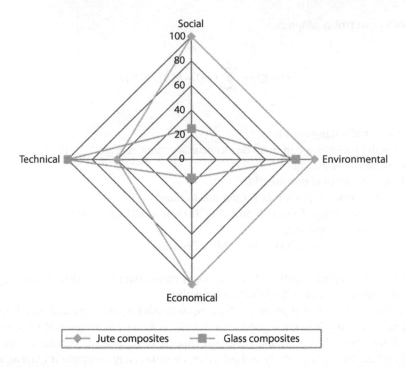

FIGURE 13.10 Performance of Bonnet aspects. (From Alves, C., Ferrão, P.M.C., Silva, A.J., Reis, L.G., Freitas, M., Rodrigues, L.B., and Alves, D.E., 2010. *Journal of Cleaner Production*, *18*(4), pp. 313–327.)

many important data available to us so we could expect that in the future more reliable LCA studies will be available to the public.

Unfortunately, the other drawback is caused by researchers and that is a fact that in the LCA, literature considered a lot of assumptions and that makes them hard to compare (Kim and Wallington 2013) and in some cases impossible. It is really important that the LCA result should be carefully interpreted as the results of an LCA could change drastically based on the methodology, assumptions, and other values and choices.

LCA still needs a lot of improvements in many of its aspects; for example, a positive impact in LCA has been missed out, like the situation that we end up emitting sulfur dioxide, this molecule reacts with sulfates and therefore mitigates the global warming effects, but it is simply not calculated. This issue is relatively easy to fix. We can calculate and reduce the amount from the global warming potential, but there are some more complicated aspects of LCA that need to be addressed and improved like creating an inventory data for noises, noise after a certain limit will affect human well-being. We need to pay more attention to odor as it is a very important factor. Imagine if we have a greener process that has a terrible odor for a city. It may not be pleasant and no corporation, no matter how powerful, is interested in making people hate them. Another serious aspect is the changes in soil quality, desertification, salination, erosion, and so on (Finkbeiner et al. 2014). These are really important as the soil needs a long time to recover. Some other aspects of LCA include scale-up for the economy which is a situation like, for example, if we start using bioethanol based engines around the world. Not only is the effect different than a small scale it may have an indirect but a huge effect on our agriculture, forestry, and other apparently nonrelated businesses (Hellweg and Milà Canals 2014). Animal well-being, unintended (and sometimes necessary?) killing of animals is very important, but we don't have an effective measure of that in our LCA toolbox yet (Finkbeiner et al. 2014). There are many more aspects in the full LCA that have not been mentioned here like losing biodiversity and scientists around the world are working on that.

Human life is so tied up with the lack of the energy that we almost forget that if we master four fundamental interactions of nature and become able to use the energy inside those forces we could change our environment back to its original state. Inside human civilization, everything can be explained with the transportation and transforming energy, almost all the aspects of our history and life can be explained by that. Therefore, if we master these two words, then we don't need to be worried about our environmental footprints. Our complicated problem of global warming will be an easy to solve problem for us and all we need to do is to move some of the excessive energy from the atmosphere to somewhere else, until then we need to be very cautious about our footprint as we currently have just one planet and unfortunately we are moving very fast beyond the safe operating boundaries. We are capable of doing a lot of things and even reversing almost all our damages, but with the current energy problem it is simply not worth it. If the energy is out of the question, the car weight is not an issue anymore, but until then lightweighting seems to be our only solution and it is really important. Lightweighting may seem a temporary solution to our emission problem because we do have an upper limit in lightweighting. If we lightweight a vehicle too much it may compromise vehicle dynamics (Koffler 2014) and lightweighting for hybrid electric vehicles is not really as good as it is in other vehicles, as it tends to capture the kinetic energy of the brake in the form of electricity (Lohse-Busch et al. 2013).

REFERENCES

Alves, C., Ferrão, P.M.C., Silva, A.J., Reis, L.G., Freitas, M., Rodrigues, L.B., and Alves, D.E., 2010. Ecodesign of automotive components making use of natural jute fiber composites. *Journal of Cleaner Production*, *18*(4), pp. 313–327.

Anandjiwala, R.D. and Blouw, S., 2007. Composites from bast fibres—Prospects and potential in the changing market environment. *Journal of Natural Fibers*, *4*(2), pp. 91–109.

Association of Plastics Manufacturers in Europe. Available on: http://web.deu.edu.tr /metalurjimalzeme/pdf/MME%204001Polymers/ELV_summary_120503.pdf.

Auto Recycling Fact sheet. http://www.erie.pa.us/Portals/0/Content/PublicWorks/recycling /FactSheetAutoRecyclingFacts.pdf.

Bare, J.C. et al., 2003. TRACI the tool for the reduction and assessment impacts. *Journal of Industrial Ecology*, *6*(3–4), pp. 49–78.

Bogue, R., 2007. Design for disassembly: A critical twenty-first century discipline. *Assembly Automation*, *27*(4), pp. 285–289.

Boland, C., DeKleine, R., Moorthy, A., Keoleian, G., Kim, H.C., Lee, E., and Wallington, T.J., 2014. A Life Cycle Assessment of Natural Fiber Reinforced Composites in Automotive Applications (No. 2014-01-1959). SAE Technical Paper.

Boland, C.S., Kleine, R., Keoleian, G.A., Lee, E.C., Kim, H.C., and Wallington, T.J., 2015. Life cycle impacts of natural fiber composites for automotive applications: Effects of renewable energy content and lightweighting. *Journal of Industrial Ecology*, 20(1), pp. 179–189.

Bos, H.L., 2004. *The Potential of Flax Fibres as Reinforcement for Composite Materials*, Eindhoven-University, Technische Universiteit Eindhoven.

Bushi, L., Skszek, T., and Wagner, D., 2015. MMLV: Life Cycle Assessment (No. 2015-01-1616). SAE Technical Paper.

Ciacci, L., Morselli, L., Passarini, F., Santini, A., and Vassura, I., 2010. A comparison among different automotive shredder residue treatment processes. *The International Journal of Life Cycle Assessment*, 15(9), pp. 896–906.

Çinar, H., 2005. Eco-design and furniture: Environmental impacts of wood-based panels, surface and edge finishes. *Forest Products Journal*, 55(11), p. 27.

Clift, R. and Brandão, M., 2008. Carbon storage and timing of emissions. University of Surrey. Centre for Environmental Strategy Working Paper (02/08).

CSA, 2014. Life cycle assessment of auto parts—Guidelines for conducting LCA of auto parts incorporating weight changes due to material composition, manufacturing technology, or part geometry, http://shop.csa.ca/en/canada/analyse-du-cycle-de-vie/spe-14040-14/invt/27036702014.

Das, S., 2011. Life cycle assessment of carbon fiber-reinforced polymer composites. *The International Journal of Life Cycle Assessment*, 16(3), pp. 268–282.

De Vegt, O.M. and Haije, W.G., 1997. Comparative environmental life cycle assessment of composite materials. Netherlands Energy Research Foundation ECN.

Diener, J. and Siehler, U., 1999. Ökologischer Vergleich von NMT-und GMT-Bauteilen. *Die Angewandte Makromolekulare Chemie*, 272(1), pp. 1–4.

Directive, E.L.V., 2000. Directive 2000/53. EC of the European Parliament and of the Council on end-of-life vehicles.

Fearnside, P.M., Lashof, D.A., and Moura-Costa, P., 2000. Accounting for time in mitigating global warming through land-use change and forestry. *Mitigation and Adaptation Strategies for Global Change*, 5(3), pp. 239–270.

Finkbeiner, M. and Hoffmann, R., 2006. Application of life cycle assessment for the environmental certificate of the Mercedes-Benz S-Class. *The International Journal of Life Cycle Assessment*, 11(4), pp. 240–246.

Finkbeiner, M., Ackermann, R., Bach, V., Berger, M., Brankatschk, G., Chang, Y.J., Grinberg, M., Lehmann, A., Martínez-Blanco, J., Minkov, N., and Neugebauer, S., 2014. Challenges in life cycle assessment: An overview of current gaps and research needs. In *Background and Future Prospects in Life Cycle Assessment* (pp. 207–258). Springer, Netherlands.

Fonseca, A.S., Nunes, M.I., Matos, M.A., and Gomes, A.P., 2013. Environmental impacts of end-of-life vehicles' management: Recovery versus elimination. *The International Journal of Life Cycle Assessment*, 18(7), pp. 1374–1385.

Gärtner, S.O. and Reinhardt, G.A., 2004. Biobased products and their environmental impacts with respect to conventional products. In *Proceedings of the 2nd World Conference on Biomass for Energy, Industry and Climate Protection*, Rome.

Goepp, V., Zwolinski, P., and Caillaud, E., 2014. Design process and data models to support the design of sustainable remanufactured products. *Computers in Industry*, 65(3), pp. 480–490.

Han, D. and Choi, H.G., Developing sustainable new drum brake system through life cycle assessment: A case of US trucks. *Automotive Industries*, 6, p. 35.

Hellweg, S. and Canals, L.M., 2014. Emerging approaches, challenges and opportunities in life cycle assessment. *Science, 344*(6188), pp. 1109–1113.

Hermann, B.G., Debeer, L., De Wilde, B., Blok, K., and Patel, M.K., 2011. To compost or not to compost: Carbon and energy footprints of biodegradable materials' waste treatment. *Polymer Degradation and Stability, 96*(6), pp. 1159–1171.

Hertwich, E.G., 2005. Consumption and the rebound effect: An industrial ecology perspective. *Journal of Industrial Ecology, 9*(1–2), pp. 85–98.

Hoogmartens, R., Van Passel, S., Van Acker, K., and Dubois, M., 2014. Bridging the gap between LCA, LCC and CBA as sustainability assessment tools. *Environmental Impact Assessment Review, 48*, pp. 27–33.

ISO, I., 2013. TS 14067: 2013: Greenhouse Gases—Carbon Footprint of Products—Requirements and Guidelines for Quantification and Communication. International Organization for Standardization, Geneva, Switzerland.

Jenseit, W., Stahl, H., Wollny, V., and Wittlinger, R., 2003. Recovery options for plastic parts from end-of-life vehicles: An eco-efficiency assessment. Oko-Institut eV. http://www.oeko. de/oekodoc/151/2003-039-en.pdf.

Jolliet, O., Cotting, K., Drexler, C., and Farago, S., 1994. Life-cycle analysis of biodegradable packing materials compared with polystyrene chips: The case of popcorn. *Agriculture, Ecosystems & Environment, 49*(3), pp. 253–266.

Joshi, S.V., Drzal, L.T., Mohanty, A.K., and Arora, S., 2004. Are natural fiber composites environmentally superior to glass fiber reinforced composites? *Composites Part A: Applied Science and Manufacturing, 35*(3), pp. 371–376.

Kasai, J., 1999. Life cycle assessment, evaluation method for sustainable development. *JSAE Review, 20*(3), pp. 387–394.

Kaufman, M., 1963. *The First Century of Plastics: Celluloid and Its Sequel.* Plastics Institute.

Kellenberger, D., Althaus, H.J., Jungbluth, N., Künniger, T., Lehmann, M., and Thalmann, P., 2007. Life cycle inventories of building products. *Final Report Ecoinvent Data* 2.0(7).

Keoleian, G.A., Spatari, S., Beal, R.T., Stephens, R.D., and Williams, R.L., 1998. Application of life cycle inventory analysis to fuel tank system design. *The International Journal of Life Cycle Assessment, 3*(1), pp. 18–28.

Khanna, V. and Bakshi, B.R., 2009. Carbon nanofiber polymer composites: Evaluation of life cycle energy use. *Environmental Science & Technology, 43*(6), pp. 2078–2084.

Khanna, V., Bakshi, B.R., and Lee, J.L., 2008. Assessing life cycle environmental implications of polymer nanocomposites. *IEEE International Symposium on Electronics and the Environment.* IEEE, San Francisco, CA.

Kim, H.C. and Wallington, T.J., 2013. Life-cycle energy and greenhouse gas emission benefits of lightweighting in automobiles: Review and harmonization. *Environmental Science & Technology, 47*(12), pp. 6089–6097.

Kim, H.J., McMillan, C., Keoleian, G.A., and Skerlos, S.J., 2010. Greenhouse gas emissions payback for lightweighted vehicles using aluminum and high-strength steel. *Journal of Industrial Ecology, 14*(6), pp. 929–946.

Kim, S. and Dale, B., 2006. Ethanol fuels: E10 or E85–life cycle perspectives (5 pp). *The International Journal of Life Cycle Assessment, 11*(2), pp. 117–121.

Kim, S., Dale, B.E., Drzal, L.T., and Misra, M., 2008. Life cycle assessment of kenaf fiber reinforced biocomposite. *Journal of Biobased Materials and Bioenergy, 2*(1), pp. 85–93.

Klink, G., Rouilloux G., Znojek B., and Wadivkar O. 2012. Plastics. The Future for Automakers and Chemical Companies. Available on: https://www.atkearney.com /documents/10192/244963/Plastics-The_Future_for_Automakers_and_Chemical _Companies.pdf/28dcce52-affb-4c0b-9713-a2a57b9d753e.

Kobayashi, O., 1997. Car life cycle inventory assessment (No. 971199). SAE Technical Paper.

Koffler, C., 2014. Life cycle assessment of automotive lightweighting through polymers under US boundary conditions. *The International Journal of Life Cycle Assessment, 19*(3), pp. 538–545.

Kumar, S. and Putnam, V., 2008. Cradle to cradle: Reverse logistics strategies and opportunities across three industry sectors. *International Journal of Production Economics, 115*(2), pp. 305–315.

La Rosa, A.D., Recca, G., Summerscales, J., Latteri, A., Cozzo, G., and Cicala, G., 2014. Bio-based versus traditional polymer composites. A life cycle assessment perspective. *Journal of Cleaner Production, 74*, pp. 135–144.

Lewis, A.M., 2013. The potential of lightweight materials and advanced engines to reduce life cycle energy and greenhouse gas emissions for ICVs and EVs using design harmonization techniques (Doctoral dissertation, University of Michigan).

Lloyd, S.M. and Lave, L.B., 2003. Life cycle economic and environmental implications of using nanocomposites in automobiles. *Environmental Science & Technology, 37*(15), pp. 3458–3466.

Lloyd, S.M., Lave, L.B., and Matthews, H.S., 2005. Life cycle benefits of using nanotechnology to stabilize platinum-group metal particles in automotive catalysts. *Environmental Science & Technology, 39*(5), pp. 1384–1392.

Lohse-Busch, H., Diez, J., and Gibbs, J., 2013. The measured impact of vehicle mass on road load forces and energy consumption for a BEV, HEV, and ICE vehicle. *SAE International Journal of Alternative Powertrains, 2*(2013-01-1457), pp. 105–114.

Luz, S.M., Caldeira-Pires, A., and Ferrão, P.M., 2010. Environmental benefits of substituting talc by sugarcane bagasse fibers as reinforcement in polypropylene composites: Ecodesign and LCA as strategy for automotive components. *Resources, Conservation and Recycling, 54*(12), pp. 1135–1144.

Mainka, H., Täger, O., Stoll, O., Körner, E., and Herrmann, A.S., 2013. Alternative precursors for sustainable and cost-effective carbon fibers usable within the automotive industry. In *Society of Plastics Engineers (Automobile Division)–Automotive Composites Conference & Exhibition.*

Marretta, L., Lorenzo, R., Micari, F., Arinez, J., and Dornfeld, D., 2012. Material substitution for automotive applications: A comparative life cycle analysis. *Leveraging Technology for a Sustainable World*, pp. 61–66.

May, J.R. and Brennan, D.J., 2003. Application of data quality assessment methods to an LCA of electricity generation. *The International Journal of Life Cycle Assessment, 8*(4), pp. 215–225.

McManus, M.C. and Taylor, C.M., 2015. The changing nature of life cycle assessment. *Biomass and Bioenergy, 82*, pp. 13–26.

Miller, L.J., Sawyer-Beaulieu, S., and Tam, E., 2014. Impacts of non-traditional uses of polyurethane foam in automotive applications at end of life. *SAE International Journal of Materials & Manufacturing, 7*(3), pp. 711–718.

Mirabile, D., Pistelli, M.I., Marchesini, M., Falciani, R., and Chiappelli, L., 2002. Thermal valorisation of automobile shredder residue: Injection in blast furnace. *Waste Management, 22*(8), pp. 841–851.

Müssig, J., 2008. Cotton fibre-reinforced thermosets versus ramie composites: A comparative study using petrochemical-and agro-based resins. *Journal of Polymers and the Environment, 16*(2), pp. 94–102.

Nampoothiri, K.M., Nair, N.R., and John, R.P., 2010. An overview of the recent developments in polylactide (PLA) research. *Bioresource Technology, 101*(22), pp. 8493–8501.

Nourreddine, M., 2007. Recycling of auto shredder residue. *Journal of Hazardous Materials, 139*(3), pp. 481–490.

O'neill, T.J., 2003. *Life Cycle Assessment and Environmental Impact of Polymeric Products* (Vol. 13). iSmithers Rapra Publishing.

PAS 2050:2011. Available on: http://shop.bsigroup.com/upload/shop/download/pas/pas2050 .pdf.

Pawelzik, P., Carus, M., Hotchkiss, J., Narayan, R., Selke, S., Wellisch, M., Weiss, M.A.R.T.I.N., Wicke, B., and Patel, M.K., 2013. Critical aspects in the life cycle assessment (LCA) of bio-based materials–Reviewing methodologies and deriving recommendations. *Resources, Conservation and Recycling*, 73, pp. 211–228.

PE International, Inc. 2012a. Life Cycle Assessment of Polymers in an Automotive Assist Step, prepared for: American Chemistry Council. http://plastics.americanchemistry .com/Education-Resources/Publications/Life-Cycle-Assessment-of-Polymers-in-an -Automotive-Assist-Step.pdf.

PE International, Inc. 2012b. Life Cycle Assessment of Polymers in an Automotive Bolster, prepared for: American Chemistry Council, http://plastics.americanchemistry.com/Education -Resources/Publications/Life-Cycle-Assessment-of-Polymers-in-an-Automotive-Bolster .pdf.

Peijs, T., Cabrera, N., Alcock, B., Schimanski, T., Loos, J., 2002. Gibson, A.G. (Ed.), In: *Proceedings of 9th International Conference on Fibre Reinforced Composites—FRC 2002*, Newcastle upon Tyne, UK (March 26–28).

Pervaiz, M. and Sain, M.M., 2003. Carbon storage potential in natural fiber composites. *Resources, conservation and Recycling*, 39(4), pp. 325–340.

Protocol, G.H.G., 2011. Product life cycle accounting and reporting standard. *World Business Council for Sustainable Development and World Resource Institute*.

Puri, P., Compston, P., and Pantano, V., 2009. Life cycle assessment of Australian automotive door skins. *The International Journal of Life Cycle Assessment*, 14(5), pp. 420–428.

Rockström, J., Steffen, W., Noone, K., Persson, Å., Chapin, F.S., Lambin, E.F., Lenton, T.M., Scheffer, M., Folke, C., Schellnhuber, H.J., and Nykvist, B., 2009. A safe operating space for humanity. *Nature*, 461(7263), pp. 472–475.

Russo, D. and Rizzi, C., 2014. Structural optimization strategies to design green products. *Computers in Industry*, 65(3), pp. 470–479.

Ryberg, M., Vieira, M.D., Zgola, M., Bare, J., and Rosenbaum, R.K., 2014. Updated US and Canadian normalization factors for TRACI 2.1. *Clean Technologies and Environmental Policy*, 16(2), pp. 329–339.

Sawyer-Beaulieu, S.S. and Tam, E.K., 2008. Constructing a Gate-to-Gate Life Cycle Inventory (LCI) of End-Of-Life Vehicle (ELV) Dismantling and Shredding Processes (No. 2008-01-1283). SAE Technical Paper.

Schexnayder, S.M., Das, S., Dhingra, R., Overly, J.G., Tonn, B.E., Peretz, J.H., Waidley, G. and Davis, G.A., 2001. Environmental evaluation of new generation vehicles and vehicle components. Engineering Science and Technology Division, Oak Ridge National Lab., US Dept. of Energy, Oak Ridge, TN.

Schmehl, M., Müssig, J., Schönfeld, U., and Von Buttlar, H.B., 2008. Life cycle assessment on a bus body component based on hemp fiber and PTP®. *Journal of Polymers and the Environment*, 16(1), pp. 51–60.

Schmidt, W.P. and Beyer, H.M., 1998. Life cycle study on a natural fibre reinforced component (No. 982195). SAE Technical Paper.

Searchinger, T., Heimlich, R., Houghton, R.A., Dong, F., Elobeid, A., Fabiosa, J., Tokgoz, S., Hayes, D., and Yu, T.H., 2008. Use of US croplands for biofuels increases greenhouse gases through emissions from land-use change. *Science*, 319(5867), pp. 1238–1240.

Shaw, J., Kuriyama, Y., and Lambriks, M., 2011. Achieving a Lightweight and Steel-Intensive Body Structure for Alternative Powertrains (No. 2011-01-0425). SAE Technical Paper.

Shen, L., Haufe, J., and Patel, M.K., 2009. Product overview and market projection of emerging bio-based plastics PRO-BIP 2009. *Report for European Polysaccharide Network of Excellence (EPNOE) and European Bioplastics*, 243.

Song, Y.S., Youn, J.R., and Gutowski, T.G., 2009. Life cycle energy analysis of fiber-reinforced composites. *Composites Part A: Applied Science and Manufacturing, 40*(8), pp. 1257–1265.

Songstad, D., Lakshmanan, P., Chen, J., Gibbons, W., Hughes, S., and Nelson, R., 2011. Historical perspective of biofuels: Learning from the past to rediscover the future. In *Biofuels* (pp. 1–7). Springer, New York.

Stans, J. and Bos, H., 2007. CO2 reductions from passenger cars, The European Parliament's Committee on the Environment. *Public Health and Food Safety (IP/A/ENVI/ FWC/2006-172/Lot 1/C2/SC1)*. (IP/A/ENVI/FWC/2006-172/Lot 1/C2/SC1), 2007.

Stasinopoulos, P. and Compston, P., 2014. A dynamical life cycle inventory of steel, aluminium, and composite car bodies-in-white. In *Sustainable Automotive Technologies 2013* (pp. 111–117). Springer International Publishing.

Steel Recycling Institute. Steel: Driving Auto Recycling Success. Accessed December 06, 2016, http://www.recycle-steel.org/steel-markets/automotive.aspx.

Stephens, R.D., Wheeler, C.S., and Pryor, M., 2001. Life cycle assessment of aluminum casting processes (No. 2001-01-3726). SAE Technical Paper.

Subic, A., Mouritz, A., and Troynikov, O., 2009. Sustainable design and environmental impact of materials in sports products. *Sports Technology, 2*(3–4), pp. 67–79.

Sullivan, J.L., Burnham, A., and Wang, M., 2010. Energy-consumption and carbon-emission analysis of vehicle and component manufacturing (No. ANL/ESD/10-6). Argonne National Laboratory (ANL).

Sullivan, J.L., Williams, R.L., Yester, S., Cobas-Flores, E., Chubbs, S.T., Hentges, S.G., and Pomper, S.D., 1998. Life cycle inventory of a generic US family sedan overview of results USCAR AMP project (No. 982160). SAE Technical paper.

Tian, J., Ni, F., and Chen, M., 2015. Application of pyrolysis in dealing with end-of-life vehicular products: A case study on car bumpers. *Journal of Cleaner Production, 108,* pp. 1177–1183.

Tipper, R., Hutchison, C., and Brander, M., 2009. A practical approach for policies to address GHG emissions from indirect land use change associated with biofuels. *Ecometrica Technical Paper TP-080212-A.*

Tufvesson, L.M. and Börjesson, P., 2008. Wax production from renewable feedstock using biocatalysts instead of fossil feedstock and conventional methods. *The International Journal of Life Cycle Assessment, 13*(4), pp. 328–338.

Tufvesson, L.M., Lantz, M., and Börjesson, P., 2013. Environmental performance of biogas produced from industrial residues including competition with animal feed–life-cycle calculations according to different methodologies and standards. *Journal of Cleaner Production, 53*, pp. 214–223.

Union, P., 2009. Directive 2009/28/EC of the European Parliament and of the Council of April 23,2009 on the promotion of the use of energy from renewable sources and amending and subsequently repealing Directives 2001/77/EC and 2003/30/EC.

USEPA 2015. Available on: http://www.epa.gov/sites/production/files/2015-04/documents /sepupdatedpolicy15.pdf.

USEPA WARM 2015. Available on: http://www3.epa.gov/warm/pdfs/Background_Overview .pdf.

Van Den Brink, R. and Van Wee, B., 1990. Passenger car fuel consumption in the recent past. Why has passenger car fuel consumption no longer shown a decrease since 1990? Workshop "Indicators of Transportation Activity, Energy and CO$_2$ Emissions," Stockholm, 1999.

Vogtländer, J.G., van der Velden, N.M., and van der Lugt, P., 2014. Carbon sequestration in LCA, a proposal for a new approach based on the global carbon cycle; cases on wood and on bamboo. *The International Journal of Life Cycle Assessment, 19*(1), pp. 13–23.

Wambua, P., Ivens, J., and Verpoest, I., 2003. Natural fibres: Can they replace glass in fibre reinforced plastics?. *Composites Science and Technology, 63*(9), pp. 1259–1264.

Weidema, B.P. and Wesnæs, M.S., 1996. Data quality management for life cycle inventories—An example of using data quality indicators. *Journal of Cleaner Production, 4*(3), pp. 167–174.

Witik, R.A., Payet, J., Michaud, V., Ludwig, C., and Månson, J.A.E., 2011. Assessing the life cycle costs and environmental performance of lightweight materials in automobile applications. *Composites Part A: Applied Science and Manufacturing, 42*(11), pp. 1694–1709.

Wötzel, K., Wirth, R., and Flake, M., 1999. Life cycle studies on hemp fibre reinforced components and ABS for automotive parts. *Die Angewandte Makromolekulare Chemie, 272*(1), pp. 121–127.

Xu, X., Jayaraman, K., Morin, C., and Pecqueux, N., 2008. Life cycle assessment of wood-fibre-reinforced polypropylene composites. *Journal of Materials Processing Technology, 198*(1), pp. 168–177.

Zah, R., Hischier, R., Leão, A.L., and Braun, I., 2007. Curauá fibers in the automobile industry–A sustainability assessment. *Journal of Cleaner Production, 15*(11), pp. 1032–1040.

Zhao, Q. and Chen, M., 2011. A comparison of ELV recycling system in China and Japan and China's strategies. *Resources, Conservation and Recycling, 57*, pp. 15–21.

14 Case Studies—Sustainable and Lightweight Automotive Parts via Injection Molding

Birat KC, Omar Faruk, Mohini Sain, and Jimi Tjong

CONTENTS

14.1 OVERVIEW OF LIGHTWEIGHT AND SUSTAINABLE MATERIALS

Materials that come from renewable or recycled sources and have the potential to reduce the weight of current automobile parts can be defined as lightweight and sustainable materials. From literature, sustainable materials are referred to as biocomposites, green composites, and bioplastics [1,2]. Some authors also call materials that utilize recycled fiber or recycled plastics sustainable material [3,4]. These classifications of materials rely on material's environmental performance during the service life of a product or commonly called life cycle analysis (LCA).

A recent market research report on sustainable materials suggests that increasing environmental awareness among consumers and the trend in the automotive industry is toward greening their product [5]. On the other hand, governmental regulations such as CAFE standard require the automaker to improve the average fuel economy of vehicles to 54.5 miles per gallon by the year 2025. As an alternative, there is a big push from the automotive industry toward vehicle lightweighting and improving vehicles' environmental footprint by utilizing renewable and recycled raw materials. To capture the global market and be at the cutting edge of the technology, automakers are utilizing new and advanced materials for increased performance, environmental benefits, and light weight [6]. Nevertheless, use of these materials cannot compromise quality and safety standards of an automaker and must be economically feasible.

In this context, use of biocomposites is expected to grow as replacement materials in automotive applications [7–9].

In automotive plastic part manufacturing, injection molding is increasingly becoming the technology of choice for low-cost and high-volume part manufacturing. Despite its faster production cycles, excellent surfaces of the products, and facile molding of complicated shapes, the current demands on close dimensional tolerance and high dimensional stability make it necessary to understand the processing of newer materials [10,11]. Since the choice of sustainable materials affects both injection molding process and tool design, the relationship between these factors on the dimensional stability and quality of parts must be validated [12,13].

According to a report from Nova Institute in Germany, the production of sustainable materials such as biocomposites in Europe (year 2012) was estimated to be 0.35 million tons and is expected to increase 20% by the year 2020 with major application decking and automotive parts [14]. In the automotive industry, use of biocomposites has the potential to reduce part weight, improve sustainable portfolio of a product, and potential energy and cost saving during manufacturing over conventional materials such as glass fiber-reinforced or mineral-filled plastics [15]. In the past, most applications of biocomposites were limited to nonstructural or semistructural applications such as automotive interiors, household items, packaging, and electronic goods [16]. A nonstructural application such as interior trim components, spare tire carrier, and storage bins was investigated by several authors [17,18]. Some examples of the automotive interior from biocomposites are shown in Figure 14.1. Other studies were conducted to determine the feasibility of biocomposites in a structural application such as body panels and headlining [19–21].

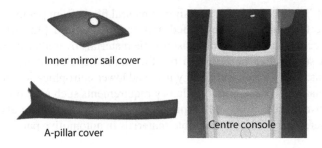

Inner mirror sail cover

A-pillar cover

Centre console

FIGURE 14.1 Automotive interior trim made from biocomposite material. (From KC, B., Panthapulakkal, S., Kronka, A., Agnelli, J.A.M., Tjong, J., and Sain, M. *J Appl Polym Sci* 2015;132:1–8.)

14.2 MATERIAL SELECTION CRITERIA

Various factors play a role in the decision-making process of selecting the right material for the right application. In a new material-based design approach that was highlighted in the multimaterial lightweight vehicle (MMLV) project from Ford Motor Company, it is changing the way materials are selected in automobiles [22]. First, it is important to understand that choice of material not only affects the design and processing but also significantly affects part performance, part cost as well as emissions during service-life and end-of-life disposal [9,23]. When it comes to selecting the right material, optimal balance of properties is required. For example, if plastics are used in an area with high exposure to chemicals, the material must exhibit good corrosion and chemical resistance while maintaining other properties such as strength and ductility [6]. And most importantly, economic restraints play a vital role in deciding the right material [24]. Nevertheless, the following factors must be taken into consideration.

- Material properties
- Processing
- Environment
- Regulatory
- Material cost and availability

Automotive materials must find the balance between material performance, cost, and its environmental impact. To date, several publications and commercial datasheets are available on sustainable materials that have good thermal, rheological, mechanical, and acoustic properties [21,25–27]. With the advance knowledge in surface chemistry and novel processing methods, it is possible to minimize issues of polymer incompatibility, water absorption, odor, and fiber distribution of sustainable materials [28,29]. In regards to renewable and recycled raw materials, there are also concerns regarding uniform quality and consistent supply to meet the automotive demand [19,30]. Cellulose fiber, the most abundant polymer in nature, can be developed from renewable sources such as wood, hemp, sisal fiber, wheat straw, rice straw, coconut fiber, and pineapple fiber. Utilization of these low-cost renewable sources to develop cellulose fiber not only makes economic sense but also it can be locally available [31]. To date, further processing of natural fibers from plants, leaf, or straw into cellulose fiber has been developed and commercialized [32–34]. In order to quantify the ecological footprint of sustainable materials over alternative materials, several LCA studies can be found in the literature [7,35,36]. Studies found lower CO_2 emissions, lower water and energy use, and lower eutrophication potential with sustainable materials. Last, both regulatory requirements such as end-of-life vehicle (ELV) directive on recycling and corporate social responsibility are also additional factors for increased used of sustainable materials in automotive parts.

14.2.1 MATERIAL PROPERTIES VERSUS PRODUCT REQUIREMENTS

Prior to selecting sustainable materials, a good understanding of product requirements is key in successfully developing a formulation that meets the design intent. Typically, most automobiles are designed for 10 years of service life (250,000 km equivalent) [37]. Material specification of each component scales down to their long-term performance over the vehicle life including environmental cycling at various heat and humidity conditions. In addition, there are specific requirements as materials are exposed to thermal, vibration, or mechanical loading. If material behavior during various loading and environmental conditions is not understood properly, product function can be compromised. For example, the automotive interior material must withstand high temperature and humidity variations (heat, cold, humidity) during its service life [25]. Although its requirement may depend on its specific location and design, loss of function from forces such as impact and scratches during its use are an important consideration when selecting materials. For exterior materials, UV resistance and long-term weathering performance are critical parameters [19,38]. Overall, materials must maintain good surface appearance during their service life. Typical material properties included in a technical datasheet with associated test standards are summarized in Table 14.1.

It is critical to understand material properties and ensure that they maintain their function during their service life [24]. For most structural designs with plastics, standard measures of material performance are tensile strength, modulus, and elongation to break [39]. Tensile strength can provide information as to what condition the material experiences catastrophic failure. The upper limit for ductile materials is yield strength at maximum load and for brittle materials is stress at break [40]. Any environment that involves stress or strains higher than these values eliminates the material for even short-term considerations. For applications where the temperature is critical, maximum short-term use temperature or HDT can possibly be used to measure softening point of the material [41]. Any material that exhibits lower HDT can be eliminated early in the selection process. However, for long-term performance characteristics, fatigue or creep resistance becomes a critical consideration [41]. The relationship between stress, time, and temperature is complicated and frequently the data needed to make good decisions about long-term behavior of a material are often not available. Sometimes, multiple data points of tensile, flexural, and impact properties at varying temperatures (typically room temperature, maximum and minimum service temperature) may be required to eliminate a lot of guesswork during material selection. Besides additional property data, it is also equally critical

TABLE 14.1
Typical Material Properties Used in Material Selection

SN	Test	Standards
1	Melt temperature	ISO 3146
2	Filler content	ISO 3451-1
3	Density	ASTM D792, ISO 1183
4	Tensile strength and modulus	ASTM D638, ISO 527
5	Flexural strength and modulus	ASTM D790, ISO 178
6	Poisson's ratio	ASTM D638, ISO 527
7	Impact strength	ASTM D256, ISO 179/180
8	Heat deflection temperature	ASTM D648, ISO 75
9	Flammability	ISO 3795, SAE J369
10	Melt flow rate	ASTM 1238, ISO 1133
11	Water absorption	ISO 62
12	Molding shrinkage	ISO 294
13	Coefficient of linear thermal expansion	ASTM E831, ISO 11359 (TMA)
14	Weathering resistance	SAE J2527
15	Heat aging performance	ASTM D573, ISO 188

to report statistical data (standard deviation and 3 sigma values) in order to determine the proper safety factor in designs.

14.3   CASE STUDIES

Three case studies with hybrid MiCel compounds and one case study with recycled carbon fiber are presented next with the emphasis on material selection and process validation of automotive parts.

14.3.1   MATERIAL SELECTION

The materials presented in this case study were developed by the Centre for Biocomposites and Biomaterials Processing (CBBP), the University of Toronto to meet or exceed specifications of existing automotive materials while offering additional attributes to designers such as lightweighting and sustainability. The materials consisted of especially engineered cellulose fiber (MiCel) extracted from trees grown in sustainably managed forests and hybridized with synthetic fiber or filler in a thermoplastic matrix (see Figure 14.2). Hybrid MiCel pellets were supplied by CBBP using a proprietary process to disperse cellulose fiber and long-inorganic filler in a thermoplastic polymer matrix (US patent US200502250009) and creating a product with high performance and uniform properties (US patent US20090314442) [32,42]. Another material discussed in this chapter is a 100% recycled material from reclaimed carbon fiber and recycled polyamide 6 (PA6). This formulation has been developed by CBBP as a substitute material for structural under-the-hood applications such as intake manifold, cam cover, and oil pans. Compared to current

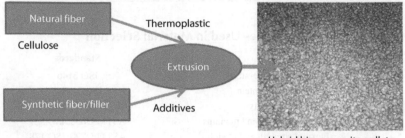

Hybrid biocomposite pellets

FIGURE 14.2   Wood cellulose fiber based black plastic pellets. (Courtesy of CBBP, University of Toronto.)

materials, MiCel hybrid composite pellets and reclaimed carbon fiber composites are expected to reduce part mass (up to 30%), improve acoustic performance, reduce manufacturing energy use, improve sustainability portfolio of a part while providing manufacturers a flexibility of utilizing existing processes and equipment [18,36].

14.3.2   CASE STUDY 1: BEAUTY SHIELD

14.3.2.1   Material Selection

A beauty shield is an under-the-hood nonstructural component that covers the top of the radiator assembly. Nevertheless, it must meet certain functions such as heat from the engine compartment and impact forces from dropped tools during servicing. Most of the current beauty shields are made using 15%–22% talc reinforced polypropylene and properties vary slightly depending on original equipment manufacturer's (OEM's) design requirements. Comparison of the commercial current material specification and MiCel composite is shown in Figure 14.3. It is clear

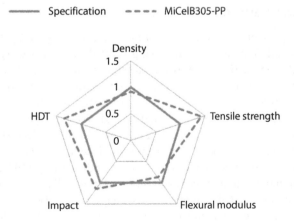

FIGURE 14.3   Relative comparison of MiCelB305-PP with the specification of an automotive beauty shield.

that MiCel hybrid composite exceeds in heat stability and impact requirements. Besides, MiCelB305-PP is 10% lighter and 15% renewable material. The cost of MiCelB305-PP was estimated at approximately US$1.08/lb based on raw material cost and 30% processing cost, which make it competitive in the current market.

14.3.2.2 Process Validation

Prior to molding trials, MiCelB305-PP pellets were dried at 80°C for 4 h (minimum). Parts were molded using 1000 ton injection molding machine with show volume of 200 oz. After 24 h, the average weight of the parts was measured to be 448.5 g (~10% lighter). Process data was also collected from current production material in the same tool and injection machine. A comparison of the two processes is shown in Table 14.2.

From Table 14.2, it is found that MiCelB305-PP required 3% longer time to fill. Nevertheless, MiCelB305-PP had 3.7% faster cycle time as a result of higher injection pressure and faster holding/packing profile. In addition, it was also molded at lower melt and nozzle temperature. It suggests that for higher volume production of MiCelB305-PP parts, significant energy and cost saving can be achieved while gaining approximately 10% weight savings.

A comparison of a beauty shield from MiCelB305-PP and current production material is shown in Figure 14.4. From the trial, surface appearance was poor for MiCelB305-PP due to tiger stripe (dull and glassy surface) defect [43]. It is commonly due to flow instabilities (often called "slip-stick phenomenon") in thermoplastics such as PP (polypropylene) and TPO (thermoplastic olefin) during filling stage of injection molding process [44]. From the literature, it has been associated with variations in process setting (melt temperature, mold temperature, and injection speed) and mold design (gate location, gate type) [44].

TABLE 14.2
Relative Process Comparison of an Automotive Beauty Shield with MiCelB305-PP and Current Material

Process Parameters	MiCelB305-PP	Current Material
Fill time (s)	1.77	1
Hold time (s)	0.60	1
Cooling time (s)	1.00	1
Cycle time (s)	0.92	1
Fill pressure (psi)	1.12	1
Pack pressure (psi)	1.16	1
Back pressure (psi)	0.36	1
Mold temperature (F)	1.18	1
Nozzle temperature (F)	0.89	1
Manifold temperatures (F)	0.90	1
Screw intensification	1.00	1
Shot size (in.)	1.77	1

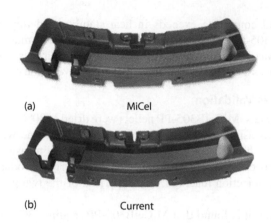

(a)   MiCel

(b)   Current

FIGURE 14.4 Beauty shield with MiCelB305-PP (a) and current material (b).

14.3.3 CASE STUDY 2: DOOR CLADDING—EXTERIOR

14.3.3.1 Material Selection

Door cladding is an automotive body component below the door panel and is designed to protect the door and body panel from stone impact and scratch defects. Typically thermoplastics are used in newer designs due to their higher energy absorption property and low-cost replacement/maintenance. In Figure 14.5, properties of MiCelB305-TPOr are compared with current door cladding material specifications which is made from talc-filled TPO (thermoplastic olefin). The comparison shows that MiCelB305-TPOr has approximately 10% lower density and contains 15% (w/w) of renewable material and 85% (w/w) of recycled material. It exceeds current material in both tensile strength and HDT performance; however, it has a slightly lower impact and flexural modulus. With the use of an impact modifier such as EPDM (ethylene-propylene-diene-terpolymer) or EPR (ethylene-propylene-rubber)

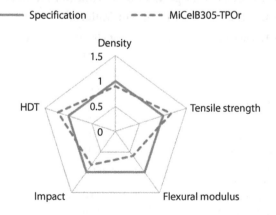

FIGURE 14.5 Relative comparison of MiCelB305-TPOr with the current material specification of door cladding.

to the current formulation, balance between stiffness and impact properties can be achieved and meet the specification [45]. Using MiCelB305-TPOr with improved impact, it is possible to make door cladding more sustainable while reducing the weight of the part by 10%.

14.3.3.2 Process Validation

In order to demonstrate the processability of MiCelB305-TPOr, door claddings were molded using 700 ton injection mold machine with dual gate (end and bottom edge gate). Prior to molding trials, MiCelB305-TPOr pellets were dried for a minimum of 4 h at 80°C. The functionality of two gates was also investigated in this trial (Table 14.3). The mold temperature (40°C), melt temperature (195°C) and hold pressure (500 psi), hold time (10 s), and back pressure (50 psi) were kept constant. For both conditions, completely filled parts were produced.

From Table 14.3, the end edge gate required a longer flow path and as a result has a longer fill time and higher pressure. The end gate also had difficulty completely filling due to end gas encapsulations and surface appearance issues as found in the beauty shield tryout. Using the bottom gate, the process was stabilized and tiger stripes were visible on the surface of the parts (Figure 14.6). Mold temperature, melt

TABLE 14.3

Comparison of Process Parameters for Door Cladding with Two Gate Configurations

Process Parameters	End Edge Gate	Bottom Edge Gate
Fill time (s)	1.8	1.73
Peak injection pressure (psi)	1925	1872
Injection speed (in/s)	2.75	3.75
Screw speed (rpm)	75	75

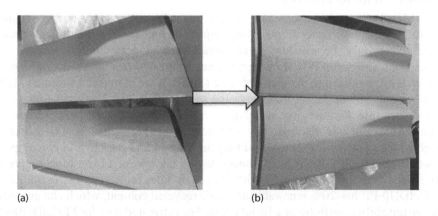

(a)   (b)

FIGURE 14.6 Door claddings with tiger stripping during the initial process (a) and no tiger stripping (b) after process optimization.

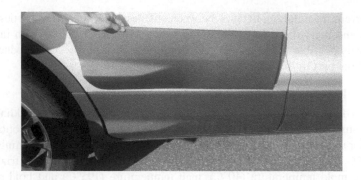

FIGURE 14.7   MiCelB305-TPOr door cladding (top) and current door cladding (bottom).

temperature, and back pressure were raised to 43.3°C, 198.89°C, and 150 psi, respectively, in order to eliminate tiger stripping.

After 24 hours of conditioning parts in ambient conditions, dimensions of the door cladding parts were measured using a coordinate measuring machine (CMM) fixture. An average gap along the profile was on average 1.3 mm, which is lower than the allowable limit of 1.5 mm per engineering specification of door cladding. Visual comparison of MiCelB305-PP was made with current production material (Figure 14.7). The MiCelB305-PP part appeared gray in color compared to the current production part which was black color. It is possible that MiCel fibers (white in color) may have influenced the surface color and suggests increased black masterbatch content to MiCel formulation to meet the surface appearance requirements of door cladding.

14.3.4   CASE STUDY 3: BATTERY TRAY

14.3.4.1   Material Selection

Under-hood plastic parts such as a battery tray must withstand heat from engine exhaust pipes, the impact from falling objects during service requirements and, more importantly, be resistant to chemicals such as acid, coolant, engine oil, brake fluid, and so on [8]. Moreover, the ease with which a material can be processed is often related to the melt flow rate (MFR) of the material. Minimum MFR requirements should be set to achieve the surface appearance of the battery tray and reduce injection pressure and injection molding cycle time [46]. Additional requirements include flammability and heat aging performance. The material selected for a battery tray is a MiCel recycled carbon and fiber hybrid biocomposites (MicelD110-PP). It has 8.69% lower density and meets all key function properties (static) to current material with 30% glass fiber reinforced PP. In addition, MiCelD110-PP has 20% renewable and 10% recycled content, which can enhance the sustainability portfolio of a battery tray. The estimated cost for MiCelD110-PP is US$1.43/lb. Comparison of properties is shown in Figure 14.8.

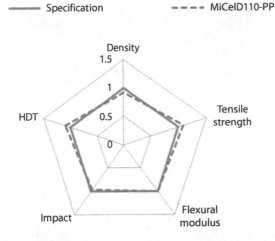

FIGURE 14.8 Relative comparison of MiCelD110-PP properties with the current material specification of a battery tray.

14.3.4.2 Process Validation

Parts were molded using 700 ton injection molding machine with the shot volume of 80 oz. Weights of the parts were measured to be 480.0 g and 494.5 g for MiCel and current material, respectively. Comparison of the two processes is shown in Table 14.4.

The average weight of the MiCelD110-PP battery tray was 10% lighter than current production part. Comparison of two parts from MiCel and current material from glass fiber and PP is given in Figure 14.9. Parts from MiCel were gray in color; however, the rib feature on the core side was less evident compared to the current production part. Dimensions of the molded battery tray were measured using a CMM

TABLE 14.4
Process Parameters for Injection Molding of an Automotive Beauty Shield with MiCelD110-PP and Current Production Material (GF30-PP)

Process Parameters	MiCelD110-PP	Current Material
Fill time (s)	1.77	1.00
Hold time (s)	0.60	1.00
Cooling time (s)	1.00	1.00
Cycle time (s)	0.92	1.00
Fill pressure (psi)	1.12	1.00
Pack pressure (psi)	1.16	1.00
Back pressure (psi)	0.36	1.00
Mold temperature (F)	1.18	1.00
Nozzle temperature (F)	0.90	1.00
Shot size (in.)	1.00	1.00

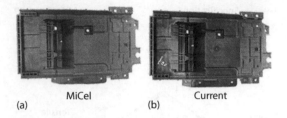

(a)   MiCel   (b)   Current

FIGURE 14.9   Comparison of battery trays made from MiCelB305-PP (a) and current material (b).

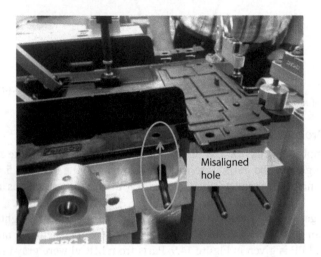

FIGURE 14.10   Battery tray in a CMM fixture for gap measurement. Circle and arrow show hole area of misalignment.

fixture. Tolerances for each hole location and datum were measured. Among all measurements taken, one of the hole features at a bolt location was misaligned (Figure 14.10). It is found that nonuniform shrinkage in the parts due to nonuniform cooling during manufacturing could result in dimensional issues in injection molding [12]. Prior to considering MiCel compounds for automotive parts, its effect on process or tool design needs to be further investigated and ensure that parts from MiCel compounds meet the dimensional requirements [27,47,48].

14.3.5   CASE STUDY 4: CAM COVER

14.3.5.1   Material Selection

As cam covers are directly bolted to the engine head, they require much higher thermal performance compared to automotive applications discussed earlier [8]. They are in constant contact with hot oil during engine peak performance and cold oil during engine shut off at sub-zero ambient conditions. It is thus critical that material must maintain its mechanical properties at both maximum and minimum service

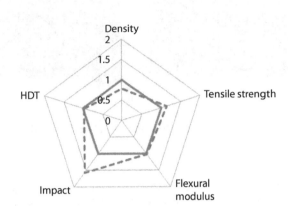

FIGURE 14.11 Comparison of RCF20-PA6r with the material specification of an engine cam cover.

temperature [49]. The material selected for cam cover was made using reclaimed carbon fiber and recycled PA6 (RCF20-PA6r). Comparison of RCF20-PA6r is shown in Figure 14.11 with cam cover specification. From the data, RCF20-PA6r has the potential to reduce weight by 22% and increase impact by approximately 56% while offering environmental benefits.

14.3.5.2 Process Validation

RCF20-PA6r was dried at 85°C for 8 hours prior to injection molding to achieve a moisture content below 0.02%. Cam covers were developed using 1000 ton injection molding press with 160 oz. shot volume. Mold temperature and melt temperature were kept similar to production material. Molding parameters comparison of RCF20-PA6r and current material are summarized in Table 14.5. Production of carbon fiber cam cover required 1.3 seconds longer time to fill and 8.84% higher injection pressure. Measurement of two parts after 24 h of conditioning at room

TABLE 14.5

Relative Process Comparison of Cam Cover with RCF20-PA6r and Current Material

Process Parameters	RCF20-PA6r	Current Material
Fill time (s)	1.54	1.00
Peak injection pressure (psi)	1.09	1.00
Hold pressing (psi)	1.25	1.00
Mold temperature (°C)	1.00	1.00
Melt temperature (°C)	1.00	1.00
Part weight (g)	0.79	1.00

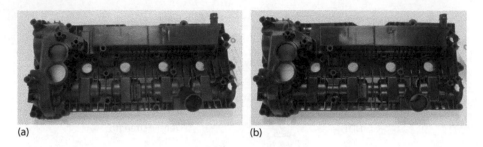

(a)   (b)

FIGURE 14.12   Cam covers from RCF20-PA6r (a) and current production material (b).

temperature suggested that RCF20-PA6 cam cover was 20.8% lighter than the current production part. The appearance of RCF20-PA6 was also comparable to the production part (Figure 14.12). Further testing such as thermal cycling and leak test is required to validate the developed material.

14.3.6   ADDITIONAL CASE STUDIES

Additional materials based on hybrid recycled carbon fiber (RCF) and MiCel were also developed by CBBP (MiCelD212-PP) for engine covers (Figure 14.13). Critical material selection parameters for engine cover were materials heat aging, sound absorption, and chemical resistance requirement. On the other hand, material parameters such as impact, heat aging, and stiffness were important parameters. In both cases, significant weight reduction (25%) was achieved while increasing renewable and recycled content on the part. Based on the visual inspection, the developed engine cover prototype does not meet the surface appearance requirements. It requires further adjustment to the formulation with the right black color masterbaches as recommended in previous tryouts. Further testing on the performance of these prototypes is underway.

FIGURE 14.13   Engine cover prototypes developed from MiCelD112-PP.

14.3.7 DESIGN AND DEVELOPMENT CHALLENGES

14.3.7.1 Material Selection

Several challenges remain in utilizing sustainable and lightweight materials in automotive parts. More importantly, these materials require a good balance of stiffness, impact, and flow properties [50]. There are also challenges relating to the availability of material database for their use in engineering design and software tools. During the early material selection process, sustainable materials data are not available to designers and product development engineers for use in design and simulation software [24]. This often results in setbacks when it comes to using materials such as biocomposites in prototype development. Most material datasheets contain properties such as tensile, flexural, impact, rheological, and HDT data. They often lack information such as thermal conductivity, shrinkage, and the coefficient of linear thermal expansion (CLTE) that are used to understand processing and injection molding tool design [51,52]. To solve this issue, a new standard (ASTM D5592-94) was developed to assist design and product engineering in utilizing new sustainable and lightweight composites in the automotive part design process [41].

14.3.7.2 Cost

Another issue is the cost of sustainable and lightweight material [24]. In most cases, the cost is directly influenced by the availability of raw materials and established supply value chain [36,53]. Despite the lower raw material cost, there is often limited information in the literature on whether compounds from these materials are cost-competitive to current alternatives. Overall, mere material substitution may not be sufficient. If a product requires redesigning for sustainable material or changes in the process, the cost of manufacturing will be higher and can translate to increased prices per part.

14.4 FUTURE OUTLOOK

Recently, more and more carbon fiber-reinforced plastics are replacing heavy metallic components in order to achieve stringent fuel economy requirements by 2020 [54]. For automakers, this translates down to cutting off 250 lb to 500 lb per vehicle every year, on average, to meet these standards (BCC Research LLC report, February 2016). More carbon-fiber based compounds are being developed by various automotive material suppliers and research institutes with an objective to replace glass fiber and metal parts and improve the sustainability of parts [55–57]. Some structural applications include powertrain components such as chain adjusters, water pump housing, and oil pans. With the current market price of carbon fiber (US$7 or higher), it is very unlikely that carbon fiber will find its application in high volume vehicles. However, given the efforts in recycling carbon fiber among several material suppliers and research groups, its prices are expected to reduce significantly and become cost-competitive for automotive applications [55]. In the future, as the technology in carbon fiber extraction from lignin-based precursor gets into the commercialization phase, the cost of carbon fiber is expected to reduce further [56].

ACKNOWLEDGMENTS

The authors thank Automotive Partnership Canada (APCJ 433821-12) for funding automotive trials. We are also thankful to researchers Dr. Suhara Panthapulakkal, Dr. Muhammad Pervaiz, Dr. Amir Beh, and Mr. Shiang Law at Centre for Biocomposites and Biomaterials Processing (CBBP), the University of Toronto for supplying the material information, and Ford Powertrain Engineering Research and Development Centre (PERDC) for coordinating trials with various automotive part suppliers.

REFERENCES

1. La Mantia, F.P. and Morreale, M. Green composites: A brief review. *Compos Part A Appl Sci Manuf* 2011;42:579–88.
2. Thakur, V.K., Singha, A.S., and Thakur, M.K. Ecofriendly biocomposites from natural fibers: Mechanical and weathering study. *Int J Polym Anal Charact* 2013;18:64–72.
3. Khalil, H.P.S.A., Bhat, A.H., and Yusra, A.F.I. Green composites from sustainable cellulose nanofibrils: A review. *Carbohydr Polym* 2011;87:963–79.
4. Bio materials, recyclables and the auto industry, Automotive World n.d. http://www.automotiveworld.com/news/electric-vehicles/66644-bio-materials-recyclables-and-the-auto-industry?highlight=natural+fiber (accessed March 20, 2012).
5. Thryft, A.R. 6 Promising new ways to make bio-based & renewable plastics. *Design News* [Online] 2016. http://www.designnews.com/author.asp?section_id=1392&doc_id=279663 (accessed January 1, 2016).
6. Park, H.S. and Dang, X.P. Development of a fiber-reinforced plastic armrest frame for weight-reduced automobiles. *Int J Automot Technol* 2011;12:83–92.
7. Faruk, O., Bledzki, A.K., Fink, H.-P., and Sain, M. Progress report on natural fiber reinforced composites. *Macromol Mater Eng* 2014;299:9–26.
8. Ozen, E., Kiziltas, A., Kiziltas, E.E., and Gardner, D.J. Natural fiber blend—Nylon 6 composites. *Polym Compos* 2013;34:544–53.
9. Ford Motor Company—Sustainability Report 2014/15. Dearborn, MI: n.d.
10. Prashantha, K., Soulestin, J., Lacrampe, M.F., Lafranche, E., Krawczak, P., Dupin, G. et al. Taguchi analysis of shrinkage and warpage of injection-moulded polypropylene/multiwall carbon nanotubes nanocomposites. *Express Polym Lett* 2009;3:630–8.
11. Nagahanumaiah, B.R. Effects of injection molding parameters on shrinkage and weight of plastic part produced by DMLS mold. *Rapid Prototyp J* 2009;15:179–86.
12. Guevara-Morales, A. and Figueroa-Lopez, U. Residual stresses in injection molded products. *J Mater Sci* 2014;49:4399–415.
13. Fuh, J.Y.H., Zhang, Y.F., Nee, A.Y.C., and Fu, M.W. *Computer-Aided Injection Mold Design and Manufacture.* New York: Marcel Dekker, 2004.
14. Carus, M. and Eder, A. Wood-Plastic Composites (WPC) and Natural Fibre Composites (NFC): European and Global Markets 2012 and Future Trends in Automotive and Construction. Hurth, Germany: 2015.
15. Martin, R.H., Giannis, S., Mirza, S., and Hansen, K. Biocomposites in challenging automotive applications, n.d.
16. Maleque, M.A., Atiqah, A., and Iqbal, M. Flexural and impact properties of kenaf-glass hybrid composite. *Adv Mater Res* 2012;576:471–4.
17. Leao, A.L., Souza, S.F., Cherian, B.M., Frollini, E., Thomas, S., Pothan, L.A. et al. Agro-based biocomposites for industrial applications. *Mol Cryst Liq Cryst* 2010;522:18/[318]–27/[327].

18. KC, B., Panthapulakkal, S., Kronka, A., Agnelli, J.A.M., Tjong, J., and Sain, M. Hybrid biocomposites with enhanced thermal and mechanical properties for structural applications. *J Appl Polym Sci* 2015;132:1–8.
19. Sathaye, A. Jute fibre based composite for automotive headlining. *SAE Intenational* 2011;01.
20. Vaidyanathan, H.P., Murty, P., and Eswara, S.P. Hybrid natural fiber composites molded auto-body panels/skins (hybrid NFPC): Processing, characterization & modeling. *SAE Int* 2011.
21. Jeoung, S.K., Lee, P., Yoo, S.E., Lee, K.D., Lee, S.N., and Han, J.K. Development process of the prototype and evaluation on the biodegradability of jute fiber/PLA fiber composites for automotive headlining. *SAE Int* 2011.
22. Jaranson, J. and Ahmed, M. MMLV: Lightweight interior systems design. *SAE Int* 2015. doi:10.4271/2015-01-1236.
23. Fatoni, R. *Product Design of Wheat Straw Polypropylene Composite*. University of Waterloo, 2012.
24. Leite, M., Silva, A., Henriques, E.A., and Madeira, J.F. Materials selection for a set of multiple parts considering manufacturing costs and weight reduction with structural isoperformance using direct multisearch optimization. *Struct Multidiscip Optim* 2015;52:635–44.
25. Mahto, K., Balaji, K.V., Adhikari, S., and Hydro, S., Industries S. Polypropylene-Starch Blend for Automotive Application 2013.
26. Maki, S. High Performance Engineered Biopolymers Via Compounding- RTP Company. Winona: n.d.
27. KC, B., Faruk, O., Agnelli, J.A.M., Lcao, A.L., Tjong, J., and Sain, M. Sisal-glass fiber hybrid biocomposite: Optimization of injection molding parameters using Taguchi method for reducing shrinkage. *Compos Part A Appl Sci Manuf* 2016;83:152–9.
28. Sain, M., Panthapulakkal, S., Law, S., and Bouillous, A. Interface modification and mechanical properties of natural fiber-polyolefin composite products. *J Reinf Plast Compos* 2005;24:121–30.
29. Dhanasekaran, S., Srinath, G., and Sathyaprasad, M. Structure, mechanical behavior and pre-treatment of natural fiber composites—A review. SAE Technical Paper 2008-28-0043, 2008. doi:10.4271/2008-28-0043.
30. Mohanty, A.K. Misra, M., Drzal, L.T. *Natural Fibers, Biopolymers, and Their Biocomposites*. Boca Raton, FL: CRC Press, 2005.
31. McGee, B. Stats Canada: Production of Field Crops in Ontario 2011 2012. http://www.omafra.gov.on.ca/english/stats/crops/estimate_metric.htm (accessed April 11, 2012).
32. Sain, M., Panthapulakkal, S., and Law, S.F. Manufacturing process for high performance short ligno-cellulosic fiber-thermoplastic composite materials. US20050225009, 2005.
33. What is THRIVE? A sustainable thermoplastic composite [Weyerhaeuser online] n.d. http://www.weyerhaeuser.com/cellulose-fibers/thrive/what-is-thrive/ (accessed January 1, 2016).
34. Cellulose Fiber Reinforced Polymer Compounds [RTP Company Online] n.d. http://www.rtpcompany.com/wp-content/uploads/2013/09/cellulose-fiber-pp.pdf (accessed January 1, 2016).
35. Boland, C. Life cycle energy and greenhouse gas emissions of natural fiber composites for automotive applications: Impacts of renewable material content and lightweighting. University of Michigan, 2014.
36. Pervaiz, M., Panthapulakkal, S., KC, B., Sain, M., and Tjong, J. Emerging trends in automotive lightweighting through novel composite materials. *Sci Res Publ* 2016:26–38.
37. Bushi, L., Skszek, T., and Wagner, D. MMLV: Life cycle assessment. *SAE Tech Pap* 2015.

38. Ford material specification WSS-M4D1002-A. 70% polypropylene copolymer (PP), 30% sisal fiber content molding compounding, exterior. IHS 2015. http://www.global spec.com/ (accessed December 4, 2015).
39. Qatu, M.S. Application of kenaf-based natural fiber composites in the automotive industry. *SAE Intenational* 2011:1–6.
40. Callister, W.D. *Materials Science and Engineering, an Introduction*. 6th ed. Hoboken, NJ: John Wiley & Sons, 2003.
41. Standard guide for material properties needed in engineering design using plastics. vol. D5592-94. 2010.
42. Sain, M., Panthapulakkal, S., and Law, S.F. Manufacturing process for hybrid organic and inorganic fibre-filled composite materials. US20090314442, 2009.
43. Iannuzzi, G., Boldizar, A., and Rigdahl, M. Characterization of flow-induced surface defects in injection moulded components—Case studies. *Annu. Trans. Nord. Rheol. Soc.*, 2009;17:9.
44. Hirano, K., Suetsugu, Y., Tamura, S., and Kanai, T. Morphological analysis of tiger-stripe and striped pattern deterioration on injection molding of polypropylene/rubber/talc blends. *J Appl Polym Sci* 2007;104:192–9.
45. Kock, C., Gahleitner, M., Schausberger, A., and Ingolic, E. Polypropylene/polyethylene blends as models for high-impact propylene—Ethylene copolymers, Part 1: Interaction between rheology and morphology. *J Appl Polym Sci* 2013:1484–96.
46. Rahman, W.A.W.A., Isa, N.M., Rahmat, A.R., Adenan, N., and Ali, R.R. Rice husk/high density polyethylene bio-composite: Effect of rice husk filler size and composition on injection molding processability with respect to impact property. *Adv Mater Res* 2009;83–86:367–74.
47. Depolo, W.S. *Dimensional Stability and Properties of Thermoplastics Reinforced with Particulate and Fiber Fillers By Dimensional Stability*. Virginia Polytechnic Institute and State University, 2005.
48. Barghash, M.A. and Alkaabneh, F.A. Shrinkage and warpage detailed analysis and optimization for the injection molding process using multistage experimental design. *Qual Eng* 2014;26:319–34.
49. Kagan, V. Forward to *Better Understanding of Optimized Performance of Welded Joints: Local Reinforcement and Memory Effects for Polyamides*. SAE Technical Paper 2001-01-0441, 2001. doi:10.4271/2001-01-0441.
50. Ramesh, M., Palanikumar, K., and Reddy, K.H. Mechanical property evaluation of sisal–jute–glass fiber reinforced polyester composites. *Compos Part B Eng* 2013;48:1–9.
51. Mehat, N.M. and Kamaruddin, S. Investigating the effects of injection molding parameters on the mechanical properties of recycled plastic parts using the Taguchi method. *Mater Manuf Process* 2011;26:202–9.
52. Youssef Ramzy, A., El-Sabbagh, M.M.A., Steuernagel, L., Ziegmann, G., and Meiners, D. Rheology of natural fibers thermoplastic compounds: Flow length and fiber distribution. *J Appl Polym Sci* 2014;131:8.
53. Mishra, S. and Sain, M. Commercialization of wheat straw as reinforcing filler for commodity thermoplastics. *J Nat Fibers* 2009;6:83–97.
54. Carbon Fibers Aid Auto Industry's Bid to Meet CAFE Standards. BCC Res LLC [Online] 2016. http://www.bccresearch.com/pressroom/pls/carbon-fibers-aid-auto-indus try%E2%80%99s-bid-to-meet-cafe-standards (accessed January 1, 2016).
55. Materials: Carbon Fiber Reinforced PP Compounds for Automotive. Plast Technol 2016. http://www.ptonline.com/products/materials-carbon-fiber-reinforced-pp-compounds-for -automotive (accessed January 1, 2016).
56. Norberg, I. Carbon fibres from kraft lignin. *KTH Royal Institute of Technology*, 2012.
57. Kadla, J., Kubo, S., Venditti, R., Gilbert, R., Compere, A., and Griffith, W. Lignin-based carbon fibers for composite fiber applications. *Carbon N Y* 2002;40:2913–20.

Index

Page numbers followed by f and t indicate figures and tables, respectively.

A

Abaca fiber PP composites
  enzyme treatment, 6
  HDT of, 5
  higher odor concentration, 6
  mechanical properties of, 5
Abrasion, property of bio-fiber thermoset
      composites, 80–81, 80f
ABS (acrylonitrile butadiene styrene), 2, 28, 29f,
      123t, 126, 356, 393t, 430t
Absorption
  energy, *see* Energy absorption
  moisture, *see* Moisture absorption
  sound, polyurethane foams, 413
Accelerated weathering tests, 111
Acetylation
  fiber composites, 14
  on flax fibers, 2, 5
  treatment, bio-fiber modification, 61
Acid anhydrides
  maleic, 15
  polyester resin from, 52
Acoustical insulation, 45, 83, 370, 384, 394t
Acrodur thermoset resin, 85f, 86
Acrylation treatment, of bio-fibers, 60, 61
Acrylonitrile butadiene styrene (ABS), 2, 28, 29f,
      123t, 126, 356, 393t, 430t
Additives
  bio-PE, properties of, 153
  biopolymers from, 140
  drop-ins with, 160
  manufacturing and curing assistance, 51
  melamine, 415
  Mg, alloying of, 336
  MMT and LDHs, 284
  in polymeric nanocomposites, 294
  rare earth, 337
  unsaturated polyester resin, 243
  use of, 117
  WPC, 105, 116
Adhesion
  epoxies for, 52
  fiber–matrix, 16, 17, 59, 119, 215, 222, 223,
      225, 233, 246, 271
  fiber surfaces and polymer matrix, 210–211
  improvement
      carbon fibers, 212, 213
      by surface treatment, 212, 213

interfacial, 6, 8, 9, 15, 20, 70, 80
  with polyurethanes, 123
  PP and jute fibers, 10
  sawdust, 12
  to UP matrix, 118
Adhesive bonding
  lightweight nanocomposite materials, 278,
      279, 285
  Mg, 350
Adler, 351
Advanced Compounding Rudolstadt (ACR),
      104
Aerostar model, 362
Aging
  bio-based oils, 194–197, 195f, 197f, 200
  heat, PA based composites, 231, 232–234,
      232f, 233f, 234f
AgriPlas, 128, 129
Agrolinz Melamine GmbH, 120
Aimplas, 128
Airlaying, 122
Alkali-lime glass
  defined, 241
  with high boron oxide, 242
Alkaline copper quaternary (ACQ)-treated
      lumber, 122
Alkali treatment
  bio-fiber composites, 60, 61, 77
  fiber modification, 60, 72
  fiber tensile strength, 67
  fillers modification by, 117
  flax fibers, 72
  flax/PP composites, 8
  hemp and kenaf fibers, 70
  jute fibers, 67, 68t, 69
  jute/vinylester composites, 68t
  silane coupling agent and, 67
  thermal-alkali treatment, 70
Alloys
  Al-Si alloys, sliding wear of, 312–315
  cost, 355
  in vehicles, applications, 304–306, 305t
      body structure, 306
      chassis applications, 305–306
      powertrain applications, 304, 305
  wear regimes and transitions in, 309–311
      mild and severe wear, 310–311, 310f
Alloys, Mg, 334–343; *see also* Magnesium

Index 497